PETROLEUM ENGINEERING
PRINCIPLES AND PRACTICE

Dedication

To our families and friends, through whose encouragement and understanding we have been able to undertake this work.

PETROLEUM ENGINEERING
PRINCIPLES AND PRACTICE

J S ARCHER AND C G WALL
Imperial College of Science and Technology, London

Graham & Trotman

First published in 1986 by
Graham and Trotman Ltd.   Graham and Trotman Inc.
Sterling House,   13 Park Avenue,
66 Wilton Road,   Gaithersburg,
London SW1V 1DE   MD 20877,
UK   USA

British Library Cataloguing in Publication Data
_____

Archer, J.S.
   Petroleum engineering : principles and practice.
   1. Petroleum engineering
   I. Title II. Wall, C.G.
   622′.3382   TN870

ISBN 0–86010–665–9
ISBN 0–86910–715–9 Pbk

© J S Archer and C G Wall, 1986

This publication is protected by international copyright law. All rights reserved. No part of this publication may be reproduced, stored in a retrieval system, or transmitted in any form or by any means, electronic, mechanical, photocopying, recording or otherwise, without the permission of the publishers.

Typeset in Great Britain by
Bookworm Studio Services, Salford
Printed and bound in Great Britain by
The Alden Press, Oxford

Contents

Preface x

Foreword ix

CHAPTER 1 Introduction

    1.1   Petroleum Engineering: A creative technology   1

CHAPTER 2 Reservoirs

    2.1   Conditions for occurrence   7
    2.2   Reservoir pressures   10
    2.3   Fluid pressures in a hydrocarbon zone   12
    2.4   Reservoir temperatures   13
    2.5   Nature of reservoir fluids   14
    2.6   Reservoir data – sources   14

CHAPTER 3 Oilwell Drilling

    3.1   Operations   20
    3.2   Costs   23
    3.3   Well completions and oilwell casing   23
    3.4   Completion   28
    3.5   Drilling fluid control   28
    3.6   Rheology of well fluids (drilling muds and cements)   29
    3.7   Formation breakdown pressures and leak off tests   30
    3.8   Data acquisition during drilling   30
    3.9   Mud fluids for core recovery   31
    3.10   Drilling optimization   32

	3.11	Turbine versus conventional rotary	33
	3.12	Special problems in drilling	33
	3.13	Completion for production	36

CHAPTER 4 Properties of Reservoir Fluids

4.1	Volumetric and phase behaviour of hydrocarbon systems	40
4.2	Applications to field systems	41
4.3	Compressibility	42
4.4	Measurement and prediction of reservoir fluid properties	43
4.5	Formation volume factors, B	49
4.6	Gas–oil ratios	51
4.7	Direct measurements – PVT analysis	52
4.8	Generalized correlations for liquid systems	54

CHAPTER 5 Characteristics of Reservoir Rocks

5.1	Data sources and application	62
5.2	Coring decisions	64
5.3	Conventional and oriented coring	66
5.4	Coring mud systems	66
5.5	Core preservation	67
5.6	Well site controls	68
5.7	Core for special core analysis	68
5.8	Core-derived data	68
5.9	Geological studies	68
5.10	Routine core analysis	69
5.11	Porosity	72
5.12	Permeability	78
5.13	Relationships between porosity and permeability	84

CHAPTER 6 Fluid Saturation: influence of wettability and capillary pressure

6.1	Equilibrium conditions	92
6.2	Laboratory measurements and relationship with reservoir systems	93
6.3	Pore size distribution	96
6.4	Capillary pressure hysteresis	97
6.5	Saturation distributions in reservoir intervals	98
6.6	Correlation of capillary pressure data from a given rock type	99

CHAPTER 7 Relative permeability and multiphase flow in porous media

7.1	Definitions	102
7.2	Fractional flow	104
7.3	Effects of permeability variation	106

CONTENTS

7.4	Wettability effects	108
7.5	Laboratory determination of relative permeability data	109
7.6	Residual saturations	111
7.7	*In situ* wettability control	112
7.8	Relative permeability from correlations	112
7.9	Validation of relative permeability data for use in displacement calculations	113
7.10	Pseudo-relative permeability in dynamic systems	115
7.11	Static pseudo-relative permeability functions	115

CHAPTER 8 Representation of volumetric estimates and recoverable reserves

8.1	In-place volume	122
8.2	Areal extent of reservoirs	122
8.3	Thickness maps	124
8.4	Lithofacies representation	125
8.5	Isoporosity maps	125
8.6	Isocapacity maps	126
8.7	Hydrocarbon pore volume maps	126
8.8	Probabilistic estimation	127
8.9	Recovery factors and reserves	128
8.10	Distribution of equity in petroleum reservoirs	130

CHAPTER 9 Radial Flow Analysis of Well Performance

9.1	Radial flow in a simple system	134
9.2	Development of the line source solution	135
9.3	Radial equations in practical units	136
9.4	Application of analytical solutions in well test methods	136
9.5	Pressure build-up analysis	139
9.6	Skin effect	140
9.7	Pressure drawdown and reservoir limit testing	142
9.8	Gas well testing	143
9.9	Well test procedures	145
9.10	Well testing and pressure analysis	150

CHAPTER 10 Reservoir Performance Analysis

10.1	Recovery from gas reservoirs	157
10.2	Primary recovery in oil reservoirs	159
10.3	Gravity segregation and recovery efficiencies	164
10.4	Material balance for reservoirs with water encroachment or water injection	165
10.5	Accuracy of the gross material balance equation	168

CHAPTER 11 Secondary Recovery and Pressure Maintenance

11.1	Displacement principles	173

	11.2	Factors influencing secondary recovery and pressure maintenance schemes	175
	11.3	Quality of injection fluids and disposal of brines	183

CHAPTER 12 Improved Hydrocarbon Recovery

12.1	Targets	191
12.2	The influence of recovery mechanism on residual oil	191
12.3	Permeability improvement	193
12.4	Miscible displacement mechanisms	194
12.5	Miscible flood applications	195
12.6	Chemical flood processes	196
12.7	Heavy oil recovery	200
12.8	Thermal energy	204
12.9	Gas condensate reservoirs	207
12.10	Volatile oil reservoirs	211

CHAPTER 13 Factors Influencing Production Operations

13.1	The production system	218
13.2	Reservoir behaviour in production engineering	220
13.3	Wellbore flow	221
13.4	Field process facilities	224
13.5	Natural gas processing	224
13.6	Crude oil processing	226
13.7	Heavy oil processing	228
13.8	Produced water treatment	228
13.9	Injection water treatment	229
13.10	Crude oil metering	229

CHAPTER 14 Concepts in Reservoir Modelling and Application to Development Planning

14.1	Models	233
14.2	Equations of multiphase flow	234
14.3	Simulator classifications	235
14.4	Simulator application	235
14.5	Reservoir description in modelling	237
14.6	Application of reservoir models in field development	248

APPENDIX 1 SPE Nomenclature and Units 257

Units	257
SPE Symbols Standard	259
Symbols alphabetized by physical quantity	267
Subscripts alphabetized by physical quantity	300

APPENDIX 2 Solutions to Examples in Text 310

INDEX 357

PREFACE

The need for this book has arisen from demand for a current text from our students in Petroleum Engineering at Imperial College and from post-experience Short Course students. It is, however, hoped that the material will also be of more general use to practising petroleum engineers and those wishing for an introduction into the specialist literature.

The book is arranged to provide both background and overview into many facets of petroleum engineering, particularly as practised in the offshore environments of North West Europe. The material is largely based on the authors' experience as teachers and consultants and is supplemented by worked problems where they are believed to enhance understanding.

The authors would like to express their sincere thanks and appreciation to all the people who have helped in the preparation of this book by technical comment and discussion and by giving permission to reproduce material. In particular we would like to thank our present colleagues and students at Imperial College and at ERC Energy Resource Consultants Ltd. for their stimulating company, Jill and Janel for typing seemingly endless manuscripts; Dan Smith at Graham and Trotman Ltd. for his perseverence and optimism; and Lesley and Joan for believing that one day things would return to normality.

John S. Archer and Colin G. Wall
1986

Foreword

Petroleum engineering has developed as an area of study only over the present century. It now provides the technical basis for the exploitation of petroleum fluids in subsurface sedimentary rock reservoirs.

The methodology in petroleum reservoir development requires the testing and evaluation of exploration and appraisal wells to discover the volume in place and productivity of compressible hydrocarbon fluids. The knowledge of the reserve-base is always insufficient as natural reservoirs are heterogeneous in their geometry and character. Petroleum engineers play a leading role in the design of recovery systems which require flexibility in well placement and the sizing of surface facilities for export processing to ensure that the products meet the specifications required for transportation by pipeline or tanker.

Petroleum fluids are complex mixtures of many hydrocarbons and currently prediction of their behaviour at reservoir pressures of up to 14 500 psia (1000 bar) and 450°F (230°C) is based on attempts to understand their thermodynamics. The prediction of fluid behaviour in heterogeneous reservoirs is aided by sophisticated mathematical modelling using powerful computers. Indeed, some of the most powerful computers available today are dedicated to the modelling of reservoir behaviour and a modern petroleum engineer must be capable of making full use of them. The petroleum engineer's responsibilities are of necessity very wide, and skills are required to design for data acquisition which will allow updating of reservoir models in the light of production experience. This may well lead to decisions on variations in the original production scheme. In all these activities safety and economy are mandatory.

The petroleum engineer is a resource manager, and has an obligation to analyse all the data available and to interpret it effectively in order to forecast the future performance of the reservoir, wells and plant. That the oil and gas industry is profitable is largely due to the emergence of petroleum engineering and the techniques which have been developed for its application.

The breadth of the interdisciplinary knowledge needed by today's petroleum engineer is ever increasing, and a study of the current literature shows the vast amount of effort now being applied to further the understanding and improve predictions of the behaviour of reservoir fluids and to increase their recovery. These studies and operations require the disciplines of physics, geology, classical engineering, mathematics and computer science. With such a spread of disciplines the availability of a text, describing the basics of petroleum engineering, is welcomed.

The authors are the present and past Heads of Petroleum Engineering in the Department of Mineral Resources Engineering of the Royal School of Mines, Imperial College, London and

FOREWORD

both have had field experience with major oil companies before joining Imperial College. The College has been a centre for the study of petroleum recovery since the early years of this century with courses in Oil Technology commencing in 1913. The Petroleum Engineering Section moved to its present location in the Mineral Resources Engineering Department at the Royal School of Mines in 1973 and currently runs undergraduate and Master of Sciences courses and has active post-graduate and post-doctoral research groups.

This book will both give students a good grounding in petroleum engineering and be valuable to the practising engineer as a comprehensive reference work. Through its extensive bibliography the reader will also be guided to more specialised branches of the petroleum engineering literature.

A.H. Sweatman
(ex-Chief Production Engineer, British Petroleum Company
and Visiting Professor
Petroleum Engineering
Imperial College, London, UK)

Chapter 1

Introduction

1.1 PETROLEUM ENGINEERING: A CREATIVE TECHNOLOGY

The function of petroleum engineering is to provide a basis for the design and implementation of techniques to recover commercial quantities of natural petroleums. It is of necessity a broadly based technology drawing upon the foundations of engineering, geology, mathematics, physics, chemistry, economics and geostatistics. As an engineering subject it is a little anomalous, in that design is based on observation of production performance and on a representation of the reservoir inferred from very limited sampling. Unlike many branches of engineering, reservoirs cannot be designed to fulfil a particular task, but rather an ill-defined naturally occurring reservoir is induced to produce some fraction of its contents for as long as is considered commercially attractive. With the passage of time and cumulative production, more information on the nature of the reservoir can be accumulated and the production methods can be modified. Petroleum engineering can thus represent an exercise in the application of uncertainty to design. A route to problem solution in petroleum engineering shown as Table 1.1 has been adapted from Timmerman [15]. The terminology of the subject contains varying degrees of confidence in the representation of the inplace and recoverable resource base. In Chapter 8 we discuss the representation of 'proven' quantities of hydrocarbon in terms of the availability of information and the existence of the technology to exploit recovery on commercially attractive terms. The economics of hydrocarbon recovery processes is inextricably linked with the practice of petroleum engineering. On a project basis, a petroleum engineer has a responsibility to present analyses of schemes that are both technically and financially attractive.

In the current climate of deeper reservoir exploration and increased exploitation of offshore reservoirs in the world's sedimentary basins (Fig. 1.1), costs of production are significant. For example, in terms of pre-tax cost of oil production from a 2000 mSS onshore well compared with a 3000 mSS offshore well, a ratio of 1:10 might be expected. Current exploration in maturing hydrocarbon provinces is centred on more subtle trapping mechanisms than structural *highs* and on smaller accumulations. The further recovery of hydrocarbons from reservoirs approaching the end of conventional development processes requires the cost-effective application of *enhanced* (EOR) or *improved* (IHR) hydrocarbon recovery processes. The exploitation of heavy oil (API) gravity less than 20° API) and of gas condensate and volatile oil reservoirs (API gravity greater than 45° API) requires special petroleum engineering effort, particularly in high-pressure or offshore reservoirs. Developments in the recovery of hydro-

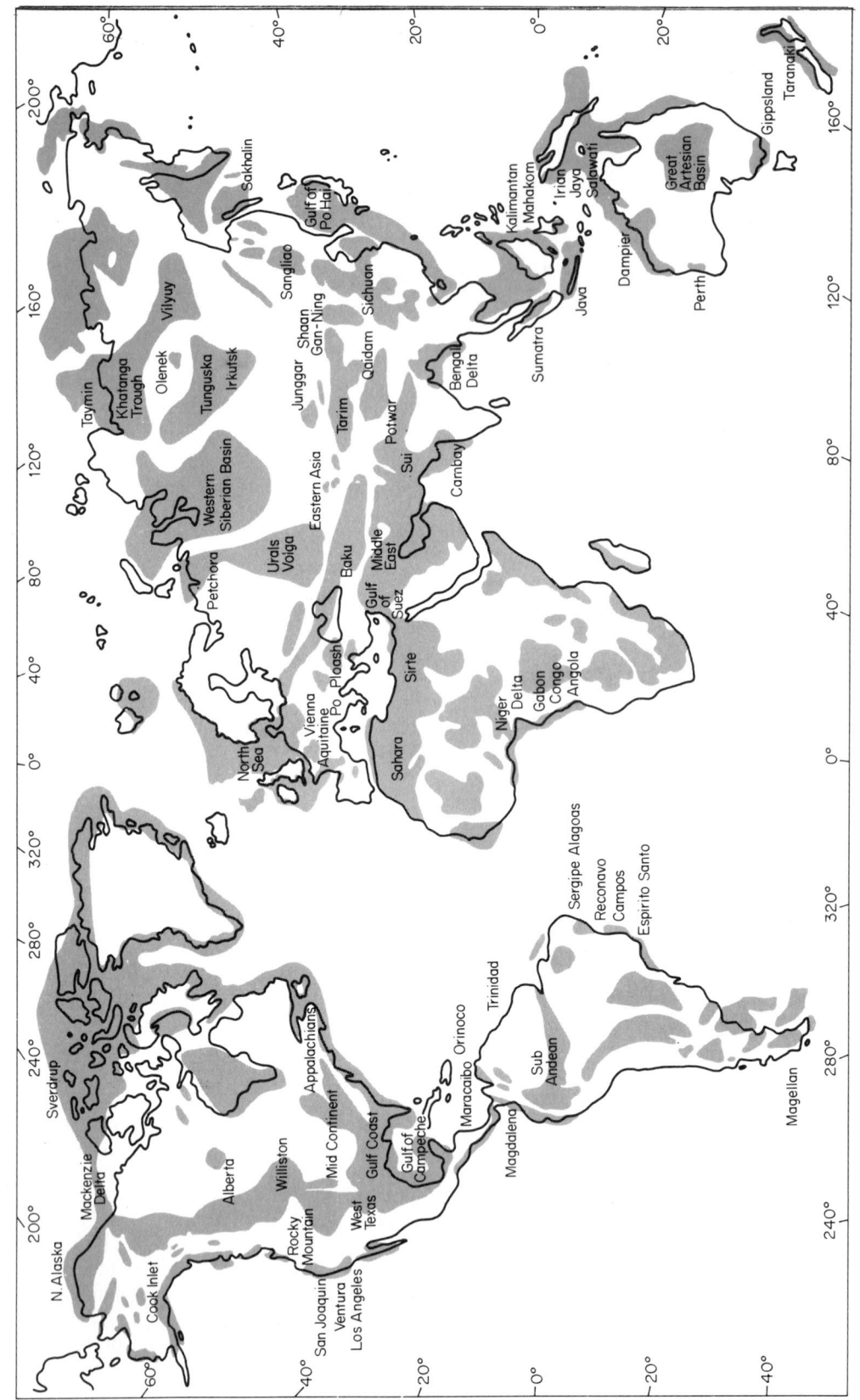

Fig. 1.1 World sedimentary basins.

1 INTRODUCTION

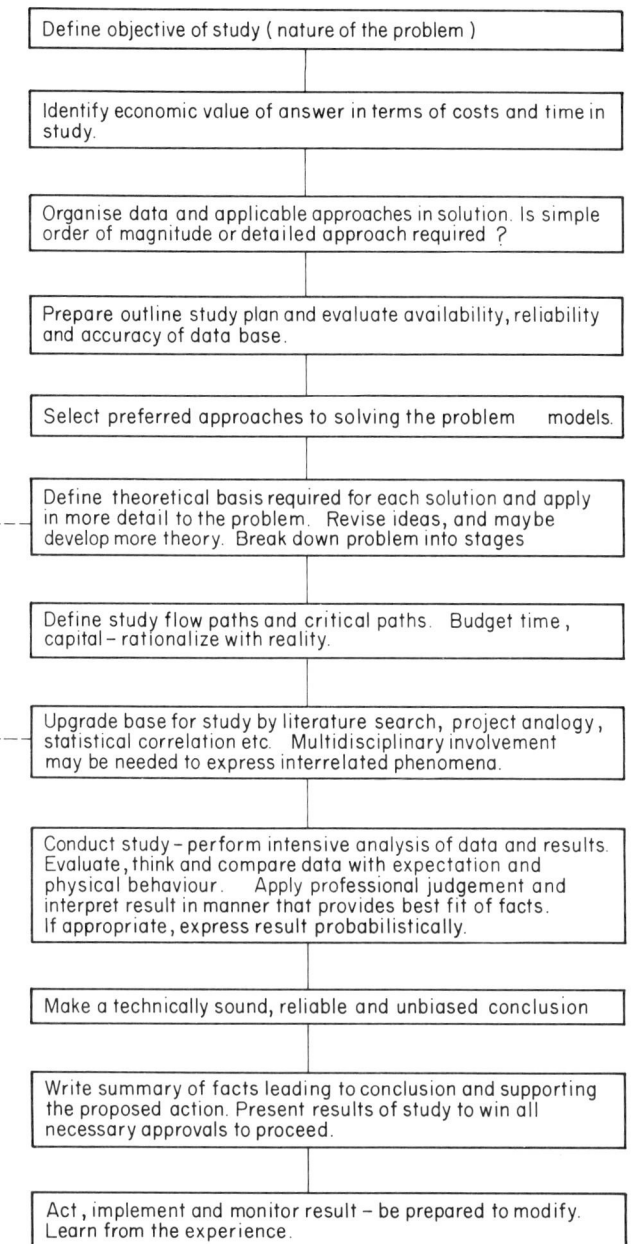

TABLE 1.1 Problem solving in petroleum engineering

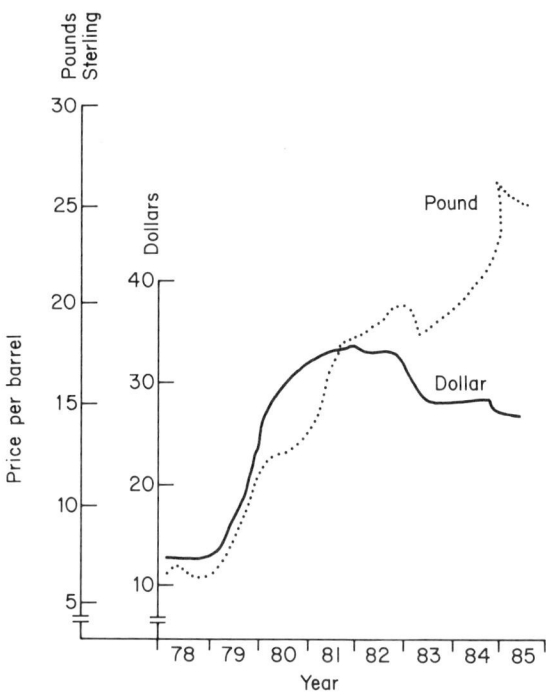

Fig. 1.2 Variation in price of average Middle East crude oil.

carbon from oil sands and oil shales requires that petroleum engineering methods are combined with the technologies of mining engineers and chemical engineers.

The current fiscal environment, particularly in NW Europe, has led to an increased economic and political awareness amongst petroleum engineers. The prices of oil in world markets is partly controlled by agreement amongst producing nations. Figure 1.2 shows the fluctuation in the average official Middle East crude oil price, compiled from figures in *Petroleum Intelligence Weekly*. The effect of the exchange rate fluctuation between the pound sterling and the US dollar is also clearly seen. The variations have an effect on exploitation policies of operating companies. The costs of reservoir development generally require an operating company to raise a substantial quantity of money from loans – the repayment terms of which are linked to a representation of reservoir production uncertainties. The cost profile for the development of an offshore oil field on the Continental Shelf (UKCS) with some 75 million barrels of recoverable oil is indicated in Fig. 1.3. The investment was over £500 million in 1985 currency, or at mid-1985 exchange rates it was over US$600 million. It is also very clear from this figure that much of the investment is exposed at least five years prior to any production revenue. This fact alone leads to a petroleum engineering design criterion of high initial production rates to shorten payout times. The development of offshore oil fields on the UKCS with recoverable reserves less than 100 million barrels should provide a greater challenge to petroleum engineers than those under development at the end of 1983, which averaged some 400 million barrels of recoverable reserves for each reservoir.

4 PETROLEUM ENGINEERING: PRINCIPLES AND PRACTICE

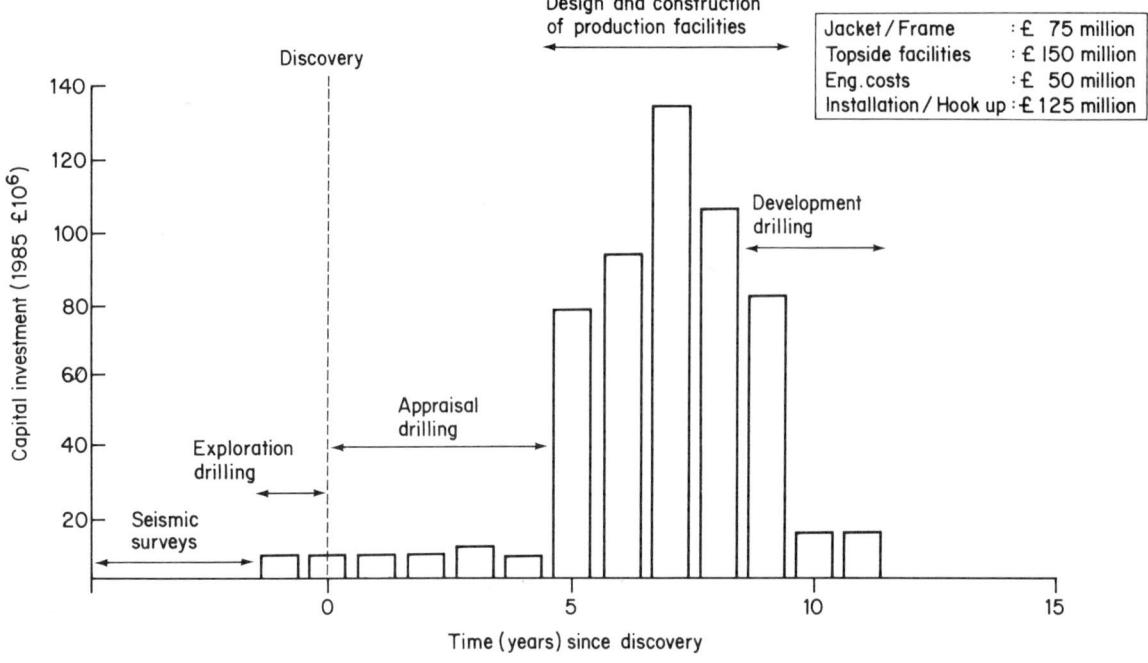

Fig. 1.3 Capital expenditure profile for hypothetical 1985 UKCS offshore oil discovery with 75 million stock tank barrels of recoverable oil.

Fig. 1.4 Offshore exploration using the semi-submersible rig Sea Conquest (Photo courtesy of BP.)

1 INTRODUCTION

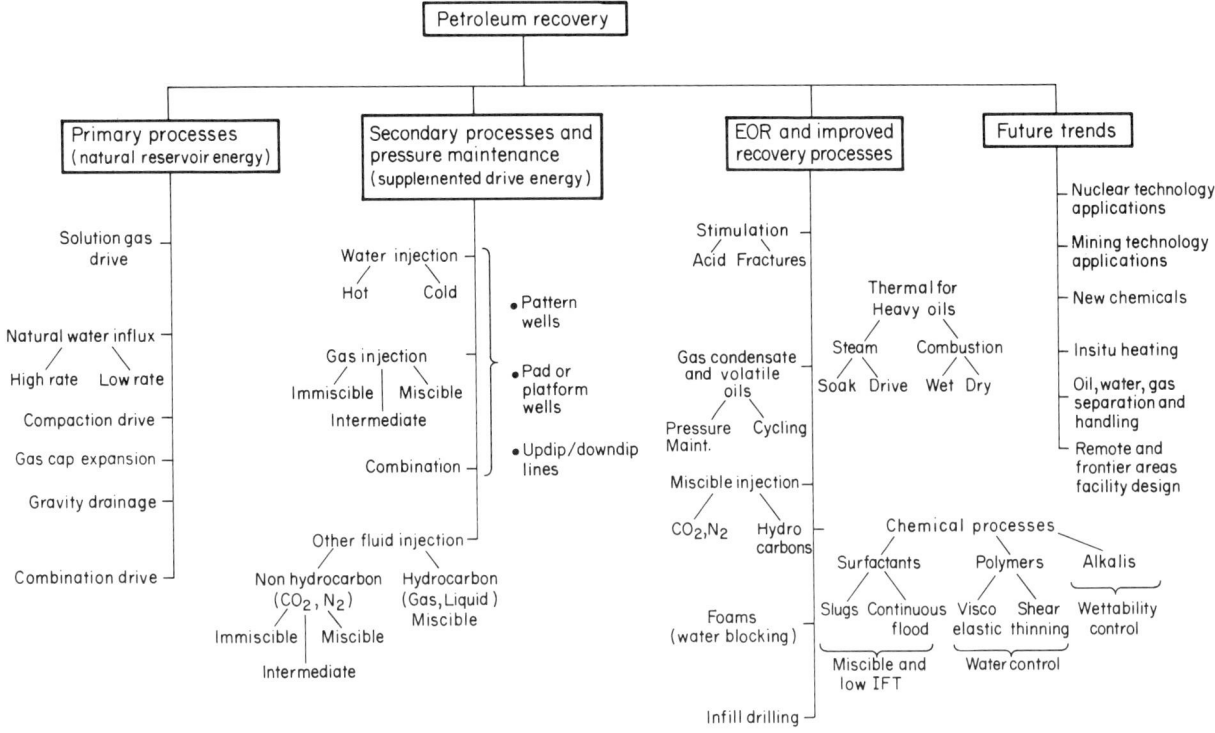

Fig. 1.5 Petroleum recovery methods.

References

[1] *International Petroleum Engineering Encyclopedia*
Pennwell Publish. Co. Tulsa (1983).
[2] British Petroleum Company plc
BP Statistical Review of World Energy (June 1984).
[3] Department of Energy
Development of the oil and gas resources of the United Kingdom (the Brown Book), Pub. D.En. 1983 and annually.
[4] Esso UK plc
Opportunities for British Industry (1984).
[5] UK Offshore Operators Association
Potential Oil and Gas Production from the UK Offshore to the Year 2000, Technical Paper (Sept. 1984).
[6] Brush, R.M. and Marsden, S.S.
Bias in engineering estimation, *JPT* (1982) 433.
[7] Marks, V.E.
Further small offshore oilfield developments, SPE Paper 12988, *Proc. Europec* (1984), 265.
[8] Underdown, D.J.
The role of taxation in optimising the exploitation of the UK continental shelf, SPE Paper 13008, *Proc. Europec* (1984), 407.
[9] Archer, J.S.
Reservoir definition and characterisation for analysis and simulation, *Proc. 11th World Pet. Cong.*, London (1983), Paper PD6(1).
[10] Perrodon, A.
Dynamics of oil and gas accumulations, *Elf Aquitaine*, Mem 5. Pau 1983.
[11] Master, C.D.
Distribution and quantitive assessment of world petroleum reserves and resources, *Proc. 11th World Pet. Cong.* (1983), Paper PD11(1).

[12] Hertz, D.B.
Risk analysis in capital investment, *Harvard Business Review* (Jan. – Feb. 1984) and in *Pet. Trans.* Reprint Series 3, SPE of AIME (1970), 176.

[13] Northern, I.G.
Investment decisions in petroleum exploration and production, *JPT* (July 1964), 727.

[14] Attanasi, E.D. and Haynes, J.L.
Economics and appraisal of conventional oil and gas (in the Western Gulf of Mexico), *JPT* (Dec. 1984), 2171.

[15] Timmerman, E.H.
Practical Reservoir Engineering, Pennwell Publishing, Tulsa (1982), 2 vols.

[16] Parra, F.R.
Financial requirements and methods of financing petroleum operations in developing countries. *Proc. UN Conf., Petroleum Exploration Strategies in Developing Countries,* The Hague (1981), Graham & Trotman 177–192.

Chapter 2
Reservoirs

2.1 CONDITIONS FOR OCCURRENCE

We may define a *reservoir* as an accumulation of hydrocarbon in porous permeable sedimentary rocks. The accumulation, which will have reached a fluid pressure equilibrium throughout its pore volume at the time of discovery, is also sometimes known as a *pool*. A hydrocarbon field may comprise several reservoirs at different statigraphic horizons or in different pressure regimes. The setting for hydrocarbon accumulation is a sedimentary basin that has provided the essential components for petroleum reservoir occurrence, namely (a) a source for hydrocarbons, (b) the formation and migration of petroleum, (c) a trapping mechanism, i.e. the existence of traps in porous sedimentary rock at the time of migration and in the migration path.

The origin of sedimentary basins and the genesis, migration and entrapment of hydrocarbons is an extensive topic covered in the geological literature[1–10] and only the essential details are reviewed here. The discovery of oil by exploration well drilling in some of the world's sedimentary basins is shown in Figs 2.1 and 2.2 in terms of intensity of exploration, maturity and exploration effort and volume by discovery. The North Sea province is seen as a relatively young exploration area in a high yield basin. From a petroleum engineering perspective it is convenient to think of sedimentary basins as accumulations of water in areas of slow subsidence into which sediments have been transported. The primary depositional processes and the nature of the sediments have a major influence on the porosity and permeability of reservoir rocks. Secondary processes, including compaction, solution, chemical replacement and diagenetic changes, can act to modify further the pore structure and geometry. With compaction, grains of sediment are subject to increasing contact and pore fluids may be expelled from the decreasing pore volume. If the pore fluids cannot be expelled the pore fluid pressure may increase. It is believed that petroleum originates from the anaerobic decomposition of fats, proteins and carbohydrates in marine or estuarine plant and animal matter, plankton and algae. This needs rapid sedimentation in organic rich waters and leads to the accumulation of organic rich clays in an anaerobic environment. Petroleum formation requires that organic source clays become *mature* by subjection to pressure and temperature. A temperature *window* in the range 140°F to 300°F, which may be correlatable with burial depth and geological time, seems to be optimal for formation of hydrocarbon mixtures classified as oils, as shown in Figs 2.3 and 2.4. Prolonged exposure to high temperatures, or shorter exposure to very high temperatures, may lead progressively to the generation of hydrocarbon mixtures characterized as condensates, wet gases and gas. The average organic content of potential source rocks is about 1% by weight. The Kimmeridge clay, the principal source rock for North Sea oil averages about 5% carbon (~ 7% organic matter) with local rich streaks greater than 40%. The hydrogen content of the organic matter should be greater than 7% by weight for potential as an oil source. It is a rule of thumb that for each percentage

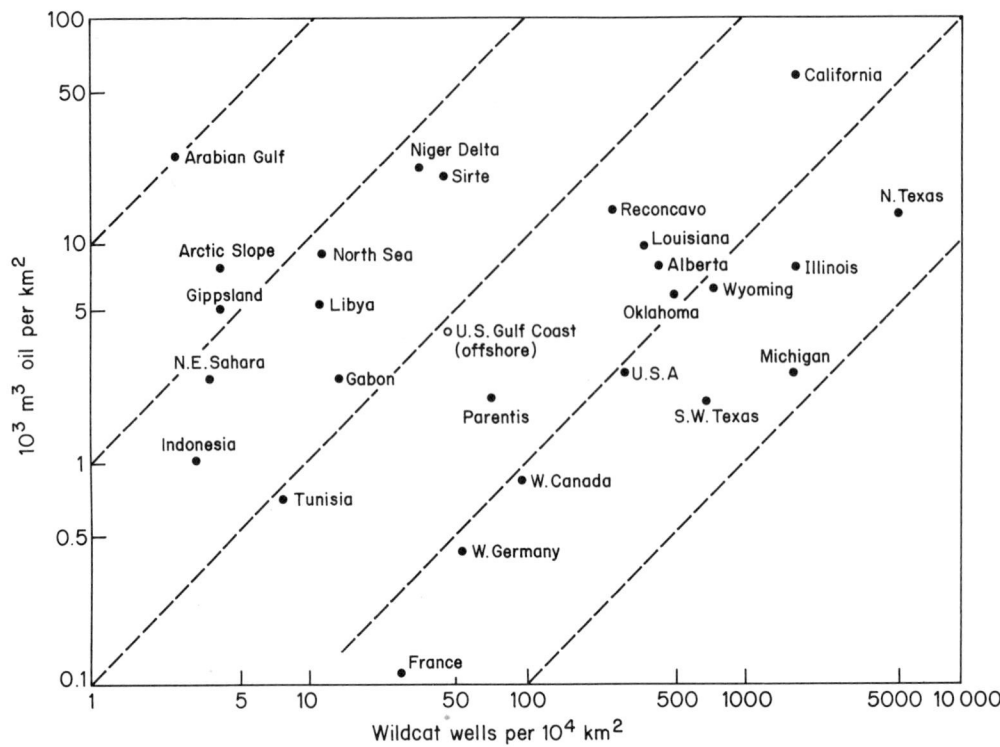

Fig. 2.1 Wildcat well success v. exploration yield and discovery (after [18]).

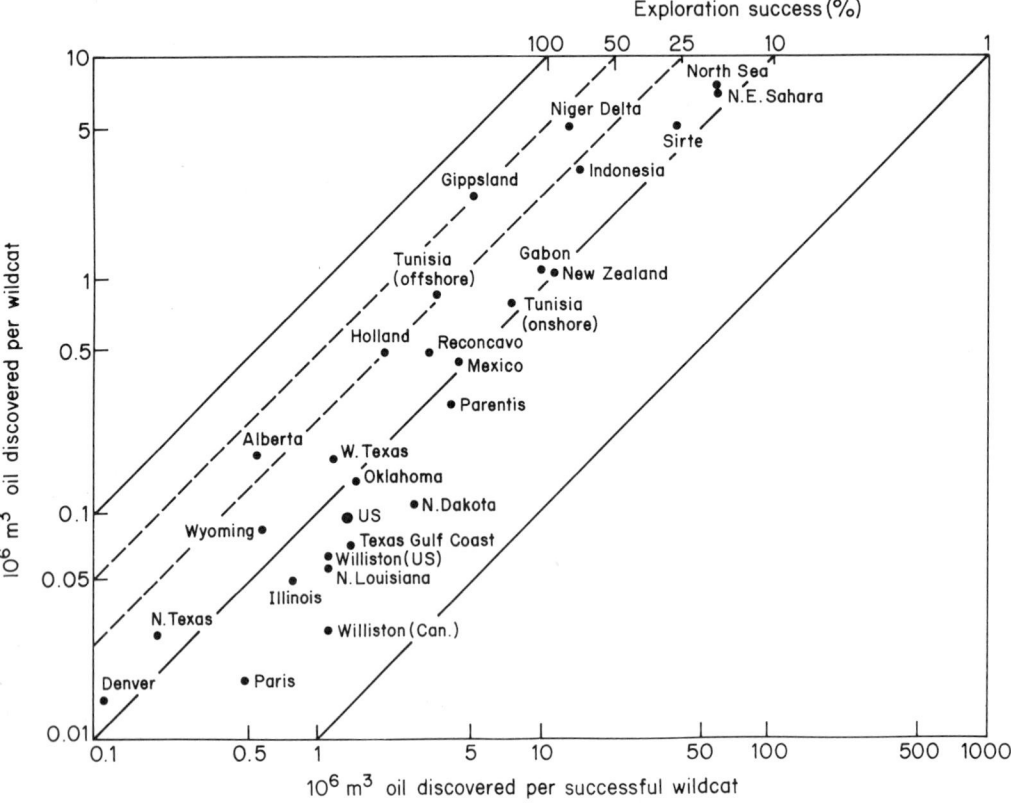

Fig. 2.2 Historical relationships between exploration intensity and yield (after [18]).

2 RESERVOIRS

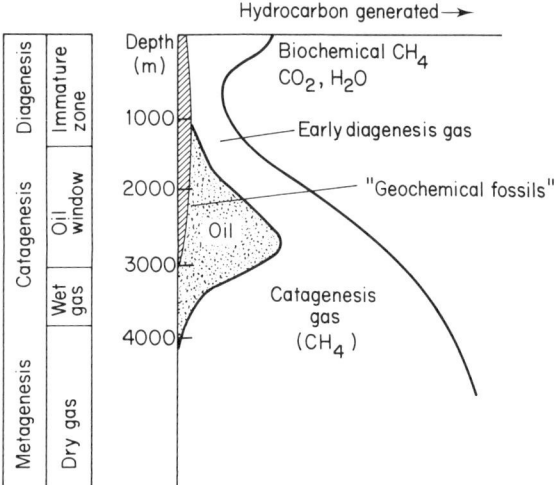

Fig. 2.3 Hydrocarbon generation for normal geothermal gradient (after [18]).

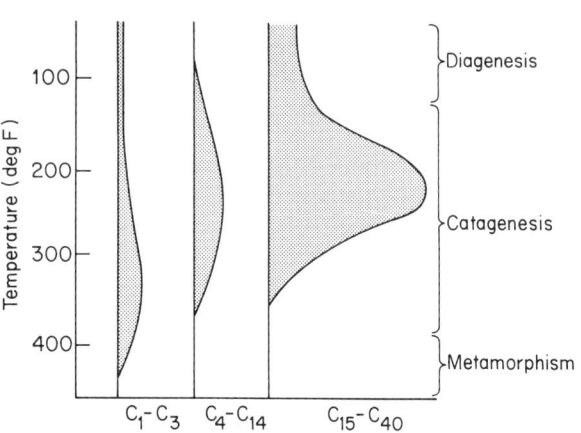

Fig. 2.4 Variation in the quantity of hydrocarbons contained in fine grained sediments as a function of temperature (after [1]).

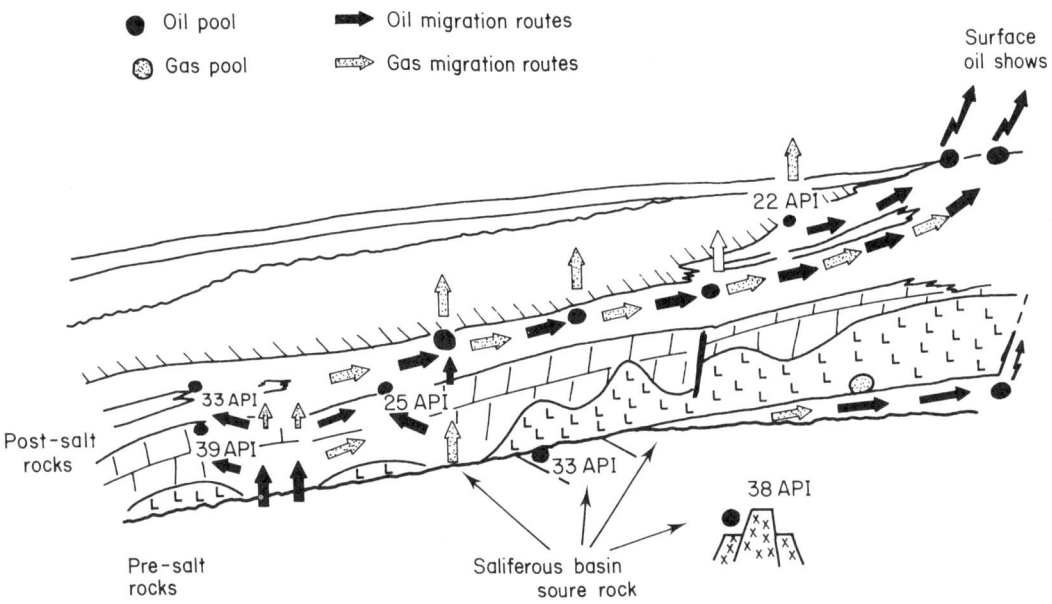

Fig. 2.5 Migration pattern model for a section of the Congo coastal basin (after [6]).

point of organic carbon in mature source rocks, some 1300–5000 cubic metres of oil per km² m (10–40 barrels of oil per acre-ft) of sediment could be generated. It is not, however, necessarily true that all the oil generated will be expelled or trapped in porous rocks.

Migration of petroleum generated from source rocks is not well understood. Since the generation of petroleum is accompanied by volume changes which can lead to high local pressures, there may well be an initiation of microfractures which provide an escape route into permeable systems such as sedimentary rocks or fault planes. The source rock microfractures are believed to heal as pressures are dissipated. The movement of petroleum may have been as a solution in water or as distinct oil or gas phases – there is no consensus on this topic. The migration process involves two main stages, namely through the source rock and then through a permeable system. In the permeable system the transport occurs under conditions of a fluid potential gradient which may take the hydrocarbon to surface or to some place where it becomes trapped. It might be assumed that less than 10% of petroleum generated in source rocks is both expelled and trapped, as shown in the example of Fig. 2.5.

The characteristic forms of petroleum trap are known as structural and stratigraphic traps, with the great majority of known accumulation being in the former style. Structural traps may be generally subdivided into anticlinal and fault traps which are described in terms of the shape of the sedimentary beds and their contacts, as shown in Figs 2.6 and 2.7.

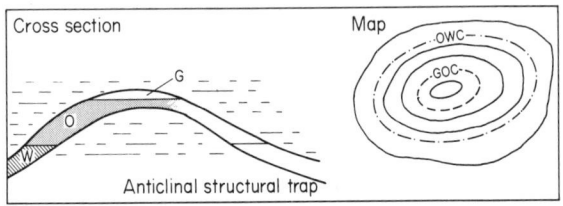

Fig. 2.6 Anticlinal structure.

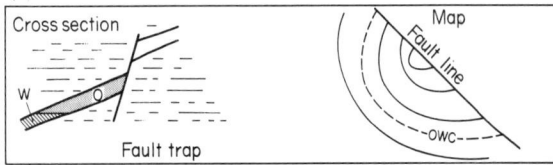

Fig. 2.7 Fault trap.

Impermeable rocks provide seals above and below the permeable reservoir rocks. At equilibrium conditions the density differences between the oil, gas and water phases can result in boundary regions between them known as fluid contacts, i.e. oil–water and gas–oil contacts.

Stratigraphic traps result when a depositional bed changes from permeable rock into fine-grain impermeable rock (Fig. 2.8). Examples of this process occur in the more distal regions and in discontinuous sands of river channels, in deeper water deltaic sediments and in the enveloping sediments of limestone reefs.

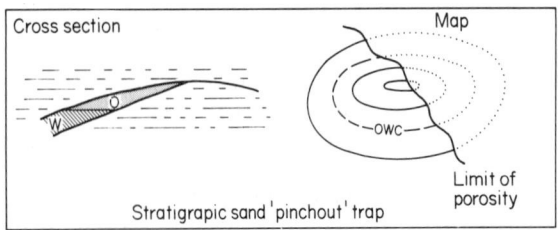

Fig. 2.8 Stratigraphic *pinch out* trap.

In the North Sea and several other sedimentary areas, a style of trapping is found which results from the truncation of inclined permeable beds by an impermeable unconformity surface (Fig. 2.9). It is debatable whether this trap should be called stratigraphic because of the trapping by fine grained sediments, or structural after the geological nature of the unconformity.

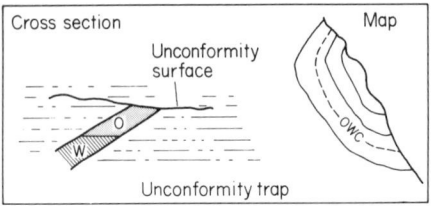

Fig. 2.9 Unconformity trap.

Many reservoirs exist as the result of a combination of structural and stratigraphic features. In the Viking Graben area of the northern North Sea, the Brent Sand reservoirs are characteristically faulted deltaic sands truncated by the Cretaceous unconformity. The seal for the Middle Jurassic is either the clays and shales of the unconformity itself or Upper Jurassic Kimmeridge shale. The down faulted mature Kimmeridge shale also provides a source for the oil as shown in Fig. 2.10.

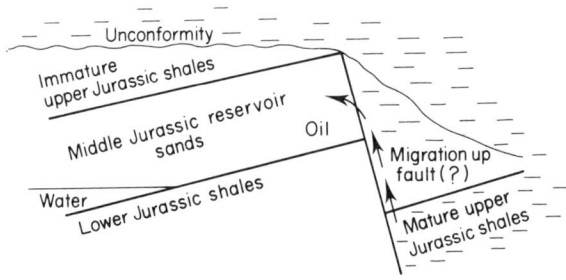

Fig. 2.10 Brent Sand trap – UKCS.

2.2. RESERVOIR PRESSURES

Hydrocarbon reservoirs are found over a wide range of present day depths of burial, the majority being in the range 500–4000 mSS. In our concept of the petroliferous sedimentary basin as a region of water into which sediment has accumulated and hydrocarbons have been generated and trapped, we may have an expectation of a regional hydrostatic gradient.

2 RESERVOIRS

That is, in a water column representing vertical pore fluid continuity the pressure at any point is approximated by the relationship

$$P_X = X \cdot G_w$$

where X is the depth below a reference datum (such as sea level) and G_w is the pressure exerted by unit height of water. The value of G_w depends on the salinity of the waters and on the temperature in the system. Fresh water exhibits a gradient of 9.79 kpa/m (0.433 psi/ft) and reservoir water systems are commonly encountered with gradients in the range 10 kpa/m (0.44 psi/ft) to 12 kpa/m (0.53 psi/ft). In reservoirs found at depths between 2000 mSS and 4000 mSS we might use a gradient of 11 kpa/m to predict pore fluid pressures around 220 bar to 440 bar as shown in Table 2.1.

Under certain depositional conditions, or because of movement of closed reservoir structures, fluid pressures may depart substantially from the normal range (see for example Magara[2]). One particular mechanism responsible for overpressure in some North Sea reservoirs is the inability to expel water from a system containing rapidly compacted shales. Abnormal pressure regimes are evident in Fig. 2.11, which is based on data from a number of Brent Sand reservoirs in the North Sea. All show similar salinity gradients but different degrees of overpressure, possibly related to development in localized basins[12].

Any hydrocarbon bearing structure of substantial relief will exhibit *abnormally* high pressures at the crest when the pressure at the hydrocarbon–water contact is normal, simply because of the lower density of the hydrocarbon compared with water.

There is a balance in a reservoir system between the pressure gradients representing rock overburden (G_r), pore fluids (G_f) and sediment grain pressure (G_g). The pore fluids can be considered to take part of the overburden pressure and relieve that part of the overburden load on the rock grains. A representation of this is

$$G_r = G_f + G_g$$

and is shown in Fig. 2.12.

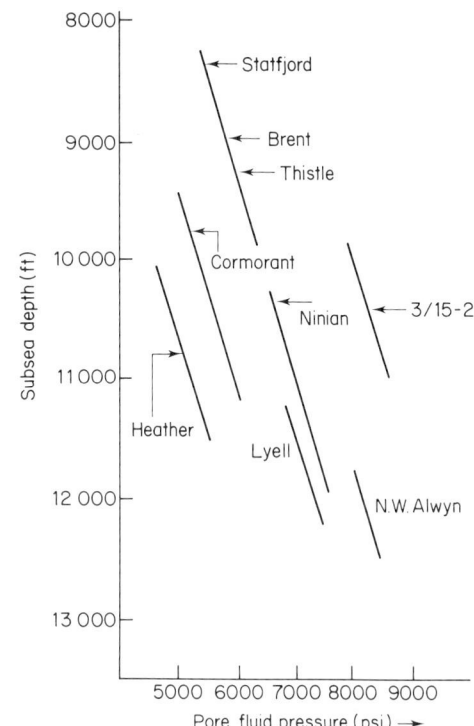

Fig. 2.11 Pressure regimes in Brent Sand reservoirs (after [12]).

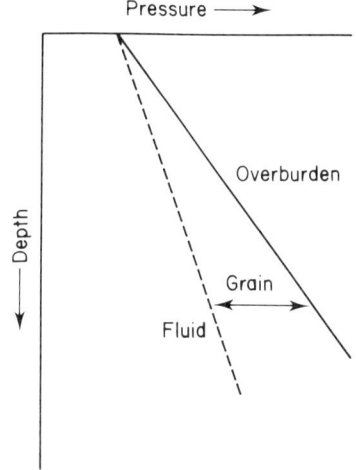

Fig. 2.12 Equilibrium gradients.

TABLE 2.1

Depth m SS (ft SS)	Gradient kPa/m (psi/ft):	Pore fluid pressure, bar (psi)		
		9.79 (0.433)	10.17 (0.45)	11.3 (0.5)
2000 (6562)		195 (2841)	203 (2953)	226 (3281)
3000 (9843)		293 (4262)	305 (4429)	339 (4921)
4000 (13 124)		392 (5683)	406 (5906)	452 (6562)

As fluid pressure is reduced, the rock grains take a proportionally increasing part of the overburden load. The magnitude of the overburden gradient is approximately 22.6 kpa/m (1 psi/ft). Abnormal fluid pressures are those not in initial fluid equilibrium at the discovery depth. Magara[2] has described conditions leading to abnormally high and abnormally low pressures. Some explanations lie in reservoirs being found at present depths higher or lower than the depth at which they became filled with hydrocarbon. This may be the result of upthrust or downthrown faulting. Overpressure from the burial weight of glacial ice has also been cited. In Gulf coast and North Sea reservoirs overpressure is most frequently attributed to rapid deposition of shales from which bound water cannot escape to hydrostatic equilibrium. This leads to overpressured aquifer–hydrocarbon systems[12].

2.3 FLUID PRESSURES IN A HYDROCARBON ZONE

We will define a fluid contact between oil and water as the depth in the reservoir at which the pressure in the oil phase (P_o) is equal to the pressure in the water phase (P_w). Reference to Fig. 2.13 shows this condition as an equilibrium condition. Strictly speaking, the position $P_o = P_w$ defines the free water level (FWL), as in some reservoirs a zone of 100% water saturation can occur above the free water level by capillarity. This effect is described further in Chapter 6. In layered sand systems which do not have equilibrium with a common aquifer, multiple fluid contacts can be found as shown in Fig. 2.14.

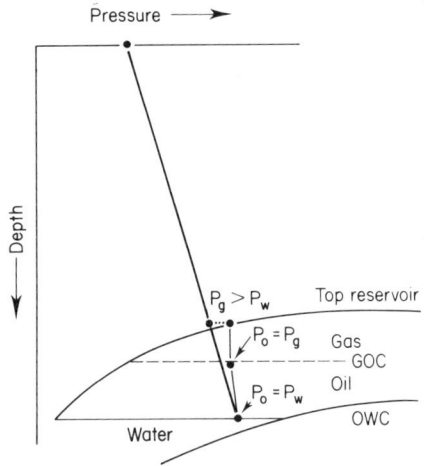

Fig. 2.13 Pressure equilibrium in a static system.

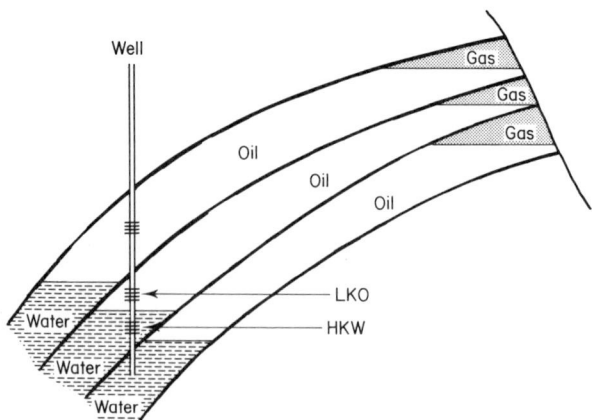

Fig. 2.14 Multiple fluid contacts showing tested interval indication of LKO and HKW.

At the water–oil contact, P_w is given by the average temperature–salinity gradient of water from the surface datum of sea level. As an equation of a straight line this is

$$P_w = X_{owc} G_w + C_1$$

where the constant C_1 can represent any degree of under- or overpressure.

At some depth, X_{owc}, we can therefore write $P_w(owc) = P_o(owc)$. Above the water–oil contact, the pressure in the oil phase is the pressure that the oil had at the water–oil contact less the density head of the oil. At any depth X_D above the water–oil contact, the pressure in the oil phase will be $P_o(X_D)$ as follows:

$$P_{oX(D)} = P_o(owc) - [\rho_o g' (X_{owc} - X_D)]$$

where ρ_o is the local oil density. At the gas-oil contact $P_o(goc) = P_g(goc)$. The pressure in the gas phase at the top of the reservoir X_T will therefore be:

$$P_g(X_{(T)}) = P_{owc} - [\rho_o g' (X_{owc} - X_{goc})] \\ - [\rho_g g' (X_{goc} - X_{(T)})]$$

Where g' is the ratio of gravitational acceleration g to the universal constant g_c. There can be significant difference at depth X_T between $P_g(X_T)$ and the calculated P_w at the same depth using the equation $P_w = X_T G_w + C_1$. This difference accounts for gas-kicks encountered sometimes during drilling operations as gas sands are penetrated.

The estimation and recognition of fluid contacts are essential in evaluating hydrocarbons in place. The placing of fluid contacts often results from

2 RESERVOIRS

consideration of information from several sources, i.e.

(a) equilibrium pressures from RFT or gradient surveys;
(b) equilibrium pressures from well tests;
(c) flow of particular fluid from particular minimum or maximum depth;
(d) fluid densities from formation samples;
(e) saturations interpreted from wireline log data;
(f) capillary pressure data from core samples;
(g) fluid saturations from recovered core.

The proving of an oil–water contact from flow tests gives rise to the terminology of LKO (lowest known oil) or ODT (oil down to) depths and HKW (highest known water) or WUT (water up to) depths. The combination of uncertainties in fluid properties for gradient extrapolation and in well test intervals means that a fluid contact is often represented as a depth range until data from several wells in a reservoir have been correlated.

A particular difficulty in hydrocarbon–water contact evaluation concerns identification in the presence of increasing shaliness. The effect of shaliness is manifest in small pore throats and high threshold capillary pressure which give high water saturation. Some of these difficulties may be resolved by capillary pressure analysis using representative core samples.

2.4 RESERVOIR TEMPERATURES

Reservoir temperature may be expected to conform to the regional or local geothermal gradient. In many petroliferous basins this is around 0.029 K/m (1.6°F/100 ft). The overburden and reservoir rock, which have large thermal capacities, together with large surface areas for heat transfer within the reservoir, lead to a reasonable assumption that reservoir condition processes tend to be isothermal.

The temperature profile from surface conditions will reflect rock property variations and can be obtained from maximum reading thermometers used with logging tools. The local geothermal gradient can be disturbed around a wellbore by drilling operations and fluids, and a Horner type analysis[14], using a suite of temperatures at a given depth from successive logging runs, can be used to obtain an indication of the undisturbed local temperature (Fig. 2.15). When temperature gradients are represented by a straight line from the surface to the reservoir interval, there may be an implied correction for water depth in offshore operations. The mean

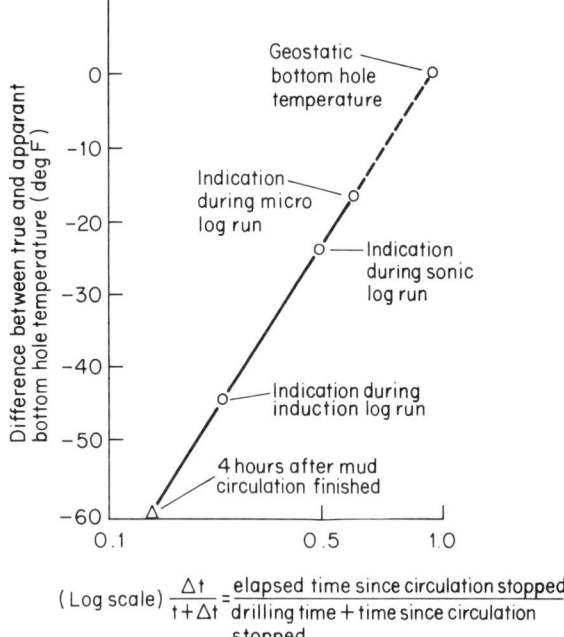

Fig. 2.15 Geostatic bottom-hole temperature from Horner analysis.

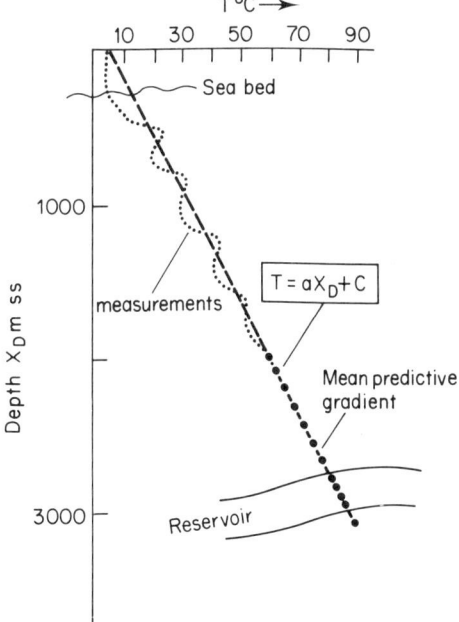

Fig. 2.16 Geostatic temperature gradient.

temperature of the North Sea has been taken as 4.5°C. As shown in Fig. 2.16, a typical geostatic temperature gradient in the reservoir interval of a northern North Sea well might be 0.029°K/m (0.016°F/ft).

2.5 NATURE OF RESERVOIR FLUIDS

Hydrocarbon accumulations are invariably associated with formation waters which may constitute extensive aquifers underlying or contiguous to the hydrocarbons, or which may exist only within the hydrocarbon bearing interval as connate or interstitial water.

If we define a phase as being a physically distinct and physically separable part of a system, then there will always be two, and frequently three, phases present within the reservoir during its producing life (oil, gas, water). The proportions, the compositions and the physical properties of the phases present may change as production proceeds, and pressures change in the essentially isothermal system. All the phases are considered compressible, although to different degrees.

The analysis of reservoir performance depends upon the prediction of the physical properties of the coexisting phases at any time and, in some cases, may require complete compositional analysis.

The volume unit of measurement in the industry is the stock tank unit, conventionally the *barrel* but frequently nowadays the cubic metre. Since stock tank oil is the result of a processing operation (gas separation), the volume resulting from unit volume of feed depends on the conditions of processing. As with gases, a volume is meaningless unless accompanied by a definition of the conditions of measurement.

By convention, stock tank conditions used in the industry are 60°F and 1 atmosphere pressure. The volumetric equivalence of one standard barrel of fluid (1 BBL) is as follows:

1 BBL = 5.615 cubic ft (ft^3)
1 BBL = 0.159 cubic metres (m^3)

A barrel at stock tank conditions of temperature and pressure is denoted STB.

The cubic foot at *standard* conditions of 60°F and 1 atmosphere has found considerable industry usage in gas volume measurement and is represented as SCF. The cubic metre at *standard* conditions of 15°C and 1 bar is represented as SCM or sm^3. The *normal* condition is sometimes used, at 0°C and 1 bar, as NCM or nm^3. In communication of data it is inevitable that recourse to unit conversions will occur.

It is most frequently necessary to relate the volume of a fluid phase existing at reservoir conditions of temperature and pressure to its equivalent volume at standard conditions. The relationship forms a term known as a *formation volume factor* and is discussed further in Chapter 4.

In the petroleum industry, oil density has long been described using an expanded inverse scale authorized by the American Petroleum Institute – the API gravity. The usual range starts with water density at 10° and rises to volatile oils and *straw coloured* condensate liquids around 60°–70°.

The relationship between API gravity and the specific gravity of the liquid (relative to water) at 60°F (SG$_{60}$) is as follows:

$$°API = \frac{141.5}{SG_{60}} - 131.5$$

or

$$SG_{60} = \frac{141.5}{(°API + 131.5)}$$

In addition to oil gravity or density, the volume of gas associated with unit volume of stock tank oil is a characterization property. Expressed as a *gas–oil ratio* or GOR, the units at a reference condition of 60°F and 1 atmosphere pressure are commonly SCF/STB and SCM/SCM. Many North Sea oils are in the region of 37° API with GORs around 600 SCF/STB.

Hydrocarbon reservoir fluids may be roughly classified as shown in Table 2.2. Although the range is continuous, the divisions are arbitrary.

Petroleum hydrocarbons consist predominantly of a series of paraffin hydrocarbons (C_nH_{2n+2}) together with some cyclic hydrocarbons (naphthenes C_nH_{2n} and aromatics C_nH_{2n-6}), but although compositional analysis is frequently taken to C_{16} or even to C_{20}, for most purposes all hydrocarbons heavier than C_6 or C_7 are frequently lumped as a composite fraction characterized by molecular weight and boiling point range. The compositions of the arbitrary classifications might typically be as those shown in Table 2.3.

2.6 RESERVOIR DATA – SOURCES

The determination of hydrocarbon in place and technically recoverable reserves requires the implementation of a data acquisition scheme. The degree of understanding of reservoir continuity and properties should improve with each well drilled but will always be a subject of uncertainty.

The data collected in the pre-development, reservoir appraisal and delineation stage needs careful planning and coordination in order to extract the maximum information. The economic justification of data acquisition is sometimes a difficult case to argue, but since production facility design and peak

2 RESERVOIRS

TABLE 2.2 Classification of hydrocarbon reservoir fluids

Fluid	API gravity	Note
Heavy oil	<20	High viscosity, high oil density and negligible gas–oil ratio. May be recently sourced or degraded black oil. At surface may form tar sands etc. Better representation with viscosity higher than say 10 cp.
Black oil	30–45	Also known as a dissolved gas oil system and constitutes majority of oil reservoirs. Critical temperatures are greater than reservoir temperature. No anomalies in phase behaviour.
Volatile oil	45–70	Very low oil specific gravities. Exists in a two-phase region. The liquid phase has very high ratios of dissolved gas to oil and the gas phase can yield a substantial part of the stock tank liquid.
Gas condensate	—	A gas phase at reservoir conditions but can undergo retrograde behaviour to yield low density oils in the reservoir.
Dry gas	—	Essentially light hydrocarbon mixture existing entirely in gas phase at reservoir conditions.

TABLE 2.3 Mole fraction compositions of hydrocarbon reservoir fluids

	Dry gas	Condensate	Volatile oil	Light black oil
C_1	0.9	0.75	0.60–0.65	0.44
C_2	0.05	0.08	0.08	0.04
C_3	0.03	0.04	0.05	0.04
C_4	0.01	0.03	0.04	0.03
C_5	0.01	0.02	0.03	0.02
C_6+		0.08	0.20–0.15	0.43
Liquid API gravity		50–75	40–60	25–45
GOR		10 000 +	3000–6000	2000

After processing:

	Tank oil	Tank gas	Separator gas
C_1	0.01	0.20	0.82
C_2	0.01	0.30	0.12
C_3	0.02	0.32	0.04
C_4	0.03+	0.14	0.01+
C_5	0.04+	0.03	0.01
C_6+	0.90	0.01	0.01

plateau production rates are calculated from reserve estimates, the narrowing of estimates in a probabilistic sense leads to greater confidence in capital commitment. Figure 2.17 shows the types of interactions using petroleum engineering and geological information during hydrocarbon exploitation.

The main sources of reservoir data, which will be amplified in later chapters, are shown in Table 2.4.

Examples

Example 2.1

The volume of Upper Jurassic source rock buried to maturation depth in the United Kingdom Continental Shelf has been estimated at 12 million km^2–m. The source rock has an average carbon content of 5% and an expected convertibility of 4500 m^3 oil per km^2–m of source rock for each percentage point of carbon. Estimate the oil in place and the technically recoverable oil in the UKCS, stating the assumptions made.

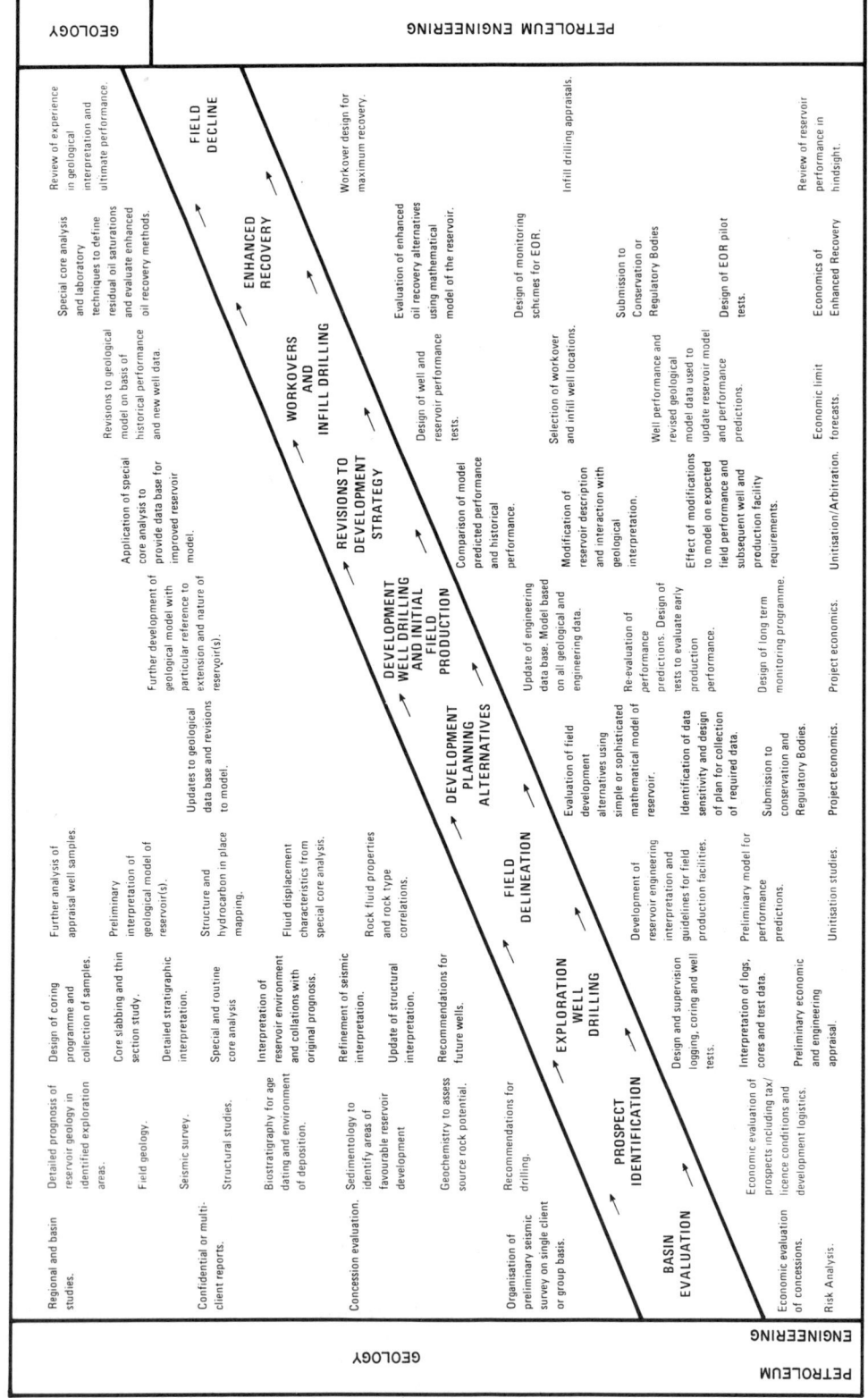

Fig. 2.17 Desired types of interaction from initial exploration through field development.

2 RESERVOIRS

TABLE 2.4 Sources of data

Source	
Drilling time logs Drill cuttings Mud logging and MWD (measurements while drilling)	These represent the earliest information available, and can help in determining intervals for coring, thicknesses of porous, hydrocarbon bearing layers, the lithology of the section, and possibly the type of hydrocarbon.
Sidewall cores	Small well-wall samples for lithology control (Fig. 2.18).
Core samples	Yield data on lithology pore structure, porosity, permeability, and may help to determine depositional environments, fluid saturations and hydrocarbon type. Special core analysis techniques will indicate recovery potential.
Bore-hole surveys: logs, wireline tests etc.	May give gross and net section thicknesses, water contacts, dips and, under favourable conditions, porosities and fluid saturations. Permeable intervals and *movable hydrocarbon* may be detected, and velocity data for seismic interpretation is obtained. The RFT tool can give valuable information on pressures and zonation. Logs may be open hole or cased hole production logs.
Well tests and fluid sampling	Hydrocarbon type and fluid samples are obtained (Fig. 2.19). Initial reservoir pressure, pressure gradients, permeability – thickness estimates, and well productivities. Generally conducted in cased hole (Fig. 2.20).

Fig. 2.18 Sidewall core operation. (Photo courtesy of BP.)

Fig. 2.19 Riser system on an offshore well allows fluids to reach surface. (Photo courtesy of BP.)

Fig. 2.20 Well testing on an offshore exploration rig. (Photo courtesy of BP.)

References

[1] Hunt, J.M.
Distribution of carbon as hydrocarbon and asphaltic compounds in sedimentary rocks, *Bull. AAPG* **61** (1977), 100.

[2] Magara, K.
Compaction and Fluid Migration, Elsevier, Amsterdam (1978).

[3] Erdman, J.Q., Morris, D.A. and Smith, J.E.
Migration, accumulation and retention of petroleum in the earth, *Proc. 8th World Pet. Cong.* (Moscow) **2** (1971), 13.

[4] Hubbert, K.
Entrapment of petroleum under hydrodynamic conditions, *Bull. AAPG* **37** (1953), 1954.

[5] Chapman, R.E.
Petroleum Geology, a Concise Study, Elsevier, Amsterdam (1973).

[6] Chiarelli, A. and Rouchet, J. du
The importance of vertical migration mechanisms of hydrocarbons, *Rev. Inst. Fr. Petrole* **32** (1977), 189.

[7] Le Blanc, R.J.
Distribution and continuity of sandstone reservoirs, *JPT* (1977), 776.

[8] Jardine, D. *et al.*
Distribution and continuity of carbonate reservoirs, *JPT* (1977), 873.

[9] Roberts, W.H. and Corpel, R.J.
Problems of petroleum migration, *Bull. AAPG Studies in Geology* **10** (1980).

[10] St. John, B.
Sedimentary basins of the world and giant hydrocarbon accumulations, *AAPG* (1980), map.

[11] UK Dept. Energy
Brown Book – Annual Report, Development of the oil and gas resources of the United Kingdom, HMSO, London (1984).

[12] Robertson Research International/ERC Energy Resource Consultants Ltd.
The Brent Sand in the N. Viking Graben, UKCS : A Sedimentological and Reservoir Engineering Study (7 vols), 1980.

[13] Jenkins, D.A.L. and Twombley, B.N.
Review of the petroleum geology of offshore NW Europe, *IMM Transactions Special Issue* (1980), 6.

[14] Horner, D.R.
Pressure build-up in wells, *Proc. 3rd World Pet. Cong.* Leiden (1951) II, 503.

[15] Grunau, H.R.
Natural gas in major basins world wide attributed to source rock type, thermal history and bacterial origin, *Proc. 11th World Pet. Cong.* (1983), Paper PD12 (3).

[16] Timko, D.J. and Fertl, W.H.
How down hole temperatures and pressures affect drilling, *World Oil* (Oct. 1972), 73.

[17] Simson, S.F. and Nelson, H.P.
Seismic stratigraphy moves towards interactive analysis, *World Oil* (March 1985), 95.

[18] Perrodon, A.
Dynamics of oil and gas accumulations, *Elf Aquitaine,* Mem 5, Pau 1983.

[19] Kassler, P.
Petroleum exploration methods, techniques and costs. *Proc. UN Conf. Petroleum Exploration Strategies in Developing Countries,* The Hague, (1981) Graham & Trotman, 103-118.

Chapter 3
Oilwell Drilling

3.1 OPERATIONS

The operation of drilling a well into a potential reservoir interval is the only way to prove the presence of hydrocarbon. Whether a well is drilled onshore or offshore is immaterial to the fundamentals of the process. A classification of wells can be made as in Table 3.1.

The main components of a well drilling operation are described with reference to Fig. 3.1. In an offshore system the drilling rig is mounted on a structure (Fig. 3.2) which may float (a drill ship or semi-submersible rig), or may be permanently or temporarily fixed to the sea bed (platform, jacket,

TABLE 3.1 Well classification

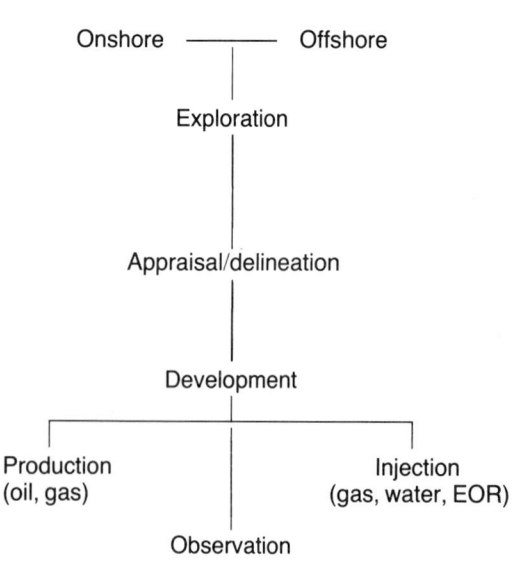

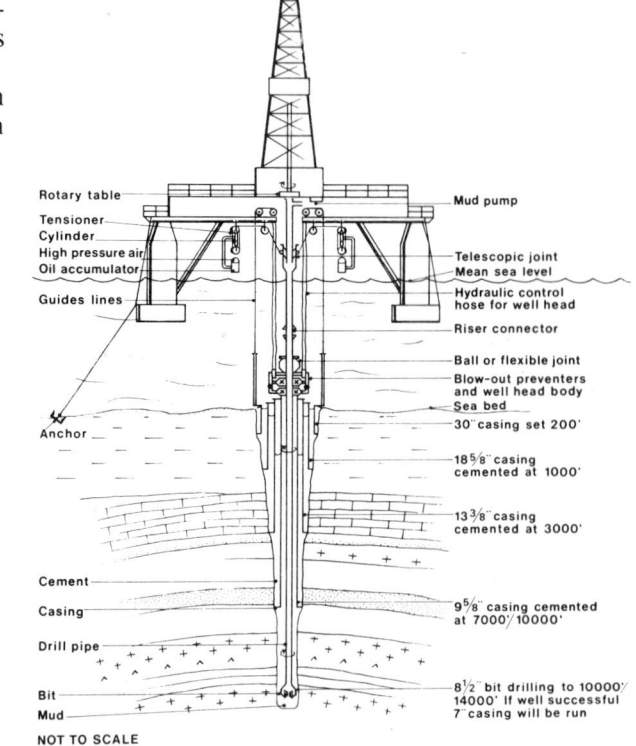

Fig. 3.1 Main components of a well drilling operation. (Photo courtesy of BP.)

3 OILWELL DRILLING

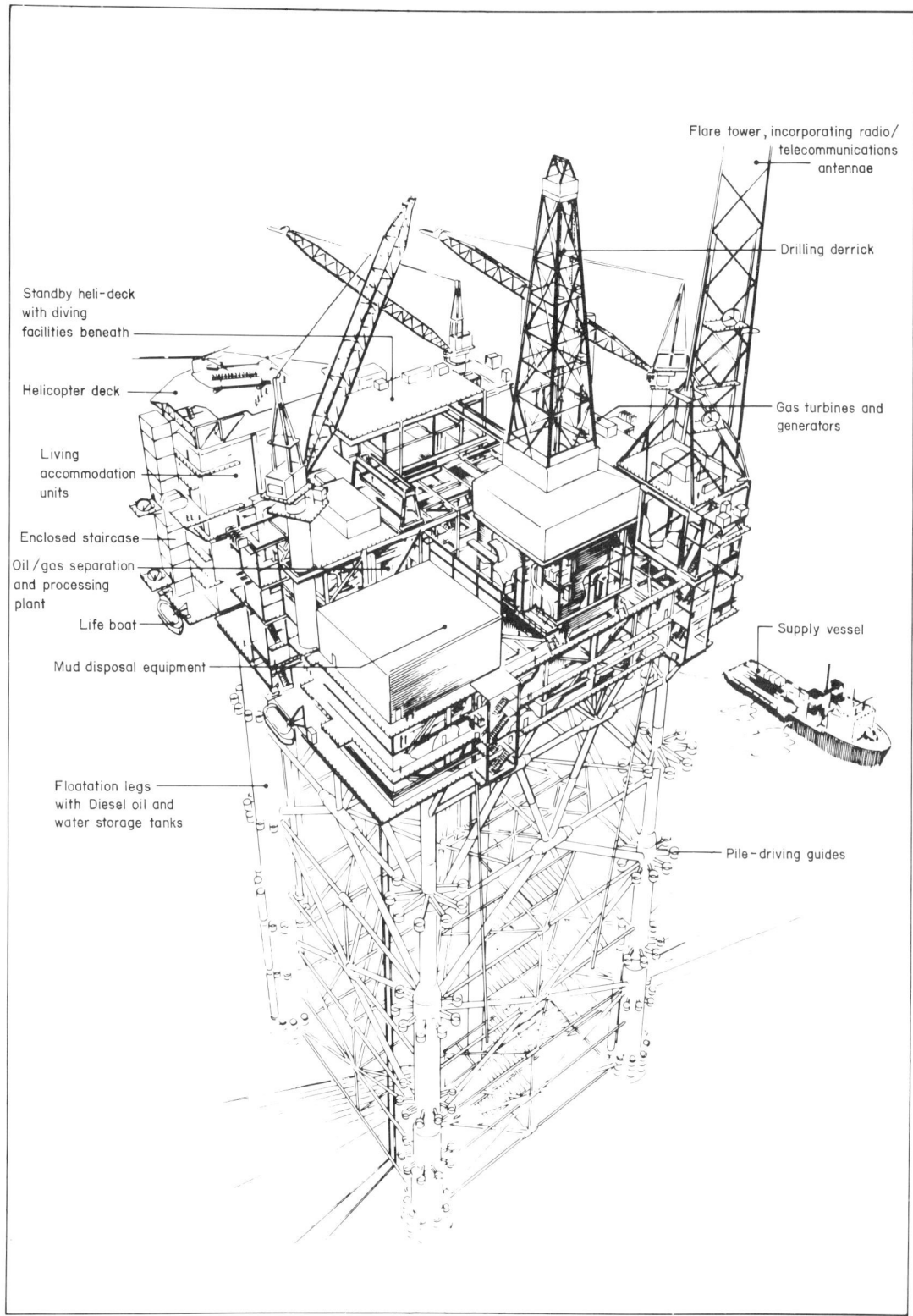

Fig. 3.2 Offshore system. (Photo courtesy of BP.)

22   PETROLEUM ENGINEERING: PRINCIPLES AND PRACTICE

Fig. 3.4 Bit types used, including diamond coring bit. (Photo courtesy of BP.)

Fig. 3.3 Large diameter and hole opening bits. (Photo courtesy of BP.)

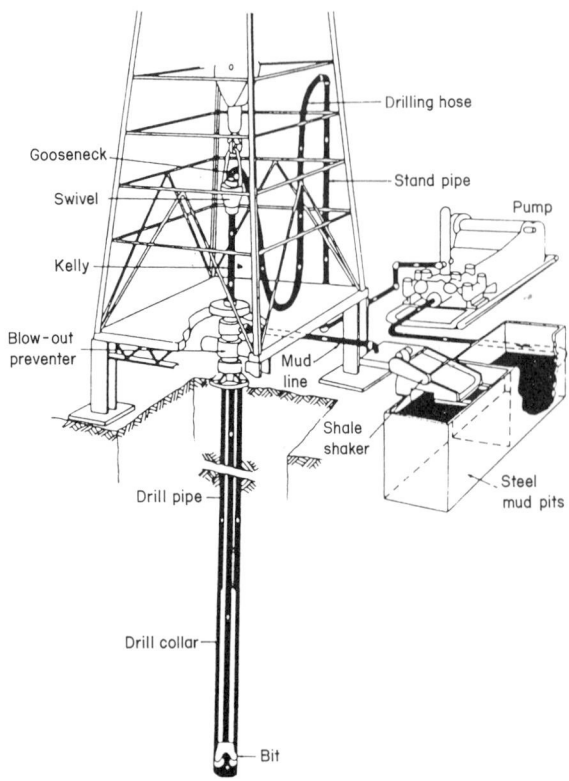

Fig. 3.5 Closed loop mud system. (Note that the mud loop is shown in black.)

jack-up rig). The drill hole is built using drill bits and steel casing for lining the drilled sections (Figs 3.3 and 3.4). The drill bits are lubricated during drilling with a fluid known as *drilling mud*, which has a composition engineered to provide (a) a density such that a pressure greater than the formation fluid pressure is maintained in the drill hole, (b) that rock cuttings are carried away from the drill bit to the surface, (c) that the drill bit is cooled. The mud may be water-based or oil-based and have components that provide particular properties needed to control the drilling. The mud system is a closed loop as can be seen in Fig. 3.5. From the mud tank the mud is

3 OILWELL DRILLING

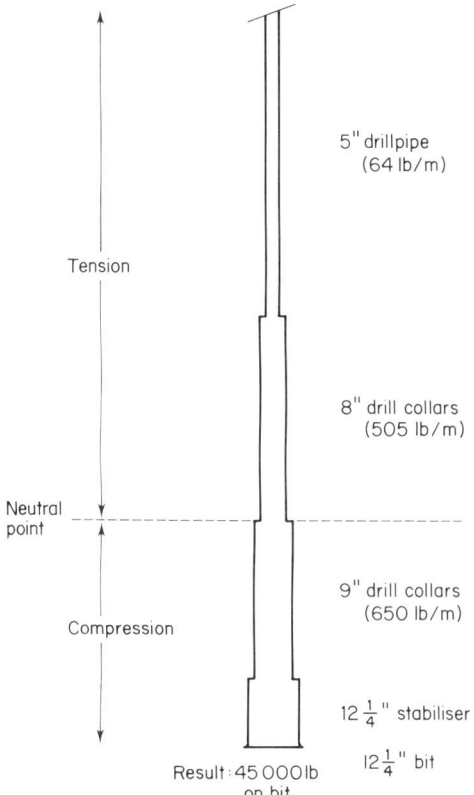

Fig. 3.6 Design basis for drilling to 3000 mSS.

pumped by a slush pump up the stand pipe and into the flexible hose. The hose is connected at a swivel joint to a hollow heavy-duty hexagonal pipe called a *kelly*. The kelly and its attached pipe are held in tension by a hoist system controlled by the *driller*. The drill pipe is connected below the kelly to provide a hollow pipe route to the drill bit. The weight acting on the drill bit is controlled by use of particularly heavy drill pipes called *drill collars* and by the tension in the upper part of the whole assembly or drill string. A neutral point in the drill string is defined by the upper part in tension from the hoist and the lower part in compression on the bit, as shown in Fig. 3.6. Rotation of the drill assembly is achieved by clamping the kelly in a bushing in a rotary table. The table speed of rotation is controlled by the driller. In some drilling operations some rotation can be achieved by use of a turbine located below the swivel. Typical drilling into sandstones at a depth of 3000 m would be achieved with a weight of some 45 000 lb on the bit and rotation at 100 rpm. In order to get to such a depth it is necessary to start with a larger hole near surface and progressively set and cement in casing, to protect the drilled hole from caving and pore fluid ingress. At 3000 m some 1000 tons of cuttings will have been removed from a typical hole. The mud circulation system ensures that fluid emerging at the bit is brought to the surface carrying the cuttings. The effluent passes onto vibrating screens which separate the larger solids from the liquid slurry, which passes to cyclone separators. The emergent cleaned mud passes into the mud tank where properties are periodically checked and chemicals added as required.

The main power requirements on a drilling rig are designed to satisfy three major functions, namely power for the hoist, power for the rotary table and power for the mud system.

3.2 COSTS

The cost of drilling operations is to a large extent dependent on well location and depth and whether the well is an exploration or a development well. The budget for a well is generally presented in a document known as an AFE (authorization for expenditure). Such a document (Table 3.2) lists in great detail all the tangible and intangible components of well costs, and guidance on its preparation can be found in the literature[12]. For current UKCS wells, a summary comparison in pounds sterling (1984) is as shown in Table 3.3.

As the most expensive phase of exploration data acquisition, and an expensive element in the development expenditure of a field, drilling operations together with well testing and well completions justify careful planning and close supervision.

When drilling exploration and appraisal wells, safety and data acquisition are the pre-eminent considerations, and fast drilling and minimum cost are subordinate (within reasonable limits).

In the case of development wells, safety is still a dominant consideration, but operations can be planned with more certainty than for wildcats. Fast drilling is important both from the point of view of time-related costs, and of the acceleration of production. Data acquisition is required within a reasonably predictable reservoir zone and so can be planned economically.

3.3. WELL COMPLETIONS AND OILWELL CASING

When drilling in relatively unknown areas, it is desirable to maintain sufficient of the hole cased and securely cemented so that a blow-out from below the casing shoe is unlikely to occur (Table 3.4). In

TABLE 3.2 AFE document

Function	Site prep./ Move on	Drill to TD	Test	Abandon/ Suspend	Work over	Total
	Days					
Rig rate						
Site survey						
Tow/Anchor						
Markers						
Mobilization						
Riser tension analysis						
Site preparation						
Anchors						
Guide bases						
Wellhead bodies						
Wellhead equipment						
Conductor						
Intermediate casing						
Deeper casing						
Casing liner						
Casing accessories						
Rock bits						
Diamond/PCD bits						
Core head						
Drilling consumables						
Mud chemicals						
Cement and additives						
Fuel oil						
Lubricants						
Catering/Accommodation						
Drilling fluid equip hire						
Drill string tool rental						
Completion tubing						
Completion tubing accessories						
Completion valves and packers						
Completion perforating and fluid						
Completion operations						
Completion Christmas trees						
Abandonment and demobilization						
Transportation						
Standby vessels						
Waste disposal						
Insurance						
Contract payments (see details below)						
Mud engineering						
Casing running						
Cement services						
Mud and drilling logs						
Turbine drilling						
Directional drilling/Survey						
Fishing						
Diving and u/w television						
Wireline logging						
Velocity calibration						
DST/Production testing						
Core barrel/Operator rental						
Sampling/Fluid analysis						
Acidization/Fracturing						
Inspection services						
Rig telecommunications						
SUB-TOTALS						

Rig: Water depth: Rotary table elevation:
Total depth target:

3 OILWELL DRILLING

TABLE 3.3 UKCS well costs (costs in millions pounds sterling, 1984)

	Exploration		Development	
	Semi-submersible (3000 mSS)	Land (2000 mSS)	N North Sea (3000 mSS)	Central North Sea (3000 mSS)
Rig	1.7	0.1	1.4	0.5
Transport	0.9	—	—	—
Contracts	0.8	0.2	0.6	0.2
Consumables	0.3	0.1	0.4	0.2
Casing & wellhead	0.3	0.2	0.6	0.4
Site preparation	0.2	0.05	—	—
Overheads	0.4	0.05	0.7	0.2
TOTAL	4.6	0.7	3.7	1.5

TABLE 3.4 Reasons for casing in a well

1. Control of well pressure
2. Wellbore *caving* prevention
3. *Freshwater* sand isolation
4. Pay zone and *thief* zone isolation
5. Confinement of production to wellbore
6. Environment for installation of production tubing

general, this will require from one third to one half of the remaining distance to the next target depth to be securely cased. In a well drilled area, knowledge of formation breakdown and fracture pressure, and of any local difficult formations – lost circulation zones, overpressured formations, caving shales etc., will enable local practices to be evaluated and implemented. Pore pressure and fracture pressures for a typical well are shown in Fig. 3.7.

When drilling through shale, the rate of penetration tends to increase with depth. In an abnormally high pressured section of hole, a sudden increase in drilling rate may occur since cuttings removal in the vicinity of the bit is aided by decrease in the mud-pore pressure differential and, in addition, shale may be softer. It is common to monitor a term known as the *d*-exponent while drilling, in order to obtain an indication of abnormal pressure. A decrease in the value of *d* indicates the possibility of abnormally high pressures.

$$d = \frac{\log\,[R/60N]}{\log\,[12W/10^6 D]}$$

where R = rate of penetration, ft/h; N = rotary speed, rpm; W = weight on bit, lb; D = hole diameter, in.

Figure 3.8, drawn from data on a well in SE Asia, shows the effect on the *d*-exponent as the abnormal pressure zone is entered at 6500 ft[16].

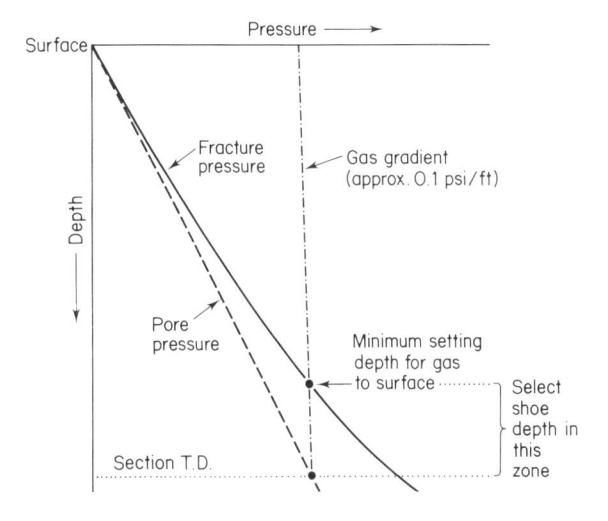

Fig. 3.7 Selection of minimum casing depth for intermediate strings.

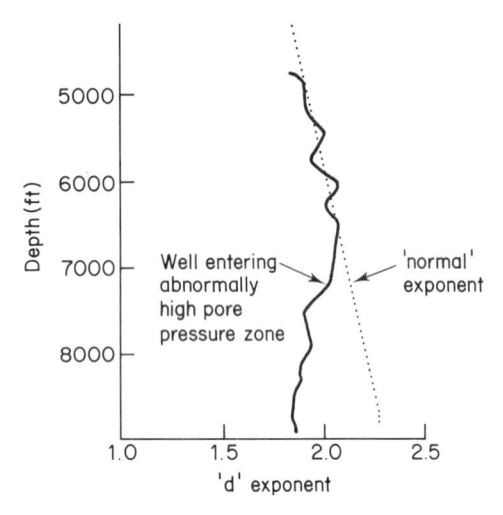

Fig. 3.8 *d*-exponent plot.

The knowledge of pore pressure is of significance in drilling and in well completion. Consider *normal* pore pressure completion as summarized in Fig. 3.7, and the reasons for casing a well as shown in Table 3.4.

To a first approximation, this means that a well with a target depth of 10 000 ft (vertical) should have an intermediate casing set at about 5000 ft (vertical). In drilling to the intermediate target of 5000 ft, a protective casing should be set at about 2000 ft. In turn, in drilling to the casing setting point at 2000 ft, a string should be set at 600–1000 ft.

In addition, in any difficult drilling situation, i.e. a new area, highly deviated wells, overpressures or lost circulation problems, the possibility should always be considered that well conditions may require the final planned casing string to be set prematurely. It is then essential to be able to continue drilling to target depth with a smaller diameter bit, and obtain the data necessary for an exploration well, or to complete a development well.

In practice, for deep, potentially high productivity wells, the smallest diameter of casing which will allow either for extended continued drilling, or for moderately high levels of production in a smaller liner is $9\frac{5}{8}$ in. A North Sea well arrangement is shown in Fig. 3.9.

This will allow continued drilling with a normal $4\frac{1}{2}$–5 in. drill string with bits up to about $8\frac{1}{2}$ in. diameter, and eliminates the problems associated

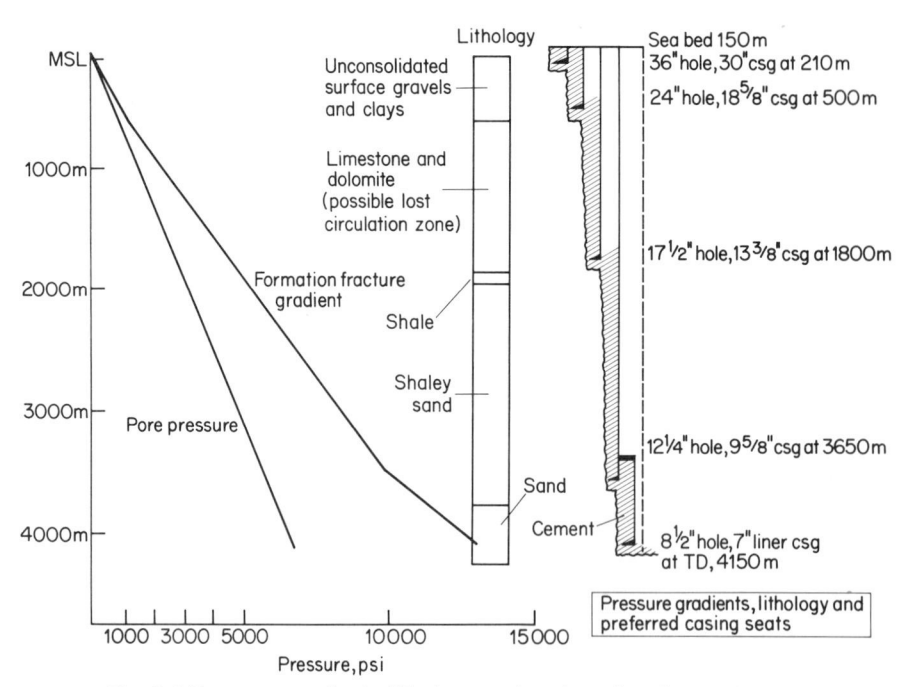

Fig. 3.9 Pressure gradients, lithology and preferred casing seats.

3 OILWELL DRILLING

(a) Burst forces

- Surface: 3500 psi
- 13 3/8" J55 68 lb/ft has effective burst rating 3100 psi
- 5000 ft: 4000 psi — Net burst at shoe 2750 psi
- 12 1/4" hole
- 10 000 ft: 4500 psi

(b) Collapse forces

- Surface — Collapse load 0 psi
- 5000 ft — Collapse load 2250 psi
- 9 5/8" N80 47 lb/ft has Collapse rating 4750 psi
- 10 000 ft — Collapse load 4500 psi

(c) Tensile forces

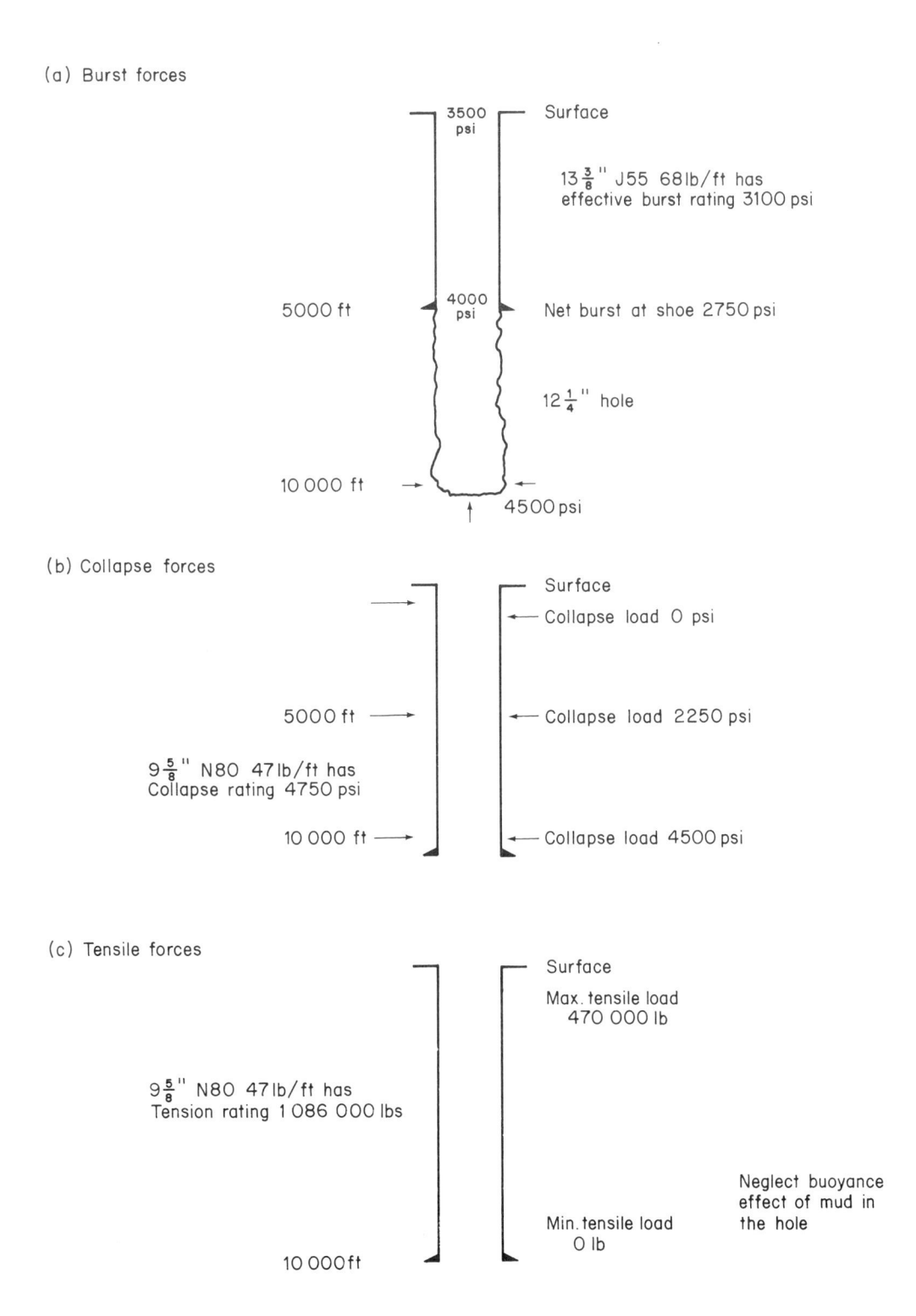

- Surface, Max. tensile load 470 000 lb
- 9 5/8" N80 47 lb/ft has Tension rating 1 086 000 lbs
- 10 000 ft, Min. tensile load 0 lb
- Neglect buoyance effect of mud in the hole

Fig. 3.10 Casing selection. (a) Burst forces, (b) collapse forces, (c) tensile forces, (d) tensile forces.

TABLE 3.5

Completion string	Hole size	Intermediate string	Hole size	Surface string	Hole size
9⅝	12¼	13⅜	17½	20	24
7	9⅝	10¾	14¾	16	20
6⅝	8½	9⅝	12¼	13⅜	17½

All sizes are in inches.

with deep drilling with 3½ in. drill pipe and small (less than 6 in.) bits.

In the largest hole possible through 9⅝ in. casing, 7 in. casing can be run as a liner, and in this or slightly smaller holes, 6⅝ in. casing will provide a fairly adequate conduit in development wells. The completion with a 9⅝ in. casing string to target depth has become a common standard for the highly deviated, highly productive wells of the North Sea, and meets all requirements of exploration and appraisal wells.

The main design criteria in casing selection involve consideration of burst, collapse and tension forces. From calculated values the overdesign safety factors are often as follows:

Burst 1.1; Collapse 1.0–0.7; Tension 1.6

Conditions in a given well environment require selection of casing according to burst, collapse and tension data in API Bulletin 5C2(S), and are shown for a hypothetical case in Fig. 3.10, which assume a 0.45 psi/ft pore pressure gradient and a 0.1 psi/ft gas column.

3.4 COMPLETION

The adoption of a particular completion size then dictates the selection of other hole and casing sizes and several representative programmes are tabulated in Table 3.5.

Even in exploration and appraisal wells which are to be abandoned, the final casing string *will* be run and cemented if hydrocarbons worth testing are encountered offshore. The greater reliability and safety of testing in casing more than offsets the cost of the casing string (unless a zone to be tested lies only just below an intermediate casing string). For development wells, a variety of completion methods are possible, the principal variations being:

- barefoot;
- screen liner uncemented;
- gravel packed liner;
- cemented liner;
- cemented casing.

The completions are illustrated in Fig. 3.11 and are summarized below.

3.4.1 Barefoot completion

This involves setting a production string in cap-rock, and drilling into the reservoir. The advantage is that there is no restriction to flow. The disadvantages are that there is no selectivity in completion, and no fluid control in completion interval.

3.4.2 Uncemented screen liner

This may be adopted for sand control – unconsolidated sand bridging on a slotted or wire wrapped perforated liner.

3.4.3 Cemented liner

This may be required when:

(a) when the flow string size is larger than can be placed in final casing size;
(b) when the *emergency* string has been run because of drilling difficulties.

3.5 DRILLING FLUID CONTROL

Drilling fluids (*muds*) are continuously circulated down the hollow drill pipe, returning through the annulus. The principal purposes of drilling fluids are to effect a primary control of formation fluid pressures, and to remove continuously from the hole the drilled rock material.

Fluids used in exploration and appraisal wells are almost universally water-based, consisting of dispersions of colloidal clays in fresh or sea water. These dispersions are stabilized with peptizing agents, and density, viscosity and filtration properties are modified and controlled by additives.

Oil-based fluids have some advantages, and are widely used in development drilling where the problems of mud salvage and re-use, and treatment and disposal of oil contaminated rock cuttings, can be handled.

In all wells, proper density control is important

3 OILWELL DRILLING

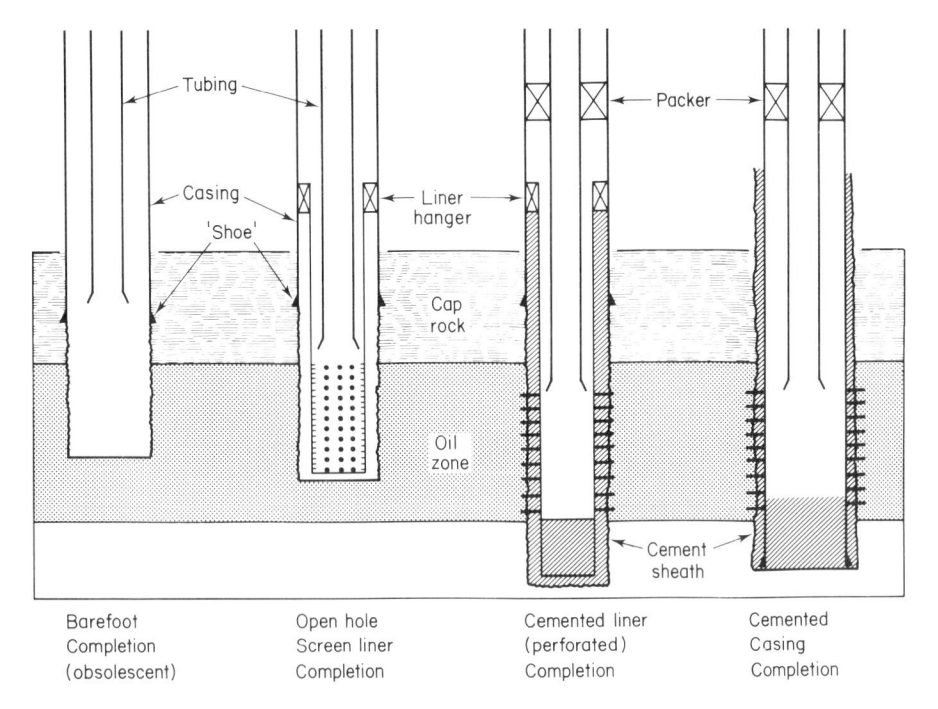

Fig. 3.11 Completion practices.

since some excess of mud pressure over formation pressure is essential for safe control of formation fluids. The excess pressure should be the minimum consistent with safe control of formation pressures (i.e. a few hundreds of psi or tens of bars) when pressures are known. Excessive pressures can lead to:

(a) lost circulation in vugular or very high porosity zones;
(b) induced fracturing and lost circulation;
(c) excessive fluid loss and thick mud cake;
(d) high differential pressure and pipe sticking;
(e) reduced rates of penetration;
(f) severe formation damage and plugged perforation etc.

Since in exploration and appraisal wells pressures are either not known or not fully established, higher mud weights are often carried than would be normal in development wells – overpressures often exceeding 1000–1200 psi (i.e. of the order of one hundred bars).

3.6 RHEOLOGY OF WELL FLUIDS (DRILLING MUDS AND CEMENTS)

The shear stress (τ) of a Newtonian fluid is directly proportional to shear rate ($\dot{v}$) and gives rise to the definition of apparent viscosity μ_a as shown in Fig. 3.12.

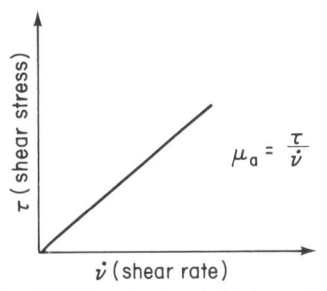

Fig. 3.12 Newtonian fluid viscosity.

For laminar flow a fluid is sheared into laminar layers, parallel to the direction of flow, with each layer moving at its specific velocity. In such a condition

$$\text{Shear rate} = \dot{v} = \frac{\text{velocity difference between two adjacent layers}}{\text{distance between the two layers}}$$

Shear stress = τ = force per unit area of the laminar layer inducing the shear

The apparent viscosity μ_a is constant at constant temperature and pressure.

The determination of the rheological properties of drilling fluids and cements is complicated for a number of reasons, including (a) the variable geometry of the well circulating system, (b) flow distortion resulting from rotation and displacement of drill string and casing, (c) the fluids are non-Newtonian and have in general a behaviour intermediate between a Bingham plastic fluid and a pseudo-plastic power law fluid, as shown in Figs 3.13 and 3.14.

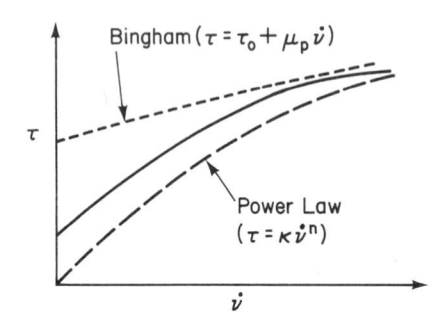

Fig. 3.13 τ–$\overset{\circ}{v}$ well fluid behaviour.

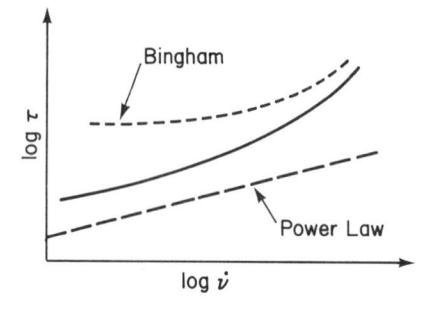

Fig. 3.14 Log τ – log $\overset{\circ}{v}$ well fluid behaviour.

3.7 FORMATION BREAKDOWN PRESSURES AND LEAK OFF TESTS

An essential factor in planning the control of any possible kick or blow-out occurring while drilling involves the avoidance of including fractures in higher formations when controlling formations at greater depths.

It is standard practice now to conduct a leak off test on formations after drilling out a casing shoe, when higher formations are fully protected. It should be emphasized that it is not the object of a test actually to break down the formation, an event which must be avoided.

In a leak off test, a plot is made of incremental mud volume pumped against pressure. When this volume–pressure relationship ceases to be linear, the formation is assumed to be in a transition from elastic to plastic deformation and to be at its yield point.

In any subsequent operations, this value of pressure at the casing shoe should not be exceeded.

3.8 DATA ACQUISITION DURING DRILLING

In exploration drilling it is imperative that no source of data should be neglected while drilling is in progress – it can never be certain that circumstances will not arise which will lead to abandonment of the well before it can be logged comprehensively by wireline.

The important sources of data while drilling is in progress are:

drilling logs;
mud logs;
MWD (measurement while drilling) logs;
MWD.

The sophisticated MWD methods are under active development and are entering more common usage.[29]

3.8.1 Drilling log

The driller's log is the most immediate information available, especially the rate of penetration under otherwise constant conditions, such as weight on bit, rotary speed and mud density.

In general, shales are harder to drill than the moderately high porosity sands and loosely consolidated sandstones that constitute good reservoir rocks. Additionally, just as high mud weights and pressures retard drilling by a *chip hold down* effect, so high formation pressures relative to the mud pressures can accelerate drilling – a reverse pressure differential leading almost to spontaneous disintegration of the formation.

Changes in rate of penetration can then frequently be correlated with sand and shale intervals, and prior indication of a porous sand interval can instigate a close examination of appropriate cuttings.

A very sudden substantial increase in drilling rate *should* lead to a cessation of drilling while a check test is made for fluid influx and a possible kick.

3.8.2 Mud logging

This involves the continuous analysis of gases extracted from the circulating mud [15] by a chromatograph and a sour gas (H_2S) detector. While background methane is always present, a change in methane concentration, and particularly increases in ethane and heavier hydrocarbons, will frequently indicate the presence of hydrocarbon bearing intervals. Given such indications, the mud itself will be tested directly for hydrocarbon content (ultraviolet light/fluorescence or total analysis by distillation for hydrocarbons).

3.8.3 Cutting logs

Provided that the travel time log between bottom hole and surface can be established, cuttings at the surface can be collected and examined for apparent porosity, permeability and hydrocarbon content, and shaly samples examined for stratigraphic and palaeontologic evidence of age. Obviously, cuttings will have been washed thoroughly by the drilling fluid stream and only residual oil traces will remain. These may, however, be detectable by examination of the solvent extract or *cut* for fluorescence under ultraviolet light.

If good indications of reservoir rock can be detected very early, e.g. drilling break, with good sandstone cuttings, good mud log indications and hydrocarbon indications in the cuttings, a decision may be made to take a complete formation core.

Figure 3.15 is reproduced from[15] to indicate a well description log for a well drilling with an 8¼ in. bit through a shaly sand series with interbedded salt layers. The drilling mud was a salt water-based mud in the upper section with a density of 1.58 g/cc and 78 000 ppm salinity. The drill string became stuck in a salt section at 8774 ft and after freeing, drilling was continued with oil-based mud. The log data shown includes from left to right the following data:

(a) cutting percentage;
(b) depth;
(c) rate of penetration in min/ft, scale 1=0 to 50 min/ft FSD;
(d) mud salinity in ppm;
(e) gas detector (units % equivalent methane) showing amount of gas in the analysed gas–air mixture coming from the degasser in the return flow line;
(f) cutting description by the wellsite geologist.

It was noted in reviewing the well history that the upper saliferous beds were drilled with non-saturated mud and that salt did not show in the cuttings brought to surface (i.e. cuttings not representative of formation which exaggerated sandstone). The constant increase in salinity, the sharp increase in drilling rate and the total disappearance of gas indication are correlatable phenomena which should have led to the suspicion of saliferous beds. The observation of substantial salt was made while circulating with the drill string stuck at 8774 ft.

3.9 MUD FLUIDS FOR CORE RECOVERY

Some companies operate a policy of coring any good indication; some operate a policy of never coring the first exploration well. The rationale of the latter policy is that after drilling, logging and possibly testing an exploration well, a precise identification can be made of the zones which are of the greatest interest for coring. Additionally, the formation pressures will be better known, and a mud programme designed to give minimum alteration of core properties can be adopted.

It is almost inevitable that a core will be heavily invaded and thoroughly flushed by mud and mud filtrate. For maximum useful reservoir information, particularly on wettability and capillary properties of the reservoir rocks, it is desirable that the drilling fluid should be as neutral as possible with respect to the reservoir rock minerals. Muds containing additives which severely affect the surface properties of minerals should be avoided.

An unchanged reservoir brine or reservoir crude would be the ideal fluid for securing unaltered specimens of reservoir rock. The production, storage and pumping of reservoir fluids will itself normally expose the fluid to some contaminants, and an oxidized crude would be severely altered. Also, brine or crude would normally not give adequate pressure control in undepleted reservoirs.

A *bland* or unreactive mud system, which is unlikely to change significantly either the wettability of reservoir rock minerals or the physical state of clay minerals within the reservoir rock, is desired. When one objective of core recovery is the evaluation of residual oil saturation, an aqueous coring fluid is necessary. To avoid changes in wettability, surfactants should be avoided as should high pH muds. To avoid clay alteration, a brine formulation similar to the connate formation water is desirable.

When an objective is the evaluation of the interstitial water saturation, an oil-based fluid is desirable, and reservoir crude with minimum active additives is a preferred base.

Viscosity and filter cake control requires the presence of thinners (peptizing agents), many of which require moderately high pH. Amongst the most common are tannin derivatives, complex phosphates and chromelignosulphonate/lignin formulations.

Additives used specifically for filter loss are CMC (a cellulose derivative), which is broadly neutral, or starch which requires a preservative, or a high pH environment or a very high salt environment.

It should be obvious that the problems of formulating a good bland coring fluid for exploration and

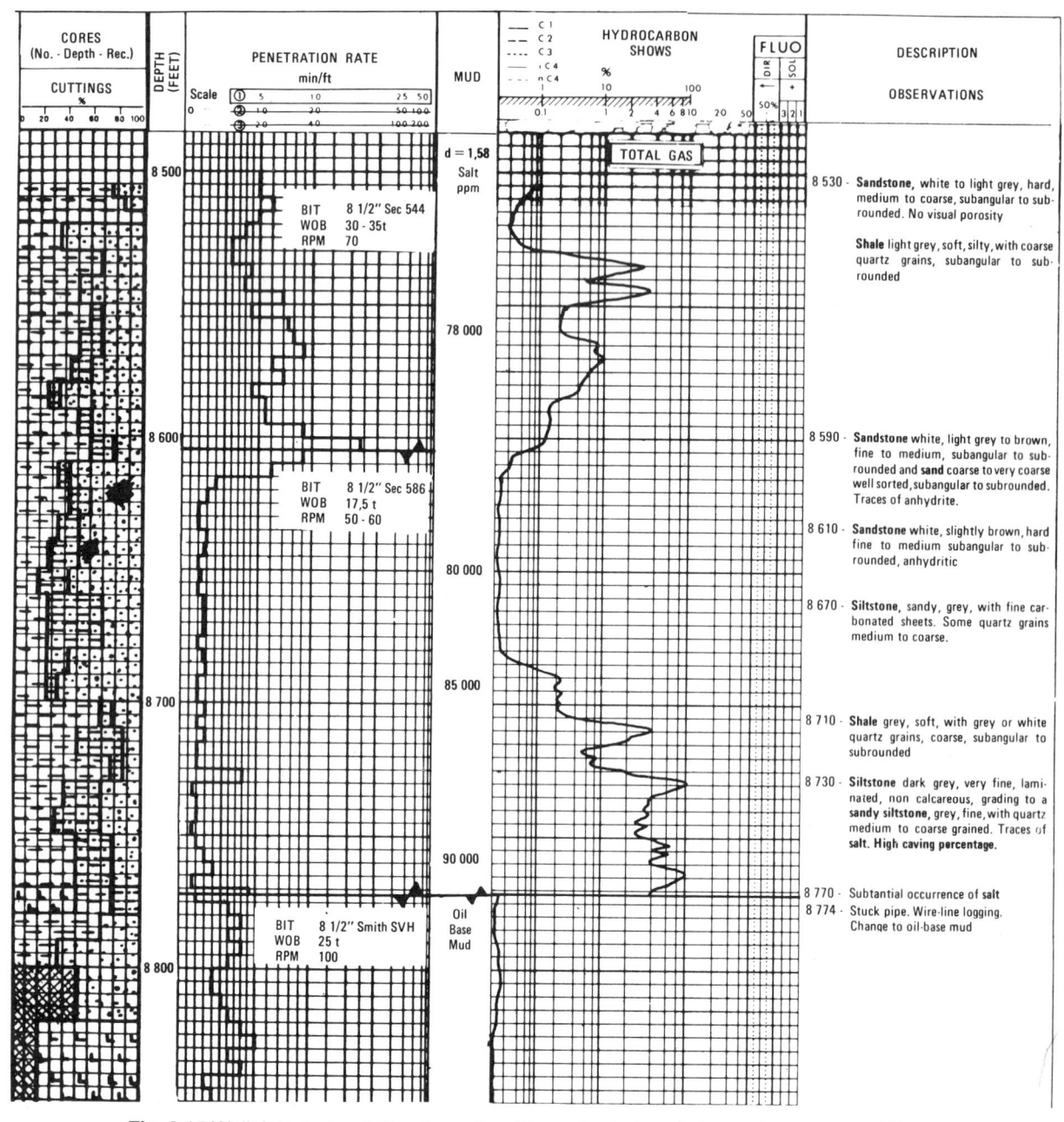

Fig. 3.15 Well data during drilling through saliferous beds in a shaly sand series (after [15]).

appraisal wells is difficult, and effective coring is frequently best conducted in development wells (though well deviation can then be a problem).

3.10 DRILLING OPTIMIZATION

The cost of drilling is such that total costs of a development can be reduced significantly if drilling costs can be reduced – the UKCS Murchison development programme managed to reduce drilling times from 50–70 days to 30–40 days for 10 000 ft TVD wells. Optimization requires a careful balance between the instantaneous drilling rate or rate of penetration, the useful life of the bit and the time taken in round trips. Conventional tricone rotary bits may last from 15 to 35 hours, by which time wear of teeth, gauge or bearings will generally have

3 OILWELL DRILLING

occurred – ideally all wearing similarly. In soft, abrasive formations, tooth and gauge wear will be a dominant factor, and it will not be worthwhile having bits with bearings of very high durability. The insert type bits used for very hard rock require very high loads for the rock crushing action necessary, and bearing failure is a dominant factor. For these bits, high quality sealed lubrication bearings are desirable to maximize bit life. Figure 3.4 shows a variety of bits in general use.

As a rule, the longest tooth (softest formation) bit which will actually cut rock without premature failure is the best overall choice.

In suitable formations – moderately hard uniform homogeneous rocks – diamond bits driven by a turbine can be extremely effective, having very long bit lives under these conditions. The rotary speeds of turbines is ill suited to the conventional tricone bit, and the combination of turbine, or the lower speed mud motor, and rock bit is restricted to some aspects of controlled directional drilling.

Diamond coring with a turbine, where only 60 ft is cored in a single trip, is the most effective coring method.

The recent developments of Stratapax bits – consisting of polycrystalline diamond compacts – may prove to be the first major advance in bit design since the development of the tricone bit.

With multiple cutting edges, the bits shear the rock in small cuttings – rather as a lathe tool or a shaper cuts – and the tool operates effectively at very low load and at moderately high speeds. The effectiveness of cut does appear to have an unusually strong dependence on the nature of the drilling fluid – in North Sea shales the bit has been especially effective in oil-based muds, but less effective than the best conventional practice in water-based fluids. On land in Holland, Stratapax bits have outperformed conventional bits in aqueous fluids. Stratapax core bits are also becoming available and could be equally effective.

3.11 TURBINE VERSUS CONVENTIONAL ROTARY

In spite of the apparent logic of generating rotary torque and power downhole where the power is needed and applied, turbine drilling has been slow to supplant conventional rotary, where the rotary torque is supplied at the surface. There have been several reasons for this:

1. Conventional tricone rotary bits cut and perform best at lower rates of revolution (± 100 rpm) than are possible with turbines (± 500 rpm). Diamond bits which can cut well at these conditions are not suited to all formations. (Stratapax bits may change this situation.)
2. The apparent power wastage, using 200–300 hp at surface to provide 10–20 useful hp at the bit, is largely irrelevant in relation to the total power requirements of a drilling rig and the power availability.
3. Turbine reliability has been suspect in the past, and turbine usage has been expensive. Consequently turbines have largely been used in special operations, and in normal drilling where conditions are known to be favourable. The special operations are particularly:

(a) diamond coring;
(b) kicking off a directional hole, where a bent sub and turbine are now the standard method;
(c) directional correction in deviated wells.

In future the combination of Stratapax and turbine may well become standard.

3.12 SPECIAL PROBLEMS IN DRILLING
3.12.1 Stuck pipe and fishing

A common cause of stuck pipe is the existence of differential pressure between borehole and formation. In particular, if a thick impermeable mud cake is allowed to build up around a pipe lying stationary on the low side of the hole, the pipe is subject to a substantial lateral thrust.

The longitudinal pull necessary to move the pipe may exceed the capacity of the rig to pull the pipe free. The capacity may be limited by:

the strength of the pipe itself;
the strength of wire rope;
the strength of derrick legs.

The remedies may be:

avoidance – use lightest possible mud weights;
  – use stabilizers or fluted drill collars or square section collars;
  – keep pipe rotating while mud cake builds up after circulation has stopped;
  – minimize delays;
after sticking – reduce pressure differential;
  – use lighter muds (wash to water or in depleted sands to oil)
  – spot lubricating and/or deflocculating chemicals around pipe at stuck section.

If the pipe remains stuck, then a fishing job results. The pipe must be backed off and a recovery string comprising

(a) pipe sub
(b) safety joint
(c) bumper sub or rotary jars
(d) drill collars
(e) drill pipe

run and an attempt made to jar the pipe free.

Where safety permits, a drill stem test string can be run, the fishing string latched in, and the pressure below the test packer reduced to reverse the pressure differential. Obviously, there must be confidence that the packer will not stick, and that the hole below the packer will not collapse completely.

When the pipe cannot be backed off for fishing, the pipe can be cut and an overshot, or a spear, used to latch on to the pipe for recovery.

3.12.2 Pressure control and well kicks

If the pressure in the borehole is reduced below the formation pressure, either through inadequate mud weight, or through lost circulation or by swabbing, then formation fluids may enter the well bore. If fluids continue to enter for any period of time, the well bore pressure will continue to fall and further fluids may enter. A series of stages can be defined and encountered:

Trip gas: a situation where after making a round trip and resuming circulation, gas is circulated out of the hole. This implies that the removal of the circulating pressure drop and the swabbing effect of the round trip have brought gas into the hole. The inference is that mud pressure and formation pressure are finely balanced.

Kick: a small influx of fluid while drilling, or possibly while making a connection. This is

Fig. 3.16 Blow out preventers (BOP) arrangement on offshore wells: (a) 21¼" BOP on board *Sea Conquest* (b) detail of BOP (c) diagram of BOP and wellhead body. (Photo courtesy of BP.)

3 OILWELL DRILLING

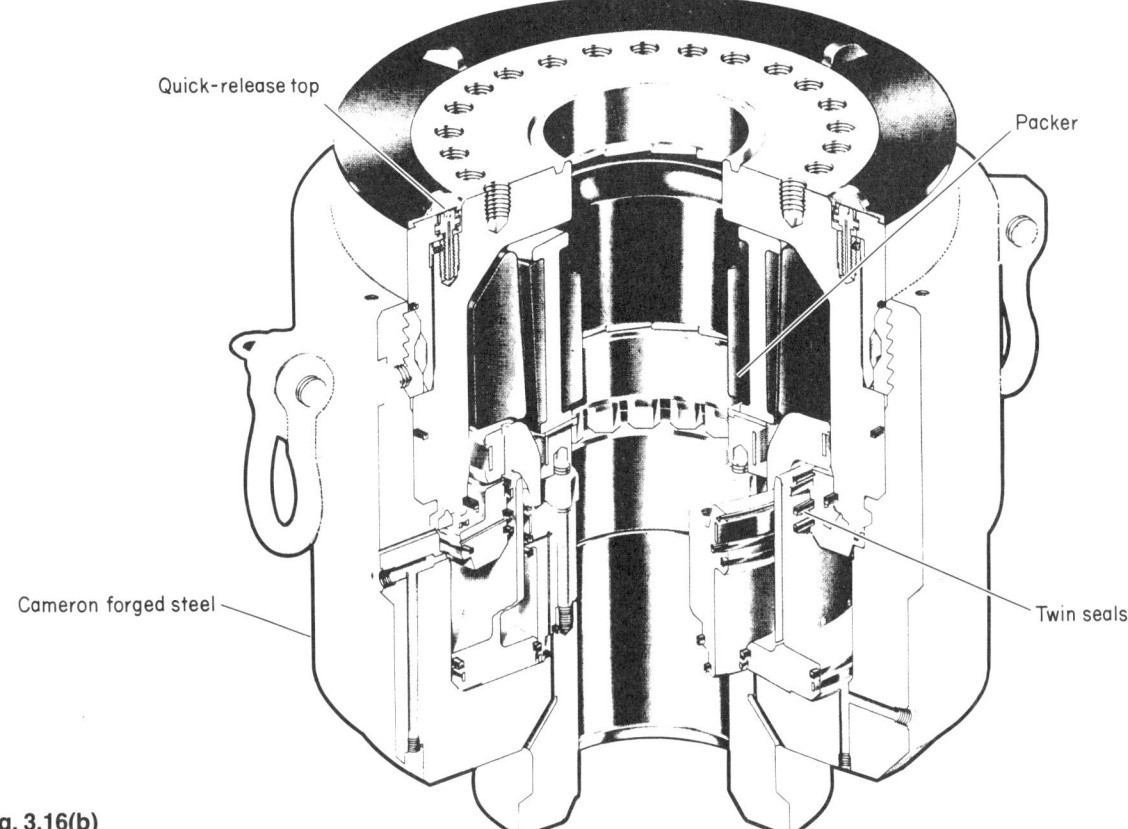

Fig. 3.16(b)

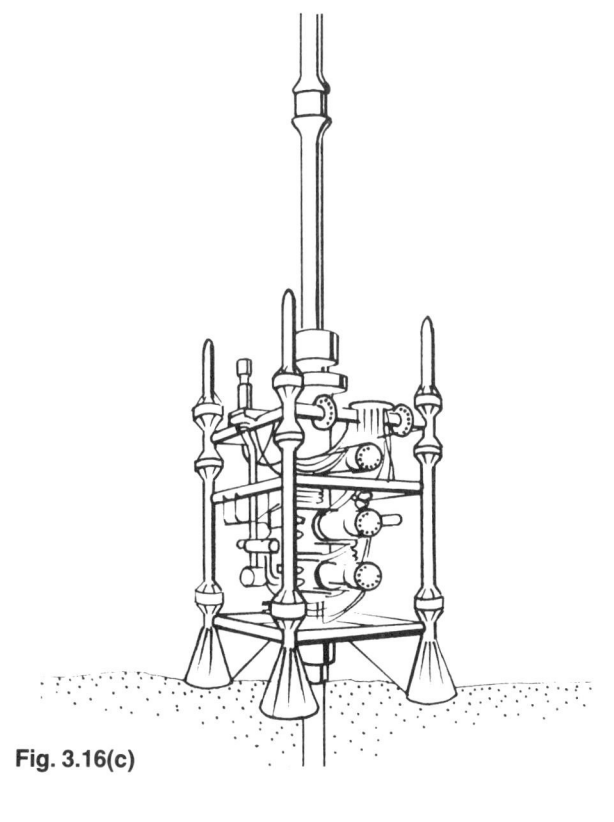

Fig. 3.16(c)

controlled by circulating against a back pressure (obtained by closing BOPs (Fig. 3.16) and circulating through a choke), preventing further fluid influx until the influx has been circulated out. The mud weight is increased to a sufficient extent to control the formation pressures.

Blow-out: a more or less uncontrolled flow of formation fluid either flowing to the surface or to a zone of natural or induced lost circulation down hole.

3.12.3 Cementation problems

Casing is run and cemented partly to protect the drilled hole from collapse, partly to isolate formations hydraulically, preventing flow of fluid from one horizon to another. Consequently, the integrity of the cement bond is of great importance.

Ideally, the cement should bond firmly to the casing itself and directly to the wall of the hole. A number of factors can prevent the attainment of a good cement bond:

Eccentricity of casing. Casing clearances in the hole can be small, and eccentricity can lead to very small clearance areas where gelled mud cannot be displaced by cement. This is a use for centralizers.

Presence of mud cake. Across all porous permeable zones, a mud cake will exist. If this mud cake is not removed, the cement *bond* will be to a mud cake, not to the formation. The tendency may be reduced by the use of scratchers, pre-wash and operation in turbulent flow.

Washed out hole. If the hole is washed out to excessive diameters, the cement may not fill the entire cross-section. This is a factor to be considered while drilling and requires careful mud control and clay stabilizing muds in the drilling phase.

Additionally, after cementing, the excess pressure due to the difference in mud and cement density (cement is at least 50% more dense than the average drilling mud) should be bled off. This prevents small contractions in the casing causing a rupture between cement and pipe.

If, after cementing, subsequent gas or water problems, or communication problems, arise, then the more difficult and expensive operations of remedial squeeze cementing may have to be undertaken.

3.13 COMPLETION FOR PRODUCTION

For flow into a cemented liner or cemented casing completion, it is necessary to perforate the casing to open a flow channel from formation to well bore.

This is effected by the detonation of shaped charges which can be placed or located selectively with a high degree of accuracy by correlation with a gamma ray log and a casing collar locator.

Intervals can be perforated selectively in one or more batch runs, with a normal perforation density of four nominal half inch shots per foot. When all perforations remain open, this will give an ideal completion, i.e. flow performance equal to or better than the theoretical uncased borehole.

3.13.1 Permanent completion

With the larger tubing strings, perforating guns can be lowered down the tubing. This enables the well to be fully completed for production with tubing and Christmas tree installed (Fig. 3.17), and the well washed to oil before perforating.

Perforation can then be carried out underbalanced, that is with a reverse pressure differential, thereby limiting plugging of perforation by mud cake. The explosive charges are, however, smaller than for full hole casing guns, and depths of penetration of the shots may sometimes be inadequate.

3.13.2 Normal completion

In this case, the perforations are made with a large size casing gun in a mud-filled hole. The larger charges give better penetration, but if mud weights are excessive, perforations may become plugged.

After perforation, tubing and packer are run and set, the BOPs flanged down and the well head and Christmas tree installed. Drilling operations on a well are completed with the installation of the Christmas tree (Fig. 3.18).

Fig. 3.17 Production tree assembly (Photo courtesy of BP).

3 OILWELL DRILLING

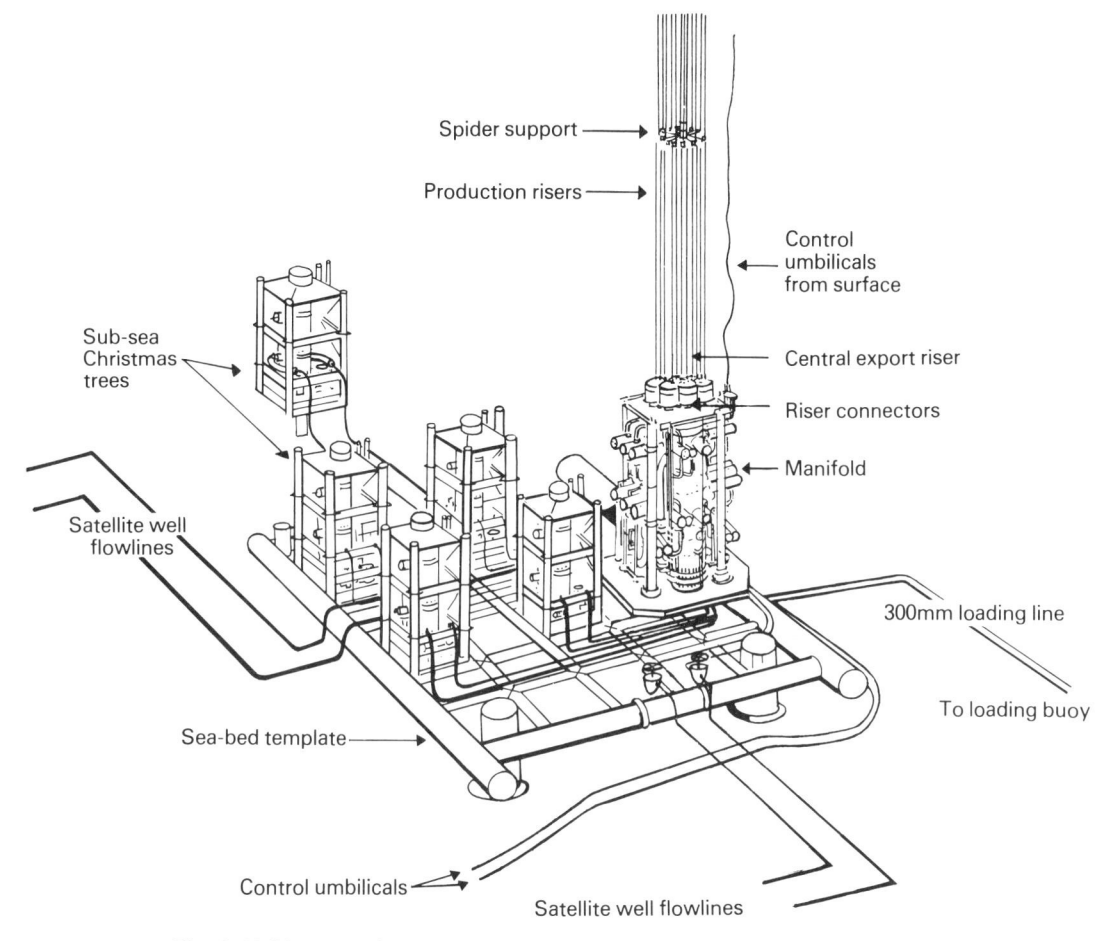

Fig. 3.18 Diagram of subsea production equipment, Buchan Field, North Sea.

Examples

Example 3.1

Design a 5½″ OD API casing string from the following grades:
  17 pounds per foot N-80 grade or P-110 grade LB threads
  20 pounds per foot N-80 grade or P-110 grade LB threads
  23 pounds per foot N-80 grade or P-110 grade LB threads

for a well in which 1.92 SG mud is left outside the casing and 1.15 SG fluid is left inside the casing. The length of string is 13 000 ft and as abnormal pressures are anticipated, assume the maximum surface pressure will be 8000 psi. The minimum section design length is 500 ft. Assume the following safety factors: Collapse = 1.125, Burst = 1.312 and Tension = 1.18. The specific gravity of the steel is taken as 7.84. Use the data of Table A 3.1 and Fig. A 3.1. (See Appendix II).

Example 3.2

In drilling through a formation to 13 000 ft the following information has been obtained:
– the pressure at 13 000 ft can be represented by an average pore pressure gradient to surface of 0.455 psi/ft
– the fracture pressure can be represented by an average gradient to surface of 0.8 psi/ft
– a gas gradient to surface of about 0.1 psi/ft can be assumed.
Specify the minimum setting depth for an intermediate string casing shoe.

References

[1] Craft, B.C., Holden, W.R. and Graves, E.D.
Well Design, Drilling and Production, Prentice-Hall, Houston.

[2] Gray, G.R. and Darley, H.C.H.
Composition and Properties of Oil Well Drilling Fluids, Gulf Publishing Company, Houston.

[3] American Petroleum Institute
API Bulletin 5C2 (1972).

[4] *Composite Catalogue of Oilfield Equipment and Services*
35th Revision (1982-83), 5 vols. Gulf Publishing Co, Houston Tx.

[5] Peden, J.M., Russel, J. and Oyeneyin, M.B.
The design and optimisation of gravel packing operations in deviated wells, SPE Paper 12997, *Proc. Europe* (1984), 321.

[6] Bruijn, H.J., Kemp, A.J. and van Donegen, J.C.
The use of MWD for turbodrill performance optimisation, SPE Paper 13000, *Proc. Europe* (1984), 353.

[7] Bailey, T.J., Bern, P.A. and McEwan, F.
Downhole rheological behaviour of low toxicity oil muds, SPE Paper 13001, *Proc. Europe* (1984), 361.

[8] Paterson, A.W. and Shute, J.P.
Experience with polycrystalline diamond compact bits in the northern North Sea, Paper EUR 339, *Proc. Europe* (1982), 575.

[9] Johnson, K.A. and Burdyulo, L.
Successful liner completion on the Muchison platform, Paper EUR 365, *Proc. Europe* (1982) 769.

[10] CSRPPGN
Blow Out Prevention and Well Control, Graham and Trotman, London (1981).

[11] ENSPM, Inst. Franc. du. Petrole
Drilling Data Handbook, Graham and Trotman, London (1978).

[12] Adams, N. and Frederick, M.
Drilling costs, *OGJ* (Dec. 6, Dec. 13, Dec. 27, 1982).

[13] Baldwin, D.D.
Drilling high angle directional wells, *Proc. 11th World Pet. Cong.* Paper PD5(2), (1983).

[14] Tanguy, D.R.
Downhole measurements while drilling, *Proc. 11th World Pet. Cong.* Paper PD5(3), (1983).

[15] ENSPM
Geological and Mud Logging in Drilling Control – Catalogue of Typical Cases, Graham and Trotman, London (1982).

[16] CSRPPGN
Drilling Mud and Cement Slurry Rheology Manual, Graham and Trotman, London (1982).

[17] van Lingen, N.H.
Bottom scavenging – a major factor governing penetration rates at depth, *Trans. AIME* **225** (1962), 187.

[18] Holster, J.L. and Kipp, R.J.
Effect of bit hydraulic horsepower on drilling rate of a PCD compact bit, *JPT* (Dec. 1984), 2110.

[19] Cooke, C.E., Kluck, M.P. and Medrano, R.
Annular pressure and temperature measurements diagnose cementing operations, *JPT* (Dec. 1984). 2181.

[20] Adams, N.
Well Control Problems and Solutions, Pennwell, Tulsa (1980).

[21] Rowlands, G.W. and Booth, N.R.
Planning for deep high pressured wells in the northern North Sea, EUR 244, *Proc. Europ. Pet. Conf.* (1980) **327**, Soc. Pet. Eng.

[22] World Oil
1985 Tubing and casing joint tables, *World Oil* (Jan. 1985), 37.

[23] Woodyard, A.H.
Risk analysis of well completion systems. *JPT* (April 1982) 713.

[24] Denholm, J.M.
Offshore drilling operations. *Trans I.Mar. E* (TM) **94** (1982) (26) 2.

[25] Mohnot, S.M.
Characterisation and control of fine particles involved in drilling, JPT (Sept 1985) 1622.

[26] Remson, D.
Planning techniques – Key to drilling efficiency. *JPT* (Sept 1985) 1613.

[27] Black, A.D., Dearing, H.L. and Dibona, B.G.
Effects of pore pressure and mild filtration on drilling rates in a permeable sandstone. *JPT* (Sept 1985) 1671.

3 OILWELL DRILLING

[28] Rabia, H.
Specific Energy as a criterion for bit selection. *JPT* (July 1985) 1225.
[29] Elliott, L.R. *et al.*
Recording downhole data while drilling, *JPT* (July 1985) 1231.
[30] Bourgoyne, A.T., and Holden, W.R.
An experimental study of well control procedures for deepwater drilling operations. *JPT* (July 1985) 1239.
[31] Warren, T.M. and Smith, M.B.
Bottomhole stress factors affecting drilling rate at depth. *JPT* (Aug. 1985) 1523.
[32] Hill, T.H., Money, R.C., and Palmer, C.R.
Qualifying drillstring components for deep drilling. *JPT* (Aug. 1985) 1511.
[33] Joosten, M.W. and Robinson, G.C.
Development of specification for christmas tree and wellhead components. *JPT* (April 1985) 655.

Chapter 4

Properties of Reservoir Fluids

As we have seen in Chapter 2, reservoir hydrocarbon fluids are mixtures of hydrocarbons with compositions related to source, history and present reservoir conditions.

While the behaviour of single component systems has no quantitative relevance to oil field systems, qualitative behaviour has some similarities, and terminology is the same.

4.1 VOLUMETRIC AND PHASE BEHAVIOUR OF HYDROCARBON SYSTEMS

Consider the pressure–specific volume relationship for a fluid at a constant temperature below its critical temperature initially held in the liquid phase at an elevated pressure. This situation is illustrated in Fig. 4.1 for propane, and generalized in Fig. 4.2.

Expansion of the system will result in large decrements in pressure for relatively small increments in volume (specific volume), due to the small compressibility of liquid systems (c), where

$$c = -\frac{1}{V}\left(\frac{\partial V}{\partial P}\right)_T$$

For most liquids over commonly encountered pressure ranges, the compressibility is independent of the pressure, i.e. compressibility is small and constant. As expansion is continued, a pressure will be reached at which some small infinitesimal gas phase will be found.

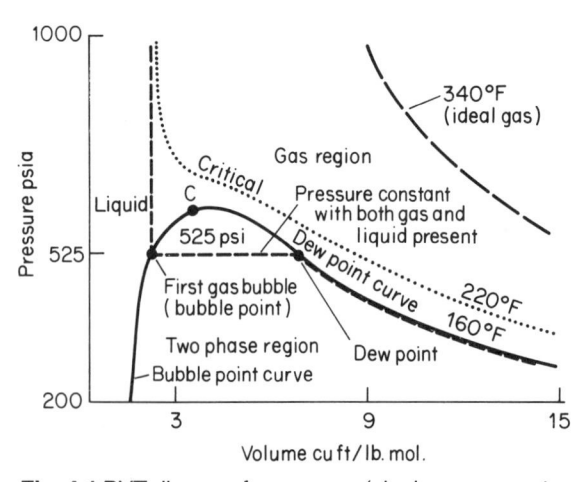

Fig. 4.1 PVT diagram for propane (single component system).

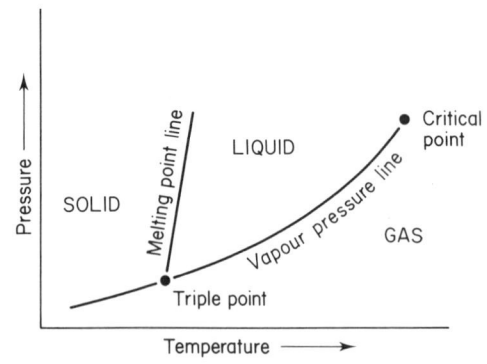

Fig. 4.2 Phase diagram for a pure single component system.

4 PROPERTIES OF RESERVOIR FLUIDS

This pressure is termed the *bubble-point* or *saturation* pressure for the temperature considered. For a pure substance, further expansion occurs at constant pressure – the vapour pressure at that temperature – the relative proportions of liquid and gas changing, until only an infinitesimal quantity of liquid is present in equilibrium with the vapour. This point on the phase envelope is termed the *dew-point*.

Expansion to lower pressures and higher specific volumes occurs in a vapour phase.

If the substance behaves as an ideal gas, then the equation $PV = nRT$ is valid, and

$$c = -\frac{1}{V}\left(\frac{\partial V}{\partial P}\right)_T = +\frac{1}{P}$$

i.e. the compressibility of an ideal gas is inversely proportional to the pressure.

A series of isotherms then generates a locus of bubble-points, and a locus of dew-points which meet at a point – the critical point – C, at which the properties of liquid and vapour become indistinguishable. The isotherm at the critical point exhibits a point of inflexion so that

$$\left(\frac{\partial P}{\partial V}\right)_{T_c} = \left(\frac{\partial^2 P}{\partial V^2}\right)_{T_c} = 0$$

Multicomponent systems exhibit slightly different phase behaviour from that of pure materials, and even simple two or three component systems may demonstrate all the phenomena associated with very complex systems, as shown in Fig. 4.3.

Expansion in the liquid phase to the bubble-point at constant temperature is similar to a pure system. Expansion through the two-phase region does not occur at constant pressure, but is accompanied by a decrease in pressure (vapour pressure) as the composition of liquid and vapour changes.

Since at any temperature the bubble-point pressure and dew-point pressure differ, the pressure–temperature relationship is not now a single line, but forms a phase envelope. On this envelope it is possible to establish a pressure above which two phases can no longer coexist – the cricondenbar – and a temperature above which two phases cannot coexist – the cricondentherm. The critical point – the pressure and temperature at which the properties of the two phases become identical – is not necessarily either one of these points, nor must it be between them.

4.2 APPLICATIONS TO FIELD SYSTEMS

4.2.1 Dry gas reservoirs

In Fig. 4.4 (a), the reservoir temperature is above the cricondentherm (maximum temperature of two-phase envelope). Consequently, decline in reservoir pressure will not result in the formation of any reservoir liquid phase.

This does not preclude the recovery of natural gas liquids from reservoir fluid as a result of change in temperature in flow lines or process facilities. There are few natural gases yielding no gas liquids, and Fig. 4.4 (b) is therefore more realistic.

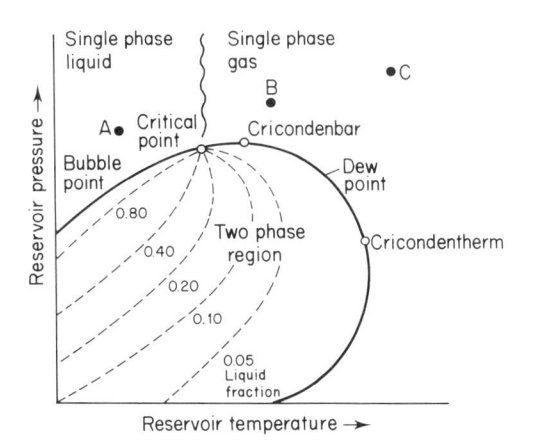

Fig. 4.3 Pressure-temperature phase diagram for multicomponent hydrocarbon reservoir fluid mixture. For isothermal production in the reservoir: position A indicates reservoir fluid found as an undersaturated oil; position B indicates reservoir fluid found as a gas condensate; position C indicates reservoir fluid found as a dry gas.

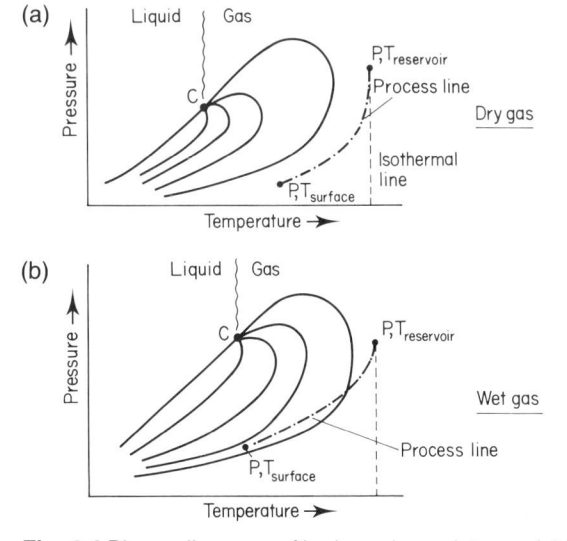

Fig. 4.4 Phase diagrams of hydrocarbon mixtures. (a) Dry gas, (b) wet gas, (c) gas condensate, (d) black oil.

Fig. 4.4 (cont.)

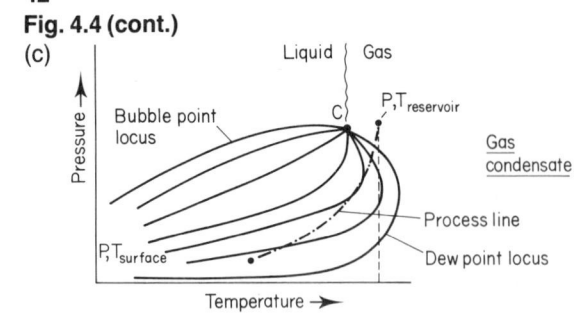

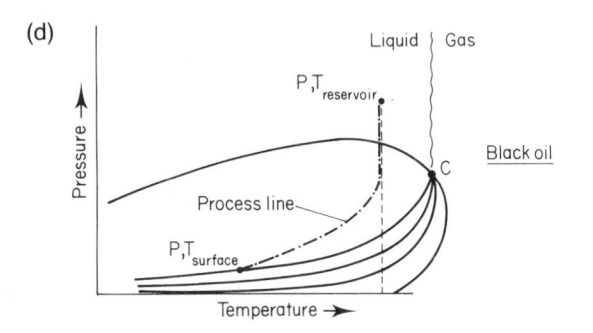

(c) [Phase diagram: Pressure vs Temperature showing Bubble point locus, Dew point locus, critical point C, P,T reservoir, P,T surface, Process line, Liquid/Gas regions — Gas condensate]

(d) [Phase diagram: Pressure vs Temperature showing P,T reservoir, P,T surface, critical point C, Process line, Liquid/Gas regions — Black oil]

4.2.2 Condensate systems

The critical temperature of the system is such that reservoir temperature is between critical and cricondentherm as shown in Fig. 4.4 (c). Under these conditions the system exhibits isothermal retrograde condensation.

At pressures above the cricondenbar, the system exists as an indeterminate *vapour* phase. Upon isothermal expansion, the phase envelope is encountered at the dew-point locus, and further reduction in pressure leads to increased condensation of a liquid phase.

If the pressure can be reduced sufficiently, the liquid phase may re-evaporate, but sufficiently low pressures may not be obtainable in practice. This phenomenon – the condensation of liquid upon decrease in pressure – is termed isothermal retrograde condensation. Another phenomenon – isobaric retrograde behaviour – can also be demonstrated, but is not of interest under the essentially isothermal conditions of petroleum reservoirs.

The liquid phase recovered (the condensate) from a condensate system, is recovered from a phase which is vapour at reservoir conditions.

This is also partly true of volatile oil systems, where the vapour phase in equilibrium with the reservoir liquid phase is particularly rich in liquefiable constituents (C_3 to C_{8+}) and a substantial proportion of stock tank liquid may derive from a reservoir vapour phase.

Contrast, the black oil – dissolved gas systems – in which the composition of gas in equilibrium with liquid is comparatively lean (except at tank conditions), the liquid recovery depends only marginally on the separated gas phase. (Nevertheless in absolute terms the maximization of liquid recovery from separator gas streams can be a valuable source of income.)

4.2.3 Volatile oil systems

These are within the two-phase region under reservoir conditions, the vapour phase corresponding to condensate compositions and conditions.

Consequently, a substantial part of the stock tank liquid is derived from a reservoir vapour phase. The concept of a system being represented by a *gas* dissolved in a liquid is then invalid.

4.2.4 Black oil systems

As shown on Fig. 4.4 (d), the critical temperature of these systems is very much higher than the reservoir temperature encountered (i.e. greater than about 373 K). Isothermal expansion from the liquid phase leads to the formation of gas at the bubble-point, and a monotonic increase in volume of the gas phase as pressure declines further. The composition of gas varies only slightly when changing conditions (except at tank conditions), the gas is relatively lean, and contributes only marginally to the separator liquid phase.

4.3 COMPRESSIBILITY

Reservoir fluids are considered compressible and, at constant reservoir temperature, we can define an isothermal compressibility as a positive term c as follows:

$$c = -\frac{1}{V}\left(\frac{\partial V}{\partial P}\right)_T$$

where V denotes original volume and P is pressure. Gas compressibility is significantly greater than those of liquid hydrocarbons, which in turn are greater than those of reservoir waters. The subscript terminology for the compressibilities of gas, oil and water is c_g, c_o, c_w. Reservoir pore volume may change with change in fluid pressure, resulting in an increased fraction of overburden being taken by reservoir rock grains. The pore volume compress-

4 PROPERTIES OF RESERVOIR FLUIDS

TABLE 4.1 Typical system compressibilities

System	Symbol	bar^{-1}	psi^{-1}
Reservoir saline waters	c_w	4.35E–05	3.0E–06
Undersaturated black oils	c_o	24.65E–05	17.0E–06
Gas at 100 bar (1450 psi)	c_g	1000E–05	689E–06
Gas at 400 bar (5800 psi)	c_g	250E–05	172E–06
Consolidated sands at 400 bar	c_f	3.63E–05	2.5E–06
Unconsolidated sands at 400 bar	c_f	14.5E–05	10E–06

ibility may be related to fluid pressure P rather than grain pressure (P_g) and treated as a positive term, c_f. This is reasonable for a constant overburden load.

$$c_f = \frac{1}{\phi}\frac{\partial \phi}{\partial P_g} = -\frac{1}{\phi}\frac{\partial \phi}{\partial P}$$

In the absence of specific information, the general order of magnitude of compressibilities is as shown in Table 4.1.

4.4 MEASUREMENT AND PREDICTION OF RESERVOIR FLUID PROPERTIES

4.4.1 Behaviour of gases

Gas is produced from gas reservoirs by expansion of the fluid from the initial reservoir pressure to some lower abandonment pressure. The recovery from the reservoir can be obtained by a mass balance on the system at the initial and end points, and the densities of the material at the two points (and at intermediate points) must be calculable or measured.

The flow behaviour, i.e. the pressure gradients developed, will depend in part on the isothermal compressibility:

$$c = -\frac{1}{V}\left(\frac{\partial V}{\partial P}\right)_T$$

while the primary processing operations of dehydration, dew-point processing and sweetening will involve other thermodynamic functions of the gas, particularly the isobaric thermal expansion coefficient

$$\alpha = \frac{1}{V}\left(\frac{\partial V}{\partial T}\right)_P$$

Both of these factors can be found by differentiating an equation of state, or can be computed from plots or tabulations of experimental data.

The possible error in these derived functions is an order of magnitude greater than the possible error in the original data (or equation of state), so that a very high degree of accuracy is necessary in equations of state (or experimental data) used for the calculation of thermodynamic functions.

The simplest equation of state is the ideal gas law:

$$PV_m = RT$$

where V_m = molar volume, or

$$PV = nRT$$

where V = volume, n = no. of moles.

The value of R, the gas constant, will depend on the system of units adopted. See Table 4.2 for examples. The inadequacy of this relation was quickly recognized, and one early modification was that of van der Waals:

$$\left(P + \frac{a}{V_m^2}\right)(V_m - b) = RT$$

where a is a factor related to, and correcting for, intermolecular forces, b is a factor related to, and correcting for, the finite volume and real geometry of a molecule.

This is a two-constant equation of state, and more and more complex equations (e.g. Beattie-

TABLE 4.2 Values of the universal gas constant

Moles	Pressure	Volume	Temperature	R
Pound	Pounds force/sq.in.	Cubic ft	°R	10.73
Pound	Atmosphere	Cubic ft	°R	0.729
Kilogram	Newton/m^2	m^3	K	8312.0

Bridgemen – five constants; Bendict-Webb-Rubin – eight constants) have been developed in attempts to improve the accuracy.

A widely used two-constant equation of state is that of Redlich-Kwong which can be arranged to a form

$$\left(P + \frac{a}{[TV_m(V_m + b)]^{0.5}}\right)(V_m - b) = RT$$

This equation has an acceptable accuracy for hydrocarbon systems over a fairly wide range of conditions, and the constants have been modified by Soave[15].

More recently, the equation of Peng and Robinson[14] has received wide acceptance

$$P = \frac{RT}{(V-b)} - \frac{a(T)}{V(V+b) + b(V-b)}$$

It may be noted at this time that Avogadro stated that under the same conditions of temperature and pressure, equal volumes of all ideal gases contain the same number of molecules. It can be shown that there are 2.73×10^{26} molecules in each pound-mole of ideal gas and that a volume of 379.4 cubic feet is occupied at 60°F and 14.7 psia by one pound molecular weight of any ideal gas. At 14.7 psia and 32°F one pound mole of gas occupies 359 cubic feet.

In this context, one mole is taken as the pounds of a component equal to its molecular weight. Since moles contain the same number of molecules (or atoms), they are used to describe system composition. At low pressure or for ideal gases, volume and mole fractions are identical.

For each ideal gas in a mixture of ideal gases, Dalton has postulated that each gas exerts a pressure equal to the pressure it would exert if it alone occupied the total volume of the mixture – that is the same as saying that the total pressure of the mixture is the sum of the partial pressures of each component, j, in the n component mixture:

$$P = \sum_{j=1}^{j=n} P_j$$

The mole fraction of the jth component is symbolized as y_j and is therefore defined as

$$y_j = \frac{P_j}{P}$$

A similar postulate by Amagat states that the sum of the partial volumes of n ideal components in a mixture is equal to the total volume under the same conditions of temperature and pressure, i.e.

$$V = \sum_{j=1}^{j=n} V_j \quad \text{and} \quad y_j = \frac{V_j}{V}$$

The apparant molecular weight of a gas mixture behaving as if it were a pure gas is defined as the sum of the product of individual component mole fractions and molecular weights

$$M_a = \sum_{j=1}^{j=n} (y_j M_j)$$

Since, for a gas, the specific gravity γ_g is defined as the ratio of the density of the gas to the density of dry air taken at the same temperature and pressure then, assuming ideal behaviour

$$\gamma_g = \frac{\text{mass gas}}{\text{volume gas}} \cdot \frac{\text{volume gas}}{\text{mass air}}$$

$$= \frac{\text{molecular weight of ideal gas (mixture)}}{\text{molecular weight of air}}$$

4.4.2 Law of corresponding states

Fluids are said to be in corresponding states when any two of the variable properties, pressure, temperature and specific volume have the same ratio to the critical values. These ratios are termed reduced values and subscripted R as follows:

$$T_R = T/T_c \,;\, P_R = P/P_c \,;\, V_R = V/V_c$$

For pure substances with simple molecules it can be shown theoretically that

$$P/P_c = f(T/T_c \,;\, V/V_c)$$

i.e. if fluids are in corresponding states then *any* dimensionless reduced property calculable from *PVT* data will be the same for those fluids, e.g. reduced densities, fugacity/pressure ratios, compressibilities.

The applicability of the law of corresponding states will depend on the phase and temperature of the substance; accuracy is greatest in the vapour

4 PROPERTIES OF RESERVOIR FLUIDS

phase, and is best for temperatures above the critical, and will also depend on the complexity and eccentricity of the molecule.

However, the law of corresponding states has been widely used in smoothing and correlating experimental data on hydrocarbons, and is used for generalized liquid phase and gas phase correlations for hydrocarbon mixtures with a considerable degree of success.

4.4.3 Pseudo-critical temperatures and pressures

The reference state for the law of corresponding states is the critical state. For most pure substances, the values of the independent variables, pressure and temperature at the critical point, have been measured and are known. For mixtures, the critical values will depend on composition, but there will not normally be a simple procedure for calculating the true critical values from composition and the values of the pure components. It has been found that the use of the true critical values of mixtures in corresponding states correlations gives less accurate results than the use of so-called *pseudo-critical* constants calculated as the mol-average values for the mixture (Kay's rule), i.e.

$$P_{pc} = \sum y_i P_{cj} \qquad T_{pc} = \sum y_i T_{cj}$$

where y_j = mol fraction of component j, P_{cj} = critical pressure of component j, T_{cj} = critical temperature of component j.

For complex hydrocarbon mixtures (e.g. the C_{7+} constituent of a system), pseudo-critical constants can be determined from specific gravity, molecular weight and boiling point of the mixture without determining the composition (see Figs 4.5 and 4.6). More complex rules for calculating critical constants have been formed, which are more accurate than Kay's rule given above, but this is generally adequ-

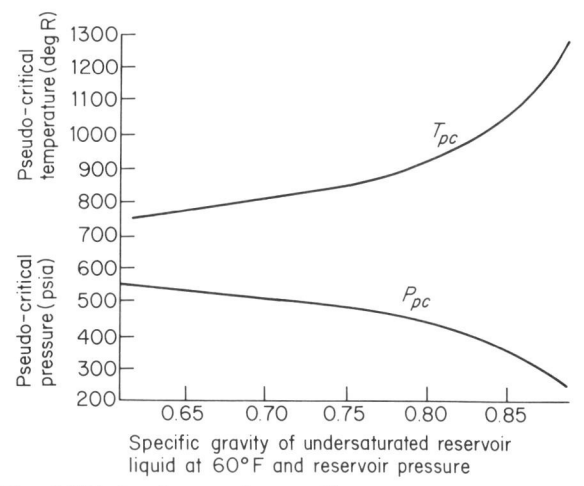

Fig. 4.5 Trube diagram for specific gravity as a function of pseudo-critical constants for undersaturated reservoir liquids (after [28]).

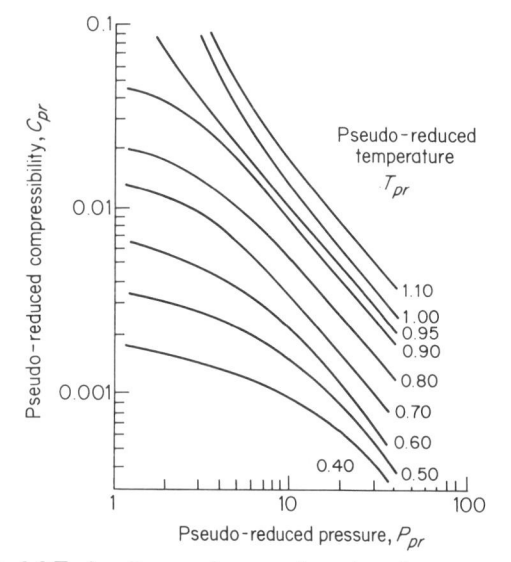

Fig. 4.6 Trube diagram for pseudo-reduced compressibility of undersaturated reservoir liquids (after [28]).

TABLE 4.3 Gas properties

Gas	Density (relative to air)	Density lb/ft³ at 60°F and 1 atmosphere	Molecular weight	P_c(psia)	T_c(°R)
Air	1.0	0.0763	28.96	547	238
Nitrogen	0.9672	0.0738	28.01	493	227
Oxygen	1.1047	0.0843	32.0	737	278
CO₂	1.5194	0.1159	944.01	1071	548
H₂S	1.1764	0.0897	34.08	1306	673
H₂	0.0696	0.0053	2.016	188	60
H₂O	—	—	18.016	3208	1165

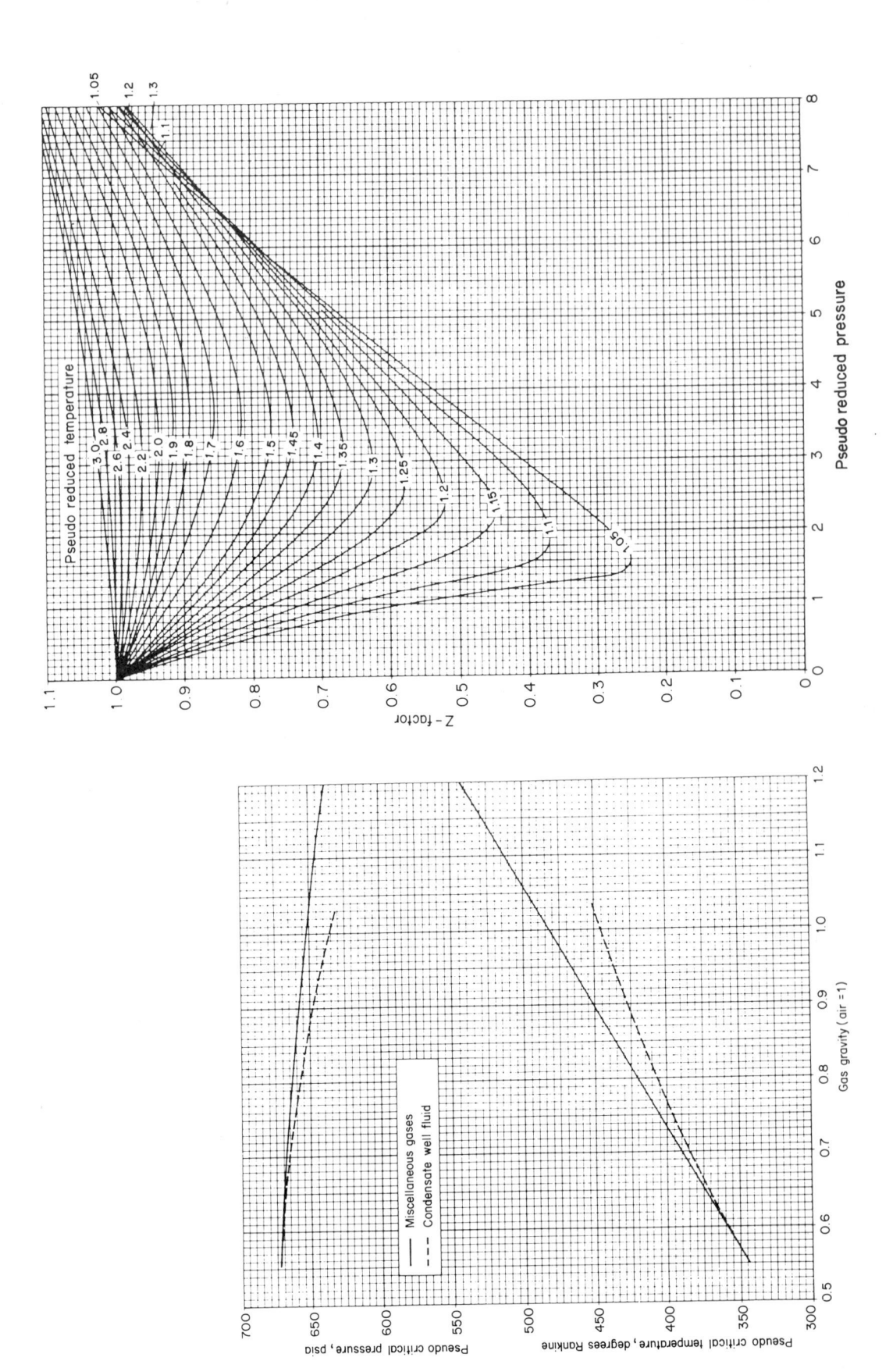

Fig. 4.7 Pseudo-critical properties and super-compressibility factor Z for natural gases (after [1]).

4 PROPERTIES OF RESERVOIR FLUIDS

TABLE 4.4 Hydrocarbon properties

Hydrocarbon	Formula	Molecular weight	Gross calorific value Btu/ft³ at s.c.	P_c(psia)	T_c(°R)	V_c(ft³/lb)
Methane	CH₄	16.042	1010	668	343	0.0993
Ethane	C₂H₆	30.06	1769	708	550	0.0787
Propane	C₃H₈	44.09	2517	616	666	0.0730
n-Butane	C₄H₁₀	58.12	3262	551	765	0.0704
i-Butane			3253	529	735	0.0725
n-Pentane	C₅H₁₂	72.15	4010	480	845	0.0690
i-Pentane			4000	489	829	0.0685
n-Hexane	C₆H₁₄	86.17	4756	437	913	0.0685
n-Heptane	C₇H₁₆	100.2	5503	397	972	0.0682
n-Octane	C₈H₁₈	114.2	6250	361	1024	0.0680
n-Nonane	C₉H₂₀	128.2	6997	332	1070	0.0673
n-Decane	C₁₀H₂₂	142.3	7742	304	1112	0.0671
Benzene	C₆H₆	78.1	3742	710	1013	0.0540
Toluene	C₇H₈	92.13	4475	596	1069	0.0570

ate for engineering accuracy with hydrocarbon systems. Critical constants for some commonly encountered components are reproduced in Tables 4.3 and 4.4, and for natural gases and condensate well fluids in Fig. 4.7.

4.4.4 Gas deviation factor Z

The ideal gas equation would predict the equality

$$\frac{PV_m}{RT} = 1$$

For real gases at pressures of more than a very few atmospheres this is not true. The behaviour of real gases can be expressed by the equation

$$PV_m = zRT$$

or

$$PV = nzRT$$

where $z = f(P,T)$ and z is termed a deviation factor (or super-compressibility factor) expressing the degree of deviation from ideality.

Because the law of corresponding states applies with satisfactory accuracy to mixtures of light hydrocarbon gases, it has been possible to correlate compressibility factors with reduced values of pressure and temperature, and these generalized correlations (the Standing-Katz correlations[1] as shown in Fig. 4.7) are widely used in approximate calculations of gas reservoir behaviour. For more accurate work, the extension and smoothing of accurate laboratory measurements by an equation of state, or direct calculation through an equation of state using the detailed composition of the gas to generate the necessary constants, will be satisfactory methods.

Hall and Yarborough[7] have used the Starling-Carnahan equation of state to calculate z, using first a Newton-Raphson iterative technique to calculate y, the reduced density from t, the reciprocal of the pseudo-reduced temperature using

$$(90.7t - 242.2t^2 + 42.4t^3) y^{(2.18+2.82t)}$$
$$- (14.76t - 9.76t^2 + 4.58t^3)y^2 + [(y + y^2 + y^3 - y^4)/(1-y^3)]$$
$$= 0.06125 p_{pr} t e^{-1.2(1-t)^2}$$

Then $zY = 0.06125 p_{pr} t e^{A-1.2(1-t)^2}$

The iterative procedure has been described by Dake[29].

Gas density, defined as mass/unit volume thus becomes at reservoir conditions:

$$\frac{M_a P}{zRT} = \rho_g$$

4.4.5 Gas viscosities

The reservoir engineer is concerned not only with the expansion behaviour of reservoir fluids, but also by flow rates and potential variations. The magnitude of flow rates and potential drops will depend directly on fluid viscosities, and in the case of gases these will depend on pressure and temperature. Again, if direct measurements of viscosity are not available, use is made of correlations based on corresponding states. The viscosities of hydrocarbon gases at atmospheric pressure are established as functions of molecular weight and temperature. The ratio of viscosity at a reduced pressure P_R and reduced temperature T_R, to the viscosity at atmospheric pressure and real temperatures T are then correlated with reduced pressure and temperature. Through these two correlations the viscosity at any given reservoir conditions can be estimated. The

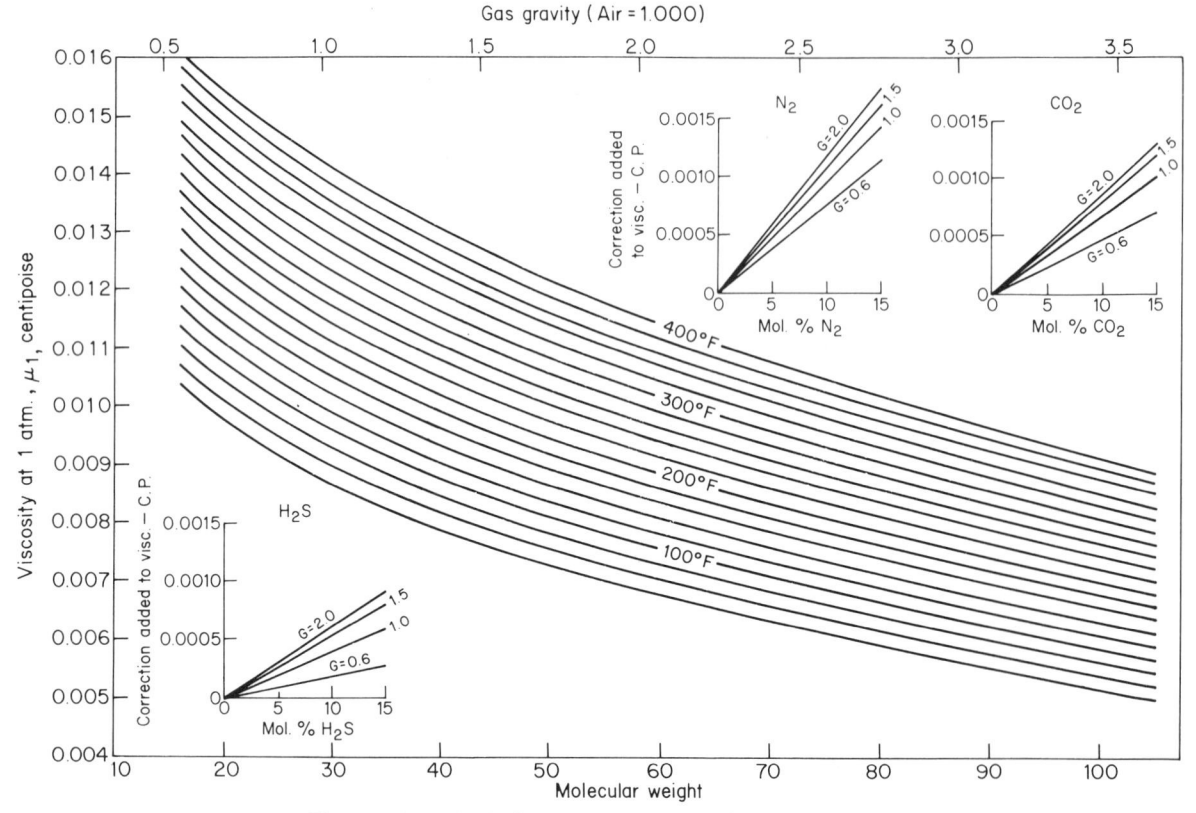

Fig. 4.8 Viscosity (μ_1) at one atmosphere for natural gases.

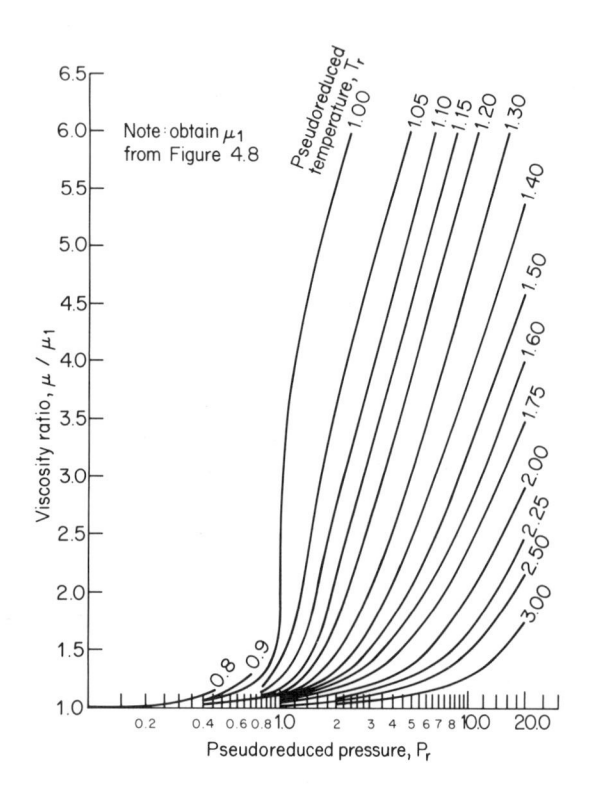

correlations, after Carr et al.[11] are shown in Figs 4.8 and 4.9.

4.4.6 Gas compressibilities

For a perfect gas:

$$PV = nRT$$

$$V = \frac{nRT}{P}$$

$$\frac{dV}{dP} = \frac{nRT}{P^2}$$

$$-\frac{1}{V}\frac{dV}{dP} = \left(\frac{-P}{nRT}\right)\left(\frac{-nRT}{P^2}\right) = \frac{1}{P} = c_g$$

Fig. 4.9 (left) Viscosity ratio for natural gases.
Note: obtain μ_1 from Fig. 4.8.

4 PROPERTIES OF RESERVOIR FLUIDS

For a real gas:

$$PV = nzRT \text{ and } z = f(P)$$
$$V = nRT\frac{z}{P}$$
$$\frac{dV}{dP} = -\left[\frac{znRT}{P^2}\right] + \frac{nRT}{P}\left(\frac{\partial z}{\partial P}\right)$$
$$-\frac{1}{V}\frac{dV}{dP} = \frac{1}{P} - \frac{1}{Z}\frac{dz}{dP} = c_g$$

so that for any ideal gas, or for real gases when the rate of change of z with p is small, compressibility can be represented by reciprocal pressure.

For real gases the gradient dz/dP is obtained by drawing a tangent to the z against pseudo-reduced pressure curve at the reservoir pseudo-reduced conditions. In this case the transformation $P = P_{pc} \cdot P_{pr}$ is used as follows:
Since

$$\frac{dz}{dP} = \frac{dP_{pr}}{dP} \cdot \frac{dz}{dP_{pr}}$$

and

$$\frac{dP_{pr}}{dP} = \frac{1}{P_{pc}}$$

then

$$\frac{dz}{dP} = \frac{1}{P_{pc}} \cdot \frac{dz}{dP_{pr}}$$

so that

$$C_g = \frac{1}{P_{pr} \cdot P_{pc}} - \frac{1}{zP_{pc}}\left(\frac{dz}{dP_{pr}}\right)$$

4.5 FORMATION VOLUME FACTORS B

Formation volume factors have been given the general standard designation of B and are used to define the ratio between a volume of fluid at reservoir conditions of temperature and pressure and its volume at a standard condition. The factors are therefore dimensionless but are commonly quoted in terms of reservoir volume per standard volume. Subscripts o, g, w are used to define the fluid phase, and i and b are often added to define initial and bubble-point conditions. Thus B_{gi} is an initial reservoir condition gas formation volume factor and B_{ob} is an oil formation volume factor at bubble-point conditions.

In general we can write

(Formation volume factor)$_{\text{gas}} = B_g$
  = f(composition, pressure, temperature)
(Formation volume factor)$_{\text{oil}} = B_o$
  = f(composition, pressure, temperature)
(Formation volume factor)$_{\text{water}} = B_w$
  = f(composition, pressure, temperature)

Defining a simple ratio does not, of course, simplify the problem of calculating the factor.

In the case of gases, this can be done through an equation of state, and this is also possible for a liquid above the saturation pressure. With liquids below the saturation pressure, the evolution of gas with decreases in pressure, and the possible dependence on composition and volume changes with pressure, the definition of simple equations of state is complicated.

4.5.1 Gas formation volume factor B_g

For a gas

$$B_g = \frac{\text{volume at operating conditions}}{\text{volume at standard conditions}}$$

$$B_g = \frac{P_o T z}{P T_o} \text{ volumes/volume}$$

For example, for reservoir conditions, 2000 psia, 585°R, $Z = 0.85$, and for reference conditions, 14.7 psia, 520°R, $z = 1$

$$B_g = 7.028 \times 10^{-3} \text{ volume/volume}$$

It is frequently convenient to work in reservoir units of barrels, and gas volumes are frequently expressed in MSCF so that

$$B_g = \frac{P_o T z}{P T_o} \times \frac{1000}{5.615} \text{ reservoir barrels/MSCF}$$

and for the above example

$$B_g = 1.2516 \text{ reservoir barrels/MSCF}$$

(In these particular units, gas formation volume factors have the same general numerical range as oil formation volume factors.)

Regrettably, some writers also use B_g to represent the reciprocal of the formation volume factor, i.e.

$$(B_g)' = \frac{\text{Volume at reference conditions}}{\text{Volume at operating conditions}}$$

but this convention will not be adopted here, and this expression will be given an alternative definition of the gas expansion factor, i.e.

$$\text{Gas expansion factor} = E = \frac{1}{B_g}$$

$$= \frac{\text{Volume at reference conditions}}{\text{Volume at operating conditions}}$$

Note that for small changes in z

$$B_g = \frac{T z P_o}{T_o} \cdot \frac{1000}{5.615} \cdot \frac{1}{P}$$

and

$$B_g \simeq \text{const} \cdot \frac{1}{P}$$

(the equation of a hyperbola) and

$$\frac{1}{B_g} = \text{constant } P$$

As shown in Fig. 4.10, a plot of E against P is approximately linear over small pressure ranges, and this can be convenient for data smoothing and interpolating.

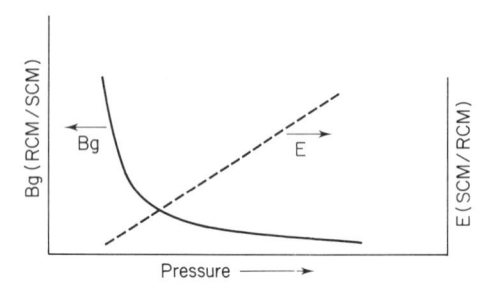

Fig. 4.10 Gas formation volume factor.

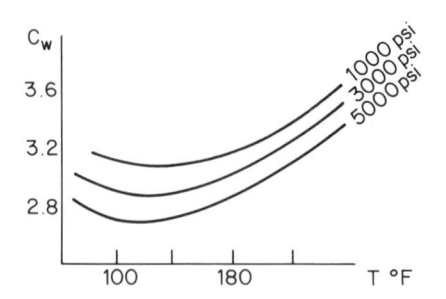

Fig. 4.11 Brine compressibility at different pressures and temperatures.

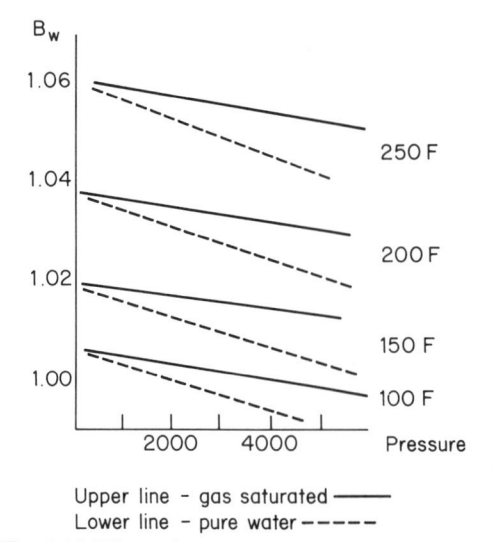

Fig. 4.12 Effect of temperature, pressure and gas saturation on water formation volume factor.

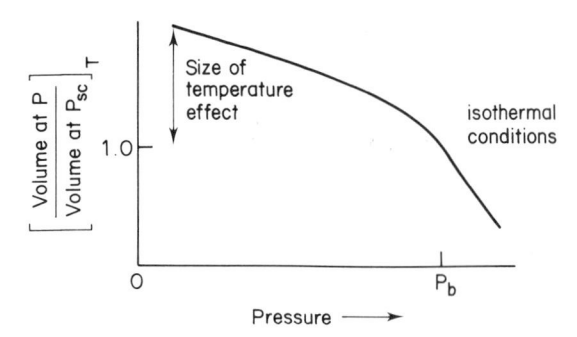

Fig. 4.13 Volume factor for water.

4.5.2 Water formation volume factor B_w

For water, gases and dissolved salts can affect the compressibility and so the formation volume factor, but this effect is often ignored. When pressure changes and water volumes are small, B_w is generally taken to be unity, and when pressure changes are large and/or water volumes are large, volume changes are calculated through compressibility, rather than formation volume factor. These data are shown in Figs 4.11 and 4.12.

The isothermal reservoir volume relationship for water containing some dissolved gas and initially existing above its bubble-point condition is shown in Fig. 4.13. It can be seen that loss in liquid volume due to evolution of gas as the pressure reduces only partially compensates for water expansion.

In general, B_w may be expressed in terms of the volume change ΔV_{wp} during pressure reduction and ΔV_{wt} during temperature reduction

$$B_w = (1 + \Delta V_{wp})(1 + \Delta V_{wt})$$

The values of ΔV terms are shown in Figs 4.14 and 4.15.

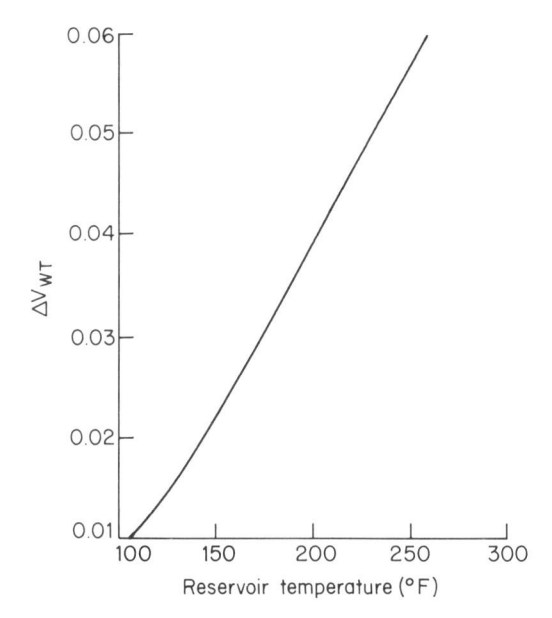

Fig. 4.14 Correction term ΔV_{wt}.

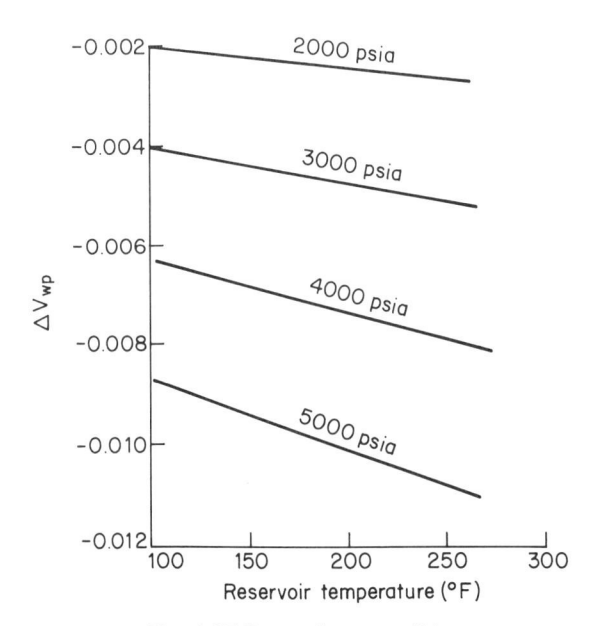

Fig. 4.15 Correction term ΔV_{wf}.

it is the total system – oil plus associated gas – that must be considered, and not simply the stock tank liquid phase. The effect of pressure on the hydrocarbon liquid and its associated gas is to induce solution of gas in the liquid until an equilibrium condition is attained. A unit volume of stock tank oil brought to equilibrium with its associated gas at reservoir pressure and temperature will almost invariably occupy a volume greater than unity (the only exception would be an oil with little dissolved gas at very high pressure).

The volume change is then a function of the partial molar volume of the gas in solution, the thermal expansion of the system with temperature change, and the compression of the liquid phase. No simple thermodynamic equation exists through which these volume changes can be calculated and formation volume factors generated. The necessary liquid properties are either directly measured or are determined from generalized correlations which can have an acceptable accuracy for the generally homologous family of components of crude oil systems.

The oil formation volume factor will be discussed in connection with the behaviour of dissolved gas systems. The shape of the B_o curve (Fig. 4.16) reflects oil compressibility while all gas stays in solution at pressures above bubble-point, and indicates liquid shrinkage when gas comes out of solution below bubble-point.

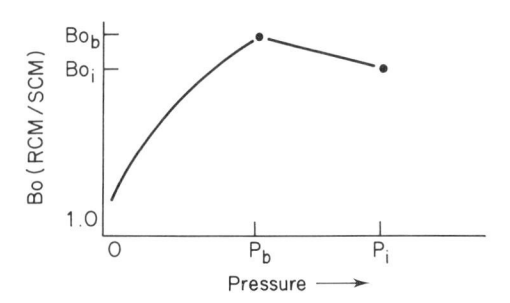

Fig. 4.16 Oil formation volume factor.

4.5.3 Oil formation volume factor B_o

Black oil, or dissolved gas systems, may conveniently be regarded as solutions of gas mixture in a liquid, the compositional changes in the gas with changing pressure and temperatures being ignored. In considering volume changes between reference conditions and reservoir (or other operating) conditions,

4.6 GAS-OIL RATIOS

The dissolved or solution gas–oil ratio, R_s, is a constant above bubble-point pressure, as shown in Fig. 4.17, and displays similar behaviour to B_o below bubble point pressure. The symbol R is generally used for gas–oil ratios and has the units of standard volumes of gas dissolved in a standard volume of liquid. When used alone, R represents an instantaneous total producing gas–oil ratio (free gas plus

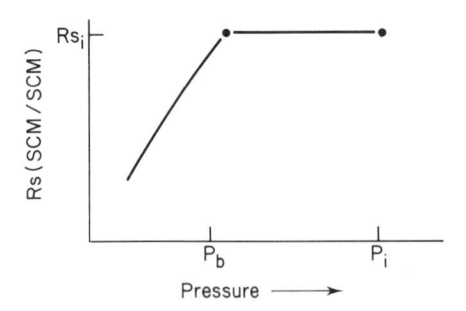

Fig. 4.17 Solution gas–oil ratio.

solution gas), while with subscript *s* the symbol R_s indicates only the dissolved gas content of the liquid. The symbol R_p indicates a cumulative ratio since start of reservoir production, and is thus total standard volume of gas produced divided by total standard volume of oil produced.

4.7 DIRECT MEASUREMENTS – *PVT* ANALYSIS

The first requirement – and difficulty – in taking measurements is that of obtaining a truly representative sample of the formation fluid.

Provided that the pressure at the bottom of a test well does not drop below the bubble-point, then a bottom-hole sample should be representative. Provided then that no leakage occurs from the sample, and that transfer from the sample vessel to the test vessel is carried out without loss of any components, then tests should be representative of samples. Multiple samples are essential. The alternative to bottom-hole sampling is recombination sampling, which is inherently less satisfactory, except perhaps for gas condensate systems.

When samples of separator oil and separator gas are recombined for test purposes, a particular problem is the choice of proportions for recombination. If a well flows under stable conditions, with a bottom-hole pressure above the bubble-point, and the separator gas–oil ratio stabilizes throughout the test, then recombination should be valid. Measured gas–oil ratios will frequently be erratic, and will occasionally vary over a range which will make recombination suspect. It must be realized that all *PVT* analysis involves a basic and unavoidable inconsistency. The thermodynamic path followed by a two-phase mixture (i.e. the changes in pressure, temperature and composition) in flowing from the reservoir to the stock tank is essentially unknown. Consequently this thermodynamic path cannot be followed or duplicated in a *PVT* analysis. In the laboratory, two processes can be followed – a constant composition (flash) separation, and a stepwise variable composition (differential) separation. For operational convenience, both these separations are carried out, and the results of the two separations are combined (in rather arbitrary fashion) to generate the data needed for material balance calculations. The experimental layout is usually similar to that indicated in Fig. 4.18.

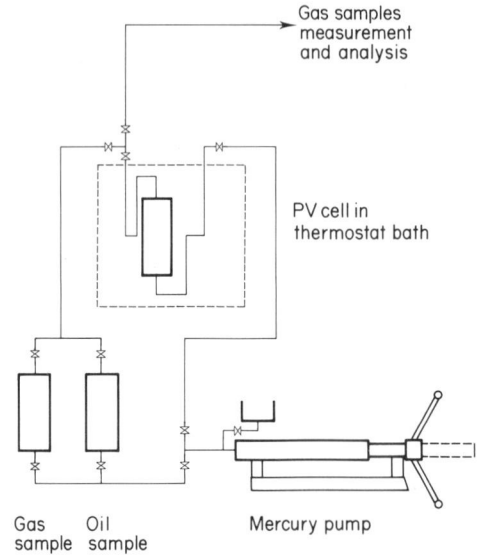

Fig. 4.18 Schematic *PVT* analysis.

In the case of black oil (dissolved gas) systems, this inconsistency is relatively unimportant, and the values generated are system properties valid within the usual range of data uncertainty. In the case of volatile oil and condensate systems, standard methods are inadequate, and process simulation by laboratory experiment is necessary to validate thermodynamic relationships.

4.7.1 Flash liberation at reservoir temperature

The fluid sample is raised to a high pressure (substantially above bubble-point) at reservoir temperature, and expanded in stages at constant composition, to give the total volume at a series of pressures.

A plot of the volumes and pressures will identify the bubble-point pressure, and data is customarily replotted in terms of relative volume V_R, as shown in Fig. 4.19.

$$V_R = \frac{\text{Volume at any arbitrary pressure}}{\text{Volume at bubble-point pressure}} = \left.\frac{V_t}{V_b}\right|_T$$

4 PROPERTIES OF RESERVOIR FLUIDS

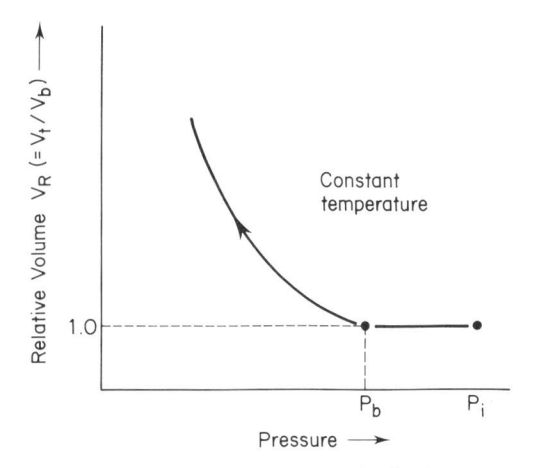

Fig. 4.19 Relative volume data by the flash process.

The initial condition is V_{Ri}.

From the plot, the compressibility above bubble-point of the liquid at reservoir temperature can be determined.

$$c_o = \frac{1 - V_{Ri}}{V_{Ri} \Delta P}$$

4.7.2 Differential liberation at reservoir temperature

Again starting above or at the bubble-point, the system is expanded in stages. In this case, the free gas phase at each stage is removed at constant pressure, and then measured by expanding to standard conditions. Consequently the expansion is not at constant composition, and the results are valid only for dissolved gas systems.

This expansion yields the solution gas–oil ratio as a function of pressure, gas expansion factors as a function of pressure, and volumes of liquid at each pressure. The liquid remaining at 1 atmosphere at reservoir temperature is termed residual oil (and its volume the residual oil volume), and cooling this to 60°F will generate a *stock tank volume* (V_{ST}).

The relative volume ratio $(V_o)_d/V_b$ is generally plotted as a function of pressure, as shown in Fig. 4.20 (a). The liberated gas ratio R_L is also plotted against pressure, as shown in Fig. 4.20 (b).

PVT reports frequently tabulate the ratio

$$\frac{(V_o)_d}{V_{ST}} = \frac{\text{liquid volume}]_{P,T}}{\text{residual liquid volume}]_{60°F}}$$

$$= \frac{V_o}{V_b} \cdot \frac{V_b}{V_{ST}}$$

as a formation volume factor. For many systems, this ratio will be sufficiently accurate (given the uncertainties in sampling) to be used in this way. The volumes of stock tank oil generated by a volume of reservoir oil is in fact dependent on the thermodynamic path followed between initial and final states.

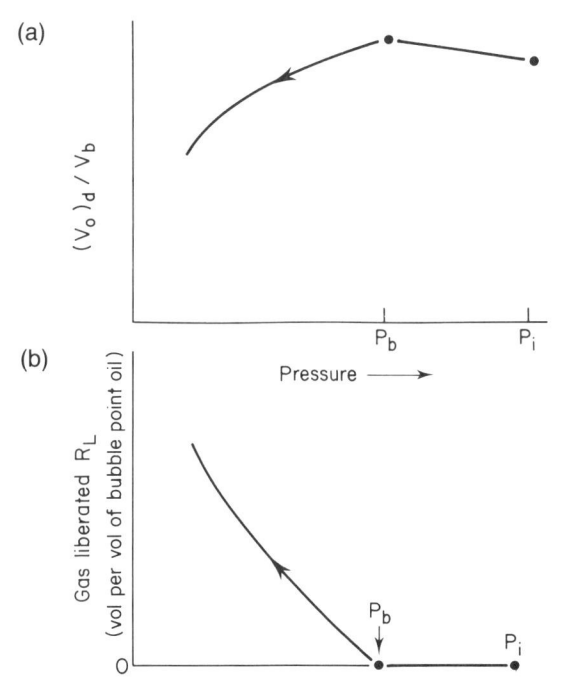

Fig. 4.20 Relative volume (a) and gas liberation (b) data by the differential process.

4.7.3 Flash separation tests

It is customary to carry out a separator test, *flashing* the bubble-point liquid to stock tank conditions through a series of intermediate stages corresponding to possible field separator conditions. The total gas evolved in this flash operation is the value taken as the total or initial solution gas–oil ratio R_{si}. This will differ from the cumulative gas released in the differential process because of the different thermodynamic path involved in cooling residual oil to standard conditions.

When this is done, the difference between residual oil (60°F) and stock tank oil obtained through a specific separation process can be corrected, although the basic inconsistency remains.

We define:

$(V_o)_d$ = Volume of oil obtained by a differential separation at any pressure, and reservoir temperature

V_b = Volume at bubble-point and reservoir temperature

$$B_o = \frac{(V_o)_d}{V_b} \cdot \frac{V_b}{(V_{ST})_F}$$

$(V_{ST})_F$ = Volume of stock tank oil obtained by flashing a volume V_b of bubble-point oil

A difficulty also arises with the definition of gas–oil ratio. If the incremental gas volumes accumulated in a differential process are summed, the results will be a total for the gas originally in solution. This is frequently referred to as initial solution GOR per unit volume of residual oil (60°F).

This volume will not be the same as the total gas obtained by flashing bubble-point oil through a specific separator process to stock tank conditions. Since the produced GOR used in material balance calculations is the separator GOR, it is customary to use the flash separator GOR as the value for R_s in material balance calculations, the differential separation values being used for the solution gas–oil ratio. Again an inconsistency remains.

4.7.4 Summary of tests

The laboratory tests can be summarized as in Table 4.5, together with the data generated.

The bubble-point oil formation volume factor B_{ob} from flash separator tests is used in the calculation of oil compressibility.

The two-phase formation volume factor B_t is calculated in consistent units from

$$B_t = \frac{V_t}{V_b} \cdot \frac{V_b}{(V_{ST})_F} = B_o + B_g \{R_{si} - R_s\}$$

The solution gas–oil ratio is thus

$$R_s = R_{si} - \left\{\frac{B_t - B_o}{B_g}\right\}$$

4.8 GENERALIZED CORRELATIONS FOR LIQUIDS SYSTEMS

4.8.1 Bubble-point pressure

The bubble-point pressure or saturation pressure is a value of considerable interest to engineers, and a value will nearly always be measured experimentally. In fields with multiple reservoirs (or where the fact of separation of horizons is not established early on), samples may not be taken for all reservoirs, and indirect approaches may be necessary.

Correlations are available for the estimation of bubble point pressure from other system properties.

$$P_b = f(\text{composition, temperature})$$

If detailed composition is known, and equilibrium ratios are available for the components at the temperature required, the bubble point can be calculated, but in general the composition for a

TABLE 4.5 Laboratory test summary

Equilibrium vaporization tests		Flash separator test
Conducted at reservoir temperature		Conducted from reservoir temperature to surface conditions
Relates oil properties at T,P to saturated oil properties at P_b and T_r		Relates oil properties at reservoir conditions to oil at stock tank conditions
Flash vaporization test	Differential liberation test	Measures:
Gas not removed Composition constant	Gas removed as it is released Composition of remaining system varies	R_{si} $\dfrac{(V_{ST})_F}{V_b} = \left(\dfrac{1}{B_{ob}}\right)$
Measures: V_t/V_b C_o	Measures: V_o/V_b R_L C_o	

4 PROPERTIES OF RESERVOIR FLUIDS

dissolved gas system is represented by the solution gas–oil ratio, the gas density and the oil density, so that

$$P_b = f(R_s, \rho_g, \rho_o, T)$$

An empirical correlation using a large amount of data, and developed by Standing[1] is

$$P_b = 18 \left[\left(\frac{R_s}{\gamma_g}\right)^{0.83} \cdot \frac{10^{0.00091T}}{10^{0.0125^{API}}} \right]$$

where $T = °F$, γ_g = gas gravity and P_b is in Psi. The relationship is shown in Fig. 4.21. Figure 4.22 shows a nomogram for an alternative method due to Lasater[16].

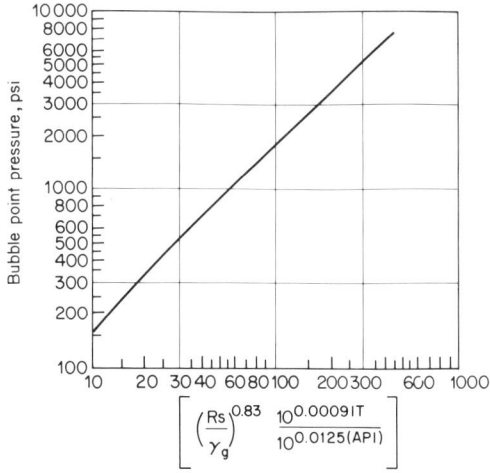

Fig. 4.21 Bubble-point correlation using Standing's data (after [1]).

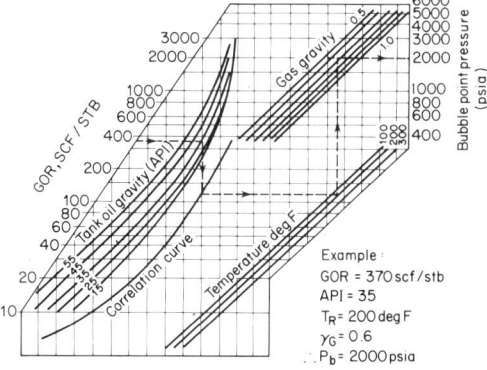

Fig. 4.22 Bubble-point pressure using the Lasater correlation (after [16]).

4.8.2 Formation volume factor

Formation volume factor for a saturated liquid can be estimated[1] from the empirical equation

$$B = 0.972 + 0.000147 F^{1.175}$$

$$F = R_s \left[\frac{\gamma_g}{(\gamma)_o}\right]^{0.5} + 1.25T$$

Evaluation of B_o over a range of pressures, when the value of solution GOR is known at one pressure, will then involve determining R_s at a series of pressures within the range by the inverse of procedure for bubble-point pressure, and the use of this value in the formation volume factor correlation.

Above the bubble-point, the formation volume factor will be given by the equation

$$B_p = B_b \{1 + c(P_b - P)\}$$
$$B_b \{1 - c(P - P_b)\}$$

and the oil compressibility within the range $P \to P_b$ will be needed for this.

Depending upon the data available, estimation for the compressibility can involve a number of cross-correlations, the final correlation being of pseudo-reduced compressibility with pseudo-reduced temperature and pseudo-reduced pressure, several previous correlations being necessary for the liquid phase.

The oil density at reservoir conditions can be estimated in appropriate units from the oil and gas densities at standard conditions and from the values at B_o and R_s at reservoir conditions, as follows:

$$\rho_o B_o = (\rho_o)_{sc} + R_s (\rho g)_{sc}$$

4.8.3 Two-phase volume factor

The total (two-phase) formation volume factor is the volume occupied at reservoir conditions by the oil and gas associated with a unit volume of stock tank oil. The parameter is designated B_t and is defined by

$$B_t = B_o + B_g (R_{si} - R_s)$$

This can be evaluated from the separate oil and gas formation volume factors at any pressure, and the solution gas–oil ratio.

Since the reservoir equations using formation volume factors involve differences in these factors, it is frequently desirable to smooth experimental or correlation derived data to improve accuracy.

Two-phase formation volume factors are most readily smoothed by the relation

$$Y = \frac{(P_b - P)}{P\left\{\dfrac{B_t}{B_b} - 1\right\}} = \frac{(P_b - P)}{P\left\{\dfrac{V_t}{V_b} - 1\right\}}$$

and the Y function is a linear function of pressure (see Fig. 4.23). This can be a valuable relationship for smoothing, correlating and extending data, especially near the bubble-point.

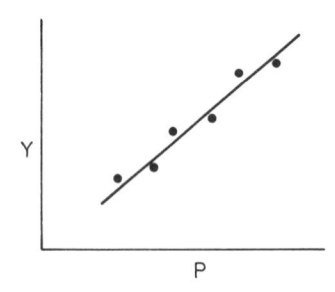

Fig. 4.23 Y-function smoothing.

4.8.4 Oil viscosity

The viscosity of a reservoir oil can be considered to be the viscosity of a *dead* oil at some reference pressure condition (e.g. tank oil), corrected for temperature and the effects of dissolved gas. It is preferable for the viscosity at the reference state to be measured, although correlations are available for this purpose.

Figure 4.24 shows the essential correlations[17]. The viscosity of dead oil at reservoir temperature and atmospheric pressure is the required start point, and a measured value at this temperature, or a measured value corrected to this temperature, is desirable. The viscosity of saturated oil is directly obtained. If the system is undersaturated, a correction for the excess pressure is needed. If no measured viscosity is available, an estimate must be made through the oil gravity.

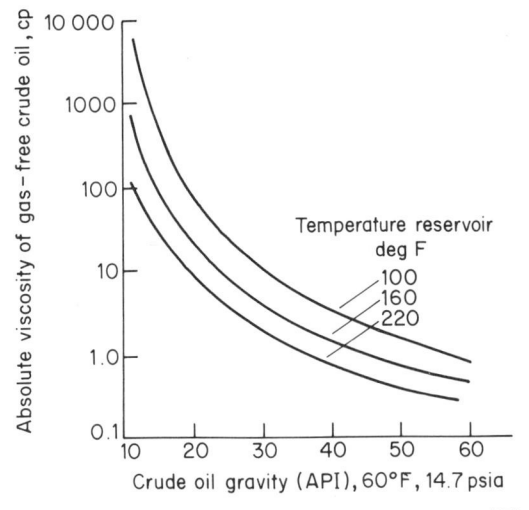

Fig. 4.24 Beal correlations for crude oil viscosity.[17]

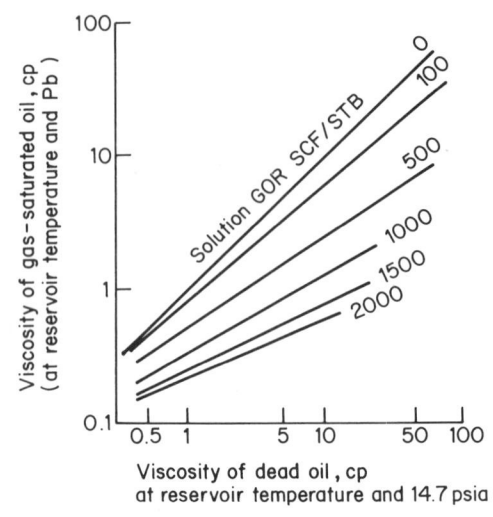

Fig. 4.24 (b)

4.8.5 Water viscosity

Water viscosity is dependent on salinity. The viscosity of various salinity brines at 1 atmosphere pressure (μ^*_T) is given in Fig. 4.25 (a) as a function of reservoir temperature[18]. At elevated pressures, the relationship

$$\mu_{P,T} = \mu^*_T \cdot f_{P,T}$$

is used where $f_{P,T}$ is obtained from the $f_{P,T}$ against T chart (Fig. 4.25 (b)) at pressures between 2000 and 10 000 psi.

4.8.6 Recent North Sea oil correlations

Using the correlation methods proposed by Standing, a number of North Sea oils have been recorrelated by Glasø[13]. The units used are oil field units with P_b the bubble-point pressure in psia; R the producing gas–oil ratio in SCF/STB; $\bar{\gamma}_g$ the average specific gravity of the total surface gases; T the reservoir temperature in °F; API is the degree API stock tank oil gravity and γ_o is the specific gravity of the stock oil. The reservoir pressure in psia is P.

For saturation pressure the relationship is

$$\log P_b = 1.7669 + 1.7447 (\log P^*_b) - 0.30218 (\log P^*_b)^2$$

where

$$P^*_b = \left[\left(\frac{R}{\bar{\gamma}_g} \right)^{0.816} \cdot \left\{ \frac{T^{0.172}}{(API)^{0.989}} \right\} \right]$$

4 PROPERTIES OF RESERVOIR FIELDS

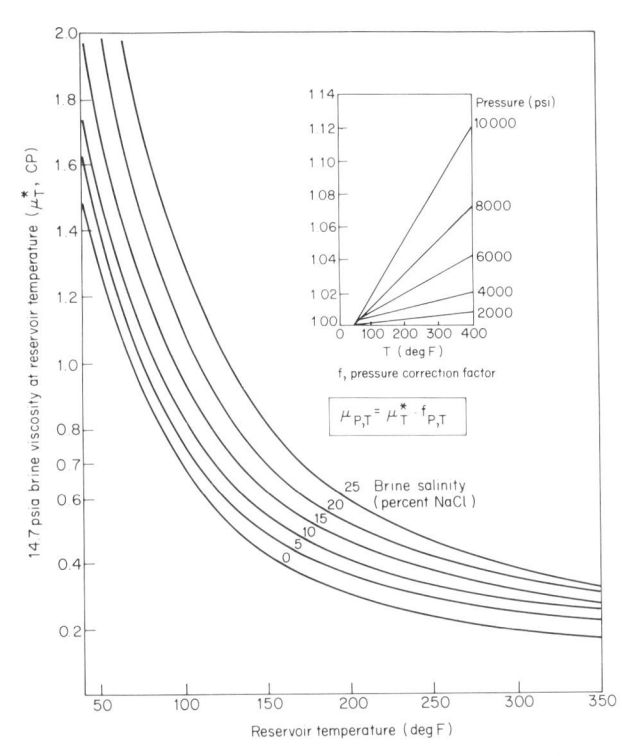

Fig. 4.25 PVT correlations for North Sea oils (after [13]).

Figure 4.26 (b) illustrates this relationship. For the two-phase flash volume factor B_t at any reservoir pressure P, the relationship between B_t and B_t^* is shown in Fig. 4.25 (c) and represents:

$$\log B_t = 0.080135 + 0.47257 (\log B_t^*) + 0.17351 (\log B_t^*)^2$$

where

$$B_t^* = R \left\{ \frac{T^{0.5}}{\bar{\gamma}_g^{0.3}} \right\} \cdot (P^{-1.1089}) \cdot \{\gamma_o^{(2.9 \times 10^{-0.00027R})}\}$$

The pressure of non-hydrocarbons, specifically CO_2, N_2, H_2S, can affect the calculated values of saturation pressure described above. The magnitude of the correction multiplier F_c, which should be applied to the calculated value, can be expressed in terms of the mole fraction y of the non-hydrocarbon component present in total surface gases, as shown in Fig. 4.26 (d), (e) and (f).

For nitrogen

$$F_c = 1.0 + \{(-2.65 \times 10^{-4} (API) + 5.5 \times 10^{-3}) T + (0.093 (API) - 0.8295)\} y_{N_2} + \{(1.954 \times 10^{-11} (API)^{4.699}) T + (0.027 (API) - 2.366)\} (y_{N_2})^2$$

For CO_2,

$$F_c = 1.0 - 693.8 (y_{CO_2}) \cdot T^{-1.553}$$

For H_2S,

$$F_c = 1.0 - (0.9035 + 0.0015 (API)) (y_{H_2S}) + 0.019 \{45 - (API)\} (y_{H_2S})^2$$

For volatile oils an exponent for the temperature of 0.130 rather than 0.172 is appropriate. The relationship between P_b^* and P_b is shown in Fig. 4.26 (a). For oil flash formation volume factor, B_{ob}, the relationship is

$$\log (B_{ob} - 1.0) = 2.91329 (\log B_{ob}^*) - 0.27683 (\log B_{ob}^*)^2 - 6.58511$$

where

where $B_{ob}^* = R \left[\dfrac{\bar{\gamma}_g}{\gamma_o} \right]^{0.526} + 0.968 T$

Examples

Example 4.1

Tabulate values of API gravity for the specific gravity range 0.70 to 0.90 in increments of 0.02.

Example 4.2

The composition and component critical values of a gas are as tabulated below:

Component	Mol.wt	Mol.fraction	Critical press (psia)	Critical temp.(°R)
C1	16	0.9	673	343
C2	30	0.05	708	550
C3	44	0.03	617	666
C4	58	0.02	551	765

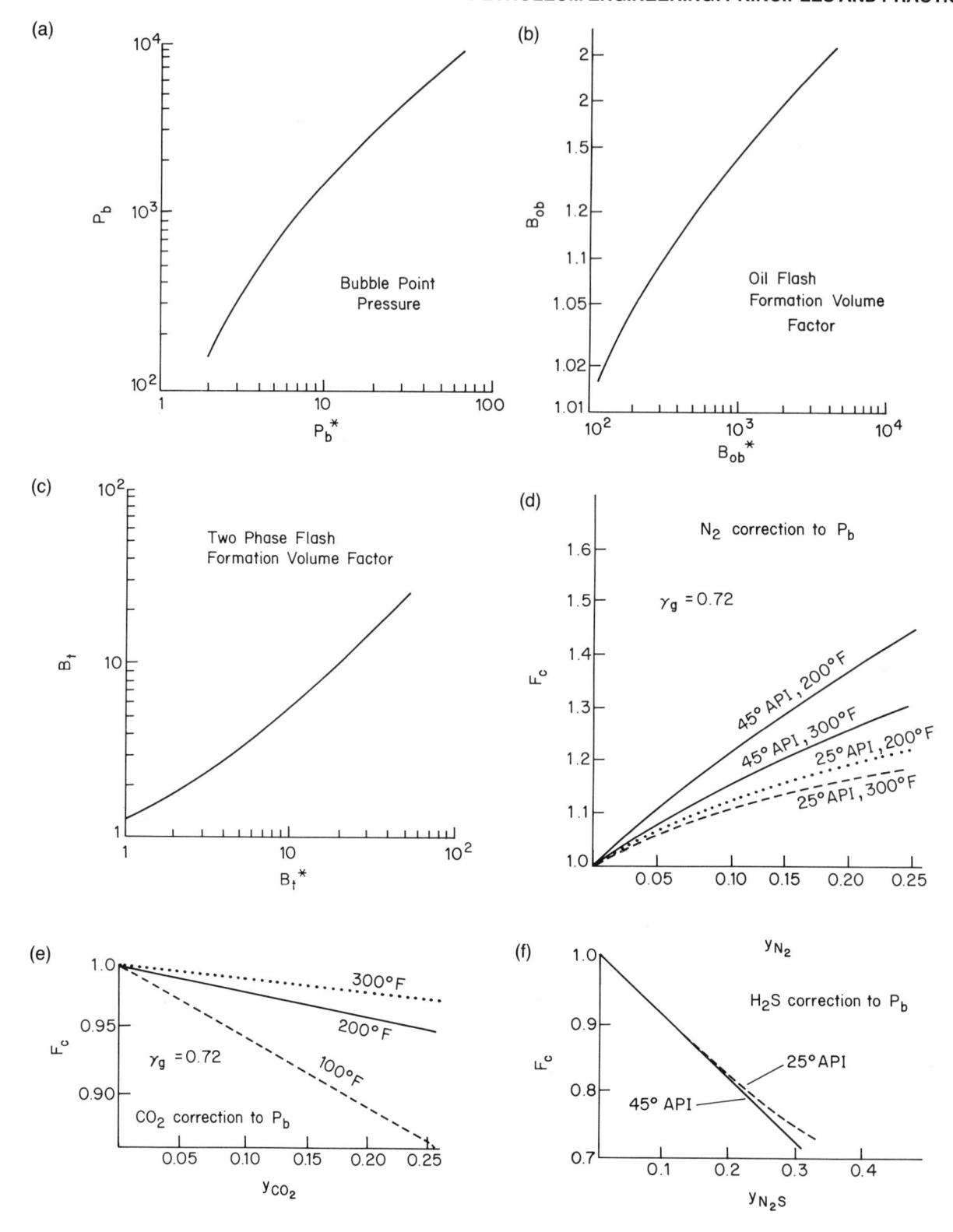

Fig. 4.26 PVT correlations for North Sea oils (after 13).

4 PROPERTIES OF RESERVOIR FLUIDS

(a) Calculate the molecular weight of the gas.
(b) Calculate the density and specific gravity relative to air at 14.7 psia and 60°F. (Mol wt. of air = 28.9)
(c) Calculate the pseudo-critical pressure and temperature.
(d) Calculate the pseudo-reduced pressure and temperature at 2000 psia and 135°F.
(e) Determine the gas deviation factor at 2000 psia and 135°F.
(f) Calculate the gas density at 2000 psia and 135°F.
(g) Calculate (both in vols/vol and in reservoir barrels/1000 scf) the gas formation volume factor at 2000 psia and 135°F.
(h) Find the gas viscosity at 2000 psia and 135°F.
(i) Calculate the compressibility of gas at 2000 psia and 135°F.
(j) If the pressure of 2000 psia is the reservoir pressure measurement at 4100 ft SS and the regional aquifer gradient is 0.44 psi/ft (below mean sea level), at what depth would a gas water contact be expected?
(k) If the crest of the structure is found to be 1000 ft above gas water contact, what density of drilling fluid will be necessary to control formation pressures at the crest (providing an over pressure of 500 psi).

Example 4.3

If a reservoir has a connate water saturation of 0.24, a gas saturation of 0.31 and compressibilities are respectively: $c_o = 10 \times 10^{-6} (\text{psi})^{-1}$; water: $3 \times 10^{-6} (\text{psi})^{-1}$; gas as calculated in 42(a); pore compressibility $c_f = 5 \times 10^{-6} (\text{psi})^{-1}$. What are:
(a) total compressibility
(b) effective hydrocarbon compressibility.

Example 4.4

(a) Using correlations find the bubble-point pressure and formation volume factor at bubble-point pressure of an oil of gravity 38° API, gas–oil ratio 750 scf/stb, gas gravity 0.70, at a temperature of 175°F.
(b) If a reservoir containing this hydrocarbon has an oil–water contact at 7000 ft s.s. and the regional hydrostatic gradient is 0.465 psi/ft, at what elevation would a gas–oil contact be anticipated?
(c) What would be the formation volume factor of the oil at a pressure 4000 psi?
(d) Find the viscosity of the oil at bubble-point pressure and 175°F. Molecular weight = 180.

Example 4.5

The following results are obtained in a PVT analysis at 200°F:

Pressure psia	4000	3000	2500	2000	1500
System vol.ml.	404	408	410	430	450

Estimate the bubble point pressure.
The system is recompressed, expanded to 2000 psia and the free gas removed at constant pressure, and is measured by expansion to 1 atmosphere.

Volume of liquid	388 ml.
Volume of gas (expanded to 1 atm., 60°F)	5.275 litres

The pressure is then reduced to 14.7 psia and the temperature to 60°F

Volume of residual liquid   295 ml.
Volume of gas (measured at 1 atm., 60°F:21 litres)

Estimate the following PVT properties:

c_o, liquid compressibility at 3000 psia
B_o at 3000 psia
B_o, B_t, R_s at 2500 psia; 2000 psia
B_g, z at 2000 psia

References

[1] Standing, M.B.
Volumetric and Phase Behaviour of Oilfield Hydrocarbons, Reinhold Publishing (1952).

[2] Clark, N.
Elements of petroleum reservoirs, *Soc. Pet. Engrs.* (1962).

[3] Standing, M.B. and Katz, D.L.
Density of natural gases, *Trans AIME* **146** (1942), 140.

[4] GPSA
Engineering Data Book, Gas Processors Suppliers Association, Tulsa (1974), 16.

[5] McCain, W.D.
The Properties of Petroleum Fluids, Pemwell., Tulsa (1973).

[6] Katz, D.L. *et al.*
Handbook of Natural Gas Engineering, McGraw-Hill Inc, New York (1959).

[7] Hall, K.R. and Yarborough, L.
How to solve equation of state for Z-factors, *OGJ* (Feb. 1974), 86.

[8] Takacs, G.
Comparisons made for computer Z-factor calculations, *OGJ* (Dec. 1976), 64.

[9] American Petroleum Institute
Recommended practice for sampling petroleum reservoir fluids, *API Pub. RP* **44** (Jan. 1966).

[10] Dodson, C.R., Goodwill, D. and Mayer, E.H.
Application of laboratory *PVT* data to reservoir engineering problems, *Trans. AIME* **198** (1953), 287.

[11] Carr, N.L., Kobyashi, R. and Burrows, D.B.
Viscosity of hydrocarbon gases under pressure, *Trans. AIME* **201** (1954), 264.

[12] Burcik, E.J.
Properties of petroleum fluids, *IHRDC* (1979).

[13] Glasø, O.
Generalised pressure-volume-temperature correlations, *JPT* **32** (1980), 785.

[14] Peng, D.Y. and Robinson, D.B.
A new two constant equation on state, *Ind. Eng. Chem. Fund* **15** (1976), 59.

[15] Soave, G.
Equilibrium constants from a modified Redlich-Kwong equation of state, *Chem. Eng. Sci.* **27** (1972), 1197.

[16] Lasater, J.A.
Bubble point pressure correlation, *Trans. AIME* **213** (1958), 379.

[17] Beal, C.
The viscosity of air, water, natural gas, crude oil and its associated gases at oil field temperatures and pressures, *Trans. AIME* **165** (1946), 94.

[18] Matthews, C.S. and Russell, D.G.
Pressure buildup and flow test in wells – Chesnut's water viscosity correlation, *SPE Monograph No. 1* (1967).

[19] Chew, J. and Connally, C.A.
A viscosity correlation for gas saturated crude oils, *Trans. AIME* **216** (1959), 23.

[20] Long, G. and Chierici, G.
Salt content changes compressibility of reservoir brines, *Pet. Eng.* (July 1961), 25.

[21] Cronquist, C.
Dimensionless *PVT* behaviour of Gulf Coast reservoir oils, *J. Pet. Tech.* (1973), 538.

[22] Firoozabadi, A., Hekim, Y. and Katz, D.L.
Reservoir depletion calculations for gas condensates using extended analyses in the Peng Robinson equation of state, *Can. J. Ch. Eng.* **56** (1978), 610.

[23] Katz, D.L.
Overview of phase behaviour in oil and gas production, *JPT* (1983), 1205.

[24] Yarborough, L.
Application of a generalised equation of state to petroleum reservoir fluids, *Adv. Chem. Series. Am. Chem. Soc.* Washington D.C. No. 182 (1979), 385.

[25] Robinson, D.B., Peng, D.Y. and Ng, H.J.
Some applications of the Peng-Robinson equation of state to fluid property calculations, *Proc. GPA,* Dallas (1977), 11.

4 PROPERTIES OF RESERVOIR FLUIDS

[26] Lohrenz, J., Bray, B.G. and Clark, C.R.
Calculating viscosities of reservoir fluids from their compositions, *Trans. AIME* **231** (1964), 1171.
[27] Clark, N.J.
Adjusting oil sample data for reservoir studies, *JPT* (Feb. 1962), 143.
[28] Trube, A.S.
Compressibility of undersaturated hydrocarbon reservoir fluids, *Trans. AIME* **210** (1957), 341.
[29] Dake, L.P.
Fundamentals of Reservoir Engineering, Elsevier Scientific, Amsterdam (1978).

Chapter 5

Characteristics of Reservoir Rocks

5.1 DATA SOURCES AND APPLICATION

To form a commercial reservoir of hydrocarbons, any geological formation must exhibit two essential characteristics. These are a capacity for storage and transmissibility to the fluid concerned, i.e. the reservoir rock must be able to accumulate and store fluids, and when development wells are drilled it must be possible for those reservoir fluids to flow through relatively long distances under small potential gradients.

Storage capacity requires void spaces within the rock, and transmissibility requires that there should be continuity of those void spaces. The first characteristic is termed porosity, the second permeability. While some estimates of reservoir rock properties can be made from electrical and radioactive log surveys, the study of core samples of the reservoir rock are always essential.

A core is a sample of rock from the well section, generally obtained by drilling into the formation with a hollow section drill pipe and drill bit. There is a facility to retain the drilled rock as a cylindrical sample with the dimensions of the internal cross-sectional area of the cutting bit and the length of the hollow section. With *conventional* equipment, this results in cores up to 10 m in length and 11 cm in diameter. It is frequently found that variation in drilling conditions and in formation rock character prevent 100% recovery of the core. In addition, the core may also be recovered in a broken condition. Friable or unconsolidated rock is frequently recovered only as loose grains, unless special core barrels are used.

The recovered core represents the record of rock type in the well section and is the basic data for interpretation of geological and engineering properties.

In general, two partially conflicting objectives must be met when taking core samples. In the first place, a careful on-site examination for hydrocarbon traces is desirable (e.g. gas bubbling or oil seeping from the core, core fluorescence on a freshly exposed surface, fluorescence and staining in solvent cuts etc.), in case an open hole drill stem test is possible and desirable. In the second place, it is desirable to preserve the core in as unchanged a condition as possible prior to laboratory evaluation. Some parts of the core should then, immediately after lithological examination and logging, be wrapped tightly in polythene or immersed in fluid and sealed into containers for transit to the laboratory. This is done in the hope that drying out of cores with changes in wettability, or changes in porosity and permeability due to washing with incompatible fresh or sea water, will be minimized. Remaining parts may then be examined for hydrocarbon traces, solvent cuts taken, and some samples washed thoroughly for detailed lithological, sedimentological and palaeontological examination. Figure 5.1 shows the kind of data obtained from recovered cores.

5 CHARACTERISTICS OF RESERVOIR ROCKS

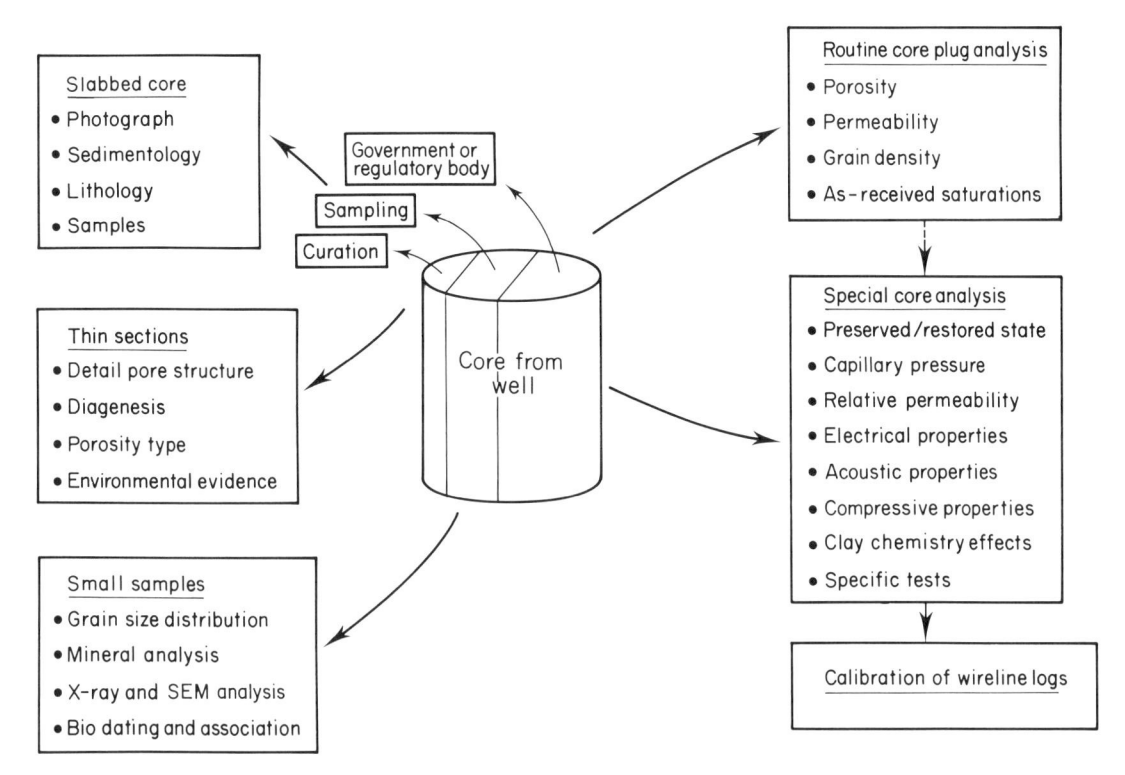

Fig. 5.1 Data obtained from cored wells.

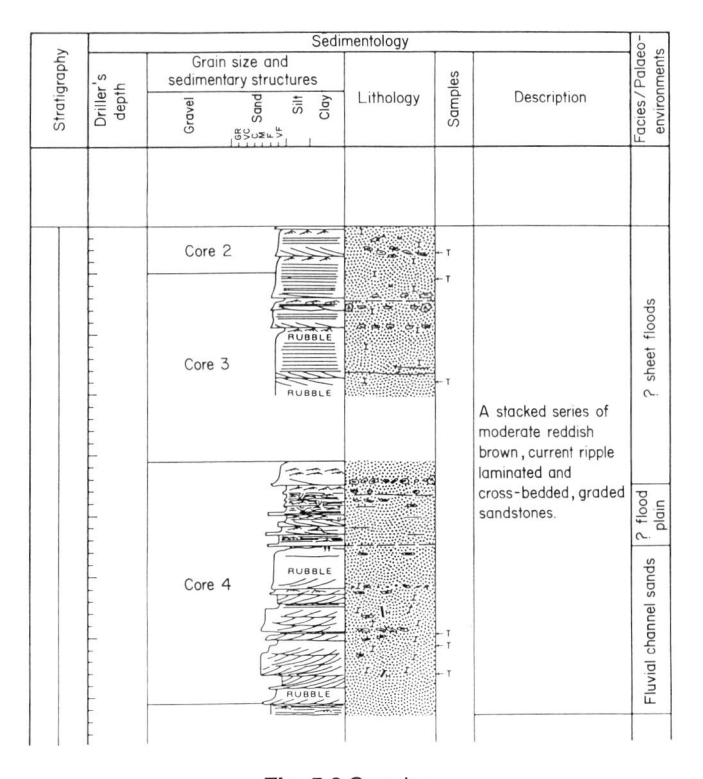

Fig. 5.3 Core log.

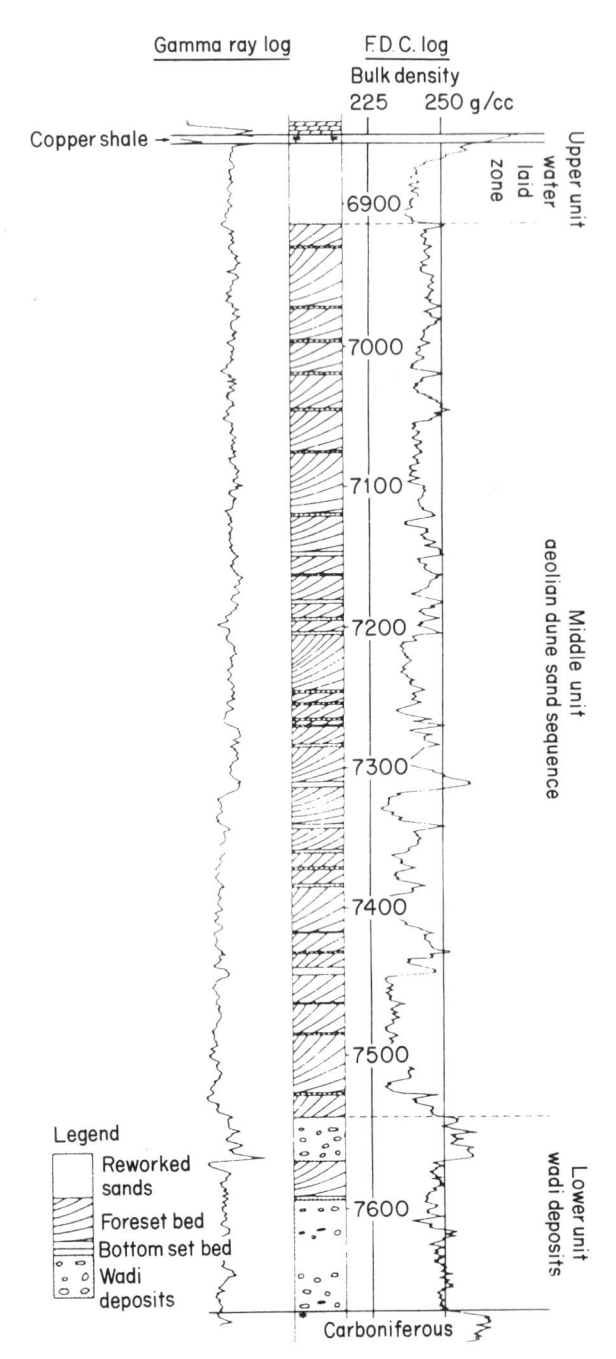

Fig. 5.2 Lithology and log character in zonation of Rotliegendes in Leman gas reservoir (UKCS) (after [36]).

5.2 CORING DECISIONS

Cores provide an opportunity to study the nature of the rock sequence in a well. They will provide a record of the lithologies encountered and can be correlated with wireline logs (Fig. 5.2). Study of the bedding character and associated fossil and microfossil record may provide an interpretation of the age and depositional environment (Fig. 5.3). Petrophysical measurements of porosity and permeability from samples of the recovered core allow quantitative characterization of reservoir properties in the well section. Samples from the recovered core are also used to study post depositional modification to the pore space (Fig. 5.4) (diagenetic studies), flow character of the continuous pore space (*special core analysis* studies) and character of recovered fluids and source rocks (geochemical studies). The diversity of information that can be obtained from recovered core implies that a number of specialists are involved in assembling a *coring* program for a new well – each specialist wishing to ensure that samples are obtained under the best possible conditions.

In general, it may be said that coring operations subdivide into two types – those conducted on exploration/appraisal wells and those on development wells. As a further generality, it is often found that the control of the coring program lies with exploration geologists for exploration/appraisal wells and with reservoir engineers for development wells. The amount of core taken is usually decided on the basis of a technical argument between data collection, technical difficulty and costs. Geologists and reservoir engineers require core for reservoir description and definition, and they may argue that the opportunity to recover samples of the reservoir is only presented once in each well. Drilling engineers tend to argue that the possibility for losing the well is increased by coring operations, and that coring adds significantly in terms of time to the cost of a well. The efficiency of a drilling operation is often measured in terms of time-related costs to move on to a location, reach a given depth, complete the well and move off to the next location. This is in direct conflict with time-consuming data retrieval and often results in coring decisions requiring management arbitration between departments.

The case for coring therefore requires a careful presentation in which the need for the information is simply explained, together with the reasons for obtaining samples under controlled conditions. The case should be supported by a time and cost analysis for the coring operation and, where possible, by an indication of the benefits in terms of incremental oil recovery (i.e. reservoir description leading to better well placement). The incremental cost of the coring may also be effectively presented as a fraction of oil value at peak production rate.

The coring of exploration wells tends to be minimal on the first well on a prospect, unless a good

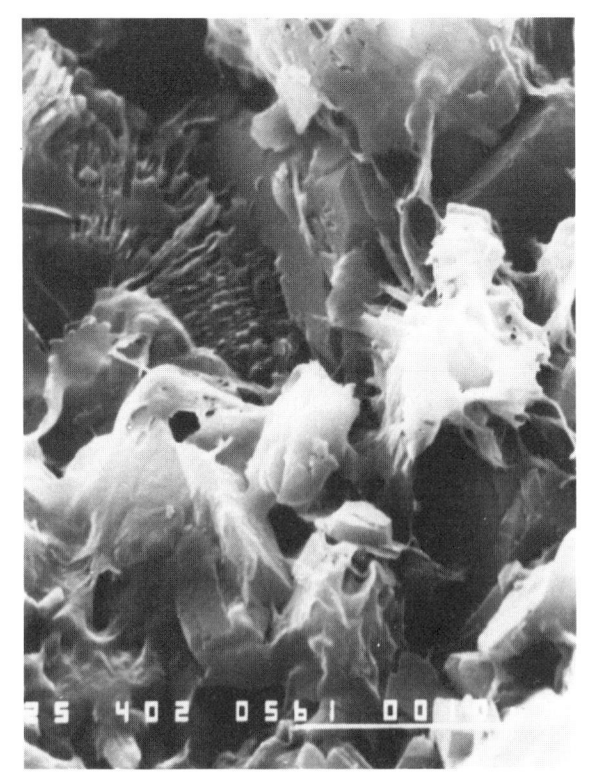

Fig. 5.4 Diagenetic modification to pore space – SEM of illite/smectite formed at the expense of kaolinite.

regional analogy exists. The second well is often designated a *type* well by the exploration department and may be extensively cored. The cored intervals will be decided on the basis of prognosis and analogy with the *discovery* well section. It is frequently found that the reservoir engineers do not fully participate in defining the core program for such appraisal wells. The opportunity exists to influence the coring mud program, the basis for sample selection for special core analysis tests, the basis for sample selection for routine core analysis, and the storage and transportation of recovered core. There may be some specialist advice from a laboratory group regarding some of these aspects.

In exploration wells the geologists have a primary concern in *describing* the core in terms of its lithological variation and defining a basis for zonation. The zonation may initially be in terms of reservoir and non-reservoir intervals, subdivided in terms of rock units. Correlation with wireline logs and between wells may be made on a preliminary basis which will be improved as sedimentological and petrographic studies proceed. Routine core analysis on plugs drilled from the core is frequently commissioned by the exploration department, without particular regard to the conditions of sample selection or preparation of the plugs. It can be important to specify the plug frequency, plug orientation, plug drilling fluids, plug cleaning processes and even the test methods for routine core analysis. These are more appropriately the responsibility of the reservoir engineer to define for the exploration department. The presentation of results may also be defined by the reservoir engineers, in combination with geological zone description.

The coring of development wells is largely controlled by reservoir engineers. In this case it is necessary for co-operation between engineers and production geologists to prepare the coring program. At this stage, reservoir zonation may be better established and cores are required by the geologists for detailed sedimentological, mineralogical and reservoir continuity studies and by engineers for attributing reservoir petrophysical properties within the more detailed zonation. The conditions of core recovery, sample selection and sample storage require particular definition since results will be affected by, for example, wettability variation.

The decisions on how many wells to core are generally taken by *management* after considerations of *need to know* technical argument and project economics. It tends to be true that the degree of reservoir continuity and the number of development wells required form an inverse relationship and that core study can provide a basis for reservoir description, which should lead to more efficient reservoir management. In many North Sea reservoirs the geological complexity can only be resolved by a combination and integration of data from many sources – core data being particularly relevant. Despite delay to first production, it is sensible to core early development wells as fully as possible in order to improve confidence in reservoir description and development plans. Coring in highly deviated wells is naturally more of a risk than in straight holes or less deviated wells, and choice of cored wells should reflect this. A target of 30% cored wells will provide reasonable reservoir control in all but the most complex geology – the coring should however be 75% complete before the reservoir has 50% of its wells.

5.3 CONVENTIONAL AND ORIENTED CORING

Conventional coring refers to core taken without regard to precise orientation – that is without reference to *true north* position. Conventional coring encompasses a range of particular core barrels and cutting head and includes:

    steel core barrel;
    plastic or fibre glass core barrel;
    rubber sleeve core barrel;
    pressure core barrel;
    sponge insert core barrel.

The core recovered from these devices does not allow visualization of rock in its exact reservoir condition orientation. This may be a limitation in sedimentological interpretation where direction of orientation has significance in predicting reservoir continuity. The orientation in fluvial deposition systems (e.g. river channels) is of particular importance in well to well correlation and may not be easily deduced from dip meter data, particularly where overbank *slump* is greater than depositional dip.

A technique that has found application in onshore coring[12] and is now achieving success in offshore coring involves the *scribing* of grooves along the axis of the core in a gyroscopically controlled orientation. The method is known as *directionally oriented* coring and requires periodic stops in the coring operation to take a measurement of orientation. The cost of oriented core offshore has recently been estimated as an incremental cost of some US $50 per metre on conventional core.

Orientation of cores is accomplished by running a conventional core barrel which has a *scribed shoe* containing three tungsten scribes (Fig. 5.5). The scribe shoe is located at the base of the inner core barrel immediately above the core catcher assembly. The scribe shoe is added to the inner barrel by replacing the inner sub with the scribe shoe sub. The top of the inner barrel is attached to the inner barrel head. A shaft extends from the inner barrel head through the safety joint and into a muleshoe attached at its top. The shaft rotates with the inner barrel. Multishot survey instruments are attached to the upper portion of the muleshoe. The survey instrument has a built-in marker which is aligned with the oriented scribe in the scribe shoe at the bottom of the barrel. The survey instrument is located within a K-monel collar to prevent any magnetic disturbance from the drill pipe.

5.4 CORING MUD SYSTEMS

It is inevitable that coring fluid will invade a porous reservoir rock to some degree. The consequence of this is that materials in the mud system, used to control viscosity, filter loss, weight etc., may cause a

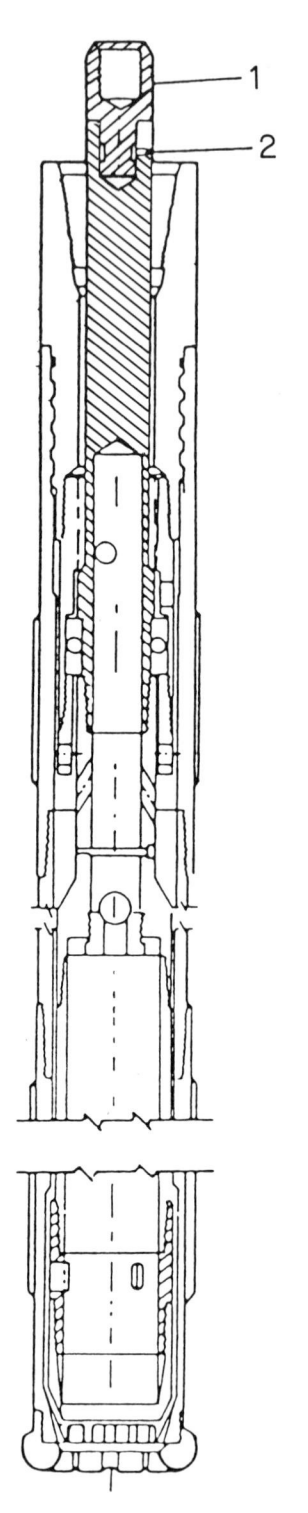

Fig. 5.5 Oriented coring: American Coldset oriented core barrel with Spenny Sun adapter (1) and scribe shoe (2) in place of conventional inner barrel sub.

5 CHARACTERISTICS OF RESERVOIR ROCKS

change in the relative affinity of the reservoir rock surface for oil and water. This change in *wettability* is manifest in the term σ cos θ, which represents the interfacial tension (σ) between oil and water under the reservoir conditions and the contact angle (θ) measured through the water phase as the angle between the oil water interface and the surface (Fig. 5.6).

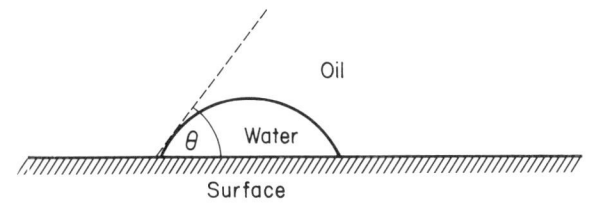

Fig. 5.6 Wetting surface.

The term σ cos θ controls the capillary forces and hence irreducible saturations in a particular rock fluid system. Changes in the term through mud system chemistry will result in the recovery of *unrepresentative* samples. It is clear that the ideal coring fluids from a sample purity point of view would be reservoir brine or reservoir crude oil, but in general these fluids would not allow adequate well control. It is therefore necessary to use a *bland* or unreactive mud system, or to otherwise demonstrate the effect of any particular mud system on the term σ cos θ.

The principal changes that might occur are those changing the wettability of the core or the physical state of *in situ* clay materials. These may change porosity and permeability as well as flow properties determined in laboratory tests.

The influence of engineers and geologists requiring core on the mud system recipe must be decided in conjunction with the drilling engineers as the primary functions of the mud are:

(a) control of subsurface pressure;
(b) lift formation cuttings;
(c) lubricate drill string;
(d) hole cleaning;
(e) formation protection.

The general limitations on mud composition dictated by formation evaluation requirements are as follows:

Unaltered wettability: Use essentially neutral pH mud, avoid use of surfactants and caustic soda. Eliminate oxidation possibilities.

Minimum filtrate invasion: Use lowest mud weight giving control of formation.

Formation damage prevention: Need compatibility with formation waters to prevent changes of clay chemistry or physical state.

Residual oil saturation: Eliminate oil base fluids. Best with formation brine composition muds. Flushing will reduce any oil saturation to just about residual, therefore use low mud circulation rate and high as practicable penetration rate.

Initial water saturation: Eliminate water in mud system. Flushing from oil *filtrate*. Best fluid is lease crude if well control permits.

The main constituents of drilling muds are classified as liquid or solid as follows:

Liquids: (1) fresh water; (2) salt water; (3) oils; (4) combination of 1–3.

Solids: (1) low gravity (approximate SG = 2.5), (a) non-reactive – sand, limestone, chert and some shales, (b) reactive – clay compound; (2) high gravity, (a) barite (approximate SG = 4.2), (b) iron ore + lead sulphide (approximate SG = 2.0).

A mud laboratory is used to evaluate the compatibility and performance of chemical additives for drilling muds planned for a particular well.

5.5 CORE PRESERVATION

The objective of core preservation is to retain the wettability condition of a recovered core sample, and to prevent change in petrophysical character. Exposure to air can result in oxidation of hydrocarbons or evaporation of core fluids with subsequent wettability change[10, 46, 51]. Retention of reservoir fluids (either oil or water) should maintain wetting character, so core may be stored anaerobically under fluid in sealed containers. Reservoir brine (or a chemically equivalent brine) will prevent ion exchange processes in interstitial clays and maintain porosity-permeability character. The use of refined oils and paraffins may cause deposition of plugging compounds, particularly with asphaltinic crude oils. Failing these techniques, the core plug may be wiped clean, wrapped in a plastic seal and foil and stored in dry ice. Usually only samples for special core analysis are stored and transported under these special conditions. The core for routine analysis, following visual inspection at the well site, is placed in boxes, marked for identification, without special care for wettability change or drying of core fluids. It is not really known whether this has any effect on the state of pore fill/replacement minerals recorded in subsequent geological analysis.

5.6 WELLSITE CONTROLS

The recovery of core at a wellsite requires care in handling core barrels and an awareness of the ultimate use of the core. Core should be wiped clean for visual inspection and marked for top and bottom and core depth in a box. Photographs of the fresh core prove invaluable in correcting later misplacements and sometimes in locating fracture zones. Core for special core analysis should be selected quickly and preserved in the agreed manner. Core for routine analysis should be dispatched quickly following the wellsite geologist's preliminary observations. Packing of transportation boxes should naturally be effective and prevent displacement of pieces.

5.7 CORE FOR SPECIAL CORE ANALYSIS

The selection of core for special core analysis is frequently a rather loose arrangement resulting from a reservoir engineering request to the wellsite geologist to preserve *some representative pieces*. While this approach may be inevitable with exploration wells, it should be possible to be more explicit during development drilling when reservoir zonation may be better understood. It is necessary to preserve samples from all significant reservoir flow intervals and these intervals must span permeability ranges. It is necessary to specify the basis for zone recognition, the amount of sample required and the conditions for preservation, transportation and storage. In cases of doubt it is preferable to preserve too much rather than too little and the geologists can always inspect the preserved core in the more controlled laboratory environment.

5.8 CORE DERIVED DATA

The non-special core analysis, i.e. non-preserved core, will generally form the bulk of the recovered core. It is now customary to pass the whole core along a conveyor belt through a device called a *Coregamma surface logger* which records a reading of natural radioactivity against equivalent downhole position of the core[4]. The readings may be compared with the gamma ray log readings obtained *in situ* in the reservoir and used to position the core pieces more precisely.

Analyses may be performed on the sample of the whole core. More usually core plugs are drilled at regular intervals (say 3–6 per metre) specified by the reservoir engineer or picked at specific intervals, and in an orientation to the whole core specified to represent bedding planes, horizontal flowpath, vertical flowpath etc. The plugs are usually about 4 cm long and are trimmed to 2.5 cm to eliminate mud invaded parts. The plug diameters are of the order 2.5–3.8 cm. These plugs are used in routine core analysis. The coolant used on the core plug drilling bit is important in many cores since it may possibly modify internal pore properties. The best coolant would be a reservoir brine, if salinity is low (danger otherwise of salt plugging), or a refined oil (as long as the crude oil is not waxy or asphaltinic (plugging)). There is a danger in using tap water in that it may change the nature of interstitial clays by modifying ionic balances.

After plug cutting, the core is usually sliced along its major axis into three slabs (known as *slabbing*). One third is designated for geological analysis, one third for curation and one third is often required by the licencing agency (e.g. the Government).

5.9 GEOLOGICAL STUDIES

The purpose of geological core study is to provide a basis for dividing the reservoir into zones and to recognize the geometry, continuity and characteristics of the various zones.

The main areas of study involve recognition of the lithology and sedimentology of the reservoir and its *vertical* sequence of rock types and grain size. This is achieved by visual observation and the result recorded as a core log. It is convenient to describe the bedding character and macrofossil character within the grain size profile, as shown in Fig. 5.3. The age of the individual rock units is usually established by association with the microfossil record (micropalynology) and this can be done as well with cuttings as with core samples. The recognition of depositional and post depositional features is achieved by core description and by microscopic observation of thin sections from cores. In addition, the fossil assemblages also provide indication of transport energy regimes (palynofacies analysis) which help support sedimentological interpretations. The environmental/depositional model of a reservoir is largely based on the observations from individual cored wells but requires correlation of data between wells and integration with other sources of informa-

5 CHARACTERISTICS OF RESERVOIR ROCKS

tion, in order to provide insight into reservoir geometry and continuity.

5.10 ROUTINE CORE ANALYSIS

5.10.1 Principles and methods

Routine core analysis is primarily concerned with establishing the variation of porosity and permeability as a function of position and depth in a well. In order to do this, samples of recovered core are subjected to measurements and the results plotted and/or tabulated. In order to provide valid analyses, different rock systems require various analytical approaches with particular names, i.e. conventional core analysis, whole core analysis, analysis of core recovered in rubber sleeve[1, 4, 5, 7, 51]. In addition, some analyses may also be performed on cuttings and sidewall cores. The techniques are reported in API booklet RP490 entitled *Recommended Practice for Core Analysis Procedure*[1].

(a) *Conventional core analysis*
This technique is applied to samples drilled from the whole core piece. Such samples are taken at regular intervals along the core and may represent a statistical sample. The samples are, however, usually biased towards the more consolidated reservoir quality intervals, because of ease of core cutting.

(b) *Whole core analysis*
The technique refers to the use of the full diameter core piece in lengths dependent only on the integrity of the core and the size of porosimeters and permeameters available. Full diameter cores are analysed only when there is reason to believe that plug samples will not reflect average properties. An example of this would be vugular carbonates where vug size may represent a significant volume of the plug sample. The use of whole core pieces tends to downgrade heterogeneous character that would be pronounced in small plugs. The cleaning of whole core sections can be difficult and time-consuming, and analysis is generally significantly more expensive than conventional core analysis.

(c) *Rubber sleeve core (also plastic/fibre glass sleeves)*
The purpose of a rubber sleeve or plastic sleeve in a core barrel is to support the core until removal in the laboratory. The technique is therefore applied particularly in formations which are friable or unconsolidated. There are several disadvantages in the method, but its utilization provides a method of at least recovering a *core* sample. The *core* cannot be visually inspected at the wellsite and is often frozen prior to transportation. Plug preparation may require frozen drilling to prevent movement of sand grains. The plug sample may be supported by some kind of sheath while in a frozen state and is often then placed in a core holder where simulated formation pressures are restored and the temperature restored to reservoir conditions.

The validity of subsequent measurements made on unconsolidated samples treated in the foregoing manner is the subject of contention, but it is clear that any grain re-orientation or lack of similarity with real reservoir overburden stresses will invalidate results.

5.10.2 Residual fluid saturation determination

In the API Recommended Practice, the methods for determining the saturation of fluids in the core as received in the laboratory include:

(a) high vacuum distillation at around 230°C (not so good for heavy oils);
(b) distillation of water and solvent extraction of oil (need to know oil gravity);
(c) high temperature (up to 650°C) retorting at atmospheric pressure (not relevant for hydrated clays in sample);
(d) combination techniques.

Sponge inserts in core barrels are sometimes used to retain reservoir fluids. The experimental techniques should give accuracy of ±5% of the true *as received* saturation condition.

If the cored interval passes through an oil–water contact this may be observable from the residual saturation data. Although the core saturations reported do not represent saturations in the reservoir, they are certainly influenced by them. Thus a rapid change in oil saturation from a relatively high volume to near zero in a similar lithology and reservoir quality interval suggests the presence of an oil–water contact. The placement of the contact will be between adjacent samples of relatively high and relatively low values, as shown in Table 5.1.

The zone appears to be of one rock type and is thus likely to have consistent capillary character. The change in core oil saturation as received in the laboratory shows a dramatic change between 2537.5 m and 2538 m from 19.6% to 1.1%. The oil–water contact would therefore be tentatively placed at the mid-point between these samples at 2537.75 m and confirmation sought from log data. The observed OWC may not be coincident with the free water level (FWL), and capillary pressure data (see Chapter 6) will be required to determine the FWL.

TABLE 5.1 Laboratory measured oil saturation in recovered core

Sample depth (m)	ϕ (%)	k (mD)	S_o (core residual) %
2536.0	30.1	1007	43.6
2536.5	28.6	897	35.7
2537.0	29.3	954	40.2
2537.5	27.9	875	19.6
2538.0	29.1	960	1.1
2538.5	29.7	1106	0.5
2539.0	28.9	984	0.9

TABLE 5.2 Presentation of routine core analysis results

Depth	Horizontal permeability milliDarcy KA	KL	Vertical permeability milliDarcy KA	KL	Helium porosity %	Saturation porosity %	Pore Saturation SO	STW	Grain dens. g/cc	Formation description
3631.35	1.4	1.0	17	14	9.6				2.69	S.ST. GY. V.F.GR. v.w. cemented w/mica
3631.70	24	20	1.4	1.0	19.9				2.67	A.A
3632.00	0.25	0.16	0.29	0.18	3.0	3.0	0.0	59.8	2.75	A.A. SILT Calcitic
3632.35	3.3	2.4	1.2	0.83	13.9				2.67	A.A. V.F.GR without calcitic
3632.70	n.p.p. rubble									
3632.80	9.8	7.6	6.2	4.7	17.3	19.7	0.0	36.6	2.66	A.A. w/org. matter
3633.35	n.p.p. rubble									
3633.70	31	26	19	15	21.5				2.68	A.A. without/org. matter
3634.00	30	25	19	15	21.2	24.5	0.0	37.9	2.67	S.ST. GY/BR F.GR sub.ang. v.w. cemented w/mica
3634.35	38	32	24	20	17.1				2.68	A.A
3634.70	85	74	62	53	18.2				2.68	A.A. w/org. matter
3635.00	17	14	13	10	20.8	20.6	0.0	39.9	2.68	S.ST. GY. V.F.GR well cemented w/mica
3635.35	28	23	21	17	22.2				2.68	A.A
3635.70	10	7.8	6.8	5.2	19.0				2.69	A.A.
3636.00	4	2.9	3.0	2.2	14.8	12.5	0.0	41.6	2.71	A.A. SILT Calcitic
3636.35	0.25	0.16	0.29	0.18	3.0				2.76	A.A
3636.70	0.25	0.16	0.29	0.18	3.8				2.77	A.A
3637.00	6.4	4.8	5.9	4.4	15.3	19.1	0.0	38.4	2.80	A.A. poor calcitic w/pyrite
3637.35	11.9	9.3	10.2	8.0	18.5				2.67	S.ST. GY/BR V.F.GR v.w. cemented w/mica
3637.70	21	17	14	11	19.6				2.67	A.A. F.GR.

5 CHARACTERISTICS OF RESERVOIR ROCKS

5.10.3 Grain density

Grain density measurements are sometimes presented in routine core analysis reports. Two methods are in use:

(a) the wet method, where a weighed sample of crushed core is placed in an unreactive refined liquid (e.g. toluene) and the displacement equated to its volume;
(b) the dry method, where a weighed uncrushed sample is placed in a Boyle's law porosimeter to determine grain volume.

The dry method is preferred.

In the grain volume determination, the effective pore volume is measured by compressing a known volume of gas at a known pressure into a core which was originally at atmospheric pressure. The grain volume is the difference between the total gas space of the void volume plus annulus and the calibrated volume of the core holder. Grain volume measurements should be reproducible to ± 0.04%.

The most usual limitation in the applicability of these measurements is their lack of representation of the bulk reservoir. This is especially so in silty or highly cemented formations.

5.10.4 Data presentation

Routine core analysis is usually presented in tabular form, as shown in Table 5.2. The rock description is provided only as a guide to character and does not pretend to be the geological sample description. Data may also be presented as a point plot against depth. It has been customary to plot permeability on a logarithmic scale, partly to present the values on a condensed scale and partly because it is often found that lithologically similar samples show a linear relationship between porosity and log permeability. For the purpose of recognizing stratification effects, it is preferred nowadays to show permeability variation with depth on a linear scale, as shown in Fig. 5.7. These data are usually not plotted in this way by service laboratories.

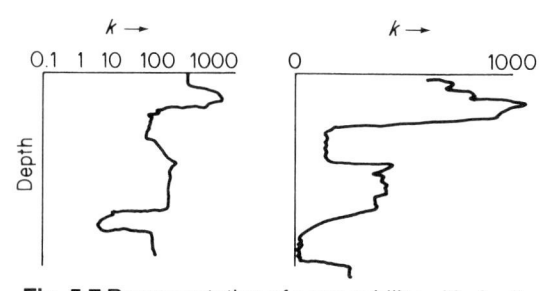

Fig. 5.7 Representation of permeability with depth.

5.11 POROSITY

Porosity is generally symbolized ϕ and is defined as the ratio of void volume to bulk volume. The void spaces in reservoir rocks are most frequently the intergranular spaces between the sedimentary particles. For regular arrangements of uniform spheres, the proportion of void space can be calculated theoretically, but this is not a useful exercise. These void spaces are microscopic in scale, equivalent to apparent diameters of voids rarely exceeding a few, or a very few, tens of microns (Fig. 5.8).

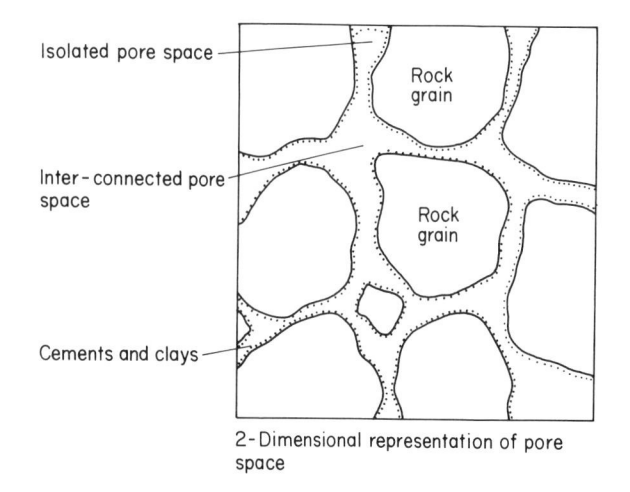

Fig. 5.8 2-D representation of pore space.

Processes subsequent to sedimentation – cementation, recrystallization, solution, weathering, fracturing – can modify substantially the proportion and distribution of void space, and in some circumstances it may be necessary to define a system as a dual porosity system having primary and secondary porosity, although since it is the physical nature of the porosity that is of interest, these may be defined as *coarse* and *fine* porosities.

The porosity of reservoir rocks may range from about 5% of bulk volume to about 30% of bulk volume, the lower porosity range normally being of interest only in dual porosity systems. The distinguishing factor between primary and secondary porosity from the reservoir engineering point of view is not the origin or mode of occurrence, but the flow capacities of the different types of porosity. It will normally only be possible to distinguish any effects of dual porosity if the *coarse* system has a flow capacity about two orders of magnitude greater than that of the *fine* system. With lesser contrasts, behaviour is virtually indistinguishable from single porosity systems with some heterogeneity. In this situation, generally only porosities greater than about 10% are likely to be of commercial interest.

Porosity may be measured directly on core samples in the laboratory, and also may be estimated *in situ* by well log analysis.

In reservoir engineering, only the interconnected porosity is of interest since this is the only capacity which can make a contribution to flow.

5.11.1 Measurement of porosity

Porosity may be determined from measurements on plugs drilled from recovered core or estimated from wireline log responses. From plug samples, the techniques of routine core analysis provide for measurement of bulk volume and either void volume or grain volume. The void volume is represented as the interconnected pore space that can be occupied by a fluid such as gas, water, oil or, for laboratory purposes, mercury. In routine core analysis, the bulk volume is usually determined either by caliper measurements or by displacement of mercury in a pycnometer (Fig. 5.9).

A technique known as the *summation of fluids* is applied to *as received* core plugs, in which the pore volume is considered equal to the sum of any oil,

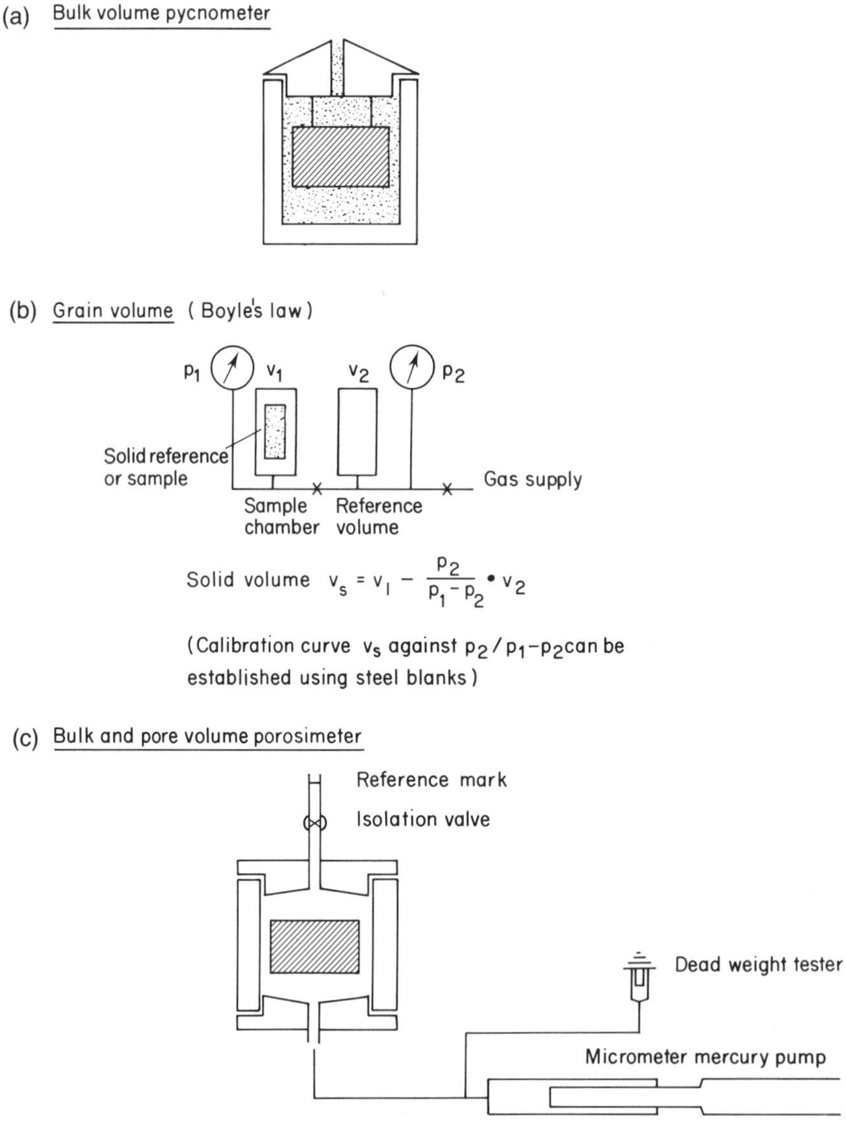

(a) Bulk volume pycnometer

(b) Grain volume (Boyle's law)

Solid volume $v_s = v_1 - \dfrac{p_2}{p_1 - p_2} \cdot v_2$

(Calibration curve v_s against p_2/p_1-p_2 can be established using steel blanks)

(c) Bulk and pore volume porosimeter

Fig. 5.9 Measurement of core plug porosity. (a) Bulk volume pycnometer; (b) grain volume (Boyle's law), calibration curve V_s against $P_2/(P_1-P_2)$ can be established using steel blanks; (c) bulk and pore volume porosimeter.

5 CHARACTERISTICS OF RESERVOIR ROCKS

water and gas occupying the sample. The method is destructive in that oil and water from a representative part of the core are determined by distillation of fluid from a crushed sample. A further representative piece is subject to mercury invasion to provide the gas filled volume. These processes are indicated in Fig. 5.10. The sum of the oil + water + gas volumes as a fraction of total bulk gives the porosity. The presence of clay bound water provides a limitation and the values of porosity are considered to represent ±0.5% around the percentage value calculated.

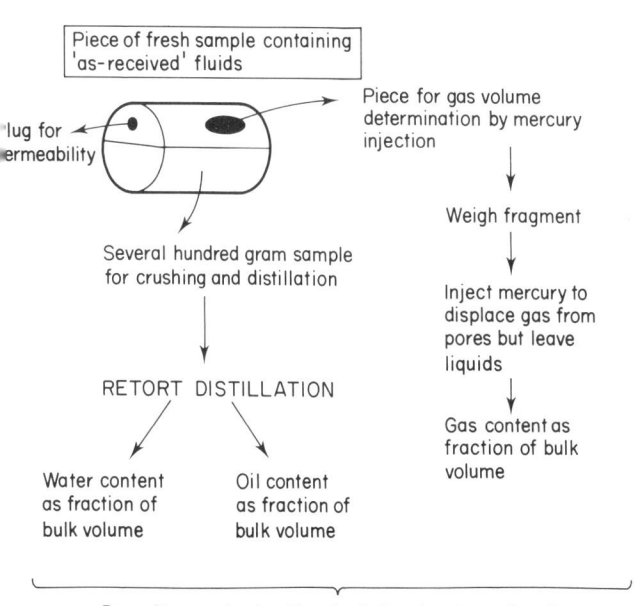

Fig. 5.10 Porosity by summation of fluids.

Non-destructive testing is generally preferred since other types of measurements are often required on a common sample. The Boyle's law method is used to provide an estimate of grain volume. In this method which has a reproducibility of about 2% of the measured porosity, a calibration curve defining the relationship between known solid volumes in a sample chamber and reference pressures and volumes is required. A schematic representation is shown in Fig. 5.9(b) and the grain volume of the sample or solid volume of say a steel cylinder is denoted by V_s. The volume of V_s in the sample chamber thus influences the observed pressure in the system compared with the pressure without the presence of a sample.

$$V_s = V_1 - \frac{P_2 V_2}{P_1 - P_2}$$

A calibration curve of V_s against P_2/P_1-P_2 is generally established using steel blanks. In precision work, the helium porosimeter using this principle has found wide acceptance.

A destructive method of porosimetry, involving injection of mercury into small and irregularly shaped sample chips or regular plugs, is designed to replace air with a measured volume of mercury. By conducting the test with small increments of mercury injection and noting the pressure required for displacement, a mercury injection capillary pressure curve and pore size distribution factor can be obtained *en route* to the porosity measurement. The maximum amount of mercury injected is equal to the pore volume of the sample. Accuracy is reported high and reproducible to one percentage point in the range of porosities of 8–40%. Dual porosity systems can be investigated by the technique if the test is conducted in equilibrium steps[4].

With low porosity, fine pore structure systems, high pressures may be necessary to approach 100% displacement, and corrections for mercury and steel vessel compressibility may become necessary, but at pressures of 6000–10 000 psi, attainable with standard equipment, most of the pore space contributing to flow is occupied. Samples are not usable for further experiments after mercury injection.

A similar, non-destructive but inherently less accurate technique of porosity measurement involves evacuation of all air from the pore spaces of a cleaned, dry weighed sample and the introduction of water into the pore space. So long as any clay minerals in the pore space remain unreactive[50], then the weight increase of the sample is directly proportional to the pore volume.

Total porosity may be obtained from crushed samples during the measurement of grain density.

Rock at reservoir conditions is subject to overburden stresses, while core recovered at surface tends to be stress relieved. It is not usual to perform routine porosity determination with anything approaching a restoration of reservoir stress, and therefore laboratory porosity values are generally expected to be higher than *in situ* values. The magnitude of the overestimation will depend upon the pore volume compressibility of the rock and the initial *in situ* porosity. In general

$$c_p = -\frac{1}{\phi} \frac{(\phi_1 - \phi_2)}{(P_1 - P_2)}$$

The form of the relationship between reservoir condition porosity ϕ_R and zero net overburden laboratory porosity ϕ_L in terms of pore volume compressibility c_p (V/V/psi) and net overburden pressure ΔP_N (= overburden pressure − fluid pressure) psi has been found for Brent sands[21] as follows:

$$\frac{\phi_R}{\phi_L} = \exp\left\{\frac{-c_P \Delta P_N}{\phi_L}\right\}$$

Inserting some typical values, say:

$\phi_L = 0.27$
$c_p = 3 \times 10^{-6}$ psi^{-1}
$\Delta P_N = 10\,000 - 4500 = 5500$ psi

$$\frac{\phi_R}{\phi_L} = \exp\left\{\frac{-3\text{E-}06\,(5500)}{0.27}\right\} = 0.94$$

The measurement of porosity on consolidated samples in routine core analysis might generally be expected to yield values of the *true* fractional porosity ±0.005, i.e. a *true* value of 27% porosity may be measured between 26.5% and 27.5% porosity.

5.11.2 Formation resistivity factor

Although when clays and shales are present the rock itself has some conductivity, in general the solid matrix can be considered non-conducting. The

Effective Overburden Pressure : 0.0 PSI

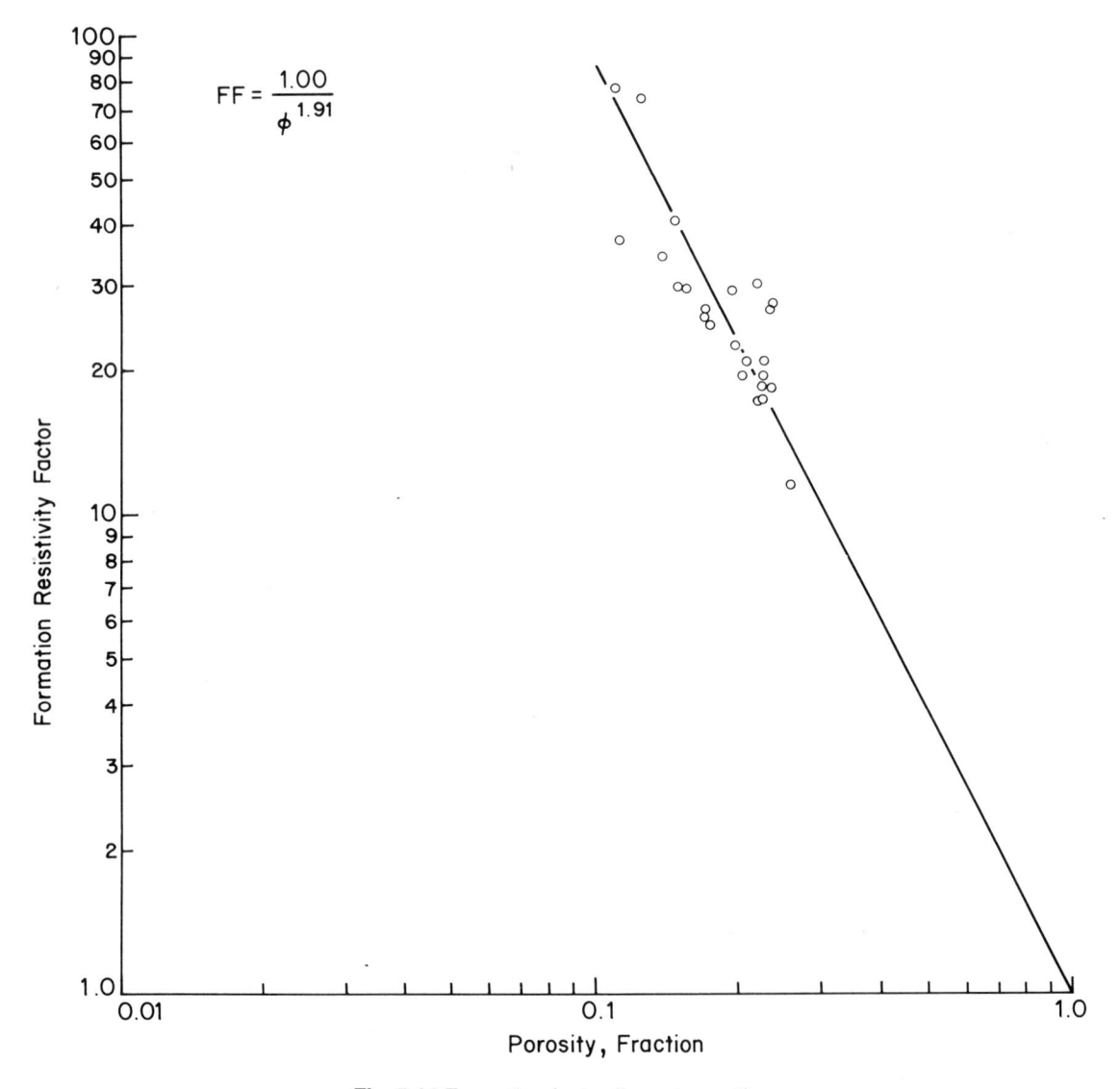

$$FF = \frac{1.00}{\phi^{1.91}}$$

Fig. 5.11 Formation factor (forced $a = 1$).

5 CHARACTERISTICS OF RESERVOIR ROCKS

electrical resistance of a rock sample fully saturated with a conducting fluid (brine) is therefore very much greater than would be the resistance of a sample of fluid of the same shape and size[3]. The ratio

$$\frac{\text{Resistivity of rock fully saturated with brine}}{\text{Resistivity of the saturating brine}} = \frac{R_o}{R_w}$$

is termed the formation resistivity factor (or formation factor, or resistivity factor) and is designated F.

Obviously, the formation resistivity factor will depend on porosity, being infinite when $\phi = 0$, and 1 when $\phi = 1$, and a relationship proposed is

$$F = \frac{a}{\phi^m}$$

where $a \approx 1$ (taken as 0.81 for sandstones, 1 for carbonates), $m \approx 2$. (This is known as the Archie Equation[3].)

The formation factor can be measured by means of an a.c. conductivity bridge, the saturated core being held between electrodes in the bridge circuit. Brine resistivity is determined by a platinum electrode dipped into the brine, forming an element of a bridge circuit.

The form of the relationship is shown in Fig. 5.11. The tests should be performed at a range of net overburden pressures, and perhaps at reservoir temperature in some cases.

5.11.3 Resistivity index

At partial brine saturations, the resistivity of rock is higher than at 100% saturation, and an index is defined:

$$I = \frac{R_t}{R_o}$$
$$= \frac{\textit{True} \text{ formation resistivity}}{\text{Resistivity of rock fully saturated with brine}}$$

This index is a function of brine saturation and to a first approximation

$$I = \frac{1}{S_w^n}$$

where $n \approx 2$. An example is shown in Fig. 5.12.

These two quantities, formation factor and resistivity index, are important in electric log validation.

5.11.4 Porosity logs

The porosity measurements from core plugs are frequently used to validate porosity interpretation from wireline logs. Most wireline logs are designed to respond, in different degrees, to lithology and porosity (Fig. 5.2) and perhaps to the fluid occupying the pore space. In making comparisons it is necessary to note that the core sample represents information essentially *at a point* and on a small scale compared to the *averaged* response of a logging tool. In heterogeneous formations, and in instances where localized mineralogy may influence response, it should be expected that overburden corrected core porosity will not match log data[39].

The main logging tools for porosity are the compensated formation density log, the neutron log and the acoustic log. In interpreting formation lithology and saturation, the gamma ray tool and variations of the induction, conductivity and spontaneous potential logs are used in addition (Fig. 5.13, after[25]). The reader is referred to the specialized log interpretation literature for particular details of their calibration, response and application[52]. As a particular example of wireline log use, the formation density tool response will be discussed further as it provides a particularly useful porosity indication in known lithology. The principle of operation concerns the scattering of gamma rays as a function of the bulk density of an environment irradiated by a gamma ray source, such as cesium

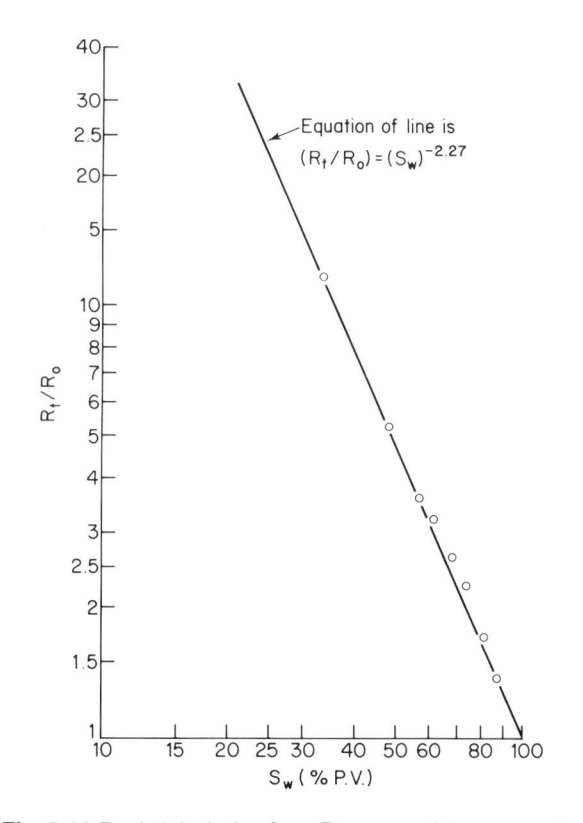

Fig. 5.12 Resistivity index for a Berea sandstone sample. Equation of line is $(R_t/R_o) = (S_w)^{-2.27}$

function of the bulk density of a formation, high densities are indicated by low count rates and low densities by high detector count rates. The count rate of a detector at a fixed distance from a source of constant intensity is very closely a sole function of the formation bulk density ρ_B.

It is usual for two detectors to be used: the short spaced one being some 6 in. from the source and the long spaced detectors 1 ft away. Gamma ray flux from the source is focused into the formation and results in about 90% of the instrument response coming from a region within 6 in. of the well bore. Both detectors are located in inserts in a tungsten carrier, the high density solid tungsten preventing unwanted photons from reaching the detectors. The detector responses are influenced by their specific length and depth, and collimation ensures that Compton scattering from formation rocks provides the prime detected energy. Radiation intensity at the detectors is measured in number of events per unit time and because all radioactive processes are statistical in nature the count rate fluctuates around an average value. An average rate is obtained by cumulating all counts for a given time and dividing by the time. As a result, the count rate does not change as abruptly as a physical boundary or as a change of character is encountered by the detector-source combination. This so-called *vertical bed resolution* is dependent on the heterogeneity of the formation and on tool design and logging speed.

The bulk density measured by the density log is the weighted average of the densities of the matrix and pore fluid such that

$$\rho_B = \rho_f \phi + \rho_{ma}(1-\phi)$$

where ρ_f is the average density of the pore fluid containing pore water, hydrocarbons and mud filtrate, and ρ_{ma} is the density of the rock matrix in the investigation region.

It is therefore clear that interpretation of the tool response requires some knowledge of the formation lithology and fluids present in the pore space as well as the heterogeneity of the investigation region. The response is invalidated in poor conditions.

In shaley formations a shale index, defined from FDC-neutron cross-plots or perhaps from gamma ray readings, is known as V_{shale} and empirically represents the fraction of shale in formations. In such shale formations the bulk density is modified for shale density ρ_{sh} as follows:

$$\rho_B = \rho_f \phi + V_{sh}\rho_{sh} + (1 - \phi - V_{sh})\rho_{ma}$$

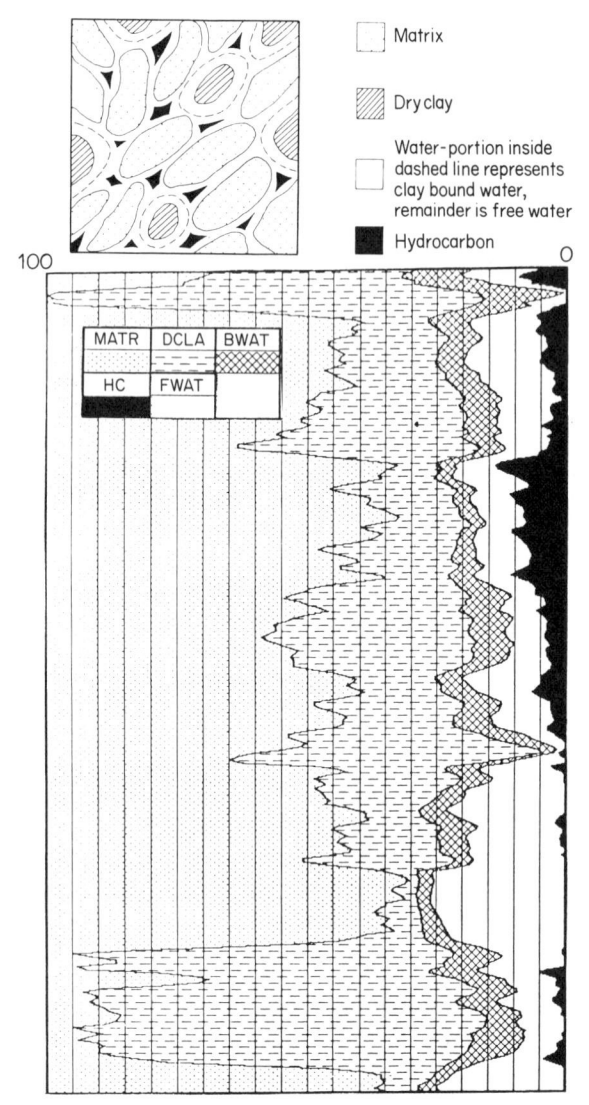

Fig. 5.13 Bulk volume interpretation from logs for a 100 ft interval of a North Sea production well (after [25]).

137 which emits 0.66 MeV gamma radiation. Gamma radiation has neither charge nor mass but may be attenuated by matter as a function of electron density. The main interaction at an energy level around 1 MeV is known as Compton scattering, whereby a photon collides with an atomic electron and some of the photon energy is imparted to the electron. A new photon of lower energy is created which in general moves in a different direction to the incident photon and which can be considered proportional to bulk density. Since the returning gamma ray intensity, measured by a Geiger-Mueller detector is an inverse exponential

5 CHARACTERISTICS OF RESERVOIR ROCKS

In correlating core compaction corrected porosity measurements with density log data, it is therefore important to recognize bed heterogeneity and boundary factors (zonation) as well as the scale of the observations. Core data in particular may not be generally representative of an interval and is influenced in practice by ease of cutting and picks of *good reservoir rock*. The nature of coring fluid influences the magnitude of ρ_f in the porosity calculation and together with ρ_{ma} provides limits on a cross-plot of core corrected porosity against ρ_B in a given zone, as shown in Figs. 5.14 and 5.15.

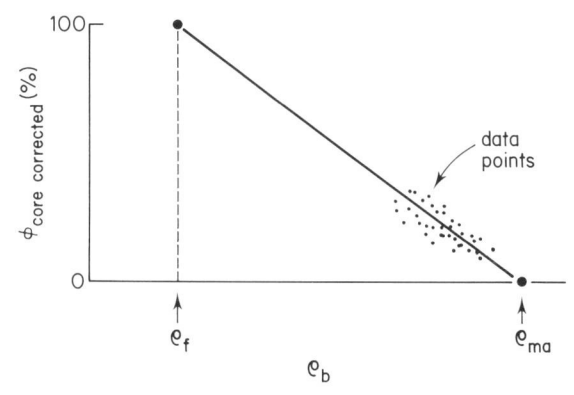

Fig. 5.14 Porosity–bulk density cross-plot.

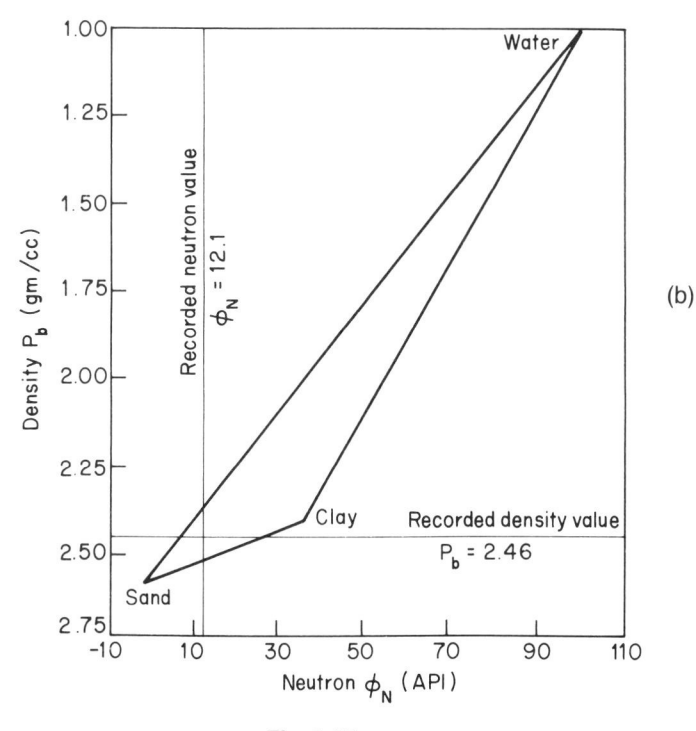

Fig. 5.15

5.11.5 Porosity distributions

Data from both core and log derived porosity interpretations may be used to provide zonal properties. The log porosity validated by core observation is the most useful working set since it will represent a continuous depth section. In contrast, most core data is discontinuous.

Sometimes it is observed that given lithologies in a reservoir with a particular depositional and diagenetic history will show a characteristic distribution of porosity in a given zone. One of the more common forms is a truncated normal distribution, as shown in Fig. 5.16. The distribution may or may not be skewed and may or may not show a trend of value with depth.

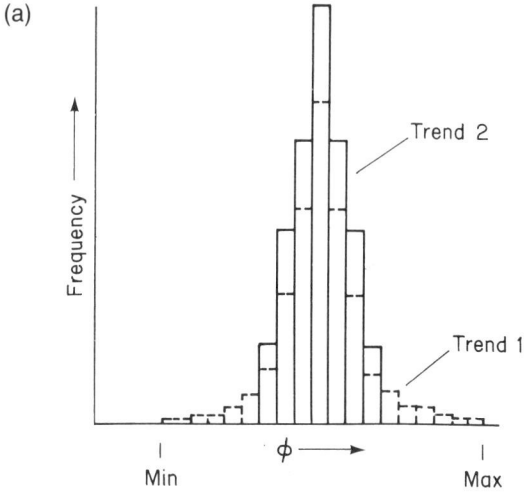

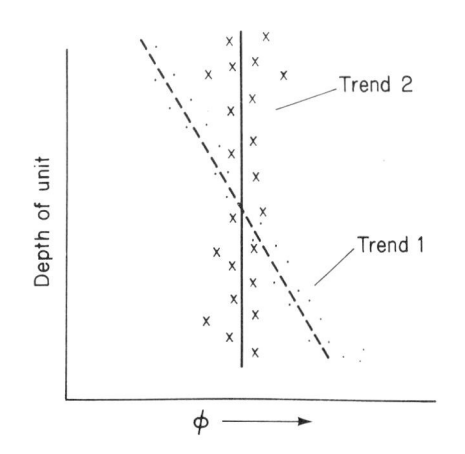

Fig. 5.16 Porosity distribution.

For a truncated normal distribution, the mean porosity of the unit is the arithmetic average. The core analysis porosity histogram is usually only a part of the log derived porosity histogram since sampling is unlikely to be statistically meaningful, and very high and very low values are sometimes missing.

Combinations of different rock units often show up with a multimodal histogram character and this requires separation into subzones, as shown in Fig. 5.17.

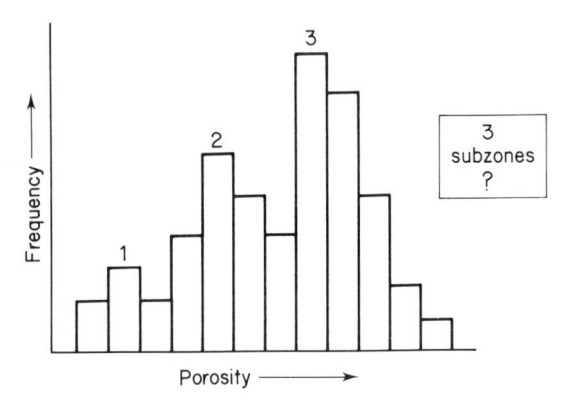

Fig. 5.17 Multimodal porosity.

5.12 PERMEABILITY

5.12.1 Fluid flow in porous media

The permeability of a rock is a measure of its specific flow capacity and can be determined only by a flow experiment. Since permeability depends upon continuity of pore space, there is not, in theory (nor in practice), any unique relation between the porosity of a rock and its permeability (except that a rock must have a non-zero porosity if it is to have a non-zero permeability). For unconsolidated rocks it is possible to establish relations between porosity, and either some measure of apparent pore diameter, or of specific surface, and permeability (e.g. Kozeny model), but these have a limited application. Again for rocks of similar lithology subjected to similar conditions of sedimentation, it may be possible to establish an approximate relation between porosity and permeability, but this is likely to be of local value only. Nevertheless, this provides one method of evaluating permeability variation from log and drill cuttings data which can be of value.

Permeability has direct analogies with thermal conductivity, electrical conductivity and diffusivity, and is defined similarly by a transport equation

$$\frac{Q}{A} = U = -\text{constant} \frac{d\Phi'}{dL}$$

i.e. a rate of transfer is proportional to a potential gradient.

Basic equations of fluid mechanics (Euler or Bernoulli equations) apply the statement of energy conservation to a flowing fluid, so that if no energy losses occur an energy balance on a unit mass of flowing fluid is

$$\frac{P_1}{\rho_1} + g'Z_1 + \frac{U_1^2}{2} = \frac{P_2}{\rho_2} + g'Z_2 + \frac{U_2^2}{2}$$

and if irreversibilities exist

$$\frac{P_1}{\rho_1} + g'Z_1 + \frac{U_1^2}{2} = \frac{P_2}{\rho_2} + g'Z_2 + \frac{U_2^2}{2} + w_f$$

The three terms can be considered to be the energy components of the fluid liable to vary during a flow process – the *pressure* energy, the kinetic energy, and the potential energy of position – and their sum can be considered to be a *potential* per unit mass of fluid.

A gradient in potential can also be defined

$$\frac{d}{dX}\left\{\frac{P}{\rho} + gZ + \frac{U^2}{2}\right\} = \frac{d\Phi}{dX}$$

and the gradient in potential is a measure of the irreversible energy losses.

If the potential terms are divided throughout by g, the acceleration due to gravity, then

$$\left\{\frac{P}{\rho g'} + Z + \frac{U^2}{2g'}\right\}$$

has the dimensions of length, and

$$\frac{d}{dX}\left\{\frac{P}{\rho g'} + Z + \frac{U^2}{2g'}\right\} = \frac{dh}{dX}$$

and

$$g'\frac{dh}{dX} = \frac{d\Phi}{dX}$$

Darcy [27] originally studied the vertical filtration of water deriving experimentally the relation

$$\frac{Q}{A} = U = -\text{constant} \frac{dh}{dL}$$

5 CHARACTERISTICS OF RESERVOIR ROCKS

where dh/dL represents a manometric gradient. Further experiments and analysis, particularly by King Hubbert [28], showed that the constant includes the fluid density and viscosity, and that the residual constant has dimensions of acceleration and rock geometry.

$$\text{constant} = (Nd^2) \cdot (\rho/\mu) \, g'$$

$$\therefore \frac{Q}{A} = U = -\left[\frac{\rho}{\mu}\right]\left[g\right]\left[\frac{dh}{dL}\right]\left[ND^2\right]$$

where N = final constant of proportionality and incorporates a shape factor, d = characteristic length dimension (e.g. mean throat diameter), ρ = density of fluid, and μ = viscosity of fluid.

Obviously, the product $g \, dh/dL$ can be related directly to the potential as defined above, and since kinetic energy changes are generally negligible

$$\frac{Q}{A} = U = -\frac{\rho}{\mu} \cdot \frac{d\Phi}{dL} \cdot (ND^2)$$

$$\therefore \Phi = \int \frac{dP}{\rho} + g'Z$$

For an incompressible fluid (or for a small pressure interval for which an average density may be used)

$$\int \frac{dP}{\rho} + g'Z = \frac{P}{\rho} + g'Z$$

and

$$\rho \frac{d\Phi}{dL} = \frac{d}{dL}\left\{P + \rho g' Z\right\} = \frac{d\Phi'}{dL}$$

and Φ' may be considered a potential per unit volume. The group constant

$$k = Nd^2 = \text{permeability}$$

is taken as the characteristic of the porous medium controlling fluid flow within the medium.

For horizontal flow $dZ/dL = 0$ and Darcy's equation can be written as

$$\frac{Q}{A} = U = -\frac{k}{\mu} \cdot \frac{dP}{dL}$$

and this is the defining equation for the measurement of permeability by flow measurement.

For the oil industry, the unit adopted is termed the Darcy, and this is defined by

$$k = \frac{Q \mu}{A \, (dP/dL)}$$

as shown in Fig. 5.18, where $Q = \text{cm}^3/\text{s}$, $\mu = \text{cp}$, $A = \text{sq. cm}$, $dp/dL = \text{atm/cm}$.

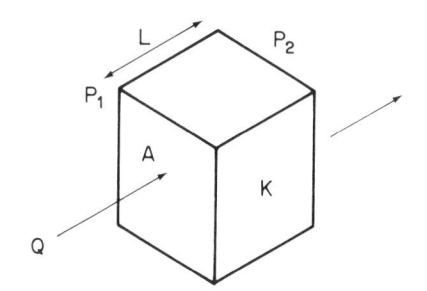

Fig. 5.18 Linear Darcy flow.

A rock has a permeability of 1 Darcy if a potential gradient of 1 atm/cm induces a flow rate of 1 cc/sec/sq cm of a liquid of viscosity 1 cp. The Darcy can be large for a practical unit, and the milliDarcy is more commonly used. With dimensions of L^2, an absolute length unit could be adopted and

$$1 \text{ Darcy} \quad 10^{-8} \, (\text{cm})^2 \, (\approx 10^{-6} \, \text{mm}^2)$$

$$1 \text{ milliDarcy} \quad 10^{-11} \, (\text{cm})^2 \, (\approx 10^{-9} \, \text{mm}^2)$$

$$(\approx 10^{-15} \, \text{m}^2)$$

5.12.2 Datum correction

The equation

$$\frac{Q}{A} = U = -\frac{k}{\mu} \cdot \frac{dP}{dL}$$

is encountered so frequently that its restriction to horizontal flow may be forgotten. It is, however, a valid equation when all pressures are corrected to a common datum level (frequently in reservoir practice a level such that equal volumes of hydrocarbon lie up dip and down dip of this datum, giving initially a correctly volumetrically weighted average pressure).

The corrected pressures are then

$$P_{1c} = P_1 + (Z - h_1) \rho g'$$

$$P_{2c} = P_2 + (Z - h_2) \rho g' = P_2 - (h_2 - Z) \rho g'$$

and the potential difference between points 1 and 2 is then as shown in Fig. 5.19.

$$\Phi' = (P_{1c} - P_{2c})$$

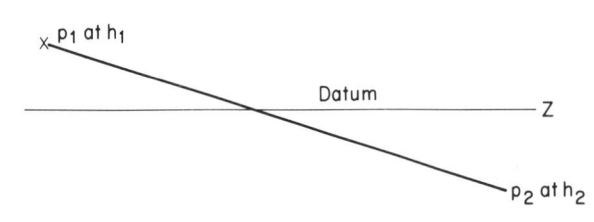

Fig. 5.19 Datum correction.

5.12.3 Linear and radial flow equations

We may now summarize the equation for steady state linear and radial flow of a single phase fluid as follows:
(a) Linear system, as shown in Fig. 5.20.

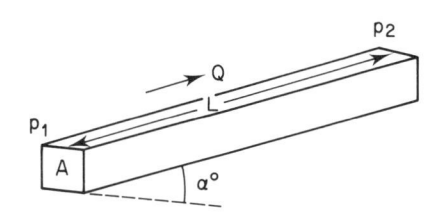

Fig. 5.20 Linear flow in dipping bed.

$$Q = \frac{kA}{\mu}\left[\frac{P_1 - P_2}{L} + \rho g' \sin \alpha\right]$$

If pressure gradient is measured in psi/ft, γ is a specific gravity relative to pure water, area is in square feet, permeability is in milliDarcies, viscosity is in centipoise and volumetric flow rate is in reservoir condition barrels per day, then we have an expression in *field units* as follows:

$$Q = 1.127 \times 10^{-3} \frac{kA}{\mu}\left[\frac{(P_1 - P_2)}{L} + 0.4335\,\gamma \sin \alpha\right]$$

(b) Radial flow system, as shown in Fig. 5.21. From Darcy's law we can write

$$Q = \frac{kA}{\mu}\frac{dP}{dr}$$

The curved surface area open to flow is $2\pi rh$ so

$$Q = \frac{2\pi kh}{\mu} r \frac{dP}{dr}$$

Integrating between the wellbore and the external boundary of the system we have

$$\int_{P_{wf}}^{P_e} dP = \frac{Q\mu}{2\pi kh} \int_{r_w}^{r_e} \frac{1}{r} dr$$

which yields

$$P_e - P_{wf} = \frac{Q\mu}{2\pi kh}\left[\ln \frac{r_e}{r_w}\right]$$

or

$$Q = \frac{2\pi kh\,(P_e - P_{wf})}{\mu \ln \{r_e/r_w\}}$$

In *field units* with Q in RB/D and length terms in feet, this becomes

$$Q = \frac{0.00708\,kh\,(P_e - P_{wf})}{\mu \ln (r_e/r'_w)}$$

In practical situations there may well be a region of altered permeability around the wellbore, which may increase or decrease the pressure drop in comparison with an unaltered system. The effect is described in terms of a skin effect S, which may be positive for reduced permeability and negative for improvement. It is incorporated into the steady state equation as follows:

$$r'_w = r_w e^{-S}$$

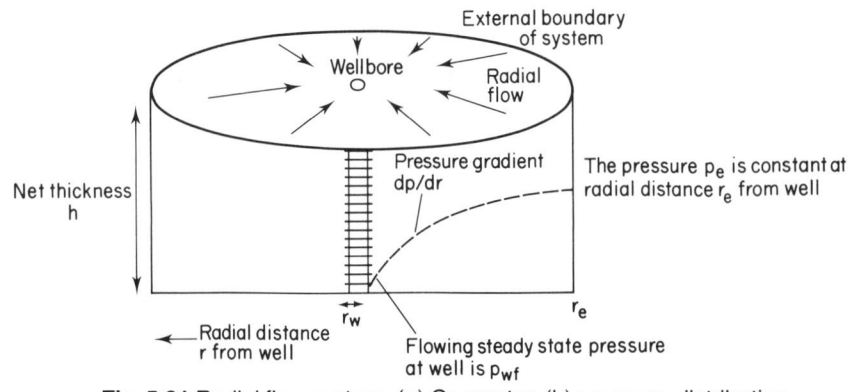

Fig. 5.21 Radial flow system. (a) Geometry, (b) pressure distribution.

5 CHARACTERISTICS OF RESERVOIR ROCKS

$$\therefore P_e - P_{wf} = \frac{Q\mu}{2\pi kh}\left\{\ln\frac{r_e}{r_w} + S\right\}$$

5.12.4 Laboratory determination of permeability

Permeability is an anisotropic property of porous rock in some defined region of the system, that is, it is directional. Routine core analysis is generally concerned with plug samples drilled normal to the long axis of the whole core. Such samples are referred to as horizontal plug samples. Sometimes these samples are specifically requested along bedding planes where it has been noted that the long axis of the whole core is not normal to the bedding plane. Any plug drilled along the long axis of the whole core is termed a vertical plug. Along with routine core analysis measurements of horizontal permeability, it is sometimes found that vertical measurements have also been specified. In carbonate reservoirs where heterogeneity is anticipated, measurements are often made on a full diameter core piece. The horizontal permeabilities of k_{max} and k_{90} are usually reported for full diameter samples (k_{90} is normal to k_{max} in horizontal plane). The permeability for horizontal laminar flow of a single fluid through a granular material has previously been given by

$$\frac{Q\mu}{A} = -k\frac{dP}{dL}$$

where the units are Q = volumetric flow rate, cm³/s; μ = fluid viscosity, cp; A = sample cross-sectional area, cm²; dP/dL = pressure gradient across sample, atm/cm; k = constant called permeability, Darcies.

For gas flow where flow rate is measured at standard conditions, the relationship between rate and pressures upstream and downstream at isothermal conditions is given by

$$P_1 Q_1 = P_2 Q_2$$

as shown in Fig. 5.22. For $P_2 = 1$ atm, then

$$Q_1 = \frac{Q_2}{P_1}$$

The linear Darcy relationship for gas flow at standard conditions thus becomes

$$Q_{sc} = \frac{kA(P_1^2 - P_2^2)}{2\mu L}$$

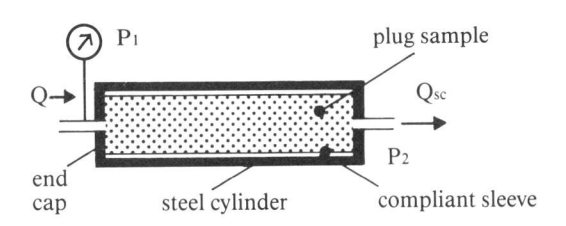

Fig. 5.22 Permeability measurement.

Dry gas is usually used (air, N_2, He) in permeability determination to minimize fluid–rock reaction, and for convenience. In the laboratory, the sample is placed in a device called a permeameter which comprises a compliant sleeve sealing the plug samples along its long axis and a steel container with end caps and pressure regulators into which the sample is placed. The region between the compliant sleeve and the inner walls of the permeameter is subjected to a confining pressure (about 200 psi) in order to prevent the flow along the long axis of the plug. Linear flow of gas through the core plug is established (flow rate proportional to pressure drop is checked). The outlet end pressure is often atmospheric. At low pressures, the permeability measured is higher than the real permeability and a correction is required. This is known as the Klinkenberg correction[6] for gas slippage and involves making several measurements of permeability at different inlet pressures. The mean permeability is determined as

$$P_m = \frac{P_1 + P_2}{2}$$

and calculated permeability is plotted against $1/P_m$ (Fig. 5.23). The true permeability is the extrapolation of the measured data on a straight line to the point $1/P_m = 0$ (i.e. infinite mean pressure). This real permeability is equivalent to the permeability that should be obtained for flow with the core saturated 100% with an unreactive liquid. For this reason the Klinkenberg corrected permeability is also called K_L (L = liquid). When corrective measurements are not made, charts are available for *typical* corrections but these may be erroneous in specific circumstances. In general, permeability corrections are unnecessary for permeabilities approaching 1 Darcy but may be in the order of 0.6 in the milliDarcy range. The nature of the liquid used in checking Klinkenberg corrections may be important in clay sensitive or reactive formations.

Using conventional equipment, permeabilities in

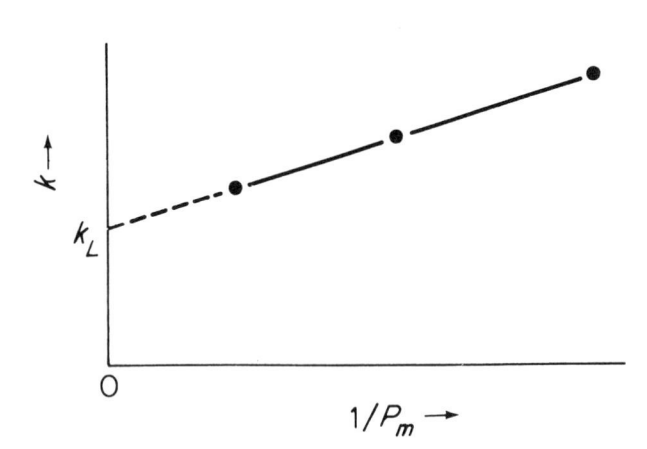

Fig. 5.23 Klinkenberg permeability correction.

the range of 1×10^{-4} D up to 20 D can be measured. The accuracy is usually within ±5% of the true value, although the measurements are poorer at high and low permeabilities. With special permeameters, values of *caprock* permeability can be determined down to around 10^{-9} D.

5.12.5 Anisotropy of permeability

While permeability as defined in petroleum usage is a property only of the rock (having the dimension L^2), the property is not necessarily identical in all size samples or orientations. The direction of greatest interest in reservoir samples is that parallel to the bedding planes – the *horizontal* permeability. The direction perpendicular to bedding planes is of considerable interest in connection with movement in the gravitational field:

- gas segregation;
- injected gas override;
- injected water tonguing;
- gravity smearing or accentuation of unstable displacement fronts.

Whenever sediments are poorly sorted, angular and irregular, the sedimentation process will ensure that *vertical* permeability will be less than horizontal permeability even in the absence of tight streaks.

The sedimentary environment may also lead to the orthogonal permeabilities in the horizontal direction also being unequal, but this is generally a lesser effect than the vertical to horizontal difference.

Consequently, in general,

$$k_x \neq k_y \neq k_z$$

with k_z generally the smallest value.

The utilization of permeability anisotropy information is very much scale and problem dependent. Core data represents microscale observation as shown in Fig. 5.24, and well test derived data is of macroscale. Application in reservoir simulation models is of intermediate scale. The representation of flow restriction is at the heart of permeability characterization and is manifest through definition of transmissibility. In the representation T with subscripts x, y, z to indicate direction, the quantification is as follows:

$$T = \frac{kA}{L} = \frac{Q\mu}{\Delta P}$$

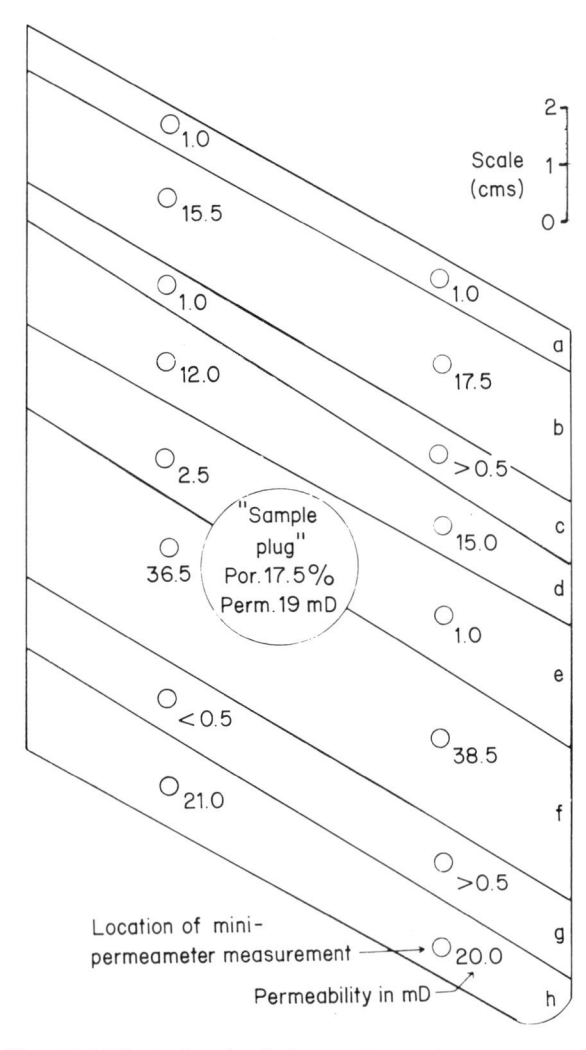

Fig. 5.24 Effect of scale of observation and measurement in permeability data from a Rotliegende aeolian sand cross bed set in the Leman gas field (after [36]).

5 CHARACTERISTICS OF RESERVOIR ROCKS

The value of k may be from appropriately scaled observation. The relationship between well test derived permeability $\bar{k}_{res}$ and core averaged permeability $\bar{k}_c$ can be represented empirically to account for scale by use of a coefficient α

$$\alpha = \frac{\bar{k}_{res}}{\bar{k}_{cor}}$$

It should, however, be remembered that in general $\bar{k}_{res}$ is an effective permeability and really should be considered as a $\bar{k} \cdot k_r$ product where k_r is a relative permeability.

5.12.6 Averaging permeabilities

Warren and Price [20] showed that the most *probable* behaviour of a heterogeneous system approaches that of a uniform system having a permeability equal to the geometric mean

$$\bar{k}_G = (k_1 \cdot k_2 \cdot k_3 \cdots k_n)^{1/n}$$

It has also been shown analytically that the mode of a log normal distribution is a geometric average.

Two simple systems can be analysed in linear geometry to determine an appropriate mean to represent an equivalent homogeneous system. These are linear beds in series and parallel with no crossflow between beds.

1. Linear beds in series

For constant flow rate we can add pressure drops, as shown in Fig. 5.25.

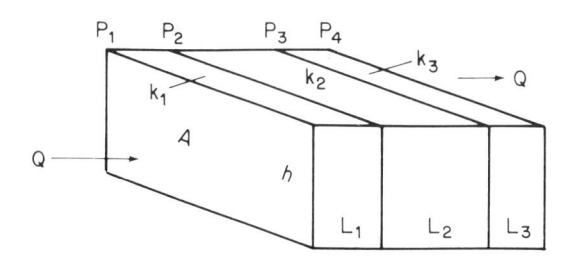

Fig. 5.25 Linear beds in series.

$$(P_1 - P_4) = (P_1 - P_2) + (P_2 - P_3) + (P_2 - P_4)$$

$$\therefore \frac{QL}{\bar{k}A} = \frac{QL_1}{k_1 A_1} + \frac{QL_2}{k_2 A_2} + \frac{QL_3}{k_3 A_3}$$

For equal A then

$$\bar{k} = L / \sum \{L_i / k_i\}$$

The appropriate mean is thus a harmonic average permeability $\bar{k}_H$.

2. Linear beds in parallel

See Fig. 5.26.

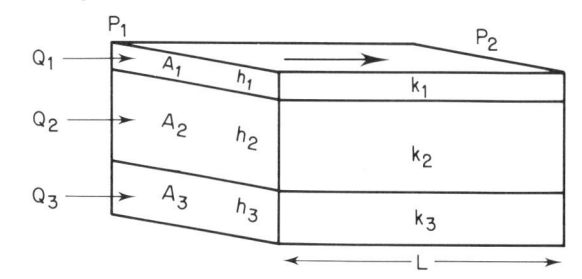

Fig. 5.26 Linear beds in parallel.

$$Q_t = Q_1 + Q_2 + Q_3$$

$$\frac{\bar{k} A (P_1 - P_2)}{\mu L} = \frac{k_1 A_1 (P_1 - P_2)}{\mu L}$$

$$- \frac{k_2 A_2 (P_1 - P_2)}{\mu L} + \frac{k_3 A_3 (P_1 - P_2)}{\mu L}$$

$$\bar{k} A = k_1 A_1 + k_2 A_2 + k_3 A_3$$

$$\bar{k} = \frac{\sum k_i A_i}{\sum A_i}$$

When all beds are the same width, then $A \propto h$ so $\bar{k}$ is an arithmetic average, $\bar{k}_A$.

$$\bar{k}_A = \frac{\sum k_i h_i}{\sum h_i}$$

This applies when beds are homogeneous such that $P_1 - P_2$ is constant in all beds at equal distances. It will not be true when water displaces oil since k_o/μ_o ahead of front is different from k_w/μ_w behind, so the pressure gradient will be different and also different between layers. Cross-flow between adjacent beds can occur unless there are permeability barriers. Cross-flow is also promoted by capillarity.

5.12.7 Permeability distributions

In a given *rock type unit*, which essentially means a unit having similarity in pore size distribution as a result of depositional and diagenetic history, it may be expected that a truncated log normal frequency of permeabilities will occur. This type of distribution was first reported in the literature by Law [13] and has been noted by others. The only source of permeability data for observing the nature of a distribution is from core analysis measurements, and there is a strong possibility that the plugs will not represent a true statistical sample of the unit. The idealized truncated log normal distribution is shown in Fig. 5.27 and has the property that the mode is equivalent to a geometric average of the sample values.

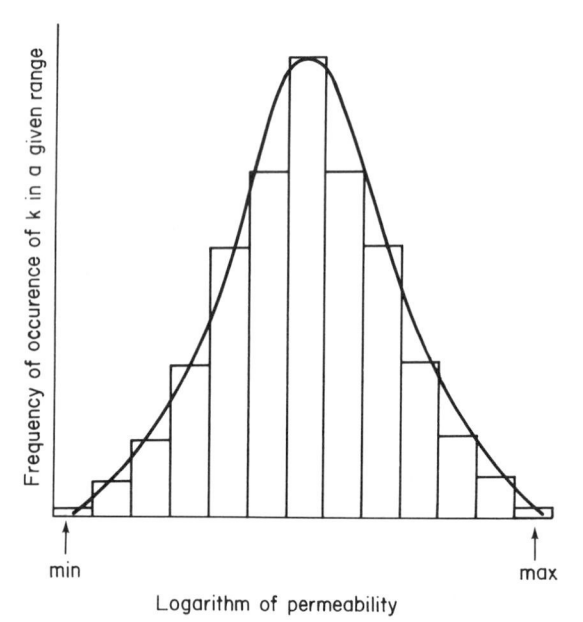

Fig. 5.27 Truncated log normal distribution.

Permeability distributions in a reservoir can be used diagnostically to aid zonation and subzonation. As shown in Fig. 5.28, the sand unit probably represents three regions. The low permeability zone may sometimes relate to diagenetic damage of pores and exist in a particular depth and/or saturation interval of a given depositional unit.

Reservoir zonation within and between wells can be aided by histogram analysis. In these circumstances, normalization of the frequency axis is recommended for ease of comparison. Statistical tests such as the Kolmogorov-Smirnoff test have found application in testing whether or not sample sets belong to a particular population [14, 30].

In zonation it is important to recognize depth and thickness trends with permeability which can be different but which can give rise to the same apparent distribution.

5.13 RELATIONSHIPS BETWEEN POROSITY AND PERMEABILITY

Porosity from validated log response provides a continuous representation of pore volume as a function of depth in a well. Core analysis data can rarely do this, so a depth record of permeability is not generally available. Empirical correlation of porosity with permeability is frequently attempted in order to provide an estimate of permeability as a function of depth. There is no theoretical relationship between porosity and permeability in natural porous systems, so any practical relationship represents a *best fit* and may be represented by a convenient mathematical relationship. The easiest relationship to test is that of a straight line and it has frequently been noted that a plot of porosity against the logarithm of permeability leads to an approximate straight line. The application of any relationship is purely in the nature of an *in–out operator* so any reasonable functional form will suffice. Some

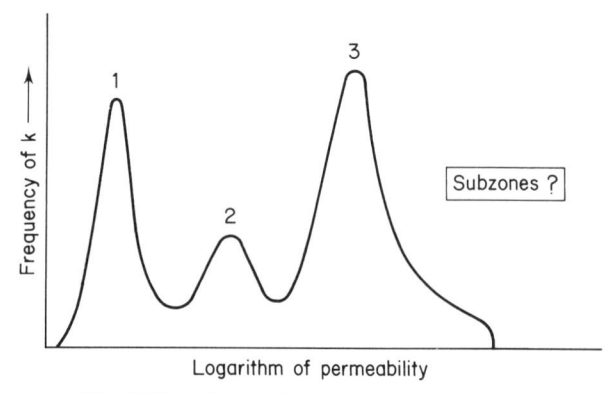

Fig. 5.28 Multimodal permeability distribution.

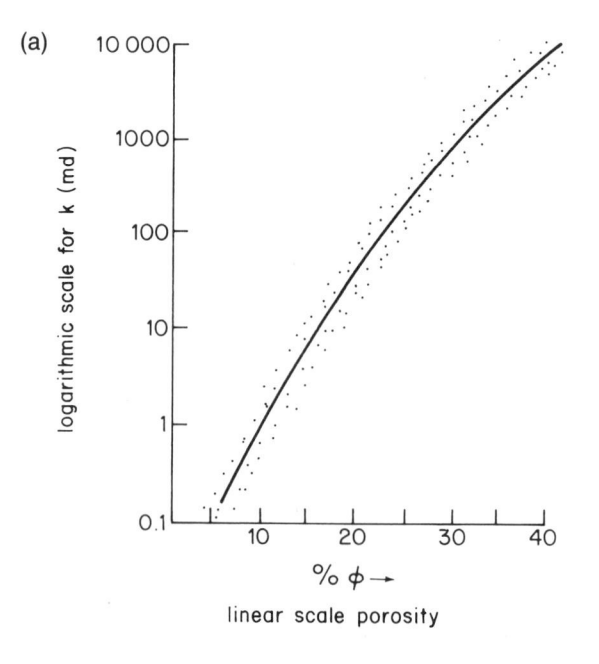

Fig. 5.29 Porosity-permeability correlations for given *rock types*. (a) Semi-log, (b) log-log, (c) semi-log multi-fit.

5 CHARACTERISTICS OF RESERVOIR ROCKS

(b)

logarithmic scale porosity

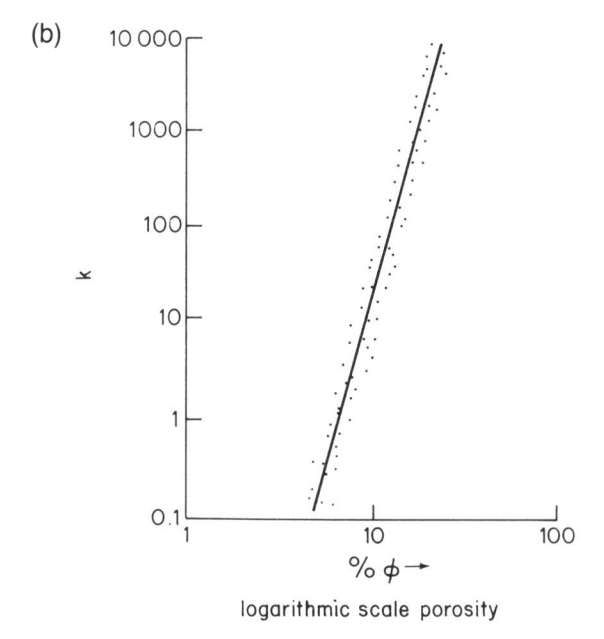

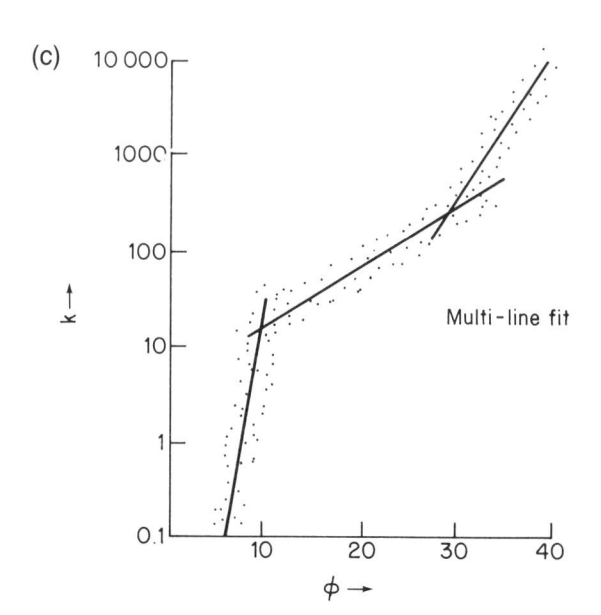

(c)

Multi-line fit

reservoir units may be represented by curves or multiple lines. As might be expected, only data from unimodal histograms are plotted for defining a particular permeability–porosity relationship. Examples of the cross-plots are shown in Fig. 5.29.

It is clear that a statistically significant volume of data is necessary to identify the relationships. It may also be apparent that the degree of fit of the equations is important as a calculated permeability is required from a porosity value at a particular depth. In poor correlations a porosity of say 25% may be used to predict a permeability of 300 mD, but be based on a data spread of permeabilities between 1500 mD and 10 mD. Since flow rate is directly proportional to permeability the potential errors are significant.

The most usual sequence of operations to provide the correlations is shown in Table 5.3.

In addition to the correlation methods outlined

TABLE 5.3 $\phi - K$ relationships

1. Depth match core and log data. Compare compaction corrected core ϕ histogram with log derived ϕ histogram and define appropriate zonation.

↓

2. For given zone plot compaction corrected core ϕ against log derived ϕ and obtain relationship. Explain anomalies.

↓

3. Plot compaction corrected porosity from core against Klinkenberg and compaction corrected core permeability (bedding plane direction) and obtain best fit.

↓

4. Define relationship between log derived *in situ* porosity and Klinkenberg and compaction corrected core permeability at common depth.

↓

5. Use each relationship in appropriate zone to predict permeability at each depth value of interest and plot permeability log.

above there are other approaches [38]. One has been the direct correlation of core corrected permeability with well test interpretation and log response informations [33]. The form of the correlating expression in a given zone is as follows:

$$k = \alpha \{f(a\rho_B + b\phi_N + c\,\Delta T + \cdots)\}$$

where $\alpha = \overline{k}_{res, welltest}/\overline{k}_{core}$; a,b,c = coefficients obtained empirically; ρ_B = formation density log response; ϕ_N = neutron log response; ΔT = acoustic log response.

Another form promoted by a logging service company requires knowledge of irreducible saturation and is based on pore geometry and log response. The Coates and Dumanoir equation [17] is a modification of one by Timur [31] and is expressed as follows:

$$k = \left\{ \frac{C}{W^4} \left(\frac{\phi}{S_{w_{irr}}} \right)^w \right\}^2$$

where

$$W^2 = 3.75 - \phi + \tfrac{1}{2} \left\{ \log \left[\frac{R_w}{R_{t_{irr}}} \right] + 2.2 \right\}^2$$

and

$$C = 465 \rho_{hc} - 188 \rho_{hc}^2 + 23$$

Examples

Example 5.1

A series of core samples from a well give the following formation factor: porosity relationships:

FRF:	30	19.3	12.5	8.4	6.0
ϕ:	0.092	0.120	0.165	0.205	0.268

Calculate the constants of the Archie equation.

In an offset well, a thick water bearing layer is encountered having a resistivity of 1.29 Ωm. The water resistivity is 0.056 Ωm. What is the porosity?

In an updip location the same formation has an apparent hydrocarbon saturation, and a true resistivity of 11.84 Ωm metres. If the exponent of the resistivity ratio equation is 2, what is the hydrocarbon saturation? What would be values for exponents of 1.8 and 2.2?

Example 5.2

Using the logs in Figs A5.2.1 and A5.2.2 (see Appendix II) together with the density–SNP crossplot in Fig. A5.2.3 evaluate the following information for the permeable zones A and B. Note these are Dresser Atlas logs in a sandy formation of a well drilled with an oil base mud – there are no resistivity logs or an SP. Assume $a = 1$, $m = 2$, $n = 2$.
(a) Tabulate log values for the zones A, B, C and shale, and plot all points on the D/N Crossplot.
(b) Determine R_W by considering only zone C.
(c) Determine R_{sh}, N_{sh} and V_{sh} from a shale zone (as included in table).
(d) Establish the clean line and the shale line for gamma ray. Read gamma ray values for zones A and B and convert to V_{sh}.
(e) Establish the shale point on a neutron–density crossplot. Evaluate V_{sh} for zones A and B.
(f) Integrate information from the two shale indicators and select the most appropriate value of V_{sh} for each level.
(g) Use this value to determine porosity.
(h) Calculate S_w at A and B using
   i) The Simandoux equation
   ii) The modified Simandoux equation
   iii) The Poupon and Leveaux equation (Indonesia Equation).

Example 5.3

A shaley sand has a porosity of 26% over an interval within which $R_t = 5$ Ωm and $R_w = 0.1$ Ωm. Laboratory analysis has shown that $Q_v = 0.3$ meq/cc. Evaluate the water saturation using the Waxman-Thomas equation. ($B = 0.046$ $\Omega^{-1}.\text{cm}^2$ meq^{-1}, $a = 1$, $m = 2$, $n = 2$).

If $R_{sh} = 1.5$ Ωm, to what value of V_{sh} is this Q_v equivalent on the modified Simandoux model?
Using these values of R_{sh} and V_{sh} calculate the water saturation from the basic Simandoux equation.

5 CHARACTERISTICS OF RESERVOIR ROCKS

Example 5.4

Darcy's law in differential form is:

$$q = - \frac{kA}{\mu} \frac{dp}{dx}$$

(a) Show that the equation for isothermal linear flow of an ideal gas in Darcy units reduces to:

$$q_{s.c} = \frac{kA(P_1^2 - P_2^2)}{2\mu L}$$

(b) If a gas flows through a core sample discharging to atmosphere at a rate of 6.2 ccs/s when a manometer upstream of the core records a pressure of 190 mm Hg, what is the permeability of the sample? Dimensions are 1 in. diameter × 1 in. length, and gas viscosity at ambient temperature is 0.018 cp.

Example 5.5

An aquifer is known to outcrop at the sea bed where the water depth is 250 ft as shown in the figure. A hydrocarbon–water contact exists at –5250 ft. The effective distance from the HWC to the outcrop is 10 miles. Under dynamic conditions, when the pressure at the HWC is 1450 psi, what is the rate of water influx in barrels per day? Assume the specific gravity of the aquifer water is 1.038. The aquifer sand has a net thickness of 65 ft, a width of 3000 ft and a permeability of 750 mD.

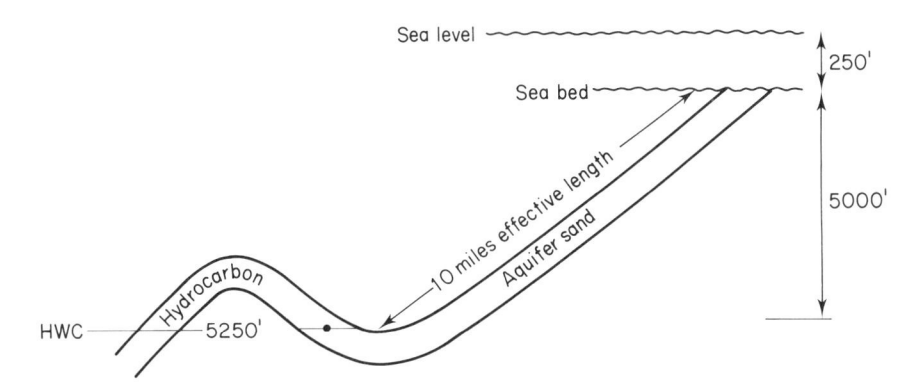

Example 5.6

The following results were obtained in flowing dry gas through a cleaned extracted dried core plug:

Core dimensions: Diameter 1 in., Length 1 in.
Gas viscosity: 0.018 cP
Atmospheric pressure: 760 mm Hg

Upstream pressure (mm Hg)	Downstream pressure	Flow rate (standard conditions) (cm³/min)
101	Atmospheric	6.4
507	Atmospheric	35.6
1520	Atmospheric	132.8

What is the permeability of the sample, after Klinkenberg correction?

Example 5.7

A core sample is saturated with brine, and mounted in a burette as shown in the accompanying diagram. When flow is started, the height of the brine above the core is as follows:

Time (s)	Height (cm)
0	100.0
100	96.1
500	82.0
1000	67.0
2000	45.0
3000	30.0
4000	20.0
5000	13.5

What is the permeability of the sample? Assume that:

Density of brine = 1.02 g/cm
Viscosity of brine = 1 centipoise $g = 981$ cm s^{-2}
1 atmosphere $\simeq 10^6$ dyne/cm^2

Example 5.8

A reservoir is bounded by three faults and an oil–water contact forming a tilted rectangular block of 3000 ft × 1000 ft × 150 ft. The oil–water contact is 1000 ft long at a depth of −5750 ft. A production rate of 1000 bbl/day of tank oil is obtained from a number of wells lying close to the upper fault boundary.

Average pressure at producing wells = 1750 psig at 5000 ft
Net sand thickness = 150 ft
Permeability to oil = 150 mD
Porosity = 26%
Viscosity of oil at reservoir conditions = 0.7 cp
Oil formation volume factor = 1.135 rb/stb
Density of oil at reservoir conditions = 50 lb/ft^3

(a) What is the pressure at the oil–water contact?
(b) If the pressure at the original oil–water contact at abandonment is 500 psig, what size of aquifer would be necessary if water drive of the reservoir were to be complete? (i.e. 1 pore volume of invading water)

Use the following relationships:

1 BBL/day = 1.84 cm^3/s
1 ft = 30.48 cm
1 atmosphere = 14.7 psi
Compressibility of water = 3×10^{-6} vols/psi

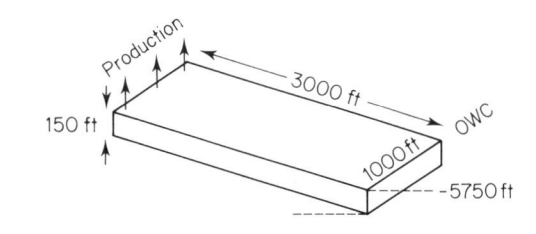

Example 5.9

Find the expressions for the average permeability of beds or zones of differing permeability when in series and in parallel, both in linear (a), (b) and radial flow (c), (d).
(e) Calculate the mean permeability for the following cases:

Zone	Depth (parallel) Length or radial extent (series)	Permeability (mD)
1	250	25
2	250	50
3	500	100
4	1000	200

References

[1] American Petroleum Institute
Recommended practice for core analysis procedure, *API RP* **40** (1960).
[2] Ryder, H.M.
Permeability, absolute, effective, measured, *World Oil* (1948), 174.
[3] Archie, G.E.
Reservoir rocks and petrophysical considerations, *AAPG Bull* **36** (1952), 278.
[4] Monicard, R.P.
Properties of Reservoir Rocks: Core Analysis, Graham and Trotman (1980).
[5] Anderson, G.
Coring and Core Analysis Handbook, Penwell, Tulsa (1975).
[6] Klinkenberg, L.J.
The permeability of porous media to liquid and gases, *API Drilling and Production Practice* (1941), 200.
[7] Amyx, J.M., Bass, D.M. and Whiting, R.L.
Petroleum Reservoir Engineering – Physical Properties, McGraw Hill (1960).
[8] Frick, T.C. (ed.)
Petroleum Production Handbook (V1, V2), SPE (1962).
[9] Buckles, R.S.
Correlating and averaging connate water saturation data, *J. Can. Pet. Tech.* **4** (1965), 42.
[10] Cuiec, L.
Study of problems related to the restoration of the natural state of core samples, *J. Can. Pet. Tech.* **16** (1977), 68.
[11] Havlena, D.
Interpretation, averaging and use of the basic geological engineering data, *J. Can. Pet. Tech.* **6** (1967), 236.
[12] Bell, H.J.
Oriented cores guide Eliasville redevelopment, *Pet. Eng. Int.* (Dec. 1979), 38.
[13] Law, J.
A statistical approach to the interstitial heterogeneity of sand reservoirs, *Trans. AIME* (1944), 202.
[14] Davis, J.
Statistics and Data Analysis in Geology, Wiley Interscience (1973).
[15] Keelan, D.K.
A critical review of core analysis techniques, *J. Can. Pet. Tech.* **11** (1972), 42.
[16] Newman, G.H.
Pore volume compressibility of consolidated friable and unconsolidated reservoir rocks under hydrostatic loading, *JPT* **25** (1973), 129.
[17] Coates, G.R. and Dumanoir, J.L.
A new approach to improved log derived permeability, *Proc. SPWLA,* 14th Ann. Symp. (1973).
[18] Teeuw, D.
Prediction of formation compaction from laboratory compressibility data, *SPEJ* (1971), 263.
[19] Testerman, J.D.
A statistical reservoir zonation technique, *JPT* (1962), 889.
[20] Warren, J.E. and Price, A.S.
Flow in heterogeneous porous media, *Trans. AIME (SPEJ)* **222** (1961), 153.
[21] RRI/ERC
The Brent Sand in the N. Viking Graben, UKCS: A Sedimentological and Reservoir Engineering Study, Vol 7 (1980).
[22] Gewers, C.W. and Nichol, L.R.
Gas turbulence factor in a microvugular carbonate, *J. Can. Pet. Tech.* (1969).
[23] Wong, S.W.
Effects of liquid saturation on turbulence factors for gas-liquid systems, *J. Can. Pet. Tech.* (1970).
[24] Firoozabadi, A. and Katz, D.L.
An analysis of high velocity gas flow through porous media, *JPT* (Feb 1979), 211.

[25] Peveraro, R.C.A. and Cable, H.W.
Determination of rock properties by quantitative processing of geophysical borehole logs, Paper EUR 273, *Proc. Europec* (1982), 29.

[26] Juhasz, I.
Normalised Qv – the key to shaley sand evaluation using the Waxman-Smits equation in the absence of core data, *Proc. SPWLA 22nd Ann. Log. Symp.* (June 1981), Z.

[27] Darcy, H.
The public fountains in the Town of Dijon, V. Dalmont, Paris (1856).

[28] Hubbert, M.K.
Darcy's law and the field equations of the flow of underground fluids, *Trans. AIME* **207** (1956), 222.

[29] Toronyi, R.M. and Ali, S.M.F.
Determining interblock transmissibility in reservoir simulators, *JPT* (1974), 77.

[30] Miller, R.L. and Kahn, J.S.
Statistical Analysis in the Geological Sciences, J. Wiley, New York (1962).

[31] Timur, A.
An investigation of permeability, porosity and residual water saturation relationships for sandstone reservoirs, *The Log Analyst* **9** (1968).

[32] Swanson, B.F. and Thomas, E.C.
The measurement of petrophysical properties of unconsolidated sand cores, *Proc. 6th Europ. Symp. SPWLA,* Paper A (1979).

[33] Allen, J.R.
Prediction of permeability from logs by multiple regression, *Proc. 6th Europ. Symp. SPWLA,* Paper M (1979).

[34] Muecke, T.W.
Formation fines and factors controlling their movement in porous media, *JPT* (1979), 144.

[35] Nagtegaal, P.J.C.
Relationship of facies and reservoir quality in Rotliegendes desert sand stones, SNS, *J. Pet. Geol.* **2** (1979), 145.

[36] Van Veen, F.R.
Geology of the Leman gas field, In *Petroleum and the Continental Shelf of NW Europe,* (Woodland, A.W., ed.), Applied Science Pub., Barking (1975), 223.

[37] Passmore, M.J. and Archer, J.S.
Thermal properties of reservoir rocks and fluids, In *Development in Petroleum Engineering – 1,* (ed. Dawe/Wilson) Elsevier Applied Science Pub., Barking (1985).

[38] Wall, C.G.
Permeability – pore size distribution correlations, *J. Inst. Pet.* **51** (1965), 195.

[39] Archer, J.S. and Hurst, A.R.
Applications of clay mineralogy in reservoir studies, *Proc. Clay Min. Conference, Cambridge* (Apr. 1985).

[40] McHardy, W.J., Wilson, M.J. and Tait, J.M.
Electron microscope and X-ray diffraction studies of filamentous illitic clay from sandstones of the Magnus field, *Clay Min. Oxford* **17** (1982), 23.

[41] Hook, J.R.
The precision of core analysis data and some implications for reservoir evaluation, *Proc. 24th SPWLA Ann. Symp.* (1983).

[42] Kruger, W.D.
Determining areal permeability distributions by calculations, *JPT* (1961), 691.

[43] Dupuy, M. and Pottier, J.
Application of statistical methods to detailed studies of reservoirs, *Rev. I.F.P.* (1963).

[44] Jennings, H.Y. and Timur, A.
Significant contributions in formation evaluation and well testing, *JPT* (1973), 1432.

[45] McLatchie, A.S., Hemstock, R.A. and Young, J.W.
The effective compressibility of reservoir rock and its effect on permeability, *Trans. AIME* **213** (1958), 386.

[46] Luffel, D.L. and Randall, R.V.
Core handling and measurement techniques for obtaining reliable reservoir characteristics, *AIME Form. Eval. Symp. Houston* (Nov. 1960).

[47] Rathmell, J.J.
Errors in core oil content data measured by the retort distillation technique, *JPT* (June 1967), 759.

[48] Pallat, N., Wilson, J. and McHardy, W.J.
The relationship between permeability and morphology of diagenetic illite in reservoir rocks, *JPT* (Dec. 1984), 2225.

[49] Trudgen, P. and Hoffman, F.
Statistically analysing core data, *SPE Proc. 41st Ann. Fall Mtg.* (1966), 1574.

5 CHARACTERISTICS OF RESERVOIR ROCKS

[50] Bush, D.C. and Jenkins, R.E.
Proper hydration of clays for rock property determinations, *JPT* (July 1970), 800.
[51] Keeland, D.
Coring, *World Oil* (March 1985), 83.
[52] Serra, O.
Fundamentals of Well Log Interpretation, (Dev. in Pet. Sci. 15A), Elsevier, Amsterdam (1984).
[53] Dewan, J.T., *Essentials of Modern Open Hole Log Interpretation,* Pennwell, Tulsa (1983).
[54] Desbrandes, R.
Encyclopedia of Well Logging, Graham and Trotman, London (1985).
[55] Hilchie, D.W.
Advanced Well Log Interpretation, D.W. Hilchie Inc., Golden, Colorado (1982).
[56] Threadgold, P.
Advances in Formation Evaluation. In *Developments in Petroleum Engineering* – 1 (Ed. Dawe–Wilson), Elsevier Applied Science, Barking (1985), 43.

Chapter 6

Fluid Saturation: Influence of Wettability and Capillary Pressure

6.1 EQUILIBRIUM CONDITIONS

The equilibrium saturation distribution in a petroleum reservoir prior to production is governed by the pore space characteristics. This happens as a result of non-wetting phase fluid (hydrocarbons) entering pore space initially occupied by wetting phase fluid (water) during migration of hydrocarbons from a source rock region into a reservoir trap. A pressure differential is required for non-wetting phase fluid to displace wetting phase fluid and this is equivalent to a minimum threshold capillary pressure and is dependent on pore size.

Capillary pressure may be defined as the pressure difference across a curved interface between two immiscible fluids. Using the example of an oil drop in a water environment (Fig. 6.1)

$$P_c = P_o - P_w$$

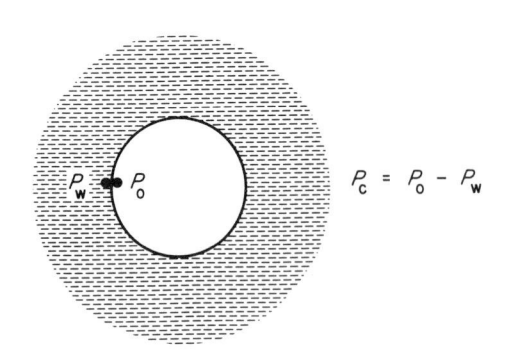

Fig. 6.1 Pressures at an interface.

The curvature of the interface suggests that the oil phase pressure P_o is greater than the water phase pressure P_w. The capillary pressure P_c is defined as the difference between the two phase pressures. By convention, the P_c term is positive for unconfined immiscible fluid pairs. The curved interface has two principal radii of curvature normal to each other, R_1 and R_2. It can be shown that the capillary pressure can also be defined in terms of these radii and in terms of the interfacial tension σ between the immiscible fluids, i.e.

$$P_c = \sigma \left(\frac{1}{R_1} + \frac{1}{R_2}\right)$$

For an immiscible fluid pair confined in a circular cross-section pore of radius r, and making the assumption that $R_1 = R_2$, we can write

$$P_c = \frac{2\sigma \cos\theta}{r}$$

where θ is the angle measured through the wetting phase (water) fluid that the surface makes at the contact with the pores wall (Fig. 6.2). The angle θ is known as the contact angle.

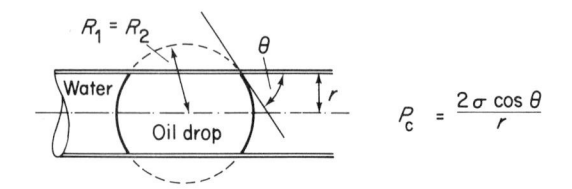

Fig. 6.2 Immiscible fluids interface in a confined capillary.

6 FLUID SATURATION

TABLE 6.1

System		Conditions	θ	σ (dynes/cm)
Wetting phase	Non-wetting phase	T = temperature P = pressure		
Brine	Oil	Reservoir, T,P	30	30
Brine	Oil	Laboratory, T,P	30	48
Brine	Gas	Laboratory, T,P	0	72
Brine	Gas	Reservoir, T,P	0	(50)
Oil	Gas	Reservoir, T,P	0	4
Gas	Mercury	Laboratory, T,P	140	480

(After [26]).

The angle θ is influenced by the tendency of one of the fluids in the immiscible pair to spread on the pore wall surface in preference to the other. The qualitative recognition of preferred spread is called a wettability preference, and the fluid which spreads more is said to be the wetting phase fluid. Contact angles are measured, by convention, through the wetting phase fluid. A table of typical fluid pairs of interest in reservoir engineering is shown in Table 6.1, together with contact angles and interfacial tensions.

The relative spreading concept applied to fluids on a surface may be used to illustrate the understanding of wettability description applied to an oil-water system in a reservoir. The main practical difficulty comes from obtaining a smooth representative pore surface at reservoir conditions of temperature and pressure on which to make measurements. In Fig. 6.3, θ is measured in the water phase to aid comparisons.

The degree of wettability exhibited depends both on the chemical compositions of the fluid pair, particularly the asphaltine content of the oil, and on the nature of the pore wall. Pure quartz sandstone or calcite surfaces are likely to be wetted preferentially by water. The presence of certain authigenic clays, particularly chamosite, may promote oil wet character. The capillary pressure forces that influence allowable saturation change in pores of a given size are thus directly influenced by wetting character. It will be seen later that fluid displacement characteristics can also be used to deduce wetting character.

6.2 LABORATORY MEASUREMENTS AND RELATIONSHIP WITH RESERVOIR SYSTEMS

Since $\sigma \cos \theta = rP_c/2$ it follows that the capillary pressure measured in any given porous system using a particular fluid pair will be related to that obtained with any other fluid pair merely by the ratio of the $\sigma \cos \theta$ terms, i.e. denoting fluid pairs by the subscripts 1 and 2

$$P_{c(1)} = P_{c(2)} \frac{(\sigma \cos \theta)_1}{(\sigma \cos \theta)_2}$$

Use is made of this relationship in conducting laboratory tests with fluids other than reservoir condition fluids. For example, air and brine with a ($\sigma \cos \theta$) value of 72 may be used to measure P_c (air-brine) in the laboratory. The relationship for P_c (reservoir oil-brine) is obtained using the appropriate value of $(\sigma \cos \theta)_{res}$ (= 26 dynes/cm).

$$P_{c(\text{res})} = P_{c(\text{lab})} \frac{(\sigma \cos \theta)_{res}}{(\sigma \cos \theta)_{lab}}$$

The migration of hydrocarbons into an initially water filled reservoir rock and the subsequent equilibrium vertical distribution of saturation is modelled in the laboratory by a non-wetting phase displacing wetting phase drainage capillary pressure test. Air and brine are frequently used as the pseudo-reservoir fluids, and the displacement is effected by increasing air pressure in a series of discrete steps in water-saturated core plugs sitting

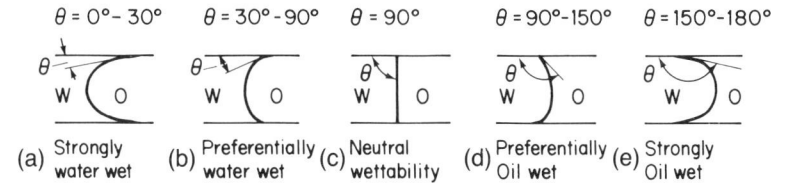

Fig. 6.3 Wetting contact angles in confined capillaries. (a) Strongly water wet, (b) preferentially water wet, (c) *neutral*, (d) preferentially oil wet, (e) strongly oil wet.

on a semi-permeable porous diaphragm. As a result of an increase in pressure (equivalent to P_c since $P_c = P_{air} - P_{water}$) the water saturation decreases and its value is established by weighing the core plug. The non-wetting phase fluid finds it easier to enter the largest pore spaces in the porous rock first since, for a given rock-fluid system,

$$P_c \propto \frac{1}{r}$$

The apparatus layout is shown in Fig. 6.4. A number of cores of similar petrophysical properties can be analysed simultaneously, with equilibrium being controlled by the core plug taking longest.

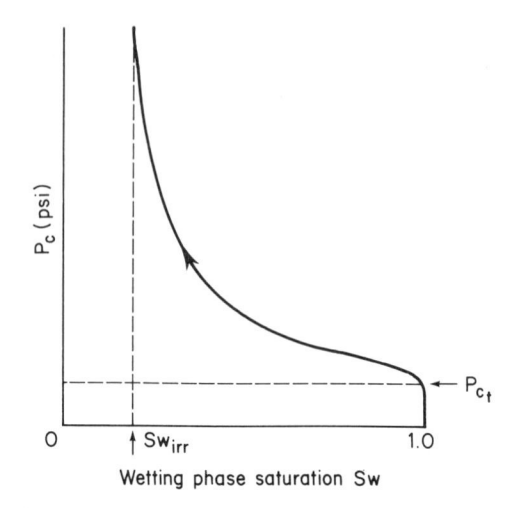

Fig. 6.5 Laboratory measurements of drainage capillary pressure.

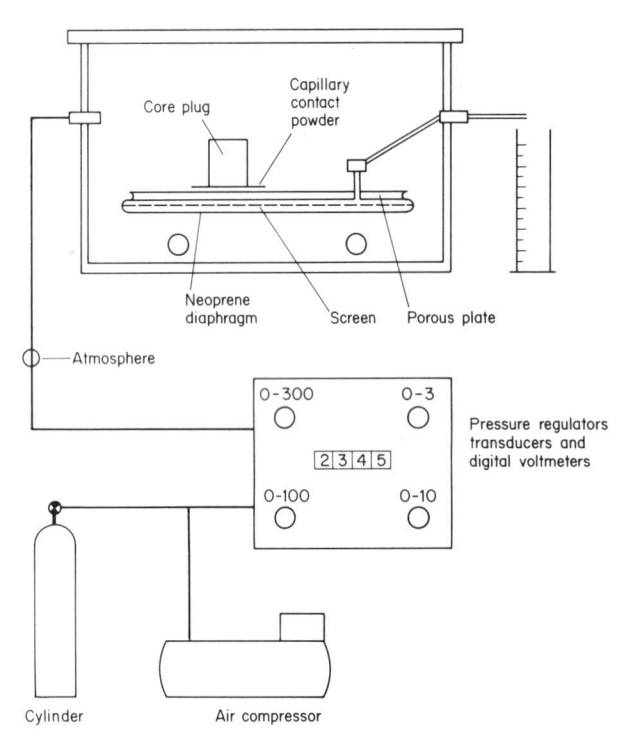

Fig. 6.4 Gas liquid drainage capillary pressure measurement. (1) Portion of liquid in saturated cores is displaced at a particular pressure level by either gas or displacing liquid. (2) Liquid saturations measured after equilibrium saturation has been reached. (3) Repetition for several successive pressure levels.

There will be a *threshold* pressure for each pore radius which has to be overcome by the applied pressure differential in order to move wetting phase fluid from that pore. The relationship between applied pressure differential (equivalent to capillary pressure) and saturation thus gives a characterization of pore size distribution. The laboratory test results may look like those shown in Fig. 6.5. The first applied pressure differential does not cause any desaturation of wetting phase and is interpreted as meaning that the threshold capillary pressure of the largest pore sizes has not been reached. Between 0.5 psi and 1.0 psi some desaturation is achieved and the minimum threshold pressure (P_{ct}) lies in this region. For pressures greater than the minimum threshold pressure, a decreasing pore size is invaded by non-wetting phase fluid until an irreducible wetting phase saturation S_{wirr} is reached and no further increase in differential pressure causes further desaturation. In laboratory tests this final irreducible saturation value is often beyond the breakdown pressure of the porous plate and is sometimes obtained by centrifuge spinning at a rotational force equivalent to about 150 psi, and measuring the quantity of any produced wetting phase. The cross-hatched region in Fig. 6.5 which lies between P_{ct} and $P_c (S_{wirr})$ is known as the transition zone region. In higher permeability reservoir rocks (500 mD) the value of P_{ct} may be indistinguishable from zero applied pressure. The physical significance of threshold pressure may be appreciated by an analogy with capillary rise of water in different bore glass tubes suspended in an open tray of water. Again, since $P_c \propto 1/r$ it will be observed that entry of the non-wetting phase should be most difficult in the smallest bore tube (highest threshold pressure), as shown in Fig. 6.6. If the density of water is denoted by ρ_w

$$P_{ct} = g'\rho_w H$$

$$P_{ct(3)} > P_{ct(2)} > P_{ct(1)}$$

6 FLUID SATURATION

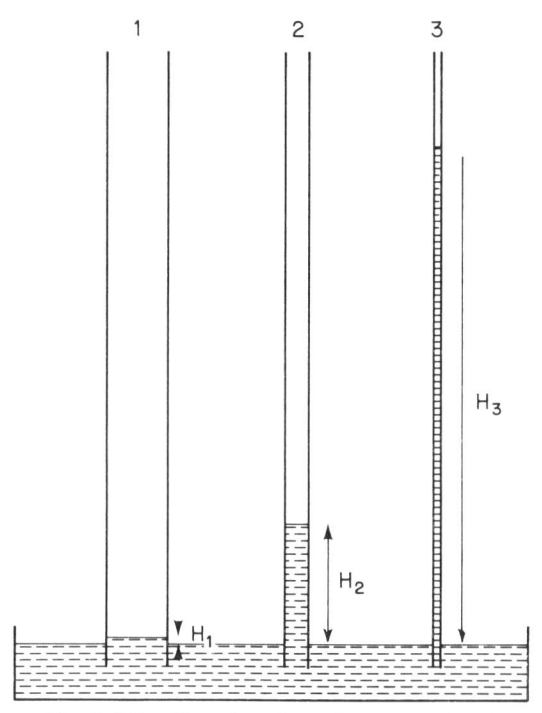

Fig. 6.6 Capillary rise above free water level.

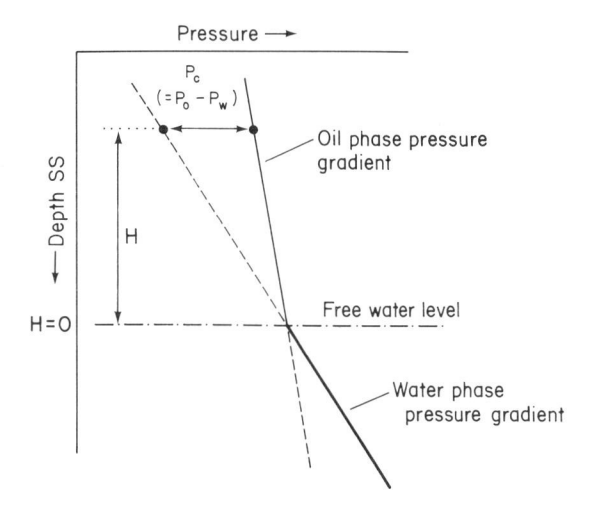

Fig. 6.7 Static pressure gradients in a homogeneous reservoir interval.

The free water level in the dish provides a convenient datum location.

The threshold capillary pressure found in reservoir rocks is proportional to the height above the free water level (FWL) datum, where a region of 100% water saturation will be found. The FWL is thus a property of the reservoir system, while an oil–water contact observed in a particular well in the reservoir will depend on the threshold pressure of the rock type present in the vicinity of the well and there may then be a zone of 100% water saturation from some height above the FWL. The relationship between height above free water level and capillary pressure is derived from consideration of the gravity-capillary pressure force equilibrium. Using the free water level as a datum and defining its position in the reservoir as the place where oil phase pressure P_o equals the water phase pressure P_w then at the FWL

$$P_o - P_w = 0 = P_{c(FWL)}$$

Figure 6.7 shows fluid gradients for the oil and water phases which are defined in terms of density of the fluids. At some height H above the free water level, which is a convenient datum and where $P_o = P_w = P_{FWL}$

$$P_o = P_{FWL} - \rho_o g/g_c \cdot H$$

and

$$P_w = P_{FWL} - \rho_w g/g_c \cdot H$$

where g/g_c or g' is the ratio of the acceleration due to gravity and the gravitational constant. In British units this ratio is unity, and in SI units is 9.81.

From our definition of capillary pressure as $P_c = P_o - P_w$, the capillary pressure at a depth equivalent to H above the free water level is given by

$$P_c = (P_{FWL} - \rho_o g/g_c H) - (P_{FWL} - \rho_w g/g_c H)$$

therefore

$$P_c = g/g_c H (\rho_w - \rho_o)$$

The saturation which exists at this height H is a rock property dependent term and is obtained from laboratory tests

$$P_c = f(S_w)$$

therefore

$$H = f(S_w)$$

Since we also have the relationship

$$P_c = \frac{2\sigma \cos \theta}{r}$$

it also follows that the saturation which occurs at height H will depend on a pore radius term r, i.e.

$$P_{c(sw)} = \frac{2\sigma \cos\theta}{r} = H_{(sw)} \cdot g/g_c \cdot (\rho_w - \rho_o)$$

therefore

$$H_{(sw)} = \frac{2\sigma \cos\theta}{r \cdot g/g_c \cdot (\rho_w - \rho_o)}$$

Similarly, the threshold height H_t which is equivalent to the height of an observed oil–water contact above FWL in a particular rock type is given by

$$H_t = \frac{g_c}{g} \cdot \frac{P_{ct}}{(\rho_w - \rho_o)}$$

The FWL depth is usually determined by noting an observed OWC in a well and conducting a drainage laboratory capillary pressure test on a rock sample from the interval to find the threshold capillary pressure, then the depth of the FWL ($= D_{FWL}$).

$$D_{FWL} = D_{OWC} + H_t$$

In frequently used oilfield units where P is in lbf/si, H is in feet and fluid densities are in units of lbm/ft^3, and g/g_c is equal to unity, then

$$P_c = \frac{H(\rho_w - \rho_o)}{144}$$

(Note: 1 g/cm^3 = 62.4 lbm/ft^3.)

The water saturation distribution in a homogeneous reservoir is shown in Fig. 6.8.

The magnitude of the threshold capillary pressure can influence the location of fluid contacts predicted by gradient intersection methods from RFT data since the tool responds to filtrate invaded zone character ($P_f = P_{hc} + P_{ct}$).

6.3 PORE SIZE DISTRIBUTION

The pore size distribution in a given rock type, which has been shown to influence saturation distribution, is usually determined using a mercury injection test. Although this test is destructive, in the sense that the sample cannot be used again, it has the advantage that high pressures can be attained and mercury, the non-wetting phase with respect to air, can be forced into very small pores. The pore size distribution function D_r is determined from the volume of mercury injected over a given pressure step [11] (Fig. 6.9).

The shapes of the capillary pressure curves can be used diagnostically to compare samples of similar rock type. Figure 6.10 shows a number of different characteristic mercury injection capillary pressure curve shapes.

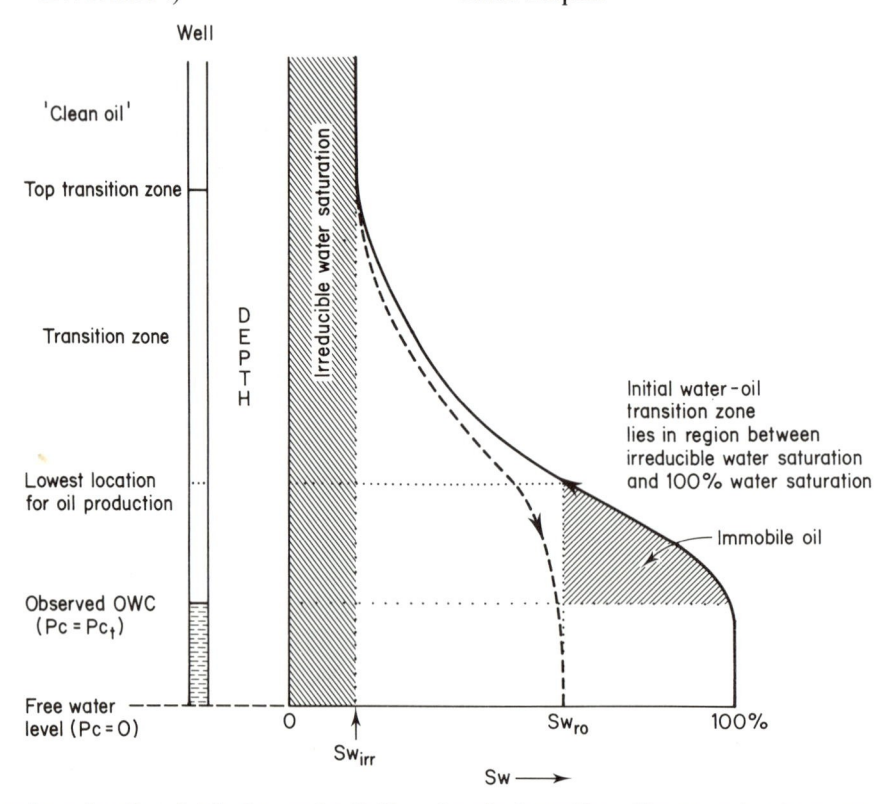

Fig. 6.8 Static water saturation distribution and definition of contacts and transition zone in a homogeneous reservoir.

6 FLUID SATURATION

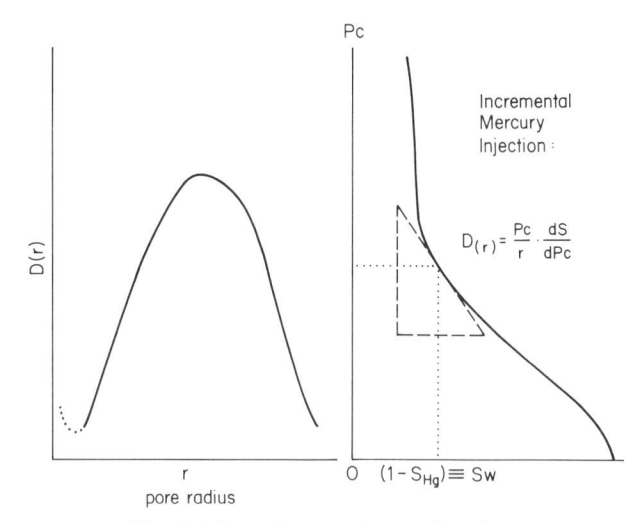

Fig. 6.9 Pore size distribution function.

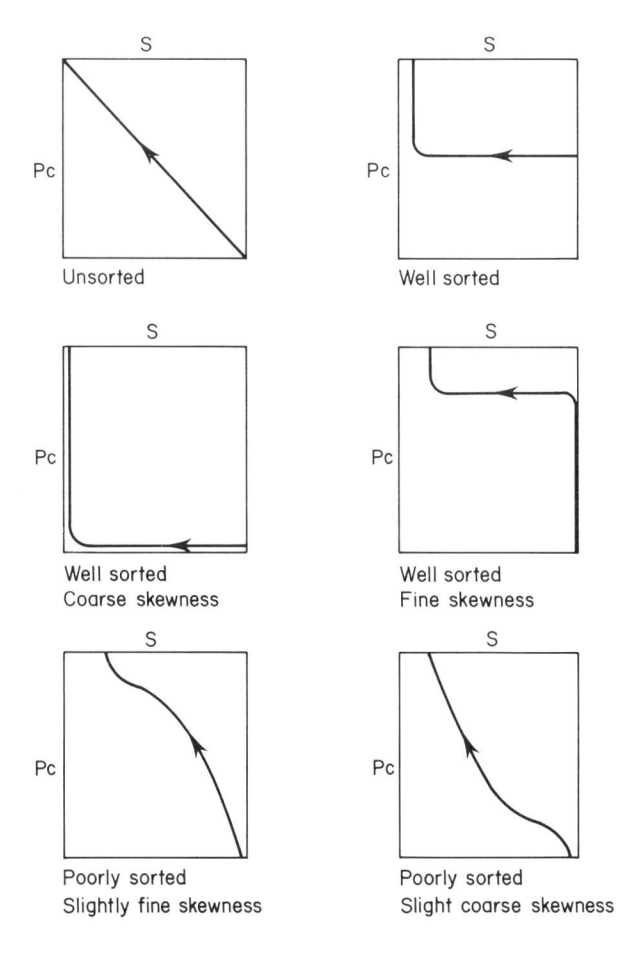

Fig. 6.10 Characteristic mercury injection capillary pressure curve shapes.

The pore size distribution would be expected to control the rate of saturation change in a given wettability system for a given phase pressure difference, and the rate would be different depending on whether phase saturation was decreasing (drainage) or increasing (imbibition). The directional effect is attributable to the threshold pressure dependency on pore radius.

6.4 CAPILLARY PRESSURE HYSTERESIS

When wetting phase pressure is increasing, the *imbibition wetting phase threshold pressure* is sometimes called a capillary suction pressure. The hysteresis phenomenon gives rise to the curve pair character shown in Fig. 6.11. The condition $P_c = 0$ in the imbibition direction effectively defines the residual non-wetting phase saturation which is therefore a property of the particular rock pore size system and should be recognized as such.

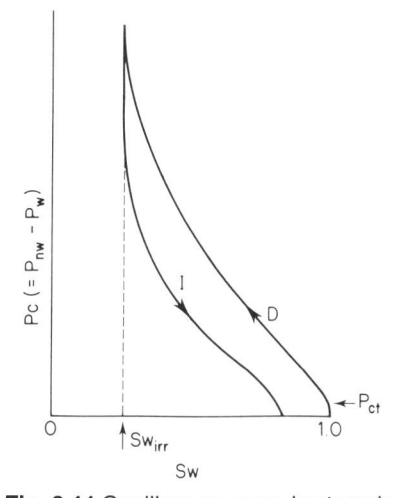

Fig. 6.11 Capillary pressure hysteresis.

The experimental difficulties in determining the definition of the imbibition direction capillary pressure curve, combined with the difficulties of using the information in reservoir simulation models, has led to the assumption by many reservoir engineers that, for practical purposes, capillary pressure hysteresis does not exist. This is close to the truth in systems without strong wetting preference and with essentially monosize pores and the magnitude of the difference $\Delta P_{c(D-I)}$ may often be negligible in comparison with viscous force pressure gradients. (This last assumption depends on whether the gradients are being compared over large distances between wells (macrosystem) or over small pore

98   **PETROLEUM ENGINEERING: PRINCIPLES AND PRACTICE**

diameter distances on a microdisplacement scale.)

For the application of hysteresis in the dynamic pressure of reservoir fluid displacement, the reader is referred to the literature [1, 5, 22, 29].

6.5 SATURATION DISTRIBUTIONS IN RESERVOIR INTERVALS

In real reservoir systems it is expected that a number of *rock type* units will be encountered. Each unit can have its own capillary pressure characteristic and the static saturation distribution in the reservoir will be a superposition of all units, as shown in Figs 6.12 and 6.13. In this example, four sand units are connected only at a common aquifer, whose free water level is denoted as FWL.

The sands are labelled 1–4 and have permeabilities k_1–k_4 with $k_1 > k_4 > k_3 > k_2$. The observed oil–water contacts representing the effects of threshold entry pressure are denoted OWC. Each sand has a capillary pressure curve, depth related to saturation, and different irreducible water saturations. A well penetrating all sands as shown will log a saturation profile as shown. Multiple oil–water contacts and transition zones which are shown can be seen to relate to appropriate portions of each sand's capil-

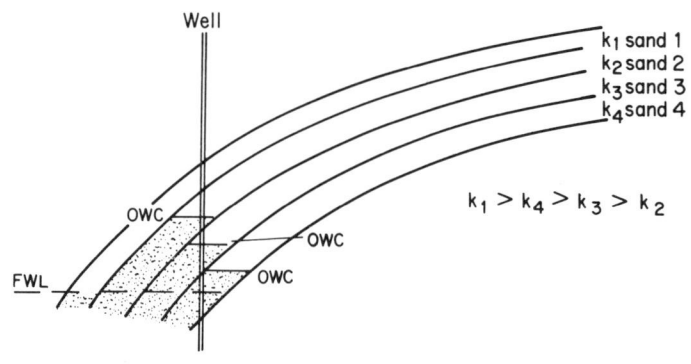

Fig. 6.12 Observed oil–water contacts and their relationship with free water level in a layered reservoir with a common aquifer.

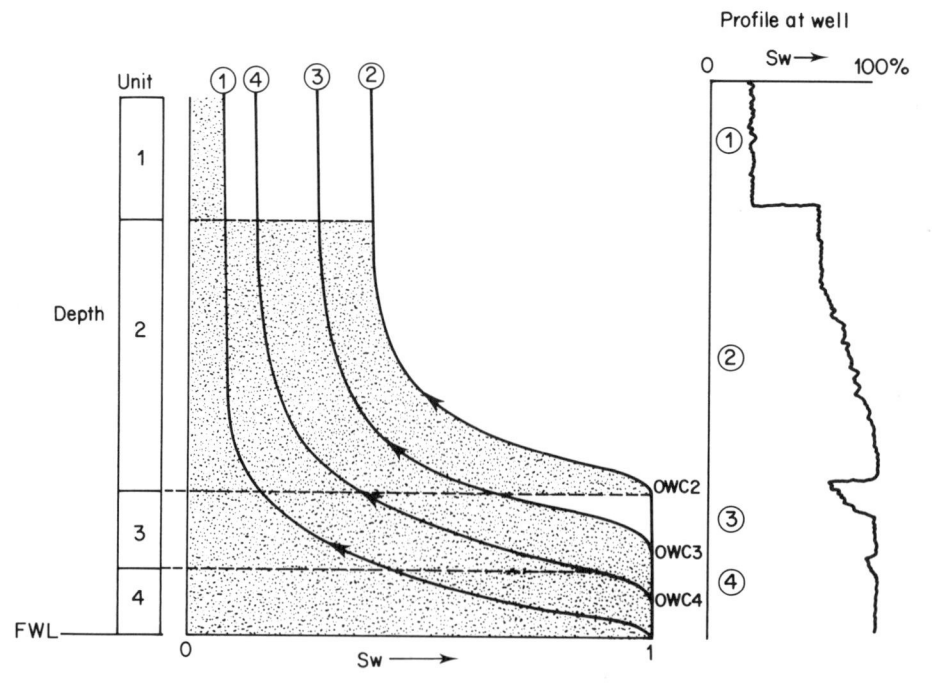

Fig. 6.13 Saturation discontinuities in a layered reservoir and an example of multiple observed oil–water contacts, but a single free water level.

6 FLUID SATURATION

lary pressure curve. This behaviour is important to recognize in correlating oil–water contacts and in the zonation of reservoirs.

6.6 CORRELATION OF CAPILLARY PRESSURE DATA FROM A GIVEN ROCK TYPE

From our definition of $P_c = 2\sigma \cos\theta/r$ where r is a mean radius, we may note that the grouping $rP_c/\sigma \cos\theta$ will be dimensionless. Since permeability has the dimension L^2 (the unit of area), then we could substitute $\sqrt{k}$ for r and maintain the dimensionless nature of the group. Leverett [4] in fact defined a dimensionless capillary pressure group in this way, with the exception that $(k/\phi)^{0.5}$ was preferred. Since capillary pressure is a function of saturation, then the dimensionless capillary pressure term (J) is also a function of saturation. Thus we may write

$$J_{(sw)} = \frac{P_{c(sw)}}{\sigma \cos\theta} \sqrt{\frac{k}{\phi}}$$

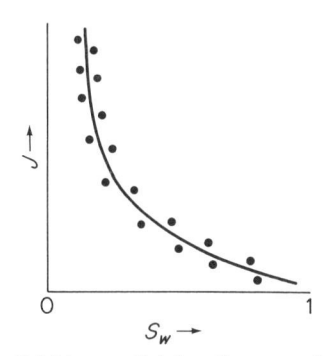

Fig. 6.14 Leverett J-function correlation.

This relationship (Fig. 6.14) will apply as a correlating group for all measurements of capillary pressure using different fluid systems, so long as the porous rocks have similar pore geometries. There will therefore be a particular correlation for given rock type, and lack of correlation can suggest the need for further zonation. Following the establishment of the correlation from representative rock samples under laboratory conditions, it is used to predict reservoir saturation distribution.

A further correlating technique makes use of an observation that in a given rock type, capillary pressure curves from samples of different permeabilities often form a family of curves, as shown in Fig. 6.15.

With this shape of curve, an approximate linearization can be made by plotting the logarithm of capillary pressure against the logarithm of permeability as iso-saturation lines, as shown in Fig. 6.16. This enables easier interpolation and regeneration of particular capillary pressure–saturation relationships to predict reservoir saturation distribution. Those curves are often obtained for sands where large permeability variations occur in a very narrow range of porosities.

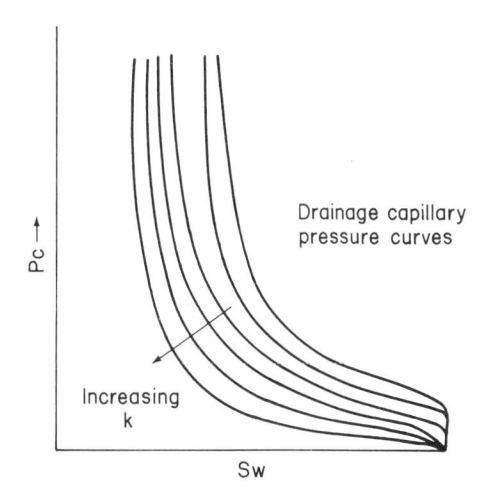

Fig. 6.15 Effect of permeability on capillary pressure in a given rock type.

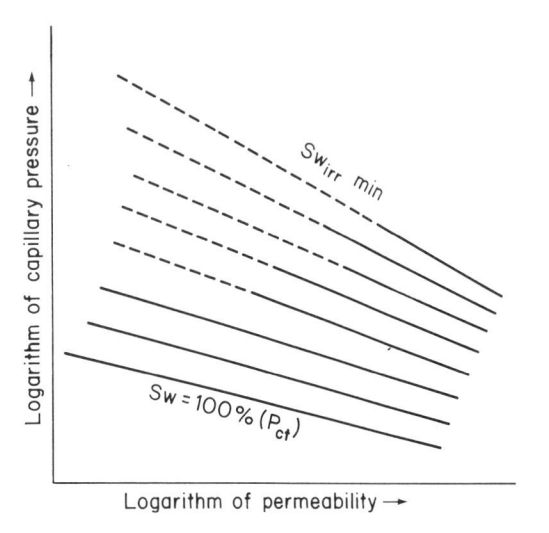

Fig. 6.16 Correlation of capillary pressure with permeability in a given rock type.

Examples

Example 6.1

An oil water capillary pressure experiment on a core sample gives the following results:

o/w capillary pressure (psia)	0	4.4	5.3	5.6	10.5	15.7	35.0
water saturation: (percent)	100	100	90.1	82.4	43.7	32.2	29.8

Given that the sample was taken from a point 100 ft above the oil–water contact, what is the expected water saturation at that elevation? If the hydrocarbon bearing thickness from the crest of the structure to the oil–water contact is 175 ft, what is the average water saturation over the interval?
($\rho_w = 64$ lbs/ft^3; $\rho_{O\text{reservoir}} = 45$ lbs/ft^3)

Example 6.2

If, in the previous example (6.1), the interfacial tension $\sigma \cos \theta$ had been 25 dyne/cm, and the permeability and porosity had been 100 mD and 18% respectively, construct the mercury capillary pressure curve for a sample of similar lithology with permeability 25 mD, porosity 13%.

Use a mercury interfacial tension of 370 dyne/cm.

Example 6.3

A drainage capillary pressure curve using an air–brine fluid pair ($\sigma \cos \theta = 72$ dyne/cm) is generated using a core plug of porosity 0.22 and permeability 150 mD. It is described as follows:

$S_{w \text{ (frac)}}$	1.0	1.0	0.9	0.8	0.7	0.6	0.5	0.4	0.3	0.2	0.2
P_c(psi)	0	1	4	6	8	9.5	11.2	13.7	16.5	23	100

It is believed that the reservoir is better represented by a porosity of 0.25 and a permeability of 500 mD. The reservoir condition oil has a specific gravity of 0.785 and formation water at reservoir conditions has a specific gravity of 1.026. The reservoir condition value of $\sigma \cos \theta$ is taken as 26 dyne/cm. Use a J-function method to generate the reservoir condition capillary pressure curve and estimate the depth relative to the free water level of the top transition zone and the observed oil–water contact.

References

[1] Morrow, N.R.
Physics and thermodynamics of capillary action in porous media, *Ind. Eng. Chem.* **62** (June 1970), 33.
[2] Muskat, M.
Physical Principles of Oil Production, McGraw Hill, NY (1949).
[3] Gregg, S.J. and Sing, K.S.W.
Adsorption, Surface Area and Porosity, Academic Press, London (1967).
[4] Leverett, M.C.
Capillary behaviour in porous solids, *Trans. AIME* **142** (1941), 152.
[5] Melrose, J.C. and Brandner, C.F.
Role of capillary forces in determining microscopic displacement efficiency for oil recovery by waterflooding, *J. Can Pet. Tech.* **13** (1974), 54.
[6] Mohanty, K.K., Davis, T.H. and Scriven, L.E.
Physics of oil entrapment in water wet rock, SPE Paper 9406, *AIME Ann. Fall Mtg.* (1980).
[7] Shull, C.G.
The determination of pore size distribution from gas adsorption data, *J. Am. Chem. Soc.* **70** (1948), 70.
[8] Pandey, B.P. and Singhal, A.K.
Evaluation of the capillary pressure curve techniques for determining pore size distribution – a network approach, *Powder Tech.* **15** (1976), 89.

[9] Kimber, O.K., Reed, R.L. and Silverberg, I.H.
Physical characteristics of natural films formed at crude oil–water interfaces, *SPEJ* (1966), 153.

[10] Donaldson, E.C., Thomas, R.D. and Lorenz, P.B.
Wettability determination and its effect on recovery efficiency, *SPEJ* (1969), 13.

[11] Ritter, L.C. and Drake, R.L.
Pore size distribution in porous material, *Ind. Eng. Chem. Fund* **17** (1945), 782.

[12] Burdine, N.T., Gournay, L.S. and Reichertz, P.P.
Pore size distribution of petroleum reservoir rocks, *Trans. AIME* **189** (1950), 195.

[13] Brunnauer, S., Emmett, P.H. and Teller, E.
The adsorption of gases in multimolecular layers, *J. Am. Chem. Soc.* **60** (1938), 309.

[14] Rose, W.R. and Bruce, W.A.
Evaluation of capillary character in petroleum reservoir rock, *Trans. AIME* **186** (1949), 127.

[15] Brooks, C.S. and Purcell, W.R.
Surface area measurements on sedimentary rocks, *Trans. AIME* **195** (1952), 289.

[16] Donaldson, E.C., Kendall, R.F., Baker, B.A. and Manning, F.S.
Surface area measurements of geologic materials, *SPEJ* **15** (1975), 111.

[17] Purcell, W.R.
Interpretation of capillary pressure data, *Trans. AIME* **189** (1950), 369.

[18] Hassler, G.L. and Brunner, E.
Measurement of capillary pressure in small core samples, *Trans. AIME* **160** (1945), 114.

[19] Slobod, R.L., Chambers, A. and Prehn, W.L.
Use of centrifuge for determining connate water, residual oil and capillary pressure curves of small core samples, *Trans. AIME* **192** (1951), 127.

[20] Schilthuis, R.J.
Connate water in oil and gas sands, *Trans. AIME* **127** (1938), 199.

[21] Brown, H.W.
Capillary pressure investigations, *Trans. AIME* **192** (1951), 67.

[22] Rapoport, L.A. and Leas, W.J.
Properties of linear waterfloods, *Trans. AIME* **198** (1953), 139.

[23] Holmes, M. and Tippie, D.
Comparisons between log and capillary pressure data to estimate reservoir wetting, SPE Paper 6856, *Ann. Fall Mtg.* (1977).

[24] Dullien, F.A. and Batra, V.K.
Determination of the structure of porous media, *Ind. Eng. Chem.* **62** (Oct. 1970), 25.

[25] Bruce, W.A. and Welge, H.
The restored state method for determination of oil in place and connate water, *OGJ* (July 26, 1947), 223.

[26] Core Labs Inc.
Special Core Analysis, Spec. Studies Section CL Inc. (April 1974).

[27] Pickell, J.J., Swanson, B.F. and Hickman, W.B.
Application of air–mercury and oil–air capillary pressure data in the study of pore structure and fluid distribution, *SPEJ* (March 1966), 55.

[28] Morrow, N.R. and Harris, C.C.
Capillary equilibrium in porous materials, *SPEJ* (March 1966), 55.

[29] Batycky, J., McCaffery, F.G., Hodgins, P.K. and Fisher, D.B.
Interpreting capillary pressure and rock wetting characteristics from unsteady-state displacement measurements, Paper SPE 9403, *Proc. 55th Ann. Fall Mtg. SPE of AIME* (1980).

[30] Melrose, J.C.
Wettability as related to capillary action in porous media, *SPEJ* (Sept. 1965), 259.

[31] Sinnokrot, A.A., Ramey, H.J. and Marsden, S.S.
Effect of temperature level upon capillary pressure curves, *SPEJ* (March 1971), 13.

[32] Mungan, N.
Interfacial phenomena and oil recovery: capillarity, In *Enhanced Oil Recovery Using Water as a Driving Fluid. World Oil* (May 1981), 149.

[33] Dunmore, J.M.
Drainage capillary pressure functions and their computation from one another, *SPEJ* (Oct. 1974), 440.

Chapter 7

Relative Permeability and Multiphase Flow in Porous Media

7.1 DEFINITIONS

Relative permeability is a concept used to relate the absolute permeability (100% saturated with a single fluid) of a porous system to the effective permeability of a particular fluid in the system when that fluid only occupies a fraction of the total pore volume.

$$k_e = k \cdot k_r$$

where k_e is the effective permeability of the phase, k_r is the relative permeability of the phase, k is the absolute permeability of the porous system.

It has been convenient to relate the relative permeability to saturation as it is observed that effective permeability decreases with decrease in the phase saturation. The relationship really expresses the Darcy flow of a two-phase or multiphase system in a porous system, i.e. it maintains a relationship for linear flow of the form

$$q_{(p)} = \frac{k_{e(p)} A}{\mu_p} \cdot \frac{d\Phi'}{dL}$$

where $q_{(p)}$ refers to phase volumetric flow rate, $k_{e(p)}$ refers to effective phase permeability, μ_p refers to phase viscosity, $d\Phi'/dL$ refers to datum corrected pressure gradient (pseudo-potential).

In two-phase systems the relationships are expressed as functions of saturation, as shown in Figs 7.1 and 7.2, where the subscripts w and nw refer to wetting and non-wetting phases respectively.

Note that $S_{w(max)}$ occurs in a two-phase system when

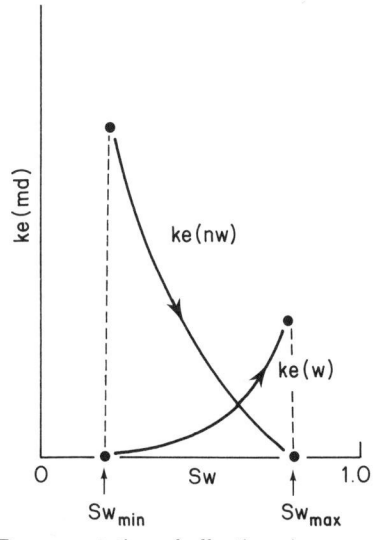

Fig. 7.1 Representation of effective phase permeability.

the non-wetting phase reaches the residual non-wetting phase saturation. Similarly, $S_{w(min)}$ represents the irreducible wetting phase saturation. The process represented in these figures is one of imbibition (i.e. increasing wetting phase saturation).

In oil–water systems in particular, the relative permeability scale is often normalised by representing relative permeability as effective permeability divided by permeability to non-wetting phase at the minimum wetting phase saturation, i.e.

$$k_r = \frac{k_e}{k_{nw\,(Sw(min))}}$$

7 RELATIVE PERMEABILITY AND MULTIPHASE FLOW

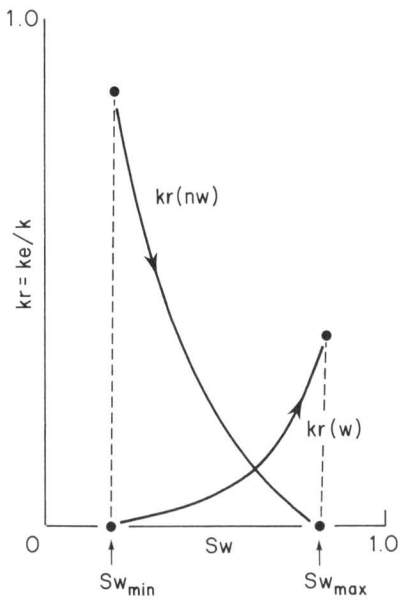

Fig. 7.2 Representation of relative permeability.

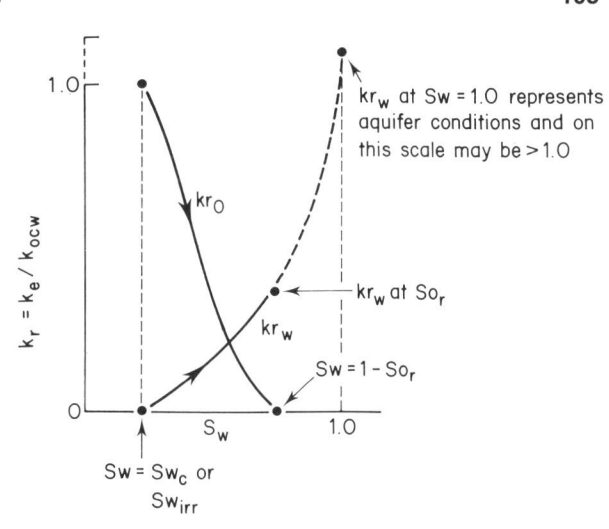

Fig. 7.3 Oil–water relative permeability (imbibition direction).

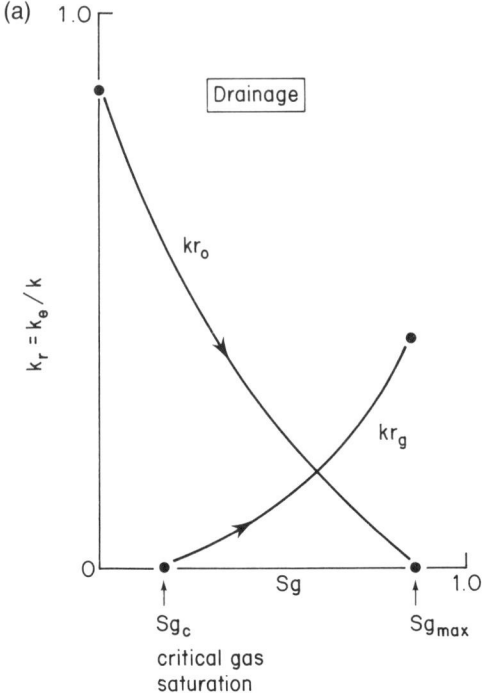

Fig. 7.4 Gas–oil relative permeability.

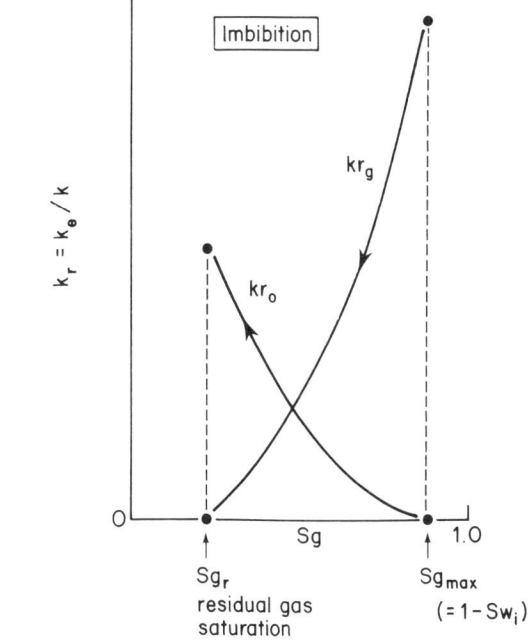

In the oil–water system this is often expressed symbolically as $k_r = k_e/k_{o_{cw}}$, where $k_{o_{cw}}$ is the oil permeability at connate water saturation, as shown in Fig. 7.3.

In the gas–oil system, the direction of displacement is particularly important as the process can represent a drainage process such as gas drive (gas displacing oil immiscibly) or an imbibition process, such as: (1) movement of an oil zone or (2) aquifer into (3) a receding depleting gas cap, as indicated in Fig. 7.4.

In gas–oil systems the third phase, water, which in

reality is always present in reservoirs is considered to stay at irreducible saturation and play no part in the displacement processes. It is therefore argued that experiments in the laboratory can be conducted with or without irreducible water present, and the effective permeabilities are correlated with total liquid saturation (S_L) rather than gas saturation. The relationship is based on

$$S_L = S_o + S_w = 1 - S_g$$

In a system where gas saturation increases from zero (a liquid drainage process) it is observed that gas does not flow until some *critical* gas saturation (S_{gc}) has been attained. This is attributed to the physical process of the gas phase becoming continuous through the system in order to flow. In liquid imbibition processes (gas saturation decreasing from a maximum initial value) the gas permeability goes to zero when the residual or trapped gas saturation (S_{gr}) is reached.

The directional aspects of relative permeability representation are often more pronounced in gas–oil systems, and modification to laboratory data may be necessary. The directional aspect may perhaps be appreciated by consideration of the difference between bedding plane gas advance towards a production well and downward gas movement vertical to the bedding plane in the vicinity of a production well, as shown in Fig. 7.5. The directional differences may be incorporated in reservoir engineering calculations by determination of frontal saturation and the use of pseudo-functions. The calculation of frontal behaviour is discussed under the heading of fractional flow analysis.

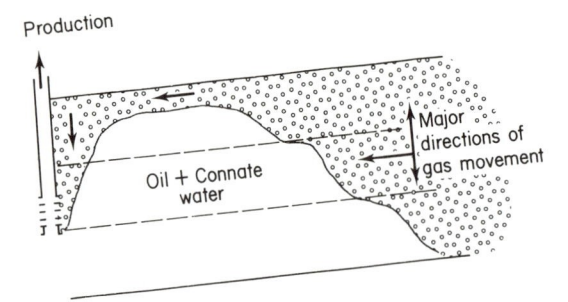

Fig. 7.5 Directional aspects of frontal gas movement.

7.2 FRACTIONAL FLOW

Phase permeability characteristics are also frequently presented in terms of permeability ratios k_w/k_o and k_g/k_o (Fig. 7.6). This is because the fractional flow of displacing fluid (f_d) at the outlet end of an incompressible linear horizontal system with no

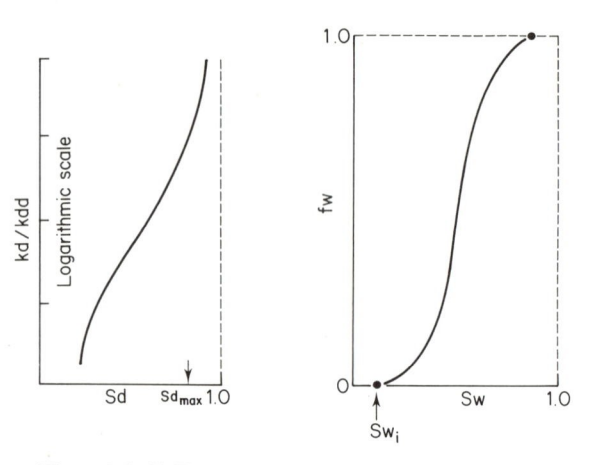

Fig. 7.6 (left) The k_d/k_{dd} ratio curve (semi-log scale).
Fig. 7.7 (right) The fractional flow curve for water displacing oil.

capillary pressure effects can be represented at different saturations in terms of mobility ratios M

$$f_d = \frac{q_d}{q_{total}} = \frac{M}{1 + M}$$

where

$$M = \frac{k_d \cdot \mu_{dd}}{\mu_d \cdot k_{dd}} = \frac{k_d}{k_{dd}} \cdot \frac{\mu_{dd}}{\mu_d}$$

and subscript d = displacing fluid, dd = displaced fluid.

The reservoir fractional flow of wetting phase displacing fluid in an oil–water system with water as the displacing fluid is therefore

$$f_w = \frac{q_w}{q_w + q_o}$$

as shown in Fig. 7.7., where q_o, q_w are reservoir condition rates.

Writing a Darcy law expression for steady state flow of each phase in a linear horizontal system we have

$$f_w = \frac{\dfrac{k\, k_{rw}}{\mu_w} \cdot A \cdot \dfrac{\Delta P_w}{L}}{\dfrac{kA}{L}\left[\dfrac{k_{r(w)}}{\mu_w}\Delta P_w + \dfrac{k_{r(o)}}{\mu_o}\Delta P_o\right]}$$

For the condition in which viscous flow forces are considerably greater than capillary forces, we can write $P_o \approx P_w$ and the expression becomes

$$f_w = \frac{1}{1 + \left[\dfrac{k_{r(o)}}{k_{r(w)}} \cdot \dfrac{\mu_w}{\mu_o}\right]}$$

7 RELATIVE PERMEABILITY AND MULTIPHASE FLOW

In tilted reservoirs of dip angle α, the fractional flow of water, using field units of RB/D for the total flow rate q_T, length terms in feet, γ as a specific gravity, viscosities in centipoise and permeabilities in milliDarcies, is given by

$$f_w = \frac{1 - \dfrac{0.001127\, k\, k_{r(o)}\, A}{q_T \cdot \mu_o}\{0.4335\,(\gamma_w - \gamma_o)\sin\alpha\}}{1 + \dfrac{\mu_w}{\mu_o}\cdot\dfrac{k_{r(o)}}{k_{r(w)}}}$$

(Note for a constant injection rate and an incompressible system $q_T = q_o + q_w = q_{INJ}$.) Typical shapes of the effective permeability curves and the f_w curve are shown in Figs 7.14 and 11.5.

Since the angle α is conventionally measured from a horizontal axis to the axis of a line in the direction of flow, it will be positive for the displacing fluid moving from downdip to updip, and will be negative for displacing fluid moving from an updip to a downdip position.

7.2.1 Analysis methods

Analysis of the fractional flow curve by the method of Buckley and Leverett [8] and Welge [10] allows the recovery performance of a homogeneous reservoir to be determined. This technique is applicable to relatively thin reservoir intervals where diffuse or dispersed flow is assumed – this means that over any part of the cross-section, the saturation of oil and water is uniform and no fluid segregation exists. This is a condition assumed in the laboratory core analysis determination of relative permeability *rock* curves.

In the Buckley-Leverett/Welge analysis of displacement of oil from a system with a uniform initial water saturation S_{wi}, a graphical technique can be utilized. A tangent drawn to the fractional flow curve from the initial water saturation has two important characteristics, as shown in Fig. 7.8. The tangent point indicates the saturation S_{wf} of the displacement front shown in Fig. 7.9. The intercept of the tangent at $f_w = 1$ indicates the average ($\overline{S}_w$) saturation behind the front up to the time of water breakthrough at the outlet end (production well) of the system. From Buckley-Leverett theory for the rate of movement of the frontal saturation, the distance X_f travelled can be related to the volume of displacing fluid injected $W_i\,(= q_i t)$ and the gradient of the fractional flow curve at the front with respect to saturation:

$$X_f = \frac{W_i}{\phi A}\cdot\frac{df_w}{dS_w}\bigg]_{S_{wf}}$$

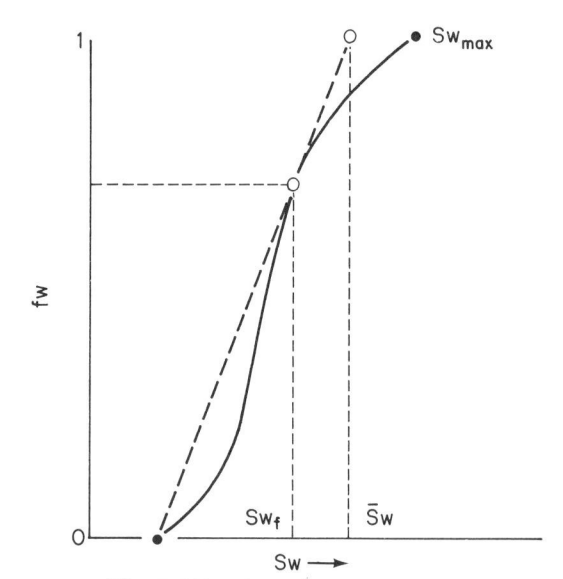

Fig. 7.8 Fractional flow analysis.

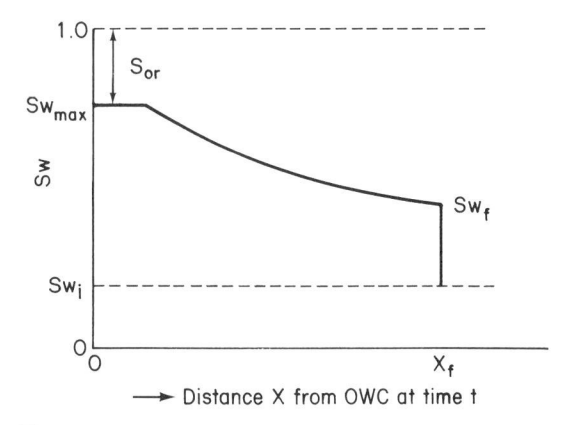

Fig. 7.9 Linear saturation profile before breakthrough.

This leads to an equivalencing for all times up to breakthrough of the saturation change behind the front to the volume of displacing fluid injected:

$$\overline{S}_w - S_{wi} = \frac{W_i}{X_f A \phi} = 1\bigg/\frac{df_w}{dS_w}\bigg]_{S_{wf}}$$

At breakthrough the distance travelled by the front X_f will equal L, the length of the system. Therefore

$$\frac{W_i}{LA\phi} = 1\bigg/\frac{df_w}{dS_w}\bigg]_{S_{wf}} = \frac{q_T \cdot t_b}{LA\phi}$$

where t_b is the breakthrough time:

$$t_b = \frac{LA\phi}{q_T}\cdot\frac{1}{df_w/dS_w}\bigg]_{S_{wf}}$$

After breakthrough of the frontal saturation at a production well for a water displacing oil system, the average water saturation in the reservoir will increase with increasing volume of injected water until a maximum value of $(1 - S_{or})$ is reached. The Welge analysis is used to calculate the average water saturation. It makes use of the fractional flow curve at a saturation greater than the frontal saturation and relates any such saturation to its fractional flow saturation gradient. It is imagined that after breakthrough, the outlet end of the system (production well) will experience an increasing water saturation with time. At any time after breakthrough, the particular outlet end saturation is defined as S_{we} and its gradient from the fractional flow curve is

$[df_w/dS_w]_{S_{we}}$

Welge demonstrated that

$$[df_w/dS_w]_{S_{we}} = \frac{1 - f_{we}}{\overline{S}_w - S_{we}}$$

This relationship at some time after breakthrough can be seen in Fig. 7.10. Since

$$\frac{q_T \cdot t_{(S_{we})}}{LA\phi} = \frac{1}{df_w/dS_w]_{S_{we}}}$$

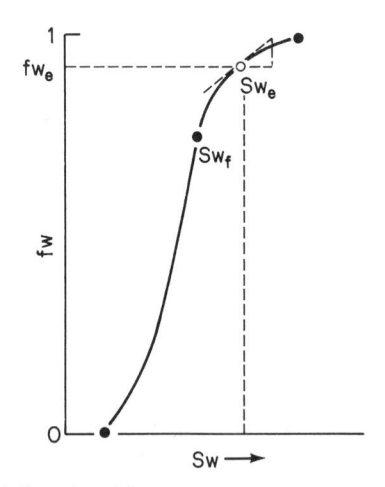

Fig. 7.10 Fractional flow gradients after breakthrough.

then the relationship between time and the attainment of a given outlet end face saturation is readily obtained, i.e.

$$t_{(Swe)} = \frac{LA\phi}{q_T} \left[\frac{1}{df_w/dS_w} \right]_{S_{we}}$$

and

$$\overline{S}_w = S_{we} + \frac{1 - f_{we}}{df_w/dS_w} \bigg]_{S_{we}}$$

The reservoir condition recovery factor after breakthrough is obtained from the ratio of oil produced to initial oil in place. In terms of saturation this will be

$$RF = \frac{\overline{S}_w - S_{wi}}{1 - S_{wi}}$$

7.3 EFFECTS OF PERMEABILITY VARIATION

Reservoirs characterized by a number of different rock types may be analysed analytically if they are totally stratified, or by numerical modelling techniques if cross-flow is significant. In both cases the effective permeability characteristics of each pore size rock type are required. Stratified reservoir analysis by analytical methods is subject to constraints regarding end point mobility ratios and the reader is referred to the original papers by Dykstra and Parsons [53, 54] and by Stiles [55]. Immiscible displacement in a system having a transition zone cannot be handled by the geometrical construction methods appropriate to a uniform initial saturation distribution, and material balance methods are needed [75]. Displacement stability can be analysed in relatively simple homogeneous linear reservoir systems, where end-point mobilities are used and the approach of Dietz [49] applied. In this case, a sharp interface is assumed to exist between the displacing and displaced fluids, that is gravity segregation dominates any capillary forces.

For a water-oil system, the gravity segregated distribution of oil and water at any distance X along the flow path can be represented as shown in Fig. 7.11 by a thickness weighted distribution. In terms of the water phase saturation S_w and the fraction n of flooded thickness, the weighted average saturation S_w'' is given by

$$S_w'' = n(1 - S_{or}) + (1 - n)S_{wi}$$

and since at this front we are dealing with end-point relative permeabilities (i.e. $k'_{ro} = k_{ro}$ at S_{wi} and $k^1_{rw} = k_{rw}$ at S_{or} then appropriate relative permeabilities at the weighted average saturation S_w'' are

$$k_{rw}'' = n \cdot k_{rw}'$$

and

$$k_{ro}'' = (1-n) k_{ro}'$$

7 RELATIVE PERMEABILITY AND MULTIPHASE FLOW

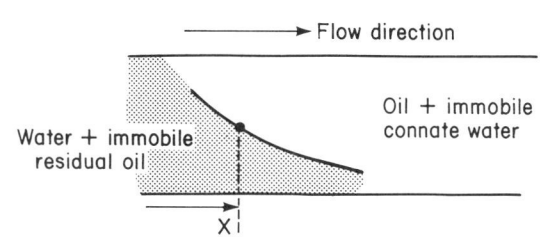

Fig. 7.11 Dietz analysis.

This leads to

$$k_{rw}'' = k_{rw}' \left[\frac{S_w'' - S_{wi}}{1 - S_{or} - S_{wi}} \right]$$

and

$$k_{ro}'' = k_{ro}' \left[\frac{1 - S_w'' - S_{or}}{1 - S_{wi} - S_{or}} \right]$$

As shown in Fig. 7.12 this indicates that a straight line relative permeability relationship with average saturation is appropriate. In this way *pseudo*-relative permeability relationships may be used to solve displacement problems in thicker sands, effectively transforming a homogeneous 2-D displacement to a 1-D problem solvable by a Buckley-Leverett/Welge technique. In such an application the fractional flow curve is generated from the k_{ro}'' and k_{rw}'' v. S_w'' relationship. The stability of displacement in a homogeneous 2-D system will depend on the mobility ratio of the fluids and the dip angle α of the reservoir. Dietz showed that the maximum rate for stable updip displacement was given, in field units of RB/D, feet, centipoise, milliDarcies and specific gravities, as

$$q_{max} = \frac{4.9 \times 10^{-4} k \cdot k'_{r(D)} \cdot A \cdot (\gamma_D - \gamma_{DD}) \sin\alpha}{\mu_D \left[\left\{ \frac{k'_{r(D)}}{k'_{r(DD)}} \cdot \frac{\mu_{DD}}{\mu_D} \right\} - 1 \right]}$$

where the subscript D refers to the displacing phase and DD to the displaced phase. Note that for downdip displacement α will be negative and that for all end-point mobility ratios less than or equal to unity the displacement is unconditionally stable and becomes piston-like.

Piston-like displacement in stratified reservoirs characterized by thin beds with no cross-flow between them can be analysed very easily by the Stiles[55] approach. In this technique, the total fractional flow of water at a producing well is determined at a number of times as each thin bed achieves water breakthrough. Since there is no cross-flow between beds, the calculation technique is facilitated by rearranging the actual beds in a sequence with the highest kh product beds at the top. For equal oil and water mobilities the pressure gradients in all beds are assumed equal and the displacing fluid is distributed between beds in proportion to the bed kh. In any bed j, therefore, the injection rate is

$$(q)_j = q_T \left\{ \frac{(kh)_j}{\sum_{j=1}^{j=n}(kh)} \right\}$$

where q_T is the total rate. The flow is considered incompressible and q_T the reservoir condition injection rate is considered equal to the total reservoir condition production rate. The pore volume of each bed between the injection and production points is

$$(PV)_j = (LA\phi)_j$$

The volume of injected fluid needed to change the saturation from S_{wi} to $S_{w(ro)}$ (or from S_{oi} to S_{or}) at breakthrough of the displacing phase is

$$(W_{ib})_j = (LA\phi(1 - S_{wi} - s_{or}))_j$$

The time taken t_b to reach this condition in the particular layer j is thus

$$(t_b)_j = \frac{(W_{ib})_j}{(q)_j}$$

As each layer reaches breakthrough, the piston displacement assumption means, for say a water–oil

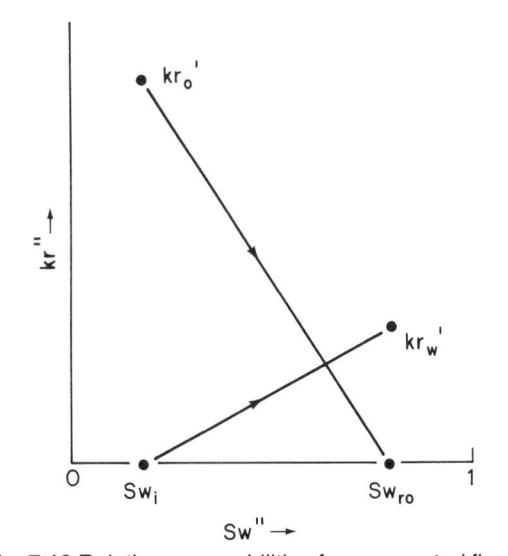

Fig. 7.12 Relative permeabilities for segregated flow.

system, that only water flows thereafter, and the layer continues to take water from the injector. In practice it may in fact be sealed off. The total production at any time t is therefore the water from any layer which has reached breakthrough, plus the oil from any layer which has reached breakthrough, plus the oil from any layer which has not yet achieved breakthrough. For an incompressible flow system this is easily calculated since the proportions at the outlet are equivalent to inlet rate distributions. At some time t in a system in which two out of n layers have reached breakthrough we have

$$q_w = (q_1 + q_2)$$
$$q_o = (q_T - q_w)$$

The fraction flow of water is given, at reservoir conditions, as

$$f_w = \frac{q_w}{q_T}$$

and the reservoir condition water–oil ratio as

$$\text{WOR} = \frac{q_w}{q_o}$$

The recovery factor at any time will be obtained by evaluating frontal positions and conducting a material balance. In any layer which has broken through, the recovery will be $(S_{oi} - S_{or})/S_{oi}$. In a layer not yet at breakthrough, the oil remaining is $(PV_j\{X(S_{or})_j + (1-X)(S_{oi})_j\}$ where X is the distance of the front from the injection location. The recovery factor for such a layer at that time is thus

$$(RF)_j \bigg]_t = \frac{(S_{oi})_j - [X(S_{or})_j + (1-X)(S_{oi})_j]}{(S_{oi})_j}$$

The overall economic recovery factor will be controlled by surface handling facilities and economic rate.

7.4 WETTABILITY EFFECTS

Wettability effects in fluid displacement are displayed by effective and relative permeability curve characteristics. As discussed previously, the wettability of a surface depends on the term $\sigma \cos \theta$, and capillary pressure controls the sequence of pore saturation change so long as viscous flow forces are not controlling. Viscous forces tend to control when the term $q_D\mu_D L/A$ in Darcy units is greater than about 2.0 [11, 76]. The effective permeability character for water wet and oil wet systems is shown in Figs. 7.13 and 7.14.

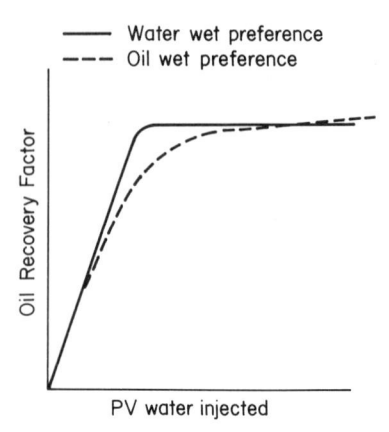

Fig. 7.13 Oil recovery efficiency.

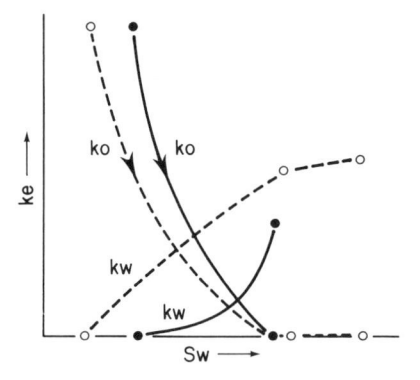

Fig. 7.14 Effect of wettability on effective permeability.

Oil wet systems tend to be characterized, in comparison with water wet systems, by:

earlier water breakthrough;
lower initial water saturations for a given pore size;
the condition $k_o = k_w$ occurs at lower values of water saturation;
less piston-like approach to a residual saturation (the S_{or} approaches zero by a film drainage mechanism);
higher values of k_w at most values of water saturation;
lower values of k_o at most values of water saturation.

The establishment of *in situ* wettability conditions[45] is therefore very important in the proper conduct of laboratory experiments, which should duplicate or account for field conditions. The most frequent laboratory measurement of wetting tendency is through the Amott test[3], but this only indicates the condition of the *as received* sample.

7.5 LABORATORY DETERMINATION OF RELATIVE PERMEABILITY DATA

Laboratory determination of effective permeability is generally conducted as a special core analysis test on representative and carefully preserved core plug samples. A reservoir condition test is conducted at reservoir pore pressure conditions and reservoir temperatures with real or simulated reservoir fluids. Such reservoir condition tests may model displacement under steady state[23], or unsteady state conditions[14], and equipment arrangements are shown in Figs. 7.15 and 7.16. Room condition relative permeability tests can be conducted at outlet end pressures of one atmosphere and at room temperature using refined oils and synthetic brines, or with gases such as air, nitrogen or helium. Gas displacement processes require a significant back pressure (say 20 bar) to facilitate flow rate interpretation.

Unsteady state relative permeability tests simulate the flooding of a reservoir with an immiscible fluid (gas or water). The determination of relative permeability is based on observation of the fractional flow of displacing phase fluid from the outlet end of the core plug and its relationship with saturation. The displacement theories of Buckley and Leverett[8] are combined with that of Welge[10] in a technique described by Johnson, Bossler and Naumann[13]. The detection of the breakthrough time of the displacing phase at the outlet core face is critical in the representation of relative permeability, and severe errors can occur with heterogeneous samples. Flow rates are determined according to the method of Rapoport and Leas[76] in order to minimize the effects of capillary pressure forces in retaining

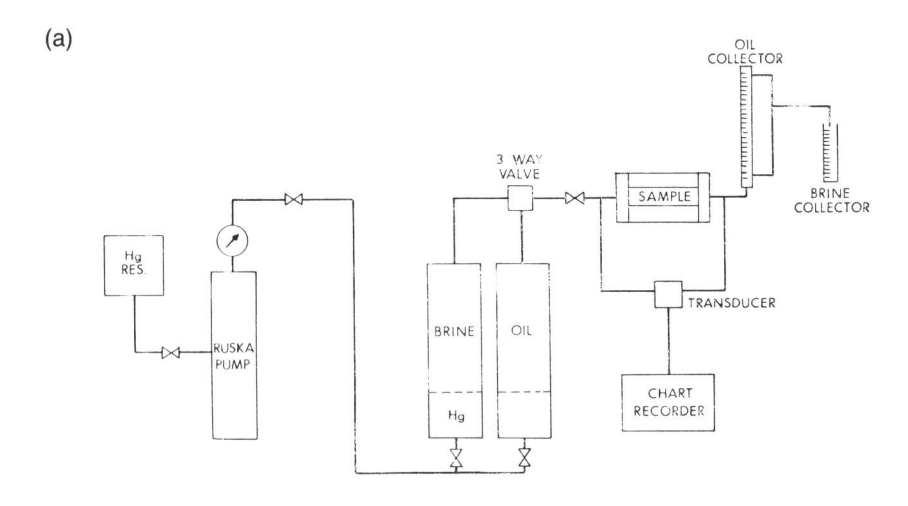

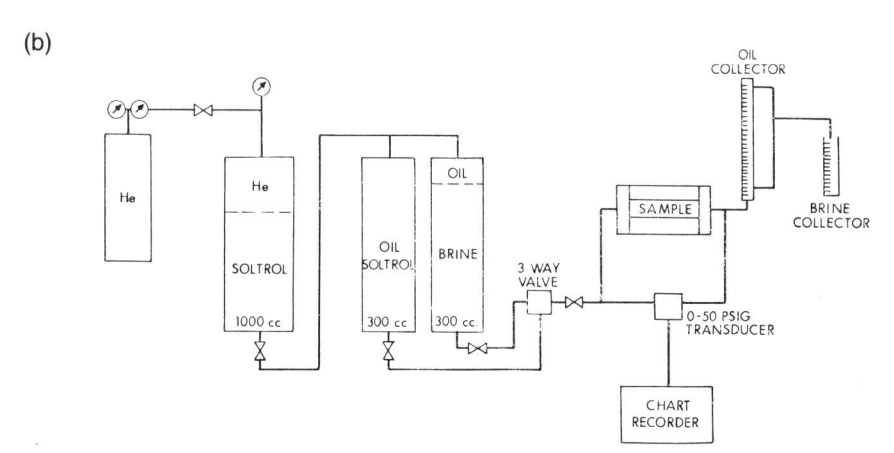

Fig. 7.15 Unsteady state relative permeability measurement: (a) constant rate; (b) constant pressure.

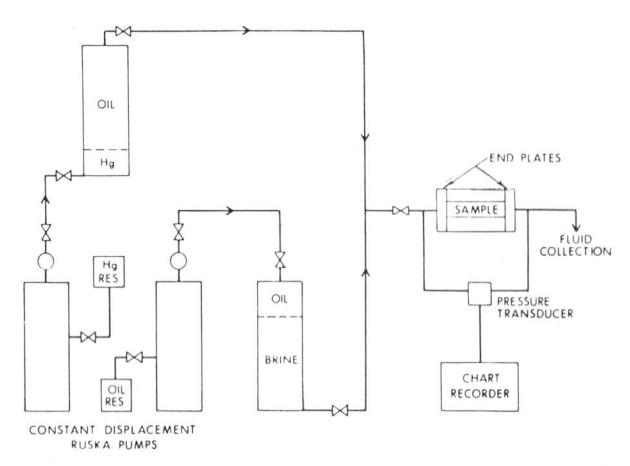

Fig. 7.16 Steady state relative permeability measurement.

wetting phase fluid at the outlet end face discontinuity. The unsteady state or dynamic displacement test is most frequently applied in reservoir analysis of strong wetting preference, and with homogeneous samples.

For reservoirs with more core-scale heterogeneity and with mixed wettability, the steady state laboratory test at reservoir conditions and with reservoir fluids is preferred. The steady state process provides simultaneous flow of displacing and displaced fluids through the core sample at a number of equilibrium ratios. At each ratio from 100% displaced phase to 100% displacing phase an equilibrium condition must be reached at which the inflow ratio of fluids equals the outflow ratio, and at which the pressure gradient between inlet and outlet is constant. At such a condition the Darcy law equation is applied to each phase to calculate effective permeability at the given steady state saturation. Capillary pressure tends to be ignored and a major difficulty is the determination of saturation at each stage. Between five and ten stages are usually needed to establish relative permeability curves.

It has become clear that the room condition tests are not necessarily a good guide to reservoir conditions behaviour. Table 7.1 shows the results of a number of different room condition and reservoir conditions unsteady state tests conducted on different small plugs of sandstones. For each test pair the injection rate and oil–water viscosity ratio was constant. The flooding efficiency ratio is defined here as the breakthrough to total oil recovery ratio for the reservoir condition test divided by the equivalent ratio for the room condition test. These data indicate the significance of temperature in that the flooding efficiency ratio appears to correlate with temperature and might be considered influenced by wettability.

In cases where reservoir condition mobility ratios are significantly greater than unity, there is some evidence to suggest that conventional short core plugs (7 cm) should not be used. To some extent this conclusion is based on the applicability of Rapoport and Leas $LV\mu_D$ core flooding criteria. In short cores, the velocity may be too high for proper imbibition processes to take place between fingers of invading water. Experimental evidence suggests that core lengths of at least 25 cm are needed to obtain consistent results. Figure 7.17 shows the effect of core length on observed breakthrough recovery at a constant $LV\mu_D$ factor in a strongly water wet outcrop sandstone. To obtain cores of lengths

TABLE 7.1 Comparison between reservoir condition and room condition waterflood tests

	Sample		Oil recovery (fraction PV)				Residual oil satn (fraction PV)		Flooding efficiency ratio
			Before BT		Between BT and WOR = 100				
No.	Reservoir cond. P_{RES} psi	T_{RES} (°F)	Room cond.	Resv. cond.	Room cond.	Resv. cond.	Room cond.	Resv. cond.	
A	3430	199	0.30	0.40	0.15	0.06	0.30	0.26	1.30
B	2100	176	0.25	0.26	0.15	0.13	0.40	0.35	1.07
C	2000	155	0.35	0.35	0.05	0.05	0.30	0.39	1.00
D	1500	128	0.30	0.27	0.12	0.25	0.40	0.35	0.73
E	500	170	0.11	0.15	0.17	0.17	0.53	0.58	1.19

7 RELATIVE PERMEABILITY AND MULTIPHASE FLOW

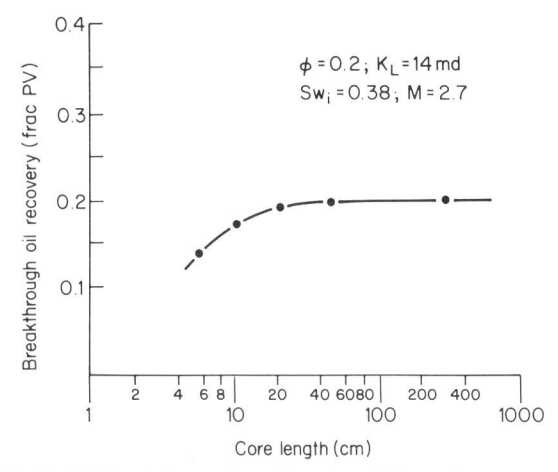

Fig. 7.17 Effect of core length on breakthrough recovery at constant $LV\mu_D$.

greater than 25 cm in the bedding plane from reservoir rock cores it is necessary to butt together small cores and construct a composite core. Composite cores should have component sections of similar petrophysical character in any representative section. When composites obey this rule, and they are maintained in capillary contact by compressive stress, Fig. 7.18 shows that they can behave as a continuous sample.

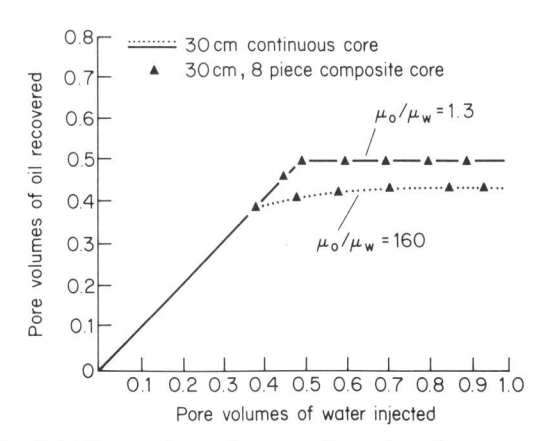

Fig. 7.18 Comparison of composite and continuous core performance with homogeneous water wet outcrop sandstone.

7.6 RESIDUAL SATURATIONS

Residual saturations tend to be dependent on pore geometry and on direction of saturation change with respect to wetting phase. In most laboratory tests, the viscous flow forces are designed to dominate capillary and gravity forces and little effect of rate on residual saturation is observed. This may not be true in the field. In gravity stabilized oil drainage by gas advance into an oil zone, residual saturations approaching single percentage figures have been claimed – this contrasts with a dynamic but otherwise equivalent wettability process of water displacing oil where residual saturation between 20% and 40% might be expected. The influence of capillary pressure is exerted at pore scale rather than at inter-well scale and, depending on wettability, can influence the location of residual fluids. The capillary number concept has been used to represent the residual oil saturation resulting from competition between viscous and capillary forces, i.e. $N_c = V\mu_D/(\sigma \cos \theta)$. Its scaling to field conditions is problematical, but some field analyses tend to support its application, as shown in Fig. 7.19. The

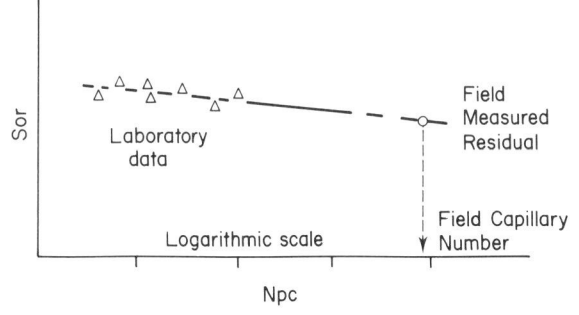

Fig. 7.19 Comparison of field and laboratory capillary numbers and residual saturations.

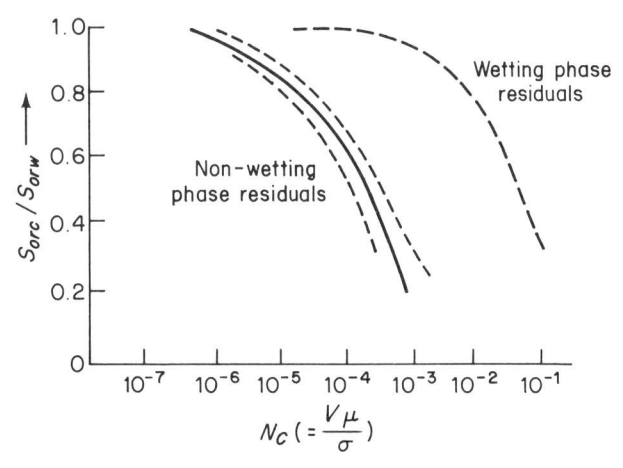

Fig. 7.20 Correlation between residual saturation ratio and capillary number (after [27]).

potential for mobilizing the residual oil saturation from conventional recovery processes is the target in improved hydrocarbon or enhanced oil recovery (IHR or EOR). Figure 7.20 shows the ratio of residual oil from chemical flood processes to that from waterflooding as a function of N_c for a range of laboratory tests [27]. The capillary number at which increased recovery starts can be used to represent mobilization. Figure 7.21 shows that when the mobilization is represented as a critical displacement ratio (units L^{-2}), then it is apparent that oil can be liberated more easily from higher permeability reservoirs. With transformation into radial coordinates it is possible to show that residual saturations in high velocity regions around a wellbore may be different from that some distance away. The critical displacement ratio for a given rate and permeability is shown in Fig. 7.22.

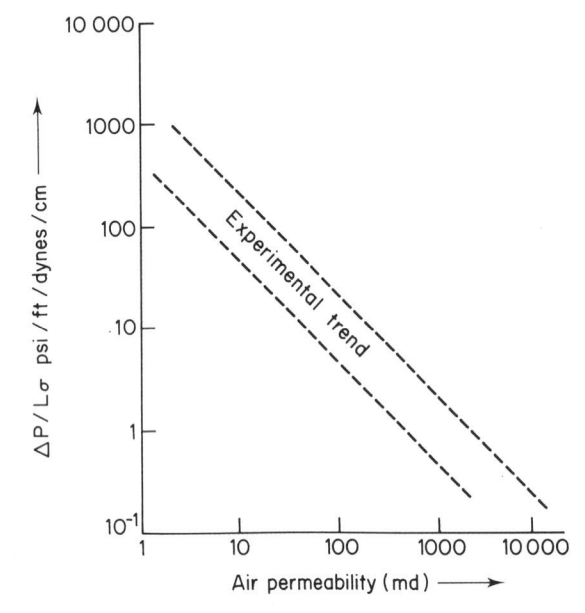

Fig. 7.21 Effect of permeability on critical displacement ratio (after [27]).

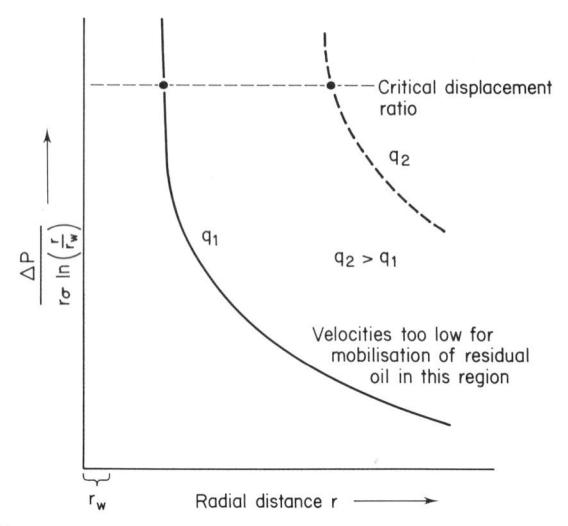

Fig. 7.22 Influence of near wellbore velocity on residual oil mobilization (after [27]).

7.7 *IN SITU* WETTABILITY CONTROL

All laboratory measurements on core samples are dependent on preservation of reservoir condition characteristics at the time of testing. It is therefore clear that any core cutting, core transportation, plug cutting, plug cleaning etc. processes which alter wettability will change effective phase permeability characteristics. Methods for restoration of *in situ* wettability conditions [45] have been proposed and are presently under scrutiny. Some restoration of wettability is claimed by conditioning cores at reservoir temperature in the presence of reservoir crude oil for some days or weeks. Both static and dynamic capillary pressure measurements can be used to demonstrate wettability change and contact angle modification. A comparison with true *in situ* wettability is, however, difficult to demonstrate.

7.8 RELATIVE PERMEABILITY FROM CORRELATIONS

In spite of the wide variety of pore structure in reservoir rocks, of preferential wettabilities between fluids and rock surfaces, and in fluid properties, normalized plots of relative permeability, $(K_o/K; K_g/K; K_w/K)$ against saturation, exhibit general similarities of form. It is, then, attractive to attempt to formulate theoretical semi-empirical, or purely empirical relationships, to assist in smoothing, extrapolating, extending (or even dispensing with) experimental measurements of effective permeability. This is particularly so since accurate, reliable, reproducible, experimental measurements are lengthy and troublesome, and the more rapid experimental techniques generally show poor reproducibility. The accuracy of approximate correlations may then be little worse than the present accuracy of the more usual measurements!

Idealized pore models have their greatest application in calculating relative permeabilities. The drainage case is conceptually the simplest, and several simple idealized flow models lead to acceptable smoothing relations. The imbibition case is more difficult to model, and gives generally less satisfactory results, but smoothing relations are not unacceptably inaccurate.

7 RELATIVE PERMEABILITY AND MULTIPHASE FLOW

7.8.1 Correlations of wetting phase permeabilities

1. Drainage case

$$K_{rw} = (S_w^*)^a$$

where $a = 4$ (Corey model), $a = 10/3$ (statistical model).

$$S_w^* = \frac{S_w - S_{wi}}{1 - S_{wi}}$$

where S_{wi} = irreducible wetting phase saturation. An alternative correlation is the Pirson model:

$$K_{rw} = S_w^3 \cdot (S_w^*)^{3/2}$$

2. Imbibition case

$$K_{rw} = (S_w^*)^4_{imb}$$

$$(S_w^*)_{imb} = (S_w^*)_{drainage} - \tfrac{1}{2}(S_w^*)^2_{drainage}$$

or

$$K_{rw} = S_w^4 \cdot (S_w^*)^{0.5}$$

7.8.2 Correlations for non-wetting phase relative permeability

1. Drainage case

$$K_{rnw} = (1 - S_w^*)^3 (1 + 2S_w^*)$$

or

$$K_{rwn} = (1 - S_w^*)(1 - (S_w^*)^{0.5} S_w^{0.5})^{0.5}$$

2. Imbibition case

$$K_{rnw} = \left\{ 1 - \frac{S_w - S_{wi}}{1 - S_{wi} - S_r} \right\}^2$$

where S_r = irreducible non-wetting phase saturation.

7.8.3 Use of correlations

With a limited amount of fairly readily determined experimental data – irreducible water saturation, residual non-wetting phase saturation, and the phase permeabilities at these saturations, these smoothing relations can be used (or modified) to generate complete relative permeability curves. Normalized end points can be adjusted using factors based on experience, i.e. $K_r = K_{rcalc} \times f$. The value of f might, for example, represent the ratio K_{ocw}/K_{abs} and be a function of permeability and/or pore geometry.

7.9 VALIDATION OF RELATIVE PERMEABILITY DATA FOR USE IN DISPLACEMENT CALCULATIONS

Laboratory derived relative permeability data reflect a number of characteristics which may be associated entirely with handling procedures and which may not be typical of *in situ* reservoir behaviour. These include:

(a) wettability change associated with coring fluids, storage, plug cutting and preparation;
(b) resaturation of plug and saturation distribution;
(c) test fluids, particularly with respect to viscosity and interfacial tension;
(d) test method, particularly with respect to steady state or unsteady state displacement mechanism and nature of the moving interface through the system with effect on distribution of residual fluid.

Recognition of these effects, together with selecting and use of relevant core plugs to represent zones of interest, can minimize the problem of applying relative permeability data in model studies.

Particular concerns in use of laboratory derived data arise from the scaling of microsize core plug displacement to reservoir simulation grid size displacement. Core plug experiments tend to be run at high viscous:capillary force ratios and the system represents a disperse or diffuse flow regime. Gravity forces are generally negligible in core plug tests. The direct application of core plug data in simulators where thick grid cells are modelled will result in improper representation of gravity forces – it can be shown that straight line relative permeability relationships are more representative of the gravity segregation of fluids. In simulators, straight line relative permeability curves do not lend themselves to stable numerical manipulation as the initial gradients to and from irreducible saturations are too steep. As a practical compromise some curvature is usually provided, as shown in Fig. 7.23.

There are no proven rules for obtaining valid relative permeability data for use in models. Laboratory experiments can explore the sensitivity of derived curves for rate effects and hysteresis in given test methods using core obtained and prepared in ways which minimize wettability alteration. The test can be conducted at reservoir conditions using reservoir (or simulated reservoir) fluids. The results of waterflood core tests can furthermore be processed using the same reservoir simulator as the reservoir model. This at least ensures that a given curve set will reproduce the fluid recovery and

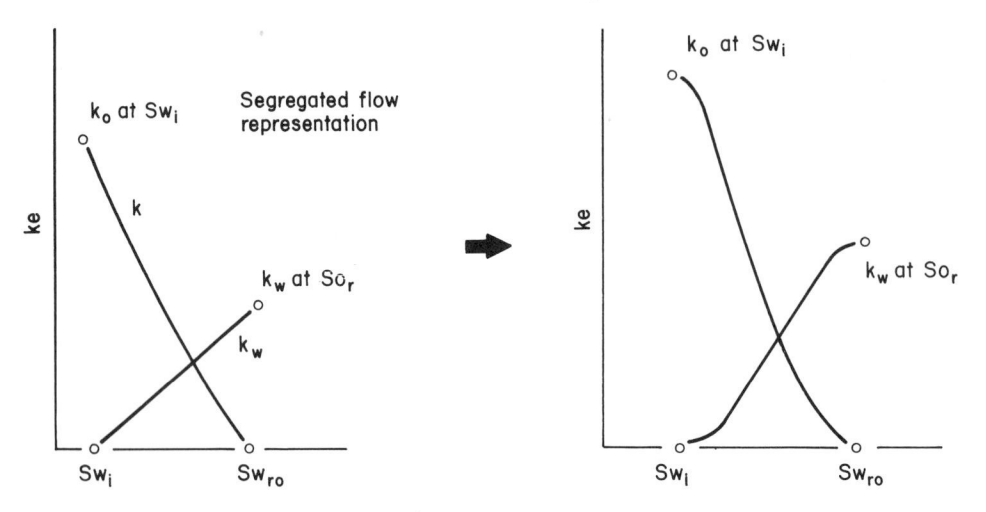

Fig. 7.23 Curvature at end points in use of straight line relative permeability curves.

pressure distribution measured in the laboratory test[30]. A flow diagram representing the use of a 1-D simulator for checking coreflood relative permeabilities is given in Fig. 7.24 and follows the Archer-Wong method.

Relative permeability in simulation models is used to determine transfer of fluid between grid cells at grid cell boundaries. In order to aid numerical stability, the relative permeability of the upstream cell block at the start of a time step is often used in effective transmissibility calculations, i.e. from Darcy law for one fluid in x-direction flowing across the boundary between Cell 1 and Cell 2 in Fig. 7.25.

$$q_o = \frac{k_{o1} A \left[\Phi'_1 - \Phi'_2\right]}{\mu_o \left\{\dfrac{Dx_1 + Dx_2}{2}\right\}}$$

The potential term (Φ') can include any capillary pressure contribution.

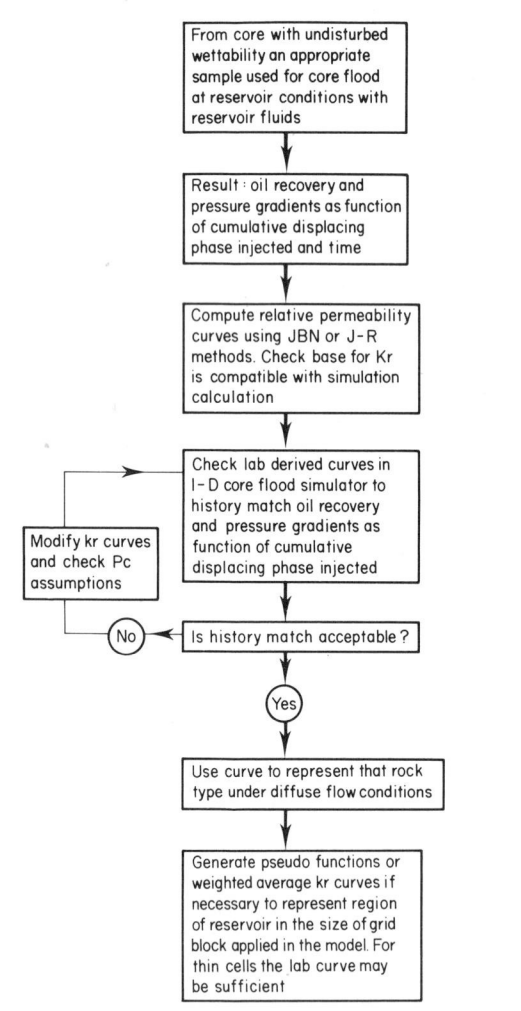

Fig. 7.24 Application of laboratory derived permeability in reservoir simulation (after [30]).

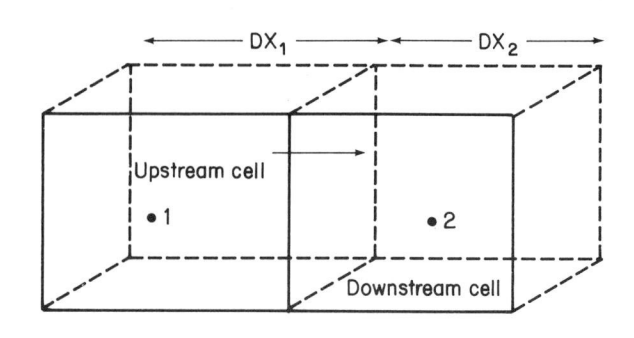

Fig. 7.25 Flow between cells.

7 RELATIVE PERMEABILITY AND MULTIPHASE FLOW

7.10 PSEUDO-RELATIVE PERMEABILITY IN DYNAMIC SYSTEMS

It is often a convenience in reservoir modelling to reduce the number of grid cells in a system in order to reduce model run costs. The flow behaviour of a reduced cell system may be matched with that of the full definition system by use of a set of *pseudo*-relative permeability curves in place of the original curves. The sole purpose of the curves is to reproduce the fluid and pressure distribution and displacement characteristics of the fine grid system in a coarse grid system. The current methodology for creating *pseudos* in a dynamic flow system is due to Kyte and Berry [58] and essentially determines the functions by summing flow rates from fine grid systems into the equivalent coarse grid and recalculating the effective permeability using Darcy's law (Fig. 7.26). The dynamic pseudo-relative permeabilities can be significantly different from calculated static pseudos or from modified pseudos obtained by history matching observed reservoir behaviour. In large field simulation, different pseudo-functions may be generated for different regions of the reservoir.

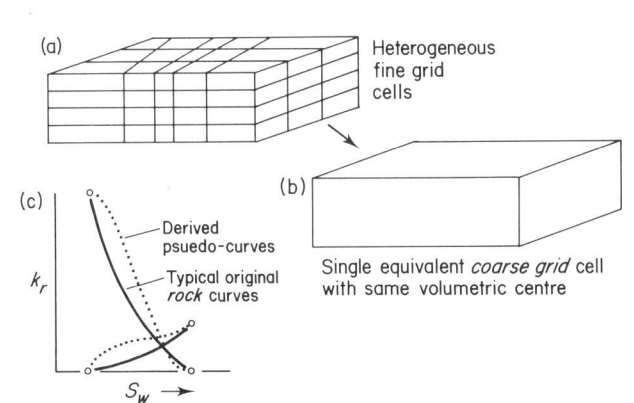

Fig. 7.26 Pseudo-relative permeability functions in coarse grid definition.

7.11 STATIC PSEUDO-RELATIVE PERMEABILITY FUNCTIONS

As a start point in many reservoir simulation problems, static pseudo-functions provide an insight into possible performance. The generation of a pseudo-curve is described for a reservoir unit with N thick layers, as shown in Fig. 7.27, in each of which segregated flow staight line relative permeability curves are assumed to apply.

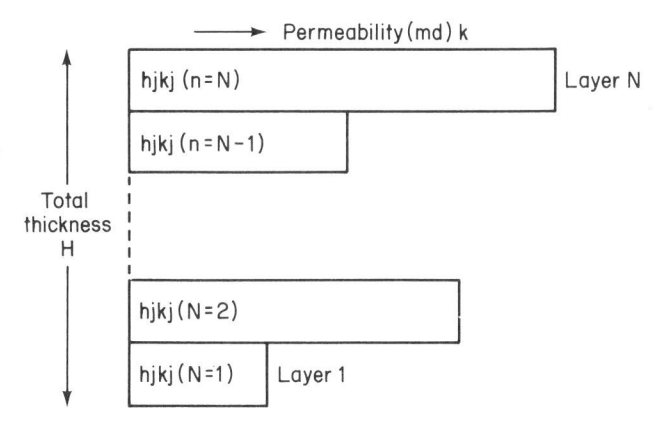

Fig. 7.27 Layer system representation for static pseudo-calculations.

The average saturation $\bar{S}_w$ as the reservoir unit approaches flood out to residual oil may be calculated assuming cross-flow through the vertical component of permeability. This is essentially an advance of bottom water, considering the position of the local oil–water contact from its initial position at the base of the bottom layer ($n = 0$) to its final position at the top of the top layer ($n = N$), as shown in Fig. 7.28.

The pseudo-functions generated will depend on position in the reservoir system and are clearly also dependent on ordering and thickness of the layers. For each condition of equilibrium oil–water contact from $n = 0$ to $n = N$ we can write

$$(\bar{S}_w)_n = \left\{ \frac{\sum_{1}^{n} h_j \phi_j (1 - S_{or_j}) + \sum_{n+1}^{N} h_j \phi_j S_{w_j}}{\sum_{1}^{N} h_j \phi_j} \right\}$$

$$(\bar{k}_{rw})_n \text{ at } (\bar{S}_w)_n = \frac{\sum_{1}^{n} h_j k_j k'_{rw_j}}{\sum_{1}^{N} h_j k_j}$$

$$(\bar{k}_{ro})_n \text{ at } (\bar{S}_w)_n = \frac{\sum_{n+1}^{N} h_j k_j k'_{ro_j}}{\sum_{1}^{N} h_j k_j}$$

The resulting pseudo-relative permeability curve is shown in Fig. 7.29, but actual shape will depend on layer ordering and reservoir character. These data can then be used in 1-D displacement calculations or in coarse simulator cells.

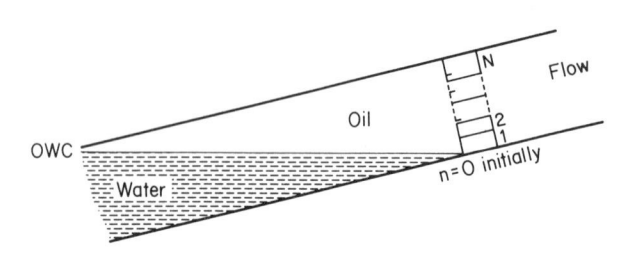

Fig. 7.28 Flow in layered system used for pseudo-calculation.

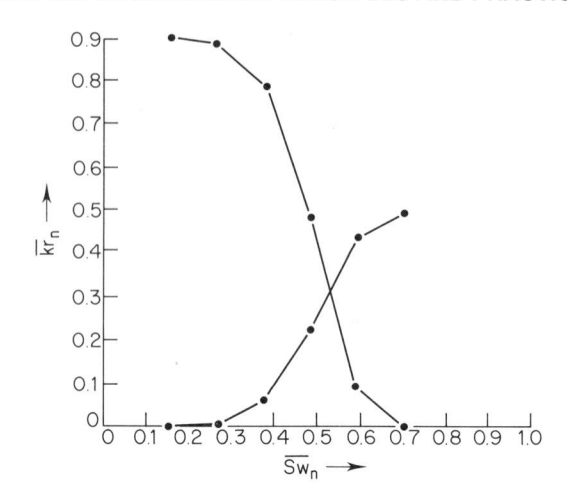

Fig. 7.29 Pseudo-curve character which might result from static calculations.

Examples

Example 7.1

The following laboratory data have been obtained from a steady state room temperature relative permeability test:

Air permeability	20 mD
Helium porosity	20%
Plug length	9 cm
Plug diameter	3.2 cm

Oil flow rate (cc/h)	Brine flow rate (cc/h)	Pressure drop psi (from transducer)	Core average water saturation (from weight change) (% PV)
90	0	49.25	15.02
75	5	91.29	19.80
60	9	109.52	25.10
45	20	123.30	32.15
30	34	137.05	40.98
15	85	164.30	54.92
0	122	147.00	68.10

The viscosity of the laboratory oil is 2 cP.
The viscosity of the laboratory brine is 1.1 cP.

Prepare the steady state relative permeability curves for this sample and comment on its characteristics.

Example 7.2

A linear horizontal sand reservoir of length 1 mile between a water injector and an oil producer is 1 mile wide and has a net thickness of 50 ft. The porosity and initial water saturation distribution are uniform and are 0.25 and 0.28, respectively. The reservoir pressure can be considered as 5000 psia and at this condition the oil formation volume factor is 1.2765 RB/STB. The oil production rate is constant prior to breakthrough

7 RELATIVE PERMEABILITY AND MULTIPHASE FLOW

at 10 000 STB/D and water injection is used to maintain reservoir pressure in the 'incompressible' system. Estimate the frontal saturation of the injection water prior to breakthrough and the time in years of water breakthrough at the production well. Further estimate the reservoir condition water cut and recovery factor one year after breakthrough assuming water injection continues at the initial rate. The end point mobility ratio has been estimated as 2.778.

The relative permeability data for the reservoir are given as follows:

S_w	k_{ro}	k_{rw}
0.28	0.90	0.00
0.30	0.80	0.02
0.35	0.427	0.05
0.45	0.25	0.17
0.55	0.10	0.38
0.60	0.03	0.52
0.65	0.00	0.70

Example 7.3

It is proposed to inject gas into an updip well of a linear geometry oil reservoir at a rate of 15×10^6 SCF/D. The reservoir is 8000 ft wide and 100 ft in net thickness and has a permeability of 800 mD. The oil has a reservoir condition viscosity of 1.8 cP; a density of 48 lb/ft^3 and a relative permeability in the presence of connate water of 0.9. The gas has a reservoir condition density of 17 lb/ft^3, a viscosity of 0.028 cP and a relative permeability in the presence of residual oil and connate water of 0.5. The reservoir dip is 10°. The gas formation volume factor is 7.5×10^{-4} RB/SCF.

Show by calculation whether you consider the gas injection stable. If the oil formation volume factor is 1.125 RB/STB what oil production rate in STB/D might be expected initially?

Example 7.4

The intitial saturation distribution and relative permeability data for a linear isolated sand reservoir subjected to water drive are as follows:

Distance from original water oil contact (ft)	S_w (% PV)	k_{rw}	k_{ro}
0	100	1.00	0.0
10	79	0.63	0.0
12	75	0.54	0.02
18	65	0.37	0.09
26	55	0.23	0.23
35	45	0.13	0.44
50	35	0.06	0.73
90	25	0.02	0.94
350	16	0.00	0.98

Calculate the fractional flow curve (f_w) for the water saturations between initial water saturation ($S_{wi} = 0.16$) and residual oil saturation ($S_{wro}) = 0.79$). Represent the initial saturation distribution graphically as a series of steps equivalent to the continuous distribution. Determine the position of the 0.79, 0.75 and 0.70 water saturations at the following times: 0.5, 1.0 and 2.0 years. The withdrawal rate from the reservoir zone is 9434 RB/D. Other relevant reservoir zone data are as follows:

Average reservoir thickness = 100 ft
Formation dip = 6°
Porosity = 21.5%
Water viscosity (res. cond.) = 0.83 cp
Reservoir oil specific gravity = 1.01

Average reservoir width = 8000 ft
Permeability = 276 mD
Oil viscosity (reservoir conditions) = 1.51 cp

Reservoir water sp. gravity = 1.05

Distance from the original *owc* to the first line of producers is 350 ft.

Determine the frontal saturation after six months production using the material balance expression:

$$\Sigma\, (s_w - s_{wi})\, \Delta x = \frac{5.615\, q_t \cdot t}{\phi A}$$

Example 7.5

Prepare the static pseudo-relative permeability curve for a five layer reservoir assuming bottom water advance. Each layer is 10 ft thick and the oil–water contact is initially at the base of the lower layer. From bottom to top the layer permeabilities are 50 mD, 500 mD, 1500 mD, 2000 mD and 2500 mD. For simplicity assume that in each layer the following properties apply:

Oil relative permeability at the initial water saturation of 15% is 0.9.

Water relative permeability at the residual oil saturation of 30% is 0.5.

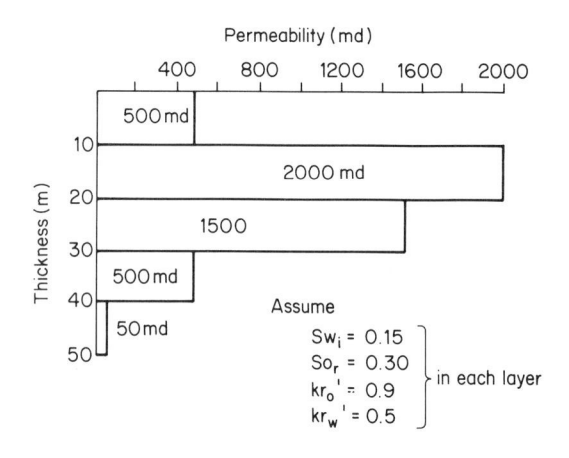

References

[1] Burdine, N.T.
Relative permeability calculations from pore size distribution data, *Trans. AIME* **198** (1953), 71.
[2] Corey, A.T.
The interrelation between gas and oil relative permeabilities, *Prod. Mon.* **19** (Nov. 1954), 34.
[3] Amott, E.
Observations relating to the wettability of porous rock, *Trans. AIME* **216** (1959), 156.
[4] Geffen, T.M., Owens, W.W., Parrish, D.R. and Morse, R.A.
Experimental investigation of factors affecting laboratory relative permeability measurements, *Trans. AIME* **192** (1951), 99.
[5] Morse, R.A., Terwilliger, P.L. and Yuster, S.T.
Relative permeability measurements on small core samples, *OGJ* (Aug 23, 1947).
[6] Osoba, J.S., Richardson, J.C., Kerver, J.K., Hafford, J.A. and Blair, P.M.
Laboratory measurements of relative permeability, *Trans. AIME* **192** (1951), 47.
[7] Jordan, J.K., McCardell, W.M. and Hocott, C.R.
Effect of rate on oil recovery by waterflooding, *OGJ* (May 13, 1957), 98.
[8] Buckley, S.E. and Leverett, M.C.
Mechanism of fluid displacement in sands, *Trans. AIME* **146** (1942), 107.

7 RELATIVE PERMEABILITY AND MULTIPHASE FLOW

[9] Cardwell, W.T.
The meaning of the triple value in non-capillary Buckley-Leverett theory, *Trans. AIME* **216** (1959), 271.

[10] Welge, H.J.
A simplified method for computing oil recovery by gas or water drive, *Trans. AIME* **195** (1952), 91.

[11] Kyte, J.R. and Rapoport, L.A.
Linear waterflood behaviour and end effects in water wet porous media, *Trans. AIME* **213** (1958), 423.

[12] Mungan, N.
Certain wettability effects in laboratory waterfloods, *J. Pet. Tech.* (Feb. 1966), 247.

[13] Johnson, E.F., Bossler, D.P. and Naumann, V.O.
Calculation of relative permeability from displacement experiments, *Trans. AIME* **216** (1959), 370.

[14] Colpitts, G.P. and Hunter, D.E.
Laboratory displacement of oil by water under simulated reservoir conditions, *J. Can. Pet. Tech.* **3** (2) (1964), 66.

[15] Kyte, J.R., Naumann, V.O. and Mattax, C.C.
Effect of reservoir environment on water–oil displacements, *J. Pet. Tech.* (June 1961), 579.

[16] Sandberg, C.R., Gournay, L.S. and Sippel, R.F.
The effect of fluid flow rate and viscosity on laboratory determinations of oil–water relative permeability, *Trans. AIME* **213** (1958), 36.

[17] Mungan, N.
Relative permeability measurements using reservoir fluids, *SPEJ* (Oct. 1972), 398.

[18] Salathiel, R.A.
Oil recovery by surface film drainage in mixed wettability rock, *JPT* (Oct. 1973), 1216.

[19] Rathmell, J.J., Braun, P.H. and Perkins, T.K.
Reservoir waterflood residual oil saturation from laboratory tests, *JPT* (Feb. 1973), 175.

[20] Jenks, L.H., Huppler, J.D., Morrow, M.R. and Salathiel, R.A.
Fluid flow within a porous medium near a diamond bit, Paper 6824, *Proc. 19th Ann. Mtg. Pet. Soc. CIM* (May 1968).

[21] Treiber, L.E., Archer, D.L. and Owens, W.W.
Laboratory evaluation of the wettability of fifty oil producing reservoirs, *SPEJ* (Dec. 1972), 531.

[22] Huppler, J.D.
Numerical investigation of the effects of core heterogeneities on waterflood relative permeabilities, *SPEJ* (Dec. 1970) 381.

[23] Braun, E.M. and Blackwell, R.J.
A steady state technique for measuring oil–water relative permeability curves at reservoir conditions, Paper SPE 10155, *Proc. 56th Ann. Fall Mtg.* (1981).

[24] Jones, S.C. and Rozelle, W.O.
Graphical techniques for determining relative permeability from displacement experiments, *SPEJ* (May 1978), 807.

[25] Craig, F.F.
The Reservoir Engineering Aspects of Waterflooding, SPE Monograph No. 3 (1971).

[26] Brown, C.E. and Langley, G.O.
The provision of laboratory data for EOR simulation, *Proc. Europ. Symp. on EOR* (1981), 81.

[27] Taber, J.J.
Research on EOR – past, present and future, In *Surface Phenomena in EOR* (Shah, D.O. ed.), Plenum Pr. (N.Y.), (1981), 13.

[28] Larson, R.G., Davis, H.T. and Scriven, L.E.
Elementary mechanisms of oil recovery by chemical methods, *JPT* (Feb. 1982), 243.

[29] Hvolboll, V.T.
Methods for accurately measuring produced oil volumes during laboratory waterflood tests at reservoir conditions, *SPEJ* (Aug. 1978), 239.

[30] Archer, J.S. and Wong, S.W.
Use of a reservoir simulator to interpret laboratory waterflood data, *SPEJ* (Dec. 1973), 343.

[31] Tao, T.M. and Watson, A.T.
Accuracy of JBN estimates of relative permeability, *SPEJ* (April 1984), 209.

[32] Sigmund, P.M. and McCaffery, F.G.
An improved unsteady state procedure for determining the relative permeability characteristics of heterogeneous porous media, *SPEJ* (Feb. 1979), 15.

[33] Collins, R.E.
Flow of Fluids Through Porous Materials, Penwell, Tulsa (1976).

[34] Brooks, R.H. and Corey, A.T.
Properties of porous media affecting fluid flow, *J. Irrig. Drain. Div., Proc. ASCE* (1966), Vol. 92, Ir. 2, 61.

[35] Torabzadeh, S.J. and Handy, L.L.
The effect of temperature and interfacial tension on water–oil relative permeabilities of consolidated sands, SPE/DOE 12689, *Proc. 4th Symp. EOR,* Tulsa (April 1984), 105.

[36] Heiba, A.A., Davis, H.T. and Scriven, L.E.
Statistical network theory of three phase relative permeabilities, SPE/DOE, 12690, *Proc. 4th Symp. EOR,* Tulsa (April 1984), 121.

[37] Ramakrishnan, T.S. and Wasan, D.T.
The relative permeability function for two phase flow in porous media – effect of capillary number, SPE/ DOE 12693, *Proc. 4th Symp. EOR,* Tulsa (April 1984), 163.

[38] Manjnath, A. and Honarpour, M.M.
An investigation of three phase relative permeability, SPE 12915, *Proc. Rocky Mt. Reg. Mtg.* (May 1984), 205.

[39] Corey, A.T., Rathjens, C.H., Henderson, J.H. and Wylie, M.
Three phase relative permeability, *Trans. AIME* **207** (1956), 349.

[40] Stone, H.L.
Estimation of three phase relative permeability and residual oil data, *J. Can. Pet. Tech.* (Oct.–Dec. 1973),53.

[41] Land, C.S.
Calculation of imbibition relative permeability for two and three phase flow, from rock properties, *SPEJ* (June 1968), 149.

[42] Dietrich, J.K. and Bondor, P.B.
Three phase oil relative permeability models, SPE 6044, *Proc. 51st Ann. Fall Mtg. SPE* (Oct. 1976).

[43] Carlson, F.M.
Simulation of relative permeability hysteresis to the non wetting phase, SPE 10157, *Proc. 56th Ann. Fall Mtg. SPE* (Oct. 1981).

[44] Killough, J.E.
Reservoir simulation with history dependent saturation functions, *SPEJ* (Feb. 1976), 37.

[45] Cuiec, L.E.
Restoration of the natural state of core samples, SPE 5634, *Proc. Ann. Fall Mtg. SPE* (Oct. 1975).

[46] Hearn, C.L.
Simulation of stratified waterflooding by pseudo relative permeability curves, *J. Pet. Tech.* (July 1971), 805.

[47] Berruin, N.A. and Morse, R.A.
Waterflood performance of heterogeneous systems, *JPT* (July 1979), 829.

[48] Coats, K.H., Dempsey, J.R. and Henderson, J.H.
The use of vertical equilibrium in two dimensional simulation of three dimensional reservoir performance, *SPEJ* (March 1971), 63.

[49] Dietz, D.N.
A theoretical approach to the problem of encroaching and by-passing edge water, *Proc. Koninkl Akad. van Wetenschappen,* (1953) Amst. V. 56–B, 83.

[50] Davis, L.A.
VHF electrical measurement of saturations in laboratory floods, SPE 8847, *Proc. 1st Jnt SPE/DoE Symp. EOR,* Tulsa (April 1980), 397.

[51] Fayers, F.J. and Sheldon, J.W.
The effect of capillary pressure and gravity on two phase flow in a porous medium, *Trans. AIME* **216** (1959), 147.

[52] Dyes, A.B., Caudle, B.H. and Erickson, R.A.
Oil production after breakthrough – as influenced by mobility ratio, *Trans. AIME* **201** (1954), 81.

[53] Dykstra, H. and Parsons, R.L.
The prediction of oil recovery by waterflood, *Sec. Rec. of oil in the U.S.,* 2nd ed. API (1950), 160.

[54] Johnson, C.E.
Predicition of oil recovery by waterflood – a simplified graphical treatment of the Dykstra-Parsons method, *Trans. AIME* **207** (1956), 345.

[55] Stiles, W.E.
Use of permeability distribution in waterflood calculations, *Trans. AIME* **186** (1949), 9.

[56] Ashford, F.E.
Computed relative permeability, drainage and imbibition, SPE Paper 2582, *Proc. 44th Ann. Fall Mtg.* (1969).

[57] Evrenos, A.I. and Comer, A.G.
Sensitivity studies of gas–water relative permeability and capillarity in reservoir modelling, *SPE 2608* (1969).

[58] Kyte, J.R. and Berry, D.W.
New pseudo functions to control numerical dispersion, *SPEJ* (Aug. 1975), 269.

[59] Dake, L.P.
Fundamentals of reservoir engineering, *Elsevier Dev. Pet. Sci* (8), (1978).

[60] Jacks, H.H., Smith, O.J. and Mattax, C.C.
The modelling of a three dimensional reservoir with a two dimensional reservoir simulator – the use of dynamic pseudo functions, *SPEJ* (June 1973), 175.

[61] Woods, E.G. and Khurana, A.K.
Pseudo functions for water coning in a three dimensional reservoir simulator, *SPEJ* **17** (1977), 251.

[62] Chappelear, J.E. and Hirasaki, G.J.
A model of oil–water coning for 2-D areal reservoir simulation, *SPEJ* (1976), 65.

[63] Koval, E.J.
A method for predicting the performance of unstable miscible displacement in heterogeneous media, *SPEJ* (June 1966), 145.

[64] Handy, L.L. and Datta, P.
Fluid distribution during immiscible displacements in porous media, *SPEJ* (Sept. 1966), 261.

[65] Higgins, R.V., Boley, D.W. and Leighton, A.J.
Unique properties of permeability curves of concern to reservoir engineers, *U.S. Bur. Mines Rept. Investig.*, RI7006 (1967).

[66] Hagoort, J.
Oil recovery by gravity drainage, *SPEJ* (June 1980), 139.

[67] Bragg, J.R. *et al.*
A comparison of several techniques for measuring residual oil saturation, SPE 7074, *Proc. Symp. Impr. Oil Rec.*, Tulsa (April 1978), 375.

[68] Deans, H.A.
Using chemical tracers to measure fractional flow and saturation *in situ*, SPE 7076, *Proc. Symp. Impr. Oil Rec.*, Tulsa (April 1978), 399.

[69] Hove, A.O., Ringen, J.K. and Read, P.A.
Visualisation of laboratory core floods with the aid of computerised tomography of X-rays, SPE 13654, *Proc. Clif. Reg. Mtg. SPE* (March 1985).

[70] McCaffery, F.G. and Bennion, D.W.
The effect of wettability on two phase relative permeability, *J. Can. Pet. Tech.* (Oct./Dec. 1974), 42.

[71] Singhal, A.K., Mekjerjee, D.P. and Somerton, W.H.
Effect of heterogeneities on flow of fluids through porous media, *J. Can. Pet. Tech.* (July–Sept. 1976), 63.

[72] Fulcher, R.A., Ertekin, T. and Stahl, C.D.
Effect of capillary number and its constituents on two phase relative permeability curves, *JPT* (Feb. 1985), 249.

[73] Melrose, J.C. *et al.*
Water–rock interactions in the Pembina field, Alberta, SPE 6049, *Proc. 51st Ann. Fall Mtg.* (1976).

[74] Archer, J.S.
Some aspects of reservoir description for reservoir modelling. *Proc. Intl. Seminar North Sea Oil and Gas Reservoirs*, Trondheim (December 1985).

[75] Slider, H.C.
Worldwide Practical Petroleum Reservoir Engineering Methods, Pennwell Books, Tulsa (1983).

[76] Rapoport, L.A. and Leas, W.J.
Properties of linear waterfloods, *Trans. AIME* **198** (1953), 139.

Chapter 8

Representation of Volumetric Estimates and Recoverable Reserves

8.1 IN-PLACE VOLUME

The volume of hydrocarbon in place in a reservoir depends on:

(a) the areal extent of the hydrocarbon region of the reservoir;
(b) the thickness of reservoir quality porous rock in the hydrocarbon region;
(c) the porosity of reservoir quality porous rock in the hydrocarbon region;
(d) the saturation of hydrocarbon in the hydrocarbon region.

This is represented in terms of average properties as follows

$$V = \bar{A}\,\bar{h}_N\,\bar{\phi}\,(1 - \bar{S}_w)$$

where $\bar{A}$ = area (avg), $\bar{h}_N$ = net thickness (gross thickness × net thickness/gross thickness) (avg), $\bar{\phi}$ = porosity (avg), $\bar{S}_w$ = water saturation (avg), V = reservoir condition volume of hydrocarbon.

At standard conditions, the volume of hydrocarbon in place is the reservoir condition volume divided by the formation volume factor. Each of the components of the volumetric equation is subject to uncertainty and spatial variation. We shall now examine the source and representation of these data and develop a probabilistic approach to volumetric estimation.

8.2 AREAL EXTENT OF RESERVOIRS

The areal extent of reservoirs are defined with some degree of uncertainty by evidence from drilled wells combined with geophysical interpretation of seismic data. The amount of well control has the main influence on the mapping and representation of reservoir structure. Maps tend to represent time stratigraphy as depositional units. Since reservoir fluids are contained in, and recovered from, permeable beds, the combination of permeable elements of a number of time stratigraphic units leads to mappable rock stratigraphic units having (a) areal extent, (b) thickness, (c) petrophysical properties.

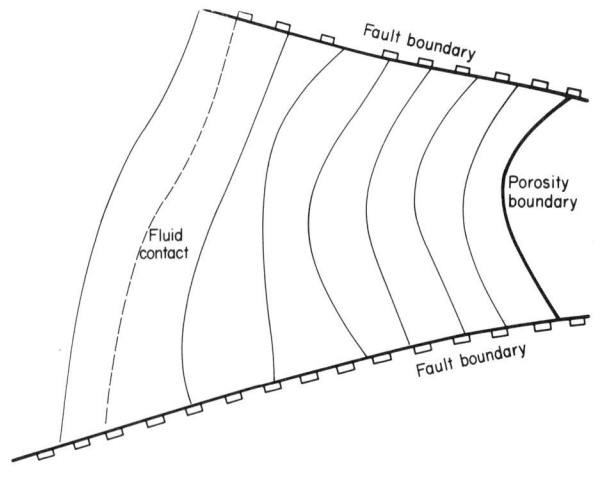

Fig. 8.1 Example of a structure contour map on top porosity.

8 VOLUMETRIC ESTIMATES AND RECOVERABLE RESERVES

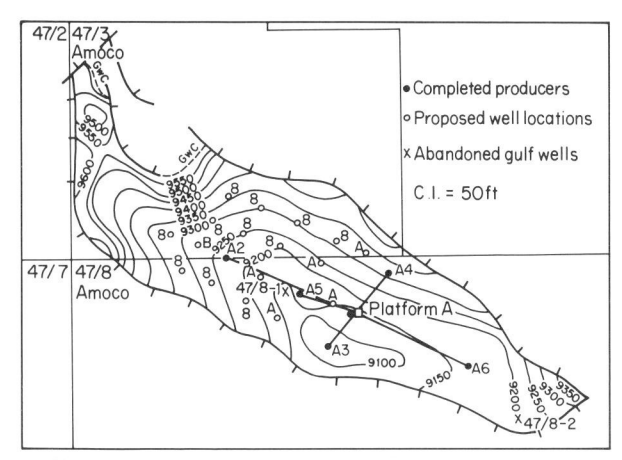

Fig. 8.3 Top sand structure map, Rough gas field[15].

Structure contour maps are used to connect points of equal elevation. It is customary to map structure at the top and base of porosity and the map indicates the external geometry of the reservoir. The map shown in Fig. 8.1 shows the boundaries as a fluid contact, a porosity limit to reservoir quality rock and fault boundaries. The contour intervals are regular and represent subsea depths. The map would be labelled *top sand, base sand* or refer to a geological age boundary. The difference in elevation between the hydrocarbon–water contact and the top of the structure is known as the closure or height of the hydrocarbon column. Structure maps of the top sands of Thistle oil reservoir[16] and the Rough gas field[15] are shown in Figs 8.2 and 8.3. A schematic cross-section of the Rough field is illustrated in Fig. 8.4[15].

The area contained within each structural contour can be measured by various mathematical techni-

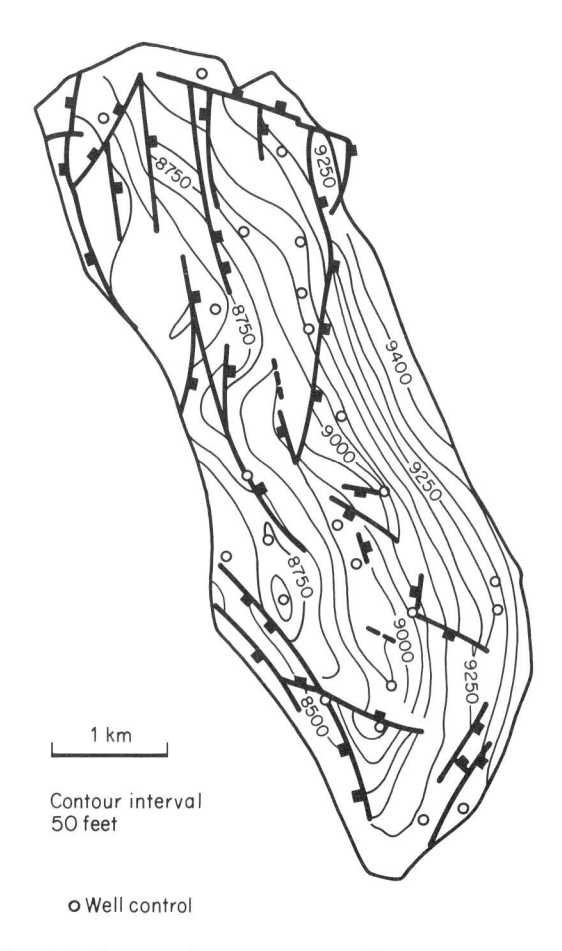

Fig. 8.2 Top sand structure map, Thistle oil reservoir. (after [16])

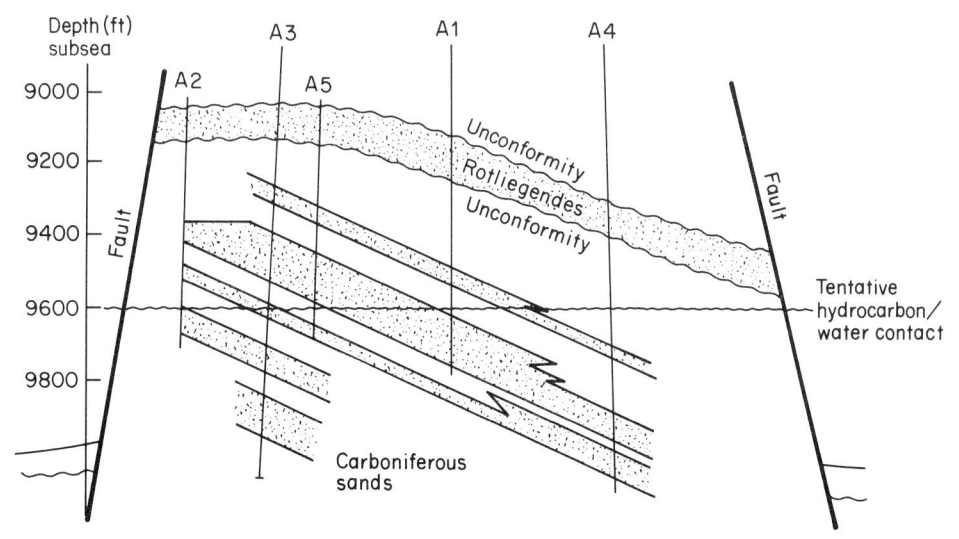

Fig. 8.4 A schematic cross-section of the Rough field. (after[15])

ques but is most usually performed by a digitizing process or with the aid of a calibrated device called a planimeter. A plot of the area contained within each contour against the contour elevation for the top and base structure maps establishes the basis for calculation of rock volume, as shown in Fig. 8.5

The area contained between the top and base structure and limited by the fluid contact can be measured from the plot by planimeter and is equivalent to the gross rock volume of the hydrocarbon interval. The validity of the maps, the scale and the accuracy of planimetering all influence the numerical value of the rock volume obtained.

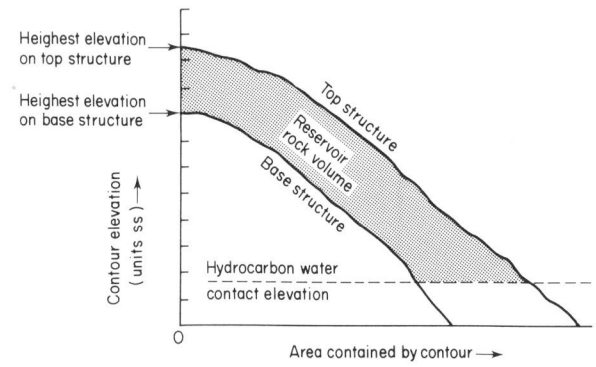

Fig. 8.5 Rock volume estimation.

8.3 THICKNESS MAPS

Contours of equi-thickness points in a reservoir can be drawn and are again influenced by geological and geophysical interpretation and well control. Where thickness represents formation thickness normal to the plane of the reservoir (true bed thickness) the contours are called *isopachs*. Often the thickness of a bed is mapped as a vertical thickness and such a contour should strictly be known as an *isochore*. A thickness mapped between top and base porosity and including non-reservoir material is known as *gross reservoir isopach*. If the impermeable beds are excluded it becomes a *net reservoir isopach*. If the thickness is measured from a zero datum of the hydrocarbon–water contact, the map would be either a net or gross pay isopach.

Figure 8.6 shows a net pay thickness isopach map and the zero contour indicates the fluid contact. Sand thickness is increasing downdip in this representation, and the sand pinches out at the top of the structure with a change in lithology to non-reservoir material. The definition of reservoir quality and non-reservoir quality material is a petrophysical definition based on core analysis, log analysis and

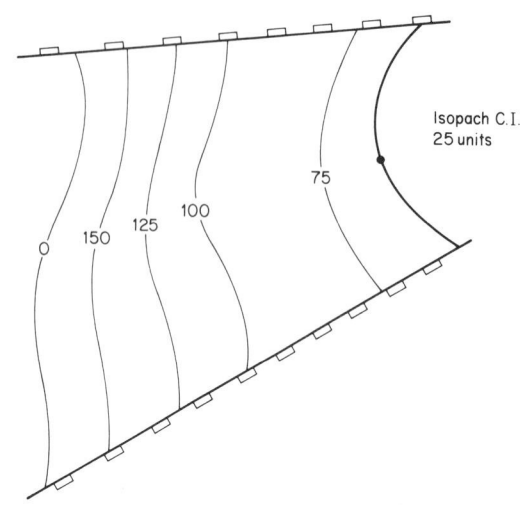

Fig. 8.6 Net pay thickness isopach.

well test interpretation. It is customary to use a porosity *cut off* in log analysis equivalent to some minimum permeability on a porosity permeability cross-plot (i.e. 8% porosity at 0.1 mD permeability). In addition, V_{clay} and water saturation are often included as additional delimiters, i.e. if $V_{clay} > 40\%$ and $S_w > 60\%$ then the rock might in a particular circumstance be considered non-reservoir quality. The net: gross ratio defines the thickness of reservoir quality rock to total thickness in a given unit. Isopach maps of sand thickness in the Rough field and the Murchison field are shown in Figs 8.7 and 8.8.

The area contained within a given isopach can be plotted against isopach value, as shown in Fig. 8.9. The area enclosed by this plot represents the net rock volume and can be obtained by planimetry.

When a reservoir is composed of a number of different rock types or sand units, and they are mapped separately, the thickness maps are known for each rock type as *isoliths*.

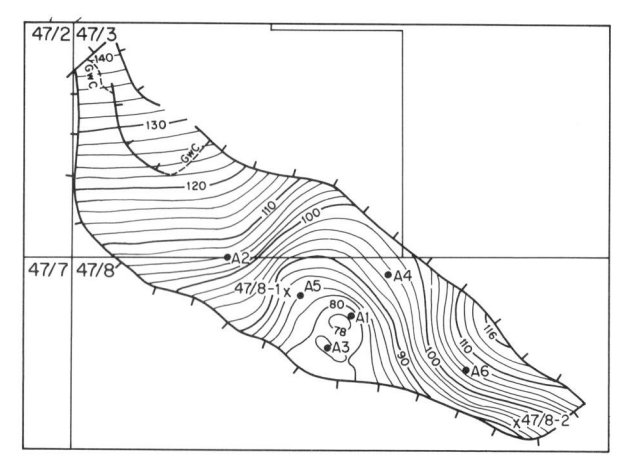

Fig. 8.7 Rough field isopach map (after [15]).

8 VOLUMETRIC ESTIMATES AND RECOVERABLE RESERVES

8.4 LITHOFACIES REPRESENTATION

Rock lithology can provide a guide in contouring, and lithofacies maps are generally presented as ratio maps, typically sand:shale or limestone:anhydrite (Fig. 8.10). The ratios follow geometric progressions (1:1, 1:2, 1:4, 1:8 etc.) and equal contour spacings do not therefore represent equal changes in lithology.

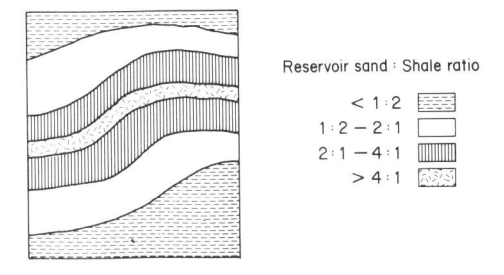

Fig. 8.10 Lithofacies mapping.

8.5 ISOPOROSITY MAPS

In a given reservoir zone or subzone, the areal variation in mean porosity may be represented. The porosity control values are thickness weighted average porosities for the zone at each well.

The shape of the isoporosity map shown in Fig. 8.11 may be obtained by application of geological modelling, by statistical techniques such as Krige mapping [17-19] or by computer controlled contour mapping! An example of an isoporosity map of the Rough gas field is shown in Fig. 8.12.

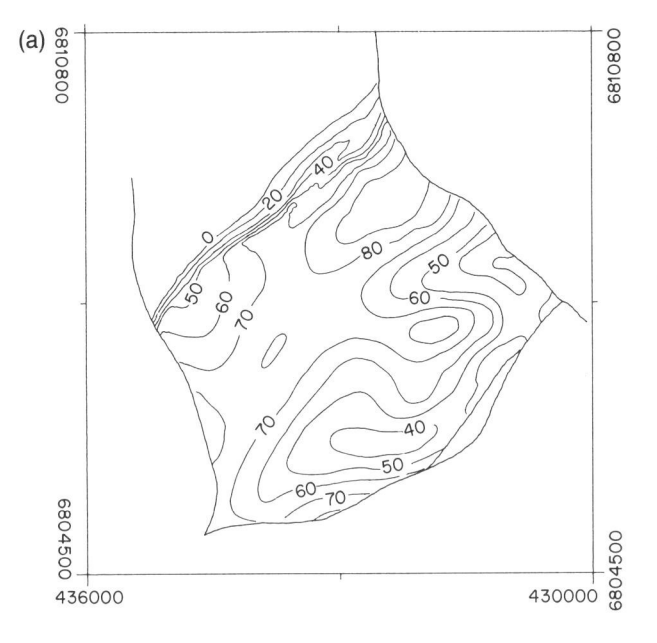

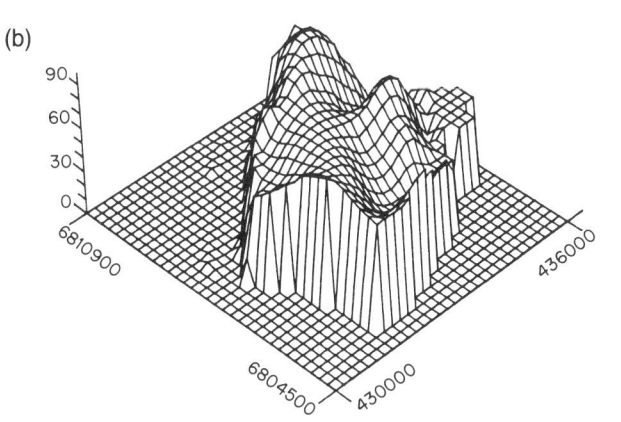

Fig. 8.8 Isopach map (a) and mesh perspective diagram (b) of Brent Sands, Murchison reservoir.

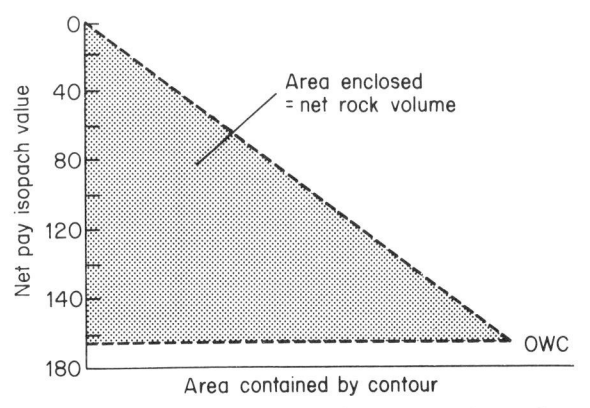

Fig. 8.9 Hydrocarbon volume from net pay isopach.

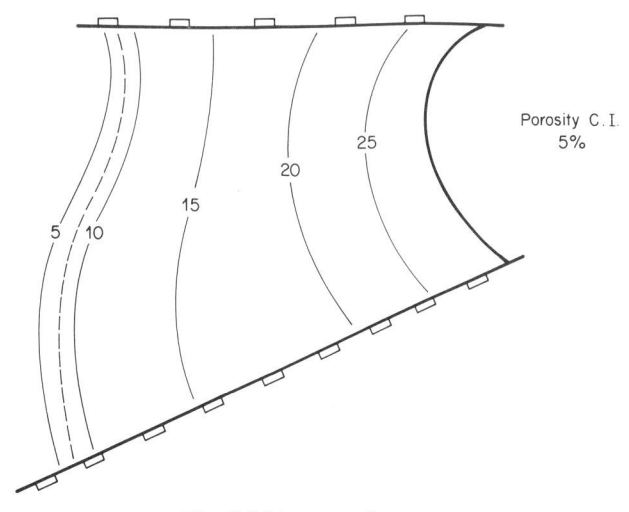

Fig. 8.11 Isoporosity map.

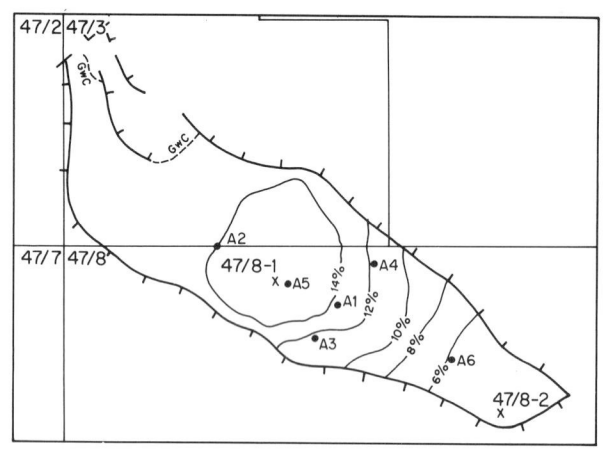

Fig. 8.12 Rough field porosity map — average gas saturation 63% (after [15]).

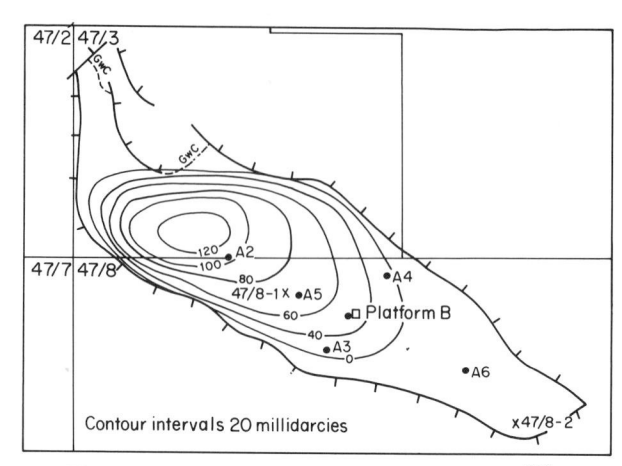

Fig. 8.14 Rough field permeability map (after [15]).

8.6 ISOCAPACITY MAPS

Isocapacity is used to denote equal values of a permeability-net thickness product, which is significant in appreciating well production capability. The product is frequently mapped instead of permeability, as permeability for a particular zone is sought as a functional relationship with porosity ($k = \text{fn}(\phi)$). Figure 8.13 shows an isocapacity map in which the absolute permeability has been obtained as an arithmetic average in the zone, i.e. at a well

$$(\bar{k}h) = h_N \frac{\sum k_i h_i}{\sum h_i}$$

The data, when modified for effective permeability can be validated by well testing. A permeability map of the Rough gas field is shown in Fig. 8.14.

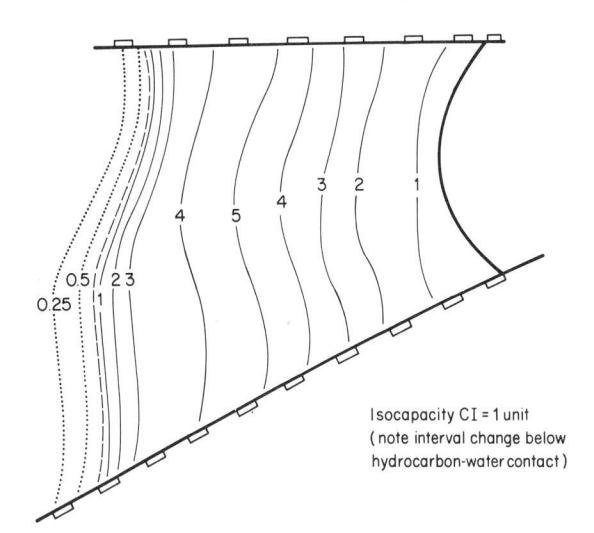

Fig. 8.13 Isocapacity map.

8.7 HYDROCARBON PORE VOLUME MAPS

As has been derived previously, an estimate of hydrocarbon pore volume can be obtained by combining the net rock volume from isopach mapping with a mean porosity $\bar{\phi}$ and a mean hydrocarbon saturation $(1 - \bar{S}_{wi})$. The porosity should be a volume weighted average, i.e.

$$\bar{\phi} = \frac{\sum \phi_j A_j h_j}{\sum A_j h_j}$$

and is obtained by dividing the reservoir into regions of constant porosity and measuring the bounding areas and mean thickness.

The isosaturation map may be derived by noting the relationship in a given rock type between irreducible water saturation and porosity. The porosities and saturations in a hydrocarbon zone are generally considered interdependent. Contouring of any isosaturation map must respect capillary transition zone characteristics and, as a consequence, *regional average* values are often of more practical use than a contoured map. Saturation should be represented as a pore volume weighted average:

$$\bar{S}_w = \frac{\sum (S_w)_j \phi_j A_j h_j}{\sum \phi_j A_j h_j}$$

An alternative to this approach is the direct mapping of hydrocarbon pore thickness (HPT) at each well control point.

HPT is characterized at a well in a given zone as $(\bar{\phi} \cdot h_N \cdot \bar{S}_h)$ where $\bar{S}_h = 1 - \bar{S}_w$. The average porosity $\bar{\phi}$ in the net thickness interval h_N is likely to be an arithmetic average, as indicated from a porosity histogram. At the hydrocarbon–water contact $\bar{S}_h$ is zero, and at the porous rock limit $\bar{\phi}$ is zero. Figure 8.15 shows the form of an HPT map. A map of HPT in the Rough gas field is indicated in Fig. 8.16. The shape of the contours might be suggested by geological controls such as structure, depositional environment and/or diagenetic modifications.

The planimetering, or any alternative way of measuring area, of the areas represented by each HPT value results in data for a plot of area against HPT, as shown in Fig. 8.17. The area under the curve generated represents the hydrocarbon pore volume of the reservoir unit.

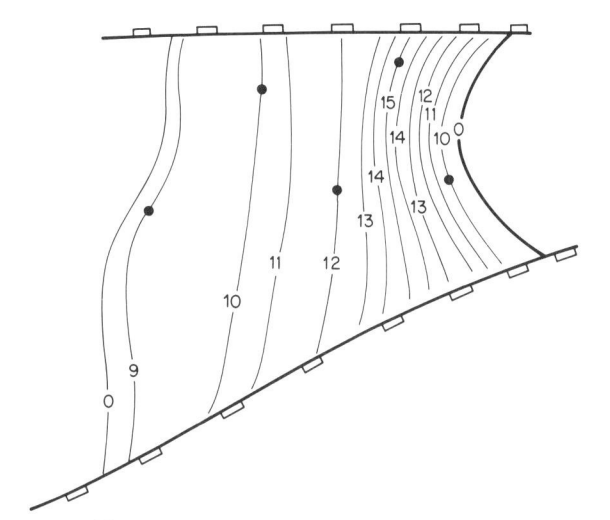

Fig. 8.15 Hydrocarbon pore thickness map.

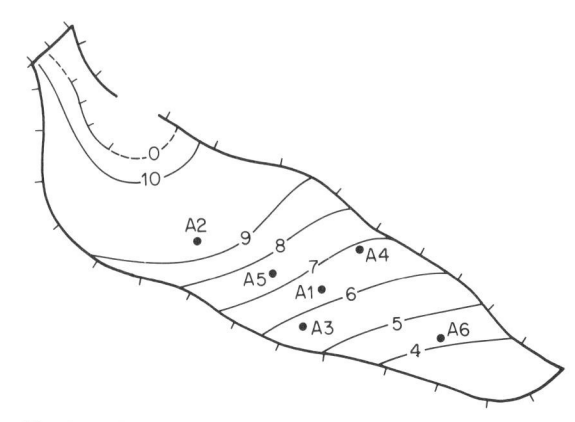

Fig. 8.16 Rough field hydrocarbon pore thickness map (after [15]).

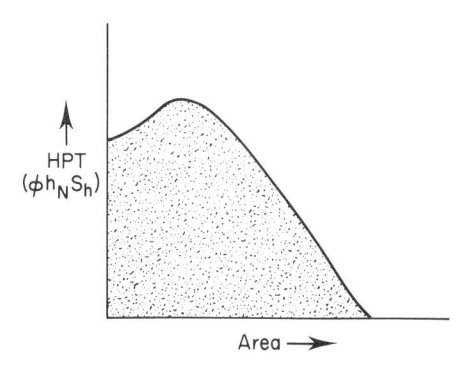

Fig. 8.17 Hydrocarbon pore volume from HPT maps (area under curve = net hydrocarbon pore volume).

8.8 PROBABILISTIC ESTIMATION

It should be apparent that there are a number of ways to map any given set of data. A numerical value of hydrocarbon pore volume thus represents one outcome of a given map combination. The presentation of hydrocarbon in place as a probabilistic estimate rather than as a deterministic value seeks to show the uncertainty of the estimate. The association of ranges and distribution shape with each of the components $A, (\phi S_h), h, N/G$ of the volumetric equation is a subjective technical exercise. One method, which has found widespread use in arriving at a probabilistic estimate, is the Monte Carlo approach [10]. Since standard condition (V_{sc}) volumes are usually needed, the reservoir volume of hydrocarbon in place is converted to a standard volume by use of an initial formation volume factor (B_{hi}) for the particular hydrocarbon, and its uncertainty is included in the estimation:

$$V_{sc} = \frac{HCPV}{B_{hi}}$$

where $HCPV$ represents reservoir condition hydrocarbon pore volume.

The shape of distributions and maximum and minimum values are generally agreed in specialist group committees, and rectangular and triangular distributions (Fig. 8.18 (a) and (b)) based on subjective assessment of reservoir characteristics are most frequently used. The distribution can easily be converted into a cumulative frequency curve which can be sampled at random. The repeated random selection of values and their probabilities from each independent variable set leads to the calculation of a large number of volumetric estimates (Fig. 8.19). The cumulative frequency of these estimates is used to show the likelihood that a given value will be at least as great as that shown, as indicated in Fig. 8.18 (c). The independent variables considered in volumetric and reserve calculations are A, h_G, N/G, (ϕS_h), B_{hi} and RF (i.e. area, gross thickness, net − gross ratio, hydrocarbon porosity, formation volume factor and recovery factor).

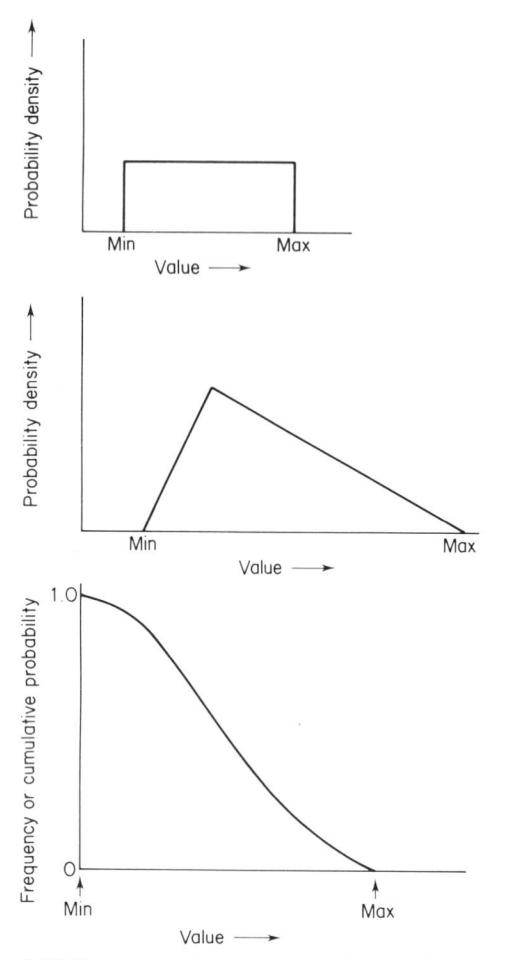

Fig. 8.18 Representation of independent variables. (a) Rectangular distribution (no preference), (b) triangular distribution (strong preference), (c) resultant distribution from multiple calculations or from multiple random sampling of distributions.

The values on the cumulative frequency graph (Figs. 8.18 and 8.21) are used to represent concepts of certainty at given levels. In common usage[1], the 90% level is known as a *proven* value, the 50% level is known as a *proven + probable* value and the 10% level is known as a *proven + probable + possible* value. Prior to drilling a well which indicates the presence of hydrocarbon, there is no *proven* value. It is also important to realize that the distribution of values can change with time as more information about a reservoir becomes available. Figure 8.20 indicates the change in volumetric estimate from pre-drilling through discovery and appraisal to early production and finally to late time depletion.

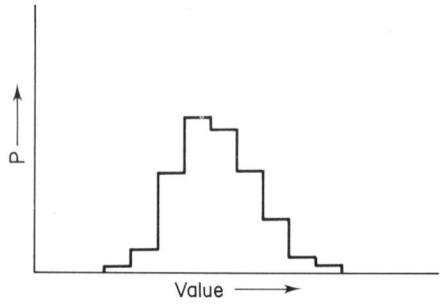

Fig. 8.19 Representation of volume in place calculations.

An alternative method attributed to Van der Laan [14] is suited to desk calculation and makes use of a number of values of each variable − 3 or 5 values to represent the probability distribution, e.g.

(1) a very optimistic value (say 0.2 chance of yielding a low value);
(2) an optimistic value (say 0.3 chance of yielding a low value);
(3) most likely value (say 0.5 chance of yielding a low value);
(4) a pessimistic value (say 0.8 chance of yielding a low value);
(5) a very pessimistic value (say 0.9 chance of yielding a low value).

Each of the five values has a probability of 1 in 5 or 20%.

All combinations of two parameters (say h and ϕ) are computed (giving 25 products), and these are reduced to five by averaging successive groups of five. The next parameter is then combined to yield a new suite of 25 values, which are again reduced to five. The process is repeated until all parameters are introduced, and a final range of five values, ranging from very pessimistic through the most likely to a very optimistic value, are then obtained.

These methods may give a clearer idea of the possible spread of results and divert attention from the potentially misleading *best estimate* value to other possibilities. When the distribution is symmetrical the *expectation value* of the field can be found at the 50% cumulative probability value.

8.9 RECOVERY FACTORS AND RESERVES

The recoverable volume of hydrocarbon from a particular reservoir will depend on reservoir rock

8 VOLUMETRIC ESTIMATES AND RECOVERABLE RESERVES

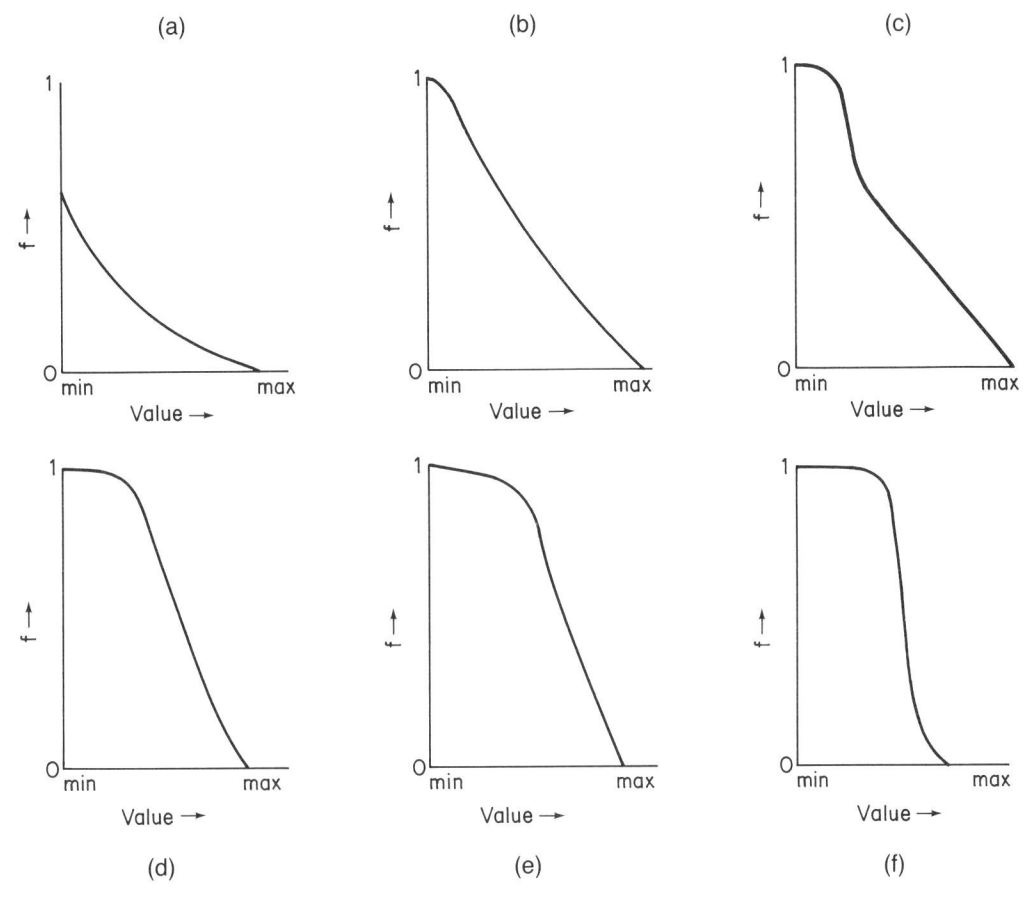

Fig. 8.20 Time (and data) variation of probabilistic estimates. (a) Pre-drilling, (b) discovery, (c) appraisal, (d) delineation/early production, (e) mature production, (f) late time depletion.

and fluid properties and continuity, as well as on economic conditions. The heterogeneity of the reservoir pore space will influence pressure gradients during dynamic displacement of hydrocarbon and leads to regions of poorer recovery than would be predicted by laboratory measured residual saturations. The fraction of original hydrocarbon in place that will be recovered, all volumes being represented at a standard condition, is known as a recovery factor. The ultimate recovery factor refers to the change in saturation of hydrocarbon from initial ($=1-S_{wi}$) to residual conditions (S_{hr}) in a completely contacted region and implies a particular recovery mechanism (i.e. waterflood, chemical flood etc.). It is not achievable throughout the reservoir.

$$RF\Big]_{ULT} = \frac{1 - S_{wi} - S_{hr}}{1 - S_{wi}}$$

The recovery factor at any stage of reservoir depletion is represented without spatial distribution in terms of the cumulative recovered hydrocarbon, $\sum q_j t_j$

$$RF = \frac{B_{hi} \sum q_j t_j}{HCPV}$$

where q_j is an interval standard condition volumetric production, t_j is an interval time, B_{hi} is a hydrocarbon initial formation volume factor.

For oil, where N is the stock tank oil in place, N_p is the cumulative stock tank oil production then, at some time t and pressure P,

$$RF = \frac{N_p}{N}\Big]_{P,t}$$

For gas, where G is the standard condition gas in place, G_p is the cumulative standard condition gas production, then, at some time t and pressure P

$$RF = \frac{G_p}{G}\Big]_{P,t}$$

Recovery factors representing a given development concept can be calculated using the techniques and methods of reservoir dynamics and represented probabilistically as a distribution. This allows a recoverable reserve estimation to be performed using a Monte Carlo technique and for the results to be represented probabilistically. For each randomly selected parameter set the calculation:

Recoverable reserve (standard conditions)
= recovery factor × hydrocarbon in place (standard conditions)

$$RR = \frac{A \cdot h_N \cdot \phi \cdot (1-S_{wi})}{B_{hi}} \cdot RF$$

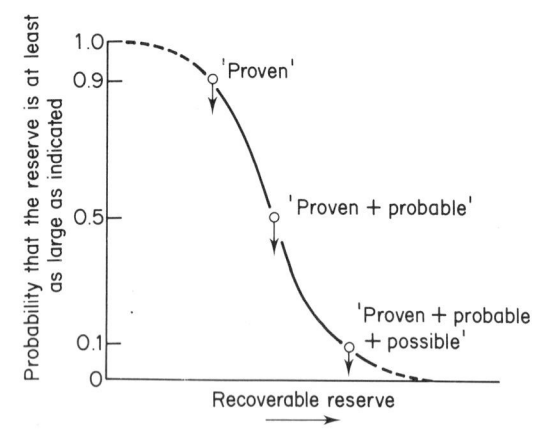

Fig. 8.21 Probabilistic representation of recoverable reserves.

The resulting cumulative frequency distribution shown in Fig. 8.21 is used with the same connotation as that of the hydrocarbon volume in place estimation, namely *proven* reserve at the 90% level, *proven + probable* reserve at the 50% level and *proven + probable + possible* reserve at the 10% level.

The recoverable reserve histogram can be used to develop a risk ratio for application in development decision making, by defining *upside* and *downside* potential of the reservoir. Development decisions are often taken assuming reserves at the 60–70% probability level.

8.10 DISTRIBUTION OF EQUITY IN PETROLEUM RESERVOIRS

It frequently occurs that the boundaries of a petroleum reservoir straddle lease lines and, in some instances, international boundaries. Under such conditions it is prudent that the owners or operators of each lease region enter into some sort of agreement to develop the reservoir in a cost and energy efficient manner. This means that the reservoir should be operated as a *unit* and the costs and revenues shared in an agreed manner. The basis for agreement depends on apportionment of equity in the unit, and the reservoir operation will then be considered *unitized*. It is appropriate that we consider unitization in this chapter as most equity formulae are based on volumetric estimation of hydrocarbons in place.

Although recoverable rather than in-place hydrocarbon may appear equitable, experience indicates that it rarely provides a basis for agreement. This is because economic recovery factors will be influenced by well density in heterogeneous reservoirs, and ultimate recovery factors may be uncertain because of their origin with core tests and the representative nature of samples. The ultimate recovery formula is also known as the movable hydrocarbon volume (MHV) formula, i.e.

$$MHV_{sc} = \frac{A \cdot h_N \cdot \phi \cdot (1-\bar{S}_{wi}-S_{hr})}{B_{hi}}$$

where A = area, h_N = net thickness, ϕ = porosity, S_{wi} = average connate water saturation, S_{hr} = average residual hydrocarbon saturation, B_{hi} = initial hydrocarbon formation volume factor. The lack of agreement between licence or lease owners over the proper representation of residual hydrocarbon saturation is the main reason why this formula is rarely applied in equity agreements.

In North Sea reservoir operation the equity determination most frequently applied in joint operating agreements is on the basis of *stock tank oil in place*. The use of this approach involves the least amount of contention but is nevertheless a major exercise involving the establishment of rules for defining net pay, fluid boundaries, transition zones, porosity, saturation and mapping techniques. Specialist subcommittees from all parties are required to formulate the bases for agreement of all parameters in the volumetric formula if unitization is to proceed. The final definition of parameters may sometimes be only quasi-technical, in that agreement is the main concern.

The use of computer models to represent a three-dimensional reservoir as a number of grid cells in which property variations are expressed is now commonplace. The summation of hydrocarbon pore volume for each grid cell leads to a deterministic evaluation of initial hydrocarbon in place (standard conditions). The mapping of the reservoir follows rules agreed in geological and geophysical subcom-

8 VOLUMETRIC ESTIMATES AND RECOVERABLE RESERVES

mittees, and the exercise may be conducted by an independent expert if there is no unit concensus. The mapping exercise provides gross reservoir rock volume as top structure and isopach (isochore) maps of each reservoir zone or layer. This of course presupposes some agreement on vertical zonation and reservoir boundaries. Figure 8.22 shows reservoir regions defined in the Dunlin field volumetric study [3].

The definitions of net thickness and porosity tend to emerge from the agreements reached in petrophysical subcommittees on log interpretation methods and core–log correlation. The compaction corrected core porosity is often taken as a standard, particularly where log interpretation involves wells drilled with both oil-based and water-based muds. The petrophysical interpretation agreement will also define the methods for interpretation of saturation from well logs. Interaction with a reservoir engineering subcommittee is necessary to reach agreement on representation of capillary transition zones and the validation of log derived saturation in a given rock type with capillary pressure data. The reservoir engineering subcommittee will usually undertake the responsibility of defining fluid contacts and will be involved in agreements regarding zonation. Reservoir engineering methods for zonation are described in Chapters 5 and 14. The fluid properties relevant to each computer grid cell allow areal and vertical variation in initial volume factors for hydrocarbons but depend on agreement reached in the reservoir engineering subcommittee on the use of fluid samples and the calculation methods for *PVT* properties (see Chapter 4). For a resultant computer model of *n* grid cells, the deterministic total hydrocarbon in place at standard conditions V_{sc} is then

$$V_{sc} = \sum_{j=1}^{j=n} \frac{A_j (h_N)_j \phi_j (S_{hc})_j}{(B_{hi})_j}$$

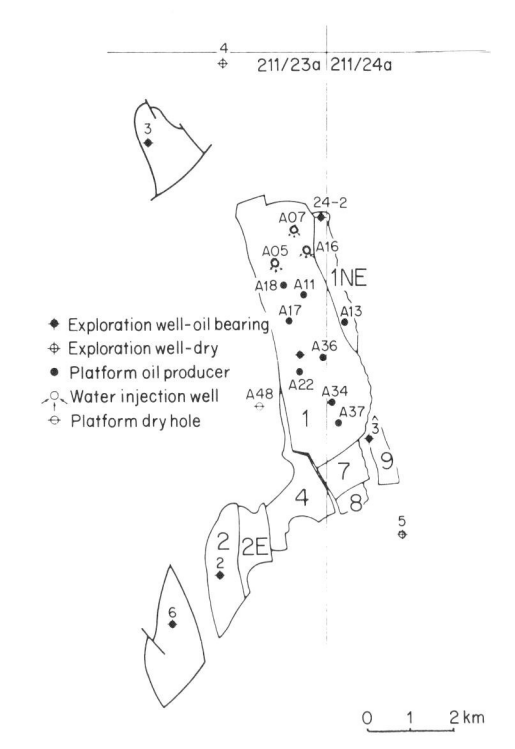

Fig. 8.22 Reservoir regions defined in the Dunlin field for volumetric calculations (after [3]).

The equity distribution between leases or licenses is obtained by arranging the computer grid system such that it follows the lease boundaries and it therefore facilitates regional subvolume calculation. The equity determination may be reviewed or redetermined at certain times in the development lifetime of the reservoir. Following any redetermination, historical costs and revenues may be reapportioned according to the nature of the Joint Operation Agreement. One owner or licencee in a unitized reservoir will agree to act as the Operator of the unit on behalf of all members.

Examples

Example 8.1

The table shows values of net sand thickness and area for a reservoir. The porosity varies linearly with sand thickness from 0.15 to 0.28, the water saturation varies hyperbolically from 1.0 to 0.33 from water contact to crest, and the oil formation volume factor is 1.355. What is the oil in place?

Table of net sand isopach values v. area within contour

Isopach value (ft)	Area within contour (acres)
350	0
300	400
250	730
200	965
150	1200
100	1520
50	2150
0	2600

NB. The general equation of hyperbola is

$$\frac{x^2}{a^2} - \frac{y^2}{b^2} = 1$$

and the relationship between x ($= S_w$, fraction) and y ($= h$, ft) can be fitted using the expression

$$h = \frac{164}{\sinh x} - 139$$

where $\sinh x$ is $\frac{1}{2}(e^x - e^{-x})$.

Example 8.2

The probabilistic distributions of reservoir properties are summarized in the following table at the cumulative frequency levels (equivalent to cumulative probability greater than a given value) of 90%, 50% and 10%. These data might correspond to minimum, most likely and maximum values. Compare the deterministic and probabilistic estimates of the recoverable reserve in stock tank barrels of oil.

Variable	Cumulative frequency level greater than given value		
	90%	50%	10%
Area (acres)	1780	2115	2450
Net thickness (ft)	225	250	275
Porosity (fraction)	0.125	0.133	0.140
Porosity-oil saturation product (fraction)	0.099	0.113	0.127
Initial oil formation volume factor (RB/STB)	1.214	1.240	1.260
Recovery factor (fraction)	0.17	0.31	0.52

References

[1] Archer, J.S.
Reservoir volumetrics and recovery factors, In *Developments in Petroleum Engineering,* (Dawe, R.A. and Wilson, D.C.,eds), Elsevier Applied Science Publishers (1985).

[2] Walstrom, J.E., Mueller, T.D. and McFarlene, R.C.
Evaluating uncertainty in engineering calculations, *JPT* (Dec. 1967), 1595.

[3] Van Rijswijk, J.J. *et al.*
The Dunlin field, a review of field development and reservoir performance to date, Paper EUR 168, *Proc. Europec* (1980), 217.

[4] Arps, J.J.
A statistical study of recovery efficiency, API Bull D14 (Oct. 1967), *Am. Pet. Inst.*

[5] Bankhead, C.C.
Processing of geological and engineering data in multi pay fields for evaluation, Pet. Trans. Reprint Series No 3, *SPE of AIME* (1970), 8.

[6] Grayson, C.J.
Bayesian analysis – a new approach to statistical decision making, Pet. Trans. Reprint Series No 3, *SPE of AIME* (1970), 215.

[7] Ryan, J.M.
Limitations of statistical methods for predicting petroleum and natural gas reserves and availability, Pet. Trans. Reprint Series No 3, *SPE of AIME* (1970), 227.

[8] Harbaugh, J.W., Doveton, J.H. and Davis, J.C.
Probability Methods in Oil Exploration, J Wiley, New York (1977).

[9] Pritchard, K.C.
Use of uncertainty analysis in evaluating hydrocarbon pore volume in the Rainbow-Zama area, *JPT* (Nov. 1970), 1357.

[10] Stoian, E.
Fundamentals and applications of the Monte Carlo method, *J. Can. Pet. Tech.* **4** (1965), 120.

[11] Archer, J.S.
Reservoir definition and characterisation for analysis and simulation, *Proc. 11th World Pet. Cong.,* London (1983), Paper PD6 (1).

[12] SPE
Standards pertaining to the estimating and auditing of oil and gas reserve information, *JPT* (July 1979), 852.

[13] Martinez, A.R. and Ion, D.C.
Classification and nomenclature systems for petroleum and petroleum reserves, *Proc. 11th World Pet. Cong.* (1983), Study Group Report.

[14] Van der Laan, G.
Physical properties of the reservoir and volume of gas initially in place, In *Proc. Symp. on the Groningen gas field,* Verhandel Konikl. Ned Geol. Mijnbouwt Genoot Geol. Ser. 25 (1968), 25.

[15] Hollis, A.P.
Some petroleum engineering considerations in the change over of the Rough gas field to the storage mode, Paper EUR 295, *Proc. Europec* (1982), 175.

[16] Hallett, D.
Refinement of the geological model of the Thistle field, In *Petroleum Geology of the Continental Shelf of North West Europe* (Illing, L.V. and Hobson, G.D.,eds), Inst. Pet., London (1981), 315.

[17] Davis, J.C.
Statistics and Data Analysis in Geology, Wiley Int., NY (1973).

[18] Krige, D.G.
Two dimensional weighted moving average trend surfaces for ore valuation, *Proc. Symp. Math. Stat. and Computer Appl. in Ore Valuation,* Johannesburg SA (1966), 13.

[19] Matheron, G.
Principles of geostatistics, *Econ. Geol.* **58** (1963), 1246.

[20] Garb, F.A.
Oil and gas reserves classification, estimation and evaluation, *JPT* (March 1985) 373.

Chapter 9

Radial Flow Analysis of Well Performance

This chapter will serve as an introduction to the subject of pressure analysis in reservoir engineering. The theoretical basis of radial flow analysis and the rudiments of data analysis for build-up and drawdown in oil and gas wells are presented. Well test procedures are briefly discussed for exploration and development wells.

9.1 RADIAL FLOW IN A SIMPLE SYSTEM

Considerations of conservation of mass, of Darcy's equation for flow, and of an equation of state for a slightly compressible liquid lead to a linearized partial differential equation of flow for a fluid flowing in a porous medium.

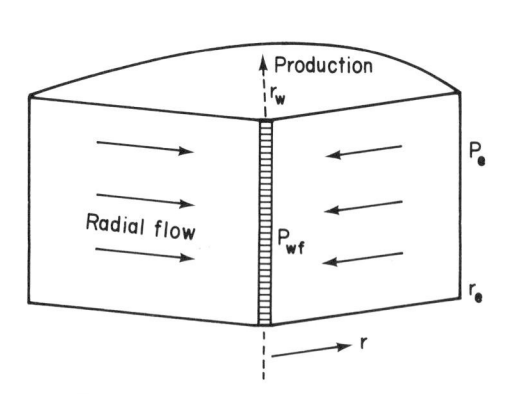

Fig. 9.1 Radial flow towards a well.

For a radial coordinate system (Fig. 9.1), with angular and vertical symmetry and isotropy, the resulting equation is

$$\frac{\partial^2 P}{\partial r^2} + \frac{1}{r}\frac{\partial P}{\partial r} = \frac{\phi \mu c}{k}\frac{\partial P}{\partial t}$$

or

$$\frac{1}{r}\frac{\partial}{\partial r}\left[r\frac{\partial P}{\partial r}\right] = \frac{\phi \mu c}{k}\frac{\partial P}{\partial t}$$

This equation is linear for the assumed conditions of constant ϕ, μ, k and small and constant compressibility. Solution is possible by Laplace transform methods (Hurst and van Everdingen [21]), and for more limited boundary conditions by applying the Boltzmann transformation:

$$S = \frac{\phi \mu c r^2}{4kt}$$

Between the limits $t = 0$ and t (when $s = x$) then:

$$\int_{P_i}^{P} dP = \frac{q\mu}{4\pi kh}\int_{\infty}^{x}\frac{e^{-s}}{S}dS$$

One solution applicable to well test analysis is the exponential integral, or line source solution [18]:

$$(P)_{r,t} = \frac{q\mu}{4\pi kh}E_i\left(-\frac{\phi \mu c r^2}{4kt}\right) + P_i$$

where, since the exponential integral of a negative argument is negative, an alternative nomenclature may be used:

9 RADIAL FLOW ANALYSIS OF WELL PERFORMANCE

$$ei(x) = -Ei(-x) = \int_x^\infty \frac{e^{-S}}{S} dS$$

where

$$x = \frac{\phi\mu c r^2}{4kt}$$

The boundary conditions necessary for this solution to apply are as follows:

(a) external boundary is at infinity, $r_e = \infty$;
(b) the rate is zero at the inner boundary r_w, and is instantaneously changed to q at time zero and maintained constant

$$(q)_{r=r_w} = 0, t \leq 0$$
$$(q)_{r=r_w} = q, t > 0$$

(c) the inner boundary $r = r_w$ is vanishingly small, $r_w \to 0$;
(d) porosity, permeability, thickness and viscosity are constant;
(e) compressibility is small and constant;
(f) pressure gradients are small;
(g) Darcy's law is valid.

In spite of these apparently severe restrictions on the use of this equation, it has wide applications and will be valid for real systems (i.e. $r_e \neq \infty$, $r_w \neq 0$) provided that:

(a) Dimensionless time for the radius at the inner boundary is greater than 5–10. The dimensionless time parameter is defined as follows:

$$t_{D_w} = \frac{kt}{\phi\mu c r_w^2} > 10$$

Since the wellbore radius is this radius in well testing, this condition is met within a few seconds or minutes of production in most cases.

(b) There is no significant pressure drop at any outer boundary. This condition will be met if the dimensionless time for the outer boundary is small, e.g.

$$t_{De} = \frac{kt}{\phi\mu c r_e^2} < 0.1$$

Since the dimensionless times are large for radii and times of practical interest, the simplified equation is valid in many situations.

9.2 DEVELOPMENT OF THE LINE SOURCE SOLUTION

This exponential integral is calculable from the series

$$ei(x) = -\log_e x - 0.5772 - \frac{x}{1} + \frac{x^2}{2\cdot 2!} - \frac{x^3}{3\cdot 3!} + \ldots$$

then, for small values of x (large values of $kt/\phi\mu cr^2$)

$$Ei(-x) \approx \log_e x + 0.5772$$

or

$$ei(x) = -0.5772 - \log_e x = -E_i(-x)$$

(The terminology $e_i(x) = -E_i(-x)$ is commonly used and values are shown in Fig. 9.2.)

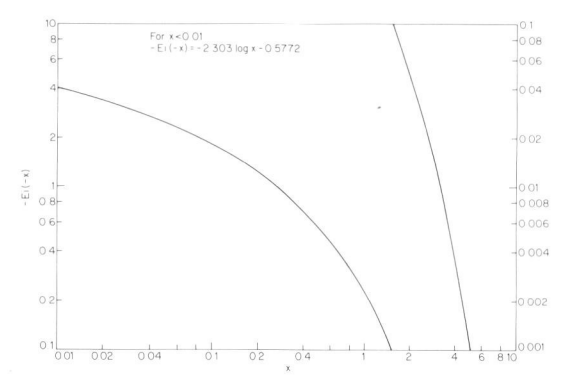

Fig. 9.2 The function $ei(x)$.

This means that:

$$P_i - P = \frac{q\mu}{4\pi kh}\left[ei\left(\frac{\phi\mu c r^2}{4kt}\right)\right]$$

$$= \frac{q\mu}{4\pi kh}\left[\log_e \frac{4kt}{\phi\mu c r^2} - 0.5772\right]$$

$$= \frac{q\mu}{4\pi kh}\left[\log_e \frac{kt}{\phi\mu c r^2} + 0.809\right]$$

Since all solutions of the diffusivity equation involve a coefficient $q\mu/kh$, a generalization of solutions is possible in the form

$$(P_i - P) = \frac{q\mu}{2\pi kh} P_D(t_D)$$

or

$$\frac{P_i - P}{[q\mu/2\pi kh]} = P_D(t_D)$$

where $q\mu/2\pi kh$ is the coefficient for steady state radial flow, and $P_D(t_D)$ is a dimensionless pressure function for the boundary conditions specified.

The dimensionless pressure function will be a function of dimensionless time and may be a simple analytical function, or a complex function requiring numerical evaluation, depending upon the complexity of the boundary conditions.

Tabulation or plots of $P_D(t_D)$ for the more common idealizations of boundary conditions are available in the literature [1,21].

9.3 RADIAL EQUATIONS IN PRACTICAL UNITS

The constant rate, radial flow equation in Darcy units is

$$(P_i - P) = \frac{q\mu}{4\pi kh}\left[\log_e \frac{kt}{\phi\mu cr^2} + 0.809\right]$$

which in field units becomes

$$(P_i - P) = \frac{70.6 q\mu B_o}{kh}\left[\log_e \frac{0.000264 kt}{\phi\mu\bar{c}_t r^2} + 0.809\right]$$

or

$$(P_i - P) = \frac{162.6 q\mu B_o}{kh}\left[\log_{10} \frac{kt}{\phi\mu\bar{c}_t r^2} - 3.23\right]$$

where p_i, p = psi, q = STB/D, μ = cp, kh = mD ft, $\bar{c}t = c_o S_o + c_w S_w + c_f$, (psi)$^{-1}$ r = ft, B_o = RB/STB.

If t is in days

$$t_D = \frac{0.006336 kt}{\phi\mu\bar{c}_t r^2}$$

9.4 APPLICATION OF ANALYTICAL SOLUTIONS IN WELL TEST METHODS

Considering the equation for a well in an infinite reservoir

$$(P_i - P) = \frac{q\mu}{4\pi kh}\left[\log_e \frac{kt}{\phi\mu\bar{c}r_w^2} + 0.809\right]$$

$$\therefore P_w = \frac{-q\mu}{4\pi kh}\log_e t \left(\frac{k}{\phi\mu\bar{c}r_w^2}\right) + \left(P_i - \frac{0.809 q\mu}{4\pi kh}\right)$$

it is apparent that if this well is maintained at a constant rate of production, then a plot of P_w against $\log_e t$ will yield a straight line, with a slope of $q\mu/4\pi kh$. If, then, the fluid viscosity is known, an estimate can be made of kh, the permeability thickness product of the formation. This estimate will be an estimate of the average kh in the area drained by the well during the test period, and consequently is representative of a very much greater reservoir volume than can be tested by coring or by wellbore survey methods.

TABLE 9.1 Characteristics of some current downhole pressure gauges (after Schlumberger)

Manufacturer	Designation	Pressure type/ Temp. type	Pressure range (psi)	Temperature range (°C)	Pres. accuracy % or psi for 10 K	Pres. resolution %fs or psi for 10 K	Temperature accuracy (°C)	Temperature resolution (°C)
Flopetrol Johnson	SSDP/SG	Strain gauge Junct. trans	10 000 15 000 (option)	0–150	±5 psi	0.02 psi	±0.5°	±0.06°
Flopetrol Johnson	SSDP/CRG	HP x-tal Plat. Res.	13 500	0–150	±0.035% ±3.5 psi	0.02 psi	±0.3°	0.03°
GRC	EMR 502 EPG 520	Capacitance Plat. Res	10 000	0–150	±0.09% FS ±9 psi	0.01 psi	±1°	±1°
Geo Services	Demeter	Strain gauge or Plat. Res	5 000 10 000 15 000	0–150	±0.04% ±4 psi	0.012% 1.2 psi	±0.3°	0.02°
Sperry Sun	MRPG MK III	Strain gauge	10 000	0–150	±0.05% ±5 psi	0.005% 0.5 psi		0.03°
Lynes	DMR 312/200	Strain gauge	5 000 10 000	0–125	±0.25%FS ±25 psi	0.025% 2.5 psi	±1°	0.03°
Lynes	DMR 314/200	Quartz	5 000	0–105	±0.05%FS ±5 psi	0.01% 1 psi	±1°	0.14°

9 RADIAL FLOW ANALYSIS OF WELL PERFORMANCE

The problem in this case is the practical one of accurate measurement of the small pressure drops normally encountered, at the fairly high absolute pressures involved. The characteristics of some current downhole pressure gauges are reproduced in Table 9.1.

Rates of pressure decline may be substantially less than 1 psi/day in extensive reservoirs with large permeabilities or thickness. Standard pressure gauges in use have an accuracy of 0.1–0.2%, but may be sensitive to pressure changes of 0.25 – 0.5 psi – and since the pressure differences are more important than the absolute value of the pressure, these can give the data required in many cases. More sensitive gauges are available which are capable of detecting pressure changes of a few hundredths of psi, but the significance of these small changes in a flowing well may be obscure. For prolonged testing, surface recording pressure gauges are available and eliminate the possible failure of clockwork or battery driven recording mechanisms.

The procedure of measuring the pressure decline

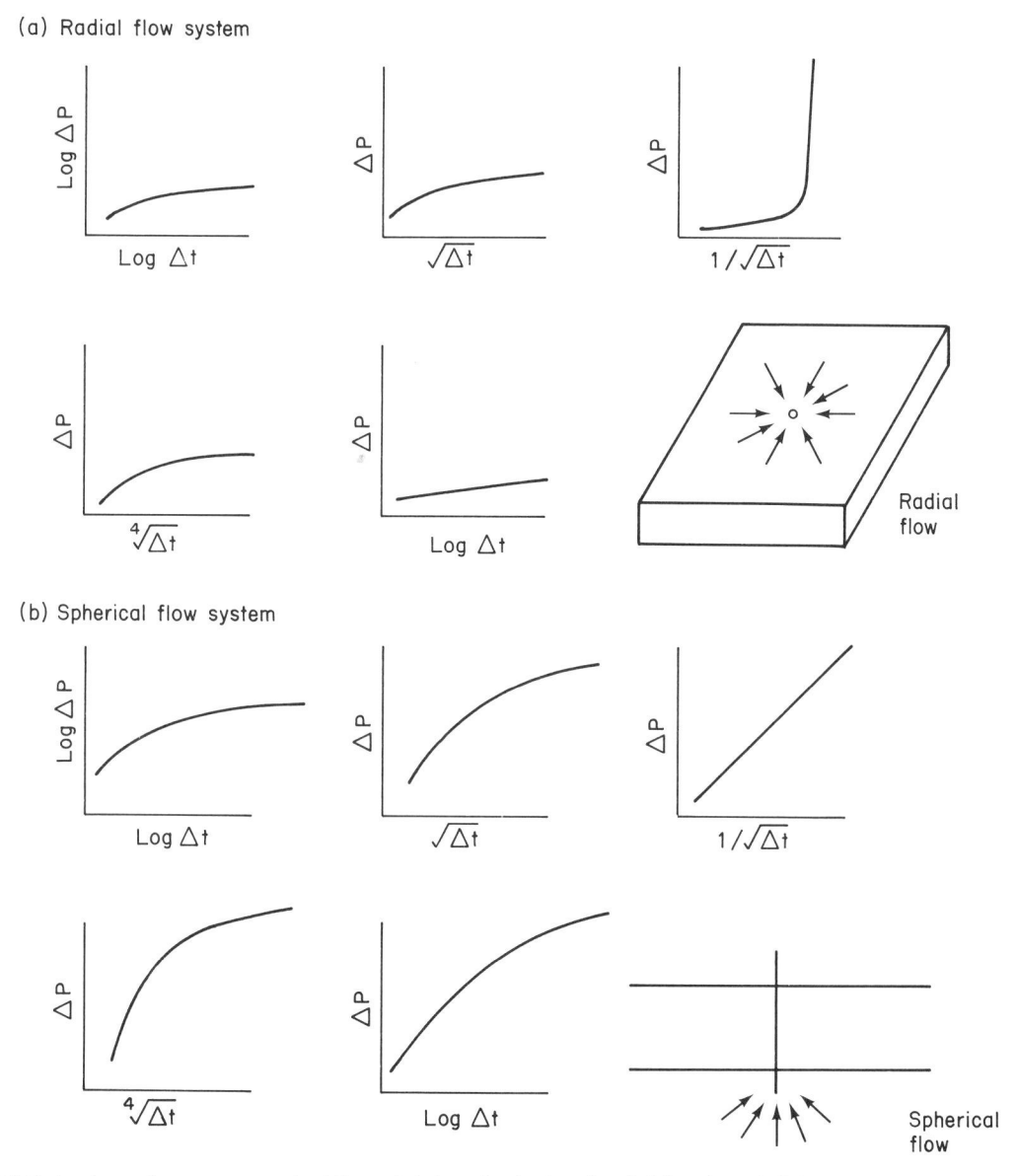

Fig. 9.3 Behaviour of pressure against time plots in spherical and radial flow in an infinite homogeneous reservoir (after [39]). (a) Radial flow system, (b) spherical flow system.

(a) Homogeneous reservoir

(b) Finite capacity fractured reservoir

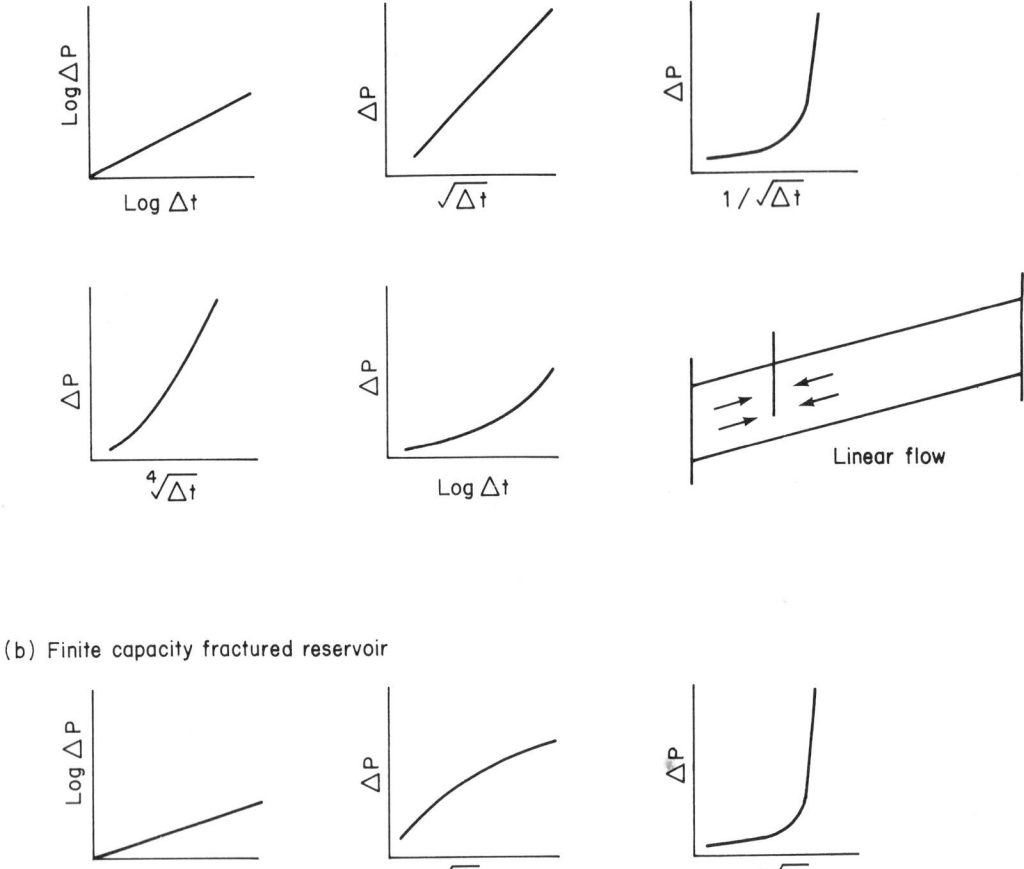

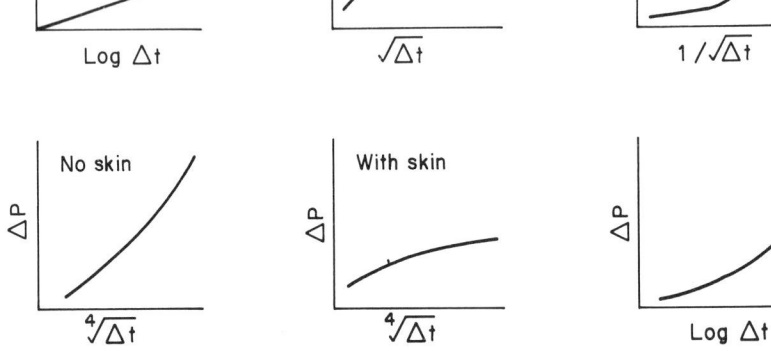

Fig. 9.4 Contrast in well test plots for assumed linear flow mechanism in infinite homogeneous and fractured reservoirs (after [39]). (a) Homogeneous reservoir, (b) finite capacity fractured reservoir.

in a producing well is known as drawdown testing and is usually undertaken for purposes other than kh measurement, which can normally be done more conveniently by pressure build-up testing.

In many reservoir geometries the pressure changes with time will indicate a number of influences:

(a) For some time following start of flow, the pressure at the wellbore is not influenced by the drainage boundary of the system, and analysis can be conducted as if the system was infinite. The solution is said to be a *transient* or early time solution.

(b) At some later time the influence of the

9 RADIAL FLOW ANALYSIS OF WELL PERFORMANCE

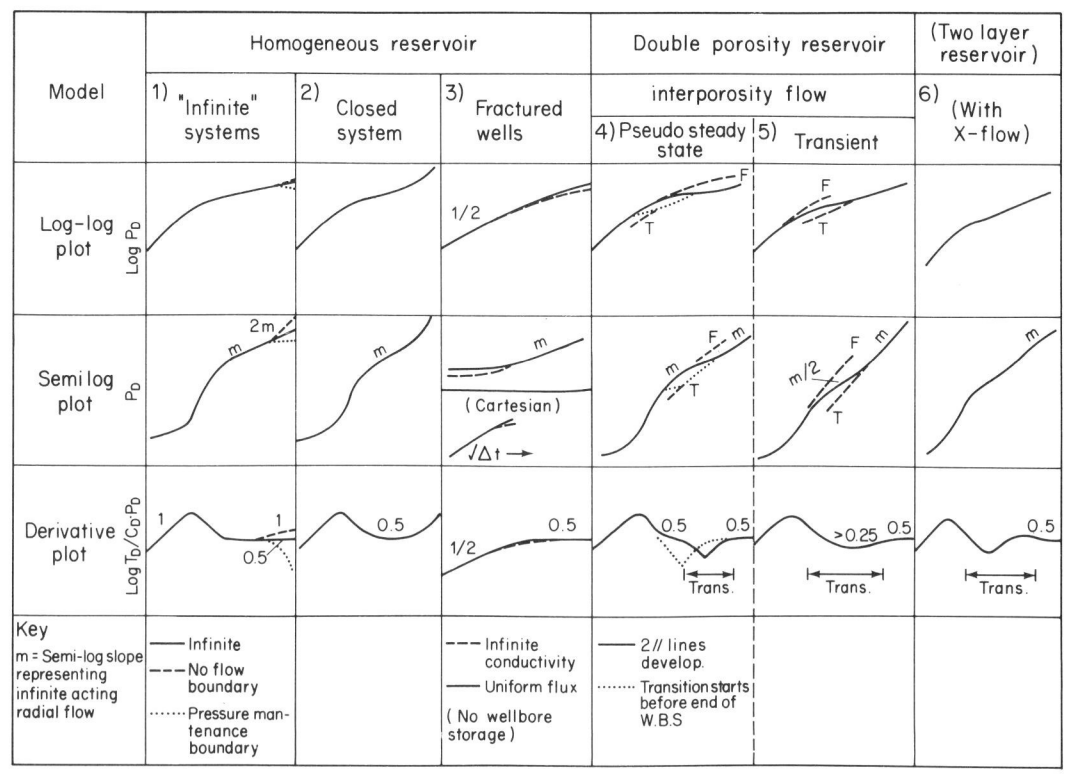

Fig. 9.5 Summary of well/reservoir model responses in different reservoir systems (after [42]).

nearest reservoir system boundary is experienced at the wellbore and the solution is said to be a *late transient* or middle time solution.

(c) When the rate of change of wellbore pressure becomes constant in time then semi-steady state stabilized flow conditions have been established.

The change from transient to semi-steady state conditions depends particularly on reservoir geometry, capacity and permeability. The *ideal* drawdown and build-up plots are only rarely seen, and interpretation is frequently difficult and ambiguous, especially in the absence of good geological and structural information. Many techniques have been proposed to identify the reason for anomalies [43,44], and Figs 9.3, 9.4 and 9.5 show some responses under different boundary conditions. Figure 9.6 is a *type-curve* showing combined pressure and pressure derivative curves [42]. This dimensionless approach is often used to match observed well response to properties of a particular reservoir type. The uncertainties in interpretation remain linked to assumptions of boundary conditions applied in solution of the original diffusivity equation.

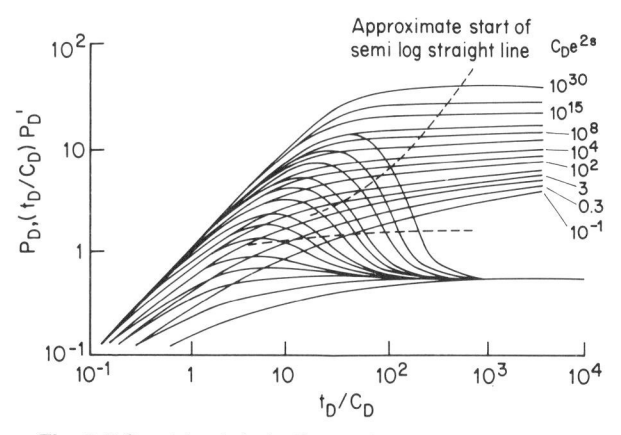

Fig. 9.6 Combined derivative and pressure type curve (after [32]).

9.5 PRESSURE BUILD-UP ANALYSIS

A pressure build-up survey involves measuring the changes in pressure which occur after a flowing well has been shut-in. If possible, the flowing pressure prior to shut-in should also be recorded.

To enable the analytical equation to be used for pressure build-up analysis, the discontinuity in the flow is removed by the technique of superposition.

If the well is considered to have flowed at rate q for a time t, prior to shut-in, then the change to zero rate at time t is effected by continuing the production, but from time t superimposing a negative rate $-q$, so that the total rate is zero, as shown in Fig. 9.7.

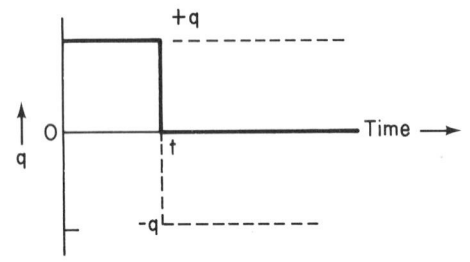

Fig. 9.7 Constant rate flow followed by constant rate shut-in.

The pressure drop at r_w due to a positive rate of production q, maintained for a time $t + \Delta t$ (Δt being the time after shut-in) is

$$\Delta P_1 = \frac{q\mu}{4\pi kh}\left[\log_e(t+\Delta t) + \log_e\frac{k}{\phi\mu\bar{c}r_w^2} + 0.809\right]$$

The pressure drop due to the superimposed negative rate of production is

$$\Delta P_2 = -\frac{q\mu}{4\pi kh}\left[\log_e \Delta t + \log_e\frac{k}{\phi\mu\bar{c}r_w^2} + 0.809\right]$$

The net pressure drop at any time $t + \Delta t$ (i.e. time after shut-in) is given by the sum of these expressions:

$$(P_i - P_{(\Delta t)}) = \frac{q\mu}{4\pi kh}\left[\log_e\frac{t+\Delta t}{\Delta t}\right]$$

Consequently a plot of $P_{(\Delta t)}$ (the pressure measured in the well at time after shut-in) against $\log_e(t+\Delta t)/\Delta t$ should give a straight line of slope $q\mu/4\pi kh$. This is known as a Horner plot (Fig. 9.8).

Conventionally, the data are plotted on semi-log paper, and the best straight line fitted to obtain a slope of m psi/$\log_{10}$ cycle. Extrapolation of the straight line to infinite shut-in time gives an extrapolated pressure P^* which, for the infinite reservoir, or for wells tested early in the life of the reservoir, has the physical significance $P^* = P_i$ (Fig. 9.9). At infinitely long shut-in time

$$\frac{t+\Delta t}{\Delta t} \to \text{unity}$$

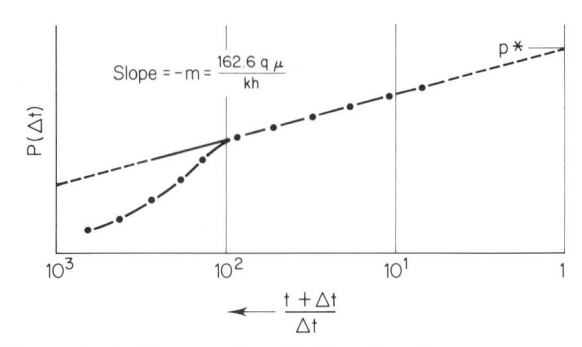

Fig. 9.8 A Horner plot (field units slope $= -m = \dfrac{162.6 q\mu B}{kh}$

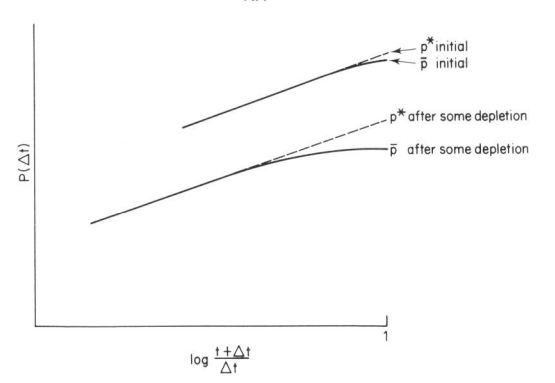

Fig. 9.9 Pressure build-up analyses at initial time and after significant reservoir production.

The permeability thickness product of the reservoir is obtained from

$$kh = \frac{162.6\, q\mu B}{m} \quad \text{milliDarcy–ft}$$

where units are field units, and m is the slope in psi/$\log_{10}$ cycle.

The use of type curves [32,44] in analysis of pressure build-up is increasingly popular, and many examples of responses can be found[1]. It is important to identify the appropriate reservoir model for type curve matching and the use of pressure derivatives reduces but does not eliminate ambiguity.

9.6 SKIN EFFECT

It is frequently found that observed pressures in the very early part of a build-up do not agree with the theoretical relation, even when corrections for after-flow production are made. Pressure drops are usually larger (rarely smaller) than theory would indicate. Better agreement is obtained if it is assumed that in the vicinity of the wellbore, some

9 RADIAL FLOW ANALYSIS OF WELL PERFORMANCE

impediment to flow exists.

Physical reasons for such an impediment are obvious and include: incomplete, inaccurate or plugged perforations; mudding off of formation or perforations by drilling fluid solids, filtrate invasion and consequent clay swelling or water blocking etc.

All these effects are local to the wellbore, affecting only a very small volume of the reservoir, and so flow within this damaged zone can justifiably be considered as steady state.

The effect of damage – the skin effect – can then be considered as a rate proportional, steady state pressure drop, given as the product of a flow rate function, and a dimensionless skin factor:

$$\Delta P_{skin} = \frac{q\mu}{2\pi kh} \cdot S = q_D \cdot S$$

where S = skin factor in a zone of *altered kh* compared to the bulk formation kh as shown in Fig. 9.10.

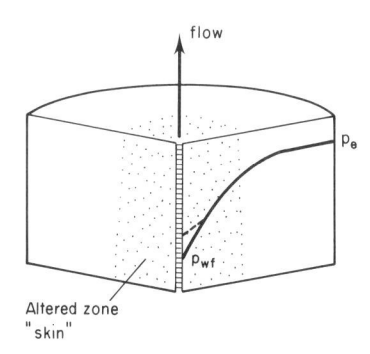

Fig. 9.10 Concept of altered zone around wellbore.

The total pressure drop for a flowing well in an infinite reservoir is then the transient solution

$$\Delta P_{total} = \frac{q\mu}{4\pi kh} \left[\log_e \frac{kt}{\phi\mu\bar{c}r_w^2} + 0.809 + 2S \right]$$

Note that uncertainty in the values of $k/\phi\mu\bar{c}r_w^2$ and S makes it difficult to calculate kh from flowing pressure data alone.

The magnitude of the skin factor can be found from a pressure build-up survey as will be shown. The equation for the pressure at the flowing well at the instant of shut-in is:

$$P_i - P_{wf} = \frac{q}{4\pi kh} \left[\log_e t + \log_e \frac{k}{\phi\mu\bar{c}r_w^2} + 0.809 + 2S \right]$$

the pressure at closed-in time is:

$$P_i - P_{(\Delta t)} = \frac{q\mu}{4\pi kh} \left[\log_e \frac{t+\Delta t}{\Delta t} \right]$$

and for small Δt (early shut-in times)

$$P_i - P_{(\Delta t)} = \frac{q\mu}{4\pi kh} \left[\log_e t - \log_e \Delta t \right]$$

$$P_{(\Delta t)} - P_{wf} = \frac{q\mu}{4\pi kh} \left[\log_e \frac{k}{\phi\mu\bar{c}r_w^2} + 0.809 + 2S \right]$$

for values of Δt such that $t + \Delta t \to t$.

Conventionally and conveniently Δt is taken as 1 and the equation manipulated to yield S, and in field, or practical, units the resulting equation used with a solution at one hour of shut-in time (P_{1h}):

$$S = 1.151 \left[\frac{P_{1h} - P_{wf}}{m} - \log_{10} \frac{k}{\phi\mu\bar{c}r_w^2} + 3.23 \right]$$

where P, P_{wf} = psi, m = slope psi per $\log_{10}$ cycle, k = mD, ϕ = fraction, μ = cp, $\bar{c}$ = (psi)$^{-1}$ and r_w = ft. For any other value of Δt (subject only to the restriction $t + \Delta t \to t$)

$$S = 1.151 \left[\frac{P_{(\Delta t)} - P_{wf}}{m} - \log_{10} \frac{k\Delta t}{\phi\mu\bar{c}r_w^2} + 3.23 \right]$$

It is possible, but not generally very useful, to interpret the skin factor as a zone of radius r_s within which the permeability is altered to some value k_s when:

$$S = \left\{ \frac{k}{k_s} - 1 \right\} \log_e \frac{r_s}{r_w}$$

but since it is not possible to assign an unambiguous value to either r_s or k_s this has limited application.

The pressure drop in the skin is given by

$$\Delta P_{skin} = \frac{q\mu}{4\pi kh} \cdot 2S \quad \text{(in Darcy units)}$$

or $0.87\, m \cdot S$ in field units when m is in psi/$\log_{10}$ cycle.

The ratio of theoretical pressure drop to the actual pressure drop may be considered an index of the flow efficiency:

$$\text{Efficiency} = \frac{P^* - P_{wf} - \Delta P_{skin}}{P^* - P_{wf}}$$

9.6.1 Negative skin factors

It is possible, after stimulation treatments, to obtain negative skin factors, and efficiencies or completion factors greater than 1. In this case:

$$\frac{q_1\mu}{4\pi(kh)_1}\left[\log_e\frac{kt_1}{\phi\mu\bar{c}r_w^2}+0.809+S_1\right]$$

$$>\frac{q_2\mu}{4\pi(kh)_2}\left[\log_e\frac{kt_2}{\phi\mu\bar{c}r_w^2}+0.809+S_2\right]$$

If the kh is apparently unchanged by the stimulation, then the physical interpretation of the situation is normally that the effective well radius has been increased, reducing the value of the log term of the right-hand side.

This effective well radius, defined as the radius of a well in a reservoir of uniform permeability giving the same pressure drop as the real system, is given by

$$r_w \text{ (effective)} = r_w e^{-S}$$

9.7 PRESSURE DRAWDOWN AND RESERVOIR LIMIT TESTING

By the time the initial disturbances due to bringing a well on production have died out, the log approximation to the exponential integral solution will be appropriate, and the flow equation will be

$$P_{wf}=P_i-\frac{q\mu}{4\pi kh}\left[\log_e\frac{kt}{\phi\mu\bar{c}r_w^2}+0.809+2S\right]$$

At early times, therefore, a plot of P_{wf} against log time will be a straight line of slope $q\mu/4kh$. If flowing well pressures are recorded, and pressure fluctuation or gauge drift and vibration do not obscure the trend, the plot may yield values of kh.

If flow is continued then at some time (usually estimated as $t_D = 0.1$ for the outer boundary of a symmetrical system), a significant pressure change will occur at the boundary. With a no-flow boundary condition, pressure will fall more rapidly than is predicted by the infinite reservoir equation. At some later time (again usually taken as $t_D = 0.3$ for the outer boundary) the reservoir is assumed to reach a pseudo steady state condition when the pressure at all points of the reservoir is falling uniformly with time and the equation for the pressure at a flowing well is obtained as follows:

since $-qdt = V_P \cdot \bar{c} \cdot dp$

then

$$P_{wf}=P_i-\frac{qt}{\pi r_e^2 h\phi\bar{c}_t}$$

and

$$\Delta P=\frac{q\mu}{2\pi kh}\left[\log_e\frac{r_e}{r_w}-\frac{1}{2}+S\right]$$

or

$$P_{wf}=\bar{P}-\frac{q\mu}{2\pi kh}\left[\log_e\frac{r_e}{r_w}-\frac{3}{4}+S\right]$$

Note that the average reservoir pressure $\bar{P}$ is

$$\bar{P}=\int_{r_w}^{r_e}dV \quad \text{and} \quad dV=2\pi rh\phi dr$$

A plot of pressure against time on linear (Cartesian) coordinate paper should be a straight line. The slope of this line is

$$\frac{dp}{dt}=\frac{-q}{\pi r_e^2 h\phi\bar{c}}$$

The pore volume is therefore

$$V_P=\frac{0.0418qB_o}{\beta\bar{c}} \quad \text{reservoir barrels}$$

where β = psi/h (slope), $\bar{c}$ = (psi)$^{-1}$ and q = stb/D. It may be noted that the presence of an aquifer can influence interpretation through its volume and compressibility, i.e. for an aquifer of volume V_w using nomenclature from Chapter 10

$$N_p B_o = N B_{o_i}\bar{c}_{oe}\Delta P + V_w\bar{c}_w\Delta P$$

Since $q_o = N_p B_o/\Delta t$, then

$$\frac{kh\Delta t}{\mu B_o(\log_e[r_e/r_w]+S-\tfrac{3}{4})} = NB_{oi}c_{oe}+V_w\bar{c}_w$$

$$= \frac{162.2\, q\, \Delta t}{m\cdot(\log_e[r_e/r_w]+S-\tfrac{3}{4})}$$

The definition of $\bar{c}$ will determine whether total pore volume or hydrocarbon pore volume is calculated. If a test can be prolonged sufficiently to reach this semi-steady state period, then the drawdown test (or reservoir limits test) can yield an estimate of hydrocarbon in place in a closed reservoir. The literature gives equations for the analysis of the intervening period (t_{De} equal to 0.1 to 0.3), but the theoretical justifications are dubious, and results

9 RADIAL FLOW ANALYSIS OF WELL PERFORMANCE

ambiguous and uncertain. However, data in this region can be used to determine minimum hydrocarbon in place since the slope of the linear scale pressure:time plot declines monotonically, and any slope taken at a time prior to semi-steady state *should* be larger (and so the pore volume estimate should be smaller) than the value calculated for the steady state condition. Approximate radial flow analysis indicates that the transient stage can be considered as pseudo steady state flow within a moving boundary (the radius of disturbance), so that observation of pressure in a well offsetting a producing test well can indicate the time when the radius of disturbance reached this point, and a pore volume (or BBLs of hydrocarbon per acre ft) can be estimated for this region. More advanced interpretation of such *interference* effects is possible, but is not considered in this text.[1]

9.8 GAS WELL TESTING

High rate gas wells are one common example of departure from the simple Darcy equation, requiring a quadratic equation for a realistic description of the pressure drop rate relationship. The basic equation for these conditions is known as the Forchheimer equation:

$$-\frac{dp}{dX} = \alpha U + \beta \rho U^2$$

and the ρU^2 term is a kinetic energy, or inertial effect correction. The usual form of equation used in field practice where flow to the wellbore is essentially radial is

$$\Delta(P^2) = A\,Q = B\,Q^2$$

where $P^2 = \bar{P}^2 - P_{wf}^2$, A is the coefficient of Darcy flow, and B is a coefficient of non-Darcy flow. B is some function of rock properties, gas properties and interval open to flow but is always treated as a grouped term.

Dividing through by Q we can transform the equation into a linear form with slope B and intercept A (Fig. 9.11):

$$\frac{\Delta(P^2)}{Q} = BQ + A$$

A single constant rate test for a well described by a non-Darcy flow equation will give kh and the extrapolated pressure, but will not define inertial effects, and may not describe accurately the productive potentials of wells.

Empirical test methods were developed for the statutory control and regulation of gas wells and

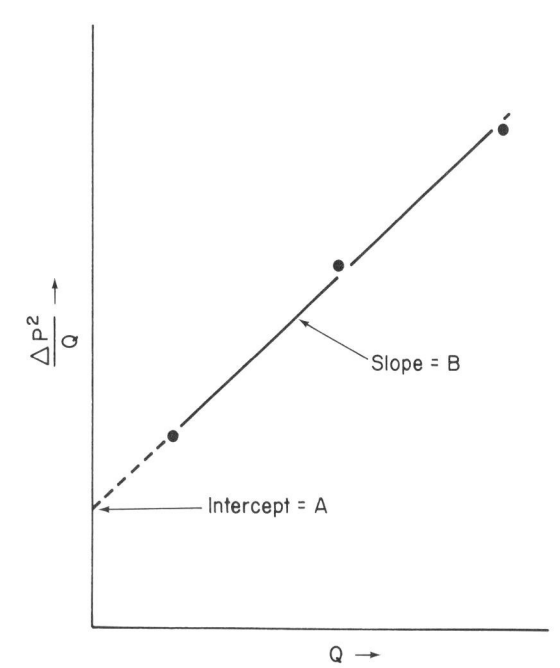

Fig. 9.11 Correlation of high rate gas test data.

reservoirs [8,11]. One advantage of these methods is that testing at more than one rate is required so that inertial effects can be investigated, although the empirical equations in use give no insight into reservoir behaviour.

The empirical relationship between rate and flowing pressure used to correlate data is known as the *back pressure equation*:

$$Q = C(P_s^2 - P_{wf}^2)^n$$

Figure 9.12 shows the log–log plot of *back pressure* test data, where Q = flow rate, standard volumes per day, P_s = static reservoir pressure, P_{wf} = stabilized flowing bottom-hole pressure, C = coefficient, n = slope, between 0.5 (turbulent) and 1.0 (laminar). The log–log slope is defined as rate/ΔP.

This equation is not strictly the result of any particular flow law, and is invalid if a very wide range of flow rates is studied. However, over moderate ranges of flow rate, test results will generate an approximate straight line when q is plotted against $(P_s^2 - P_{wf}^2)$ on log–log scales, as shown in Fig. 9.12. If the flow were true Darcy flow of an ideal gas under steady state condition, the exponent n (the *slope* of the log–log plot) would be unity, with the coefficient C given as

$$C = \text{constant} \cdot \frac{kh}{\mu z T \log \cdot [r_e/r_w]}$$

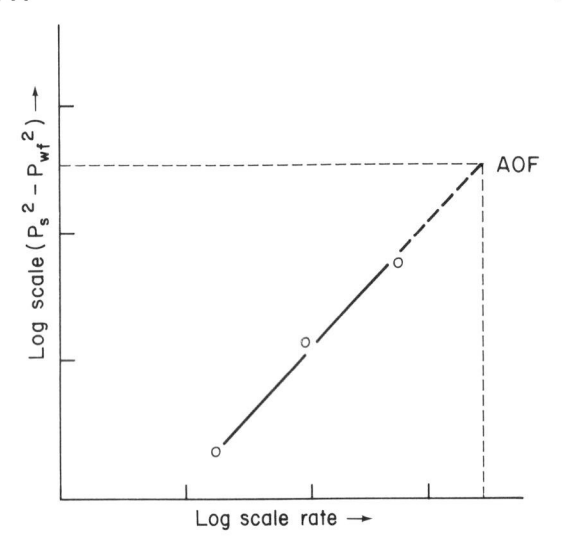

Fig. 9.12 Back pressure test analysis.

If the flow were fully turbulent the exponent n would be 0.5. Values of n between 0.5 and 1 are generally taken as indicating the existence of some inertial effect, and this will frequently be the case, but a combination of a Darcy flow equation and a power series expression for z can lead to a value of exponent other than 1.

The standard test procedure involves producing wells at a constant rate, until the flowing bottom-hole pressure approaches stabilization, and measuring P_{wf}. The well can then be allowed to build-up to static pressure, and a test carried out at a new rate; this may be repeated for three or four rates.

The results are then plotted on log–log scales, and the line extrapolated to the value $(P_s^2 - P_{wf}^2) = P_s^2$ (i.e. the theoretical flow rate corresponding to this value is termed the absolute open flow potential). The test itself may be termed an AOF test – or *absolute open flow test*.

9.8.1 Isochronal testing

The requirement of stabilized flow and full build-up between flow rates may require very prolonged testing periods in low permeability reservoirs, and modified standardized test procedures have been developed to reduce the test duration [25].

The isochronal test uses the flowing pressures after identical flow durations at each rate, the pressures not necessarily being the stabilized pressure. Between each rate, the well should return to the fully built-up conditions $((dp/Pdt) < 0.2\%/h)$ but with short flowing periods the build-up periods are also short. Unless stabilized conditions are attained in the flow periods chosen, the log–log plot of the data will not be coincident with the true back pressure curve, but will be parallel to it. Consequently, it is desirable for a test at one flow rate to be prolonged until stabilized conditions are attained, this point when plotted giving the position of the back pressure curve.

9.8.2 Modified isochronal testing

A further modification may be used, in which identical flow and shut-in periods are adopted. In this case, the last value of shut-in pressure before flow is used instead of P_s, the static pressure in the term $(P_s - P_{wf})$, and this procedure may further reduce the time necessary for testing in low permeability formations.

9.8.3 Analysis of multirate data

The empirical back pressure equation

$$Q = C(P_s^2 - P_{wf}^2)^n$$

is not especially helpful in predicting reservoir characteristics or in analysing the components of pressure drops, although it may be useful in characterizing well performance.

When multirate data are available it is more useful to revert to one of the basic flow equations in field units:

$$\bar{P}^2 - P_{wf}^2 = \frac{1422 Q \mu z T}{kh} \left[\log_e \frac{r_e}{r_w} - \tfrac{3}{4} + S \right]$$

or

$$\bar{P}^2 - P_{wf}^2 = \frac{1422 Q \mu z T}{kh} \left[\log_e \frac{0.472 \, r_e}{r_w} + S \right] + BQ^2$$

(semi-steady state)

$$P_i^2 - P_{wf}^2 = \frac{1422 Q \mu z T}{kh} \left[\tfrac{1}{2}(\log_e t_D + 0.809) + S \right] + BQ^2 \quad \text{(transient)}$$

$$\Delta(P^2) = AQ + BQ^2$$

where

$$A = \frac{1422 \mu z T}{kh} \left[\log_e \frac{0.472 \, r_e}{r_w} + S \right] \quad \text{semi-steady state}$$

9 RADIAL FLOW ANALYSIS OF WELL PERFORMANCE

or

$$\frac{1422\mu z T}{kh}\left[\tfrac{1}{2}(\log_e t_D + 0.809) + S\right] \quad \text{(transient)}$$

and B = non-Darcy coefficient, Q = MSCF/D, k = mD, h, r = ft, μ = cp, P = psia, T = °R (= 460 + °F).

It is more appropriate[34] to use the real gas pseudo-pressure $m(P)$ defined as

$$m(P) = 2\int_{P_b}^{P} \frac{P\,dP}{\mu z}$$

and either $\Delta(m(P))^2$ or $m(\bar{P}) - m(P_{wf})$ correlated with rate:

$$m(\bar{P}) - m(P_{wf})$$
$$= \frac{1422TQ}{kh}\left[\log_e \frac{r_e}{r_w} - \tfrac{3}{4} + S\right] + FQ^2$$

$$= \frac{1422TQ}{kh}\left[\log_e \frac{r_e}{r_w} - \tfrac{3}{4} + S + DQ\right]$$

where

$$D = \frac{Fkh}{1422T}$$

and DQ is known as the rate dependent skin factor[3].

This provides one method of isolating the non-Darcy effect, but does not enable the skin effect to be separated from the Darcy effect. Because of the localized nature of the non-Darcy effect, it is possible to regard this as a local additional rate dependent skin effect, rearranging the equation as

$$\frac{\Delta P^2}{\bar{\mu} z} \text{ or } \left[m(\bar{P}) - m(P_{wf})\right] \text{ or } \Delta\left(m(P)\right)^2$$
$$= \frac{1422TQ}{kh}\left[\log_e \frac{0.472 r_e}{r_w} + S'\right]$$

The analysis of multiple rate for the apparent skin effect S' will result in a set of rate dependent data, conforming to the equation

$$S' = S + DQ$$

where

$$D = \frac{kh}{1422T} \cdot B$$

It must be remembered that the skin factor calculated from a constant rate test will generally involve a non-Darcy component unless the rate is very low, and care should be exercised in determining a B correlation. It is perfectly possible for a stimulated well with a *true* negative skin to exhibit large positive apparent skin factors.

9.9 WELL TEST PROCEDURES

Tests of the flowing behaviour of discovery and appraisal wells are necessary, both to help in determining reservoir parameters, and in evaluating the productivity of individual wells, upon which will depend the number of wells needed to obtain a specified rate of production.

Because of cost considerations, the first tests on an exploration or appraisal well may be made by means of drill stem tests, without the well being completely equipped for continuing production.

9.9.1 Drill stem testing

A drill stem test is a temporary completion of a well, enabling the well to be brought on production without a production flow string and wellhead, for pressure surveys to be made, and the well then killed prior to abandonment or permanent completion. The drill string is customarily used as the flow string (Fig. 9.13) (hence the term drill stem test), although in some cases the test tool assembly may be run on a tubing string, kept for this purpose. A review of currently used equipment and techniques is referenced in [45].

Drill stem tests may be run in open hole, thereby eliminating the need to run a casing string, but the uncertainty in obtaining a good packer seat, the greatly increased possibility of sticking pipe and losing the hole, and the lack of selectivity in testing, severely limit the utility of open-hole tests, and tests in cased holes are very much to be preferred. When testing in cased holes, a number of intervals can be perforated for test at one time, and by use of a retrievable bridge plug and the testing tool, intervals can be tested selectively in a series of drill stem tests. Progressive testing of increasingly thick intervals allows assessment of productivity and zone contributions.

9.9.2 Testing tools and assemblies

There are the essential components in an assembly of testing tools.

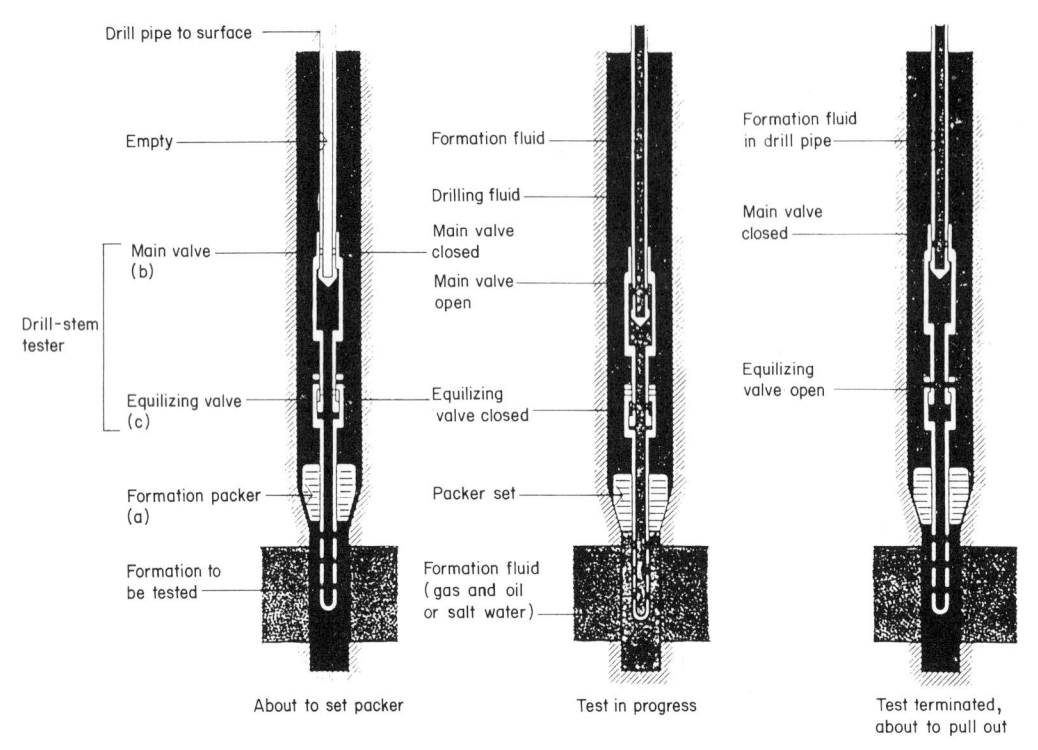

Fig. 9.13 Drill stem testing.

1. Packer

The interval to be tested must be exposed to a reduced pressure in order to induce flow. The mud column in the annulus must then be isolated from the test interval, by means of a packer.

2. Tester valve

This is the main valve in the tool, which enables the drill pipe to be run wholly or partially empty of drilling fluid.

When the test packer is set, opening of the test valve exposes the test interval to the low pressure of the drill pipe bore, inducing flow. The valve is operated by vertical movement of the drill pipe, against a spring holding the valve normally closed. A hydraulic control system delays the opening of the valve when drill pipe weight is applied so that the drill pipe weight can be used for setting the packer without opening the valve, and the valve will not open with a temporary hold up when running in the hole.

Valves operated by changes in annulus pressure are increasingly used instead, imposing less restriction to flow than the mechanically operated types. They can be operated reasonably reliably under the conditions of a floating drilling vessel.

3. Choke assembly

On a first test of a formation when pressures and potentials are unknown, a bottom-hole choke will generally be run. This serves to throttle the flow of the well and dissipate some of the pressure of the system, relieving the wellhead assembly of excessive pressures. Since the initial flow period will displace drilling fluid (possibly gelled), mud cake, perforating debris etc., which may plug narrow restrictions in valves and chokes, some screening system is necessary at fluid inlets.

When pressures are known, chokes may be omitted completely, or a selected size run to control pressures and rates.

4. Reverse circulating sub

Since the tester valve will be closed when the pipe is pulled, fluid would be released only as stands are broken on the derrick floor, if there were no means of opening the drill pipe bore to the annulus. The reverse circulation sub provides this connection and eliminates the necessity of spilling formation fluids (since these can be circulated out before pulling), and also enables the pipe to be pulled essentially dry.

9 RADIAL FLOW ANALYSIS OF WELL PERFORMANCE

5. Safety joints and jars

These are not essential to the test operation, but assist in the recovery of the test string, in case the packer or any other part of the tool becomes stuck. The safety joint enables the test string to be detached at some point above the packer.

6. Samplers

Particularly when flow does not reach the surface, it may be desirable to obtain a fluid sample at a pressure, as close as possible to bottom-hole pressure. Samplers are available which close top and bottom valves at the end of the final flow period. (Note that this is a flowing sample and does *not necessarily* correspond to a true sample of reservoir fluid.)

7. Pressure gauges

Any type of pressure gauge [15,24,47] can be run with a test string, but a test will always use two gauges, which should be carefully calibrated gauges of known accuracy, commonly with one gauge in the flow stream, and with one *blanked off,* that is contained within a chamber with ports to the wellbore, but with no flow through the chamber. Table 9.1 shows some current downhole pressure gauge characteristics. Figure 9.14 shows the principle of the rugged Amerada gauge with mechanical clocks.

8. Cushion

The water cushion is a column of water contained within the drill pipe above the tester valve, so that the test string is not run completely empty.

The main object of the cushion is to ensure that some back pressure is maintained on the formation immediately after the tester valve is opened. If a moderately high pressure formation were instantaneously exposed to a pressure effectively atmospheric, an explosively violent period of flow could result, leading to plugging of, or damage to, tools, and possible damage to unconsolidated formations. A secondary concern in very deep high pressure wells is that of preventing collapse of the test string due to the unbalanced annulus mud column pressure.

As an alternative to a water cushion, the test string can be pressure charged with nitrogen to any predetermined pressure, giving a variable cushion effect as the nitrogen is bled off.

9.9.3 Test procedures, data analysis

Accurate interpretation of test results requires continuous monitoring of all events, so that as much data as is possible on flow and pressure can be subsequently reconstructed.

Immediately after the valve is opened, flow may be detected by a blow of air at the surface, as formation fluids flow into the well. The first signifi-

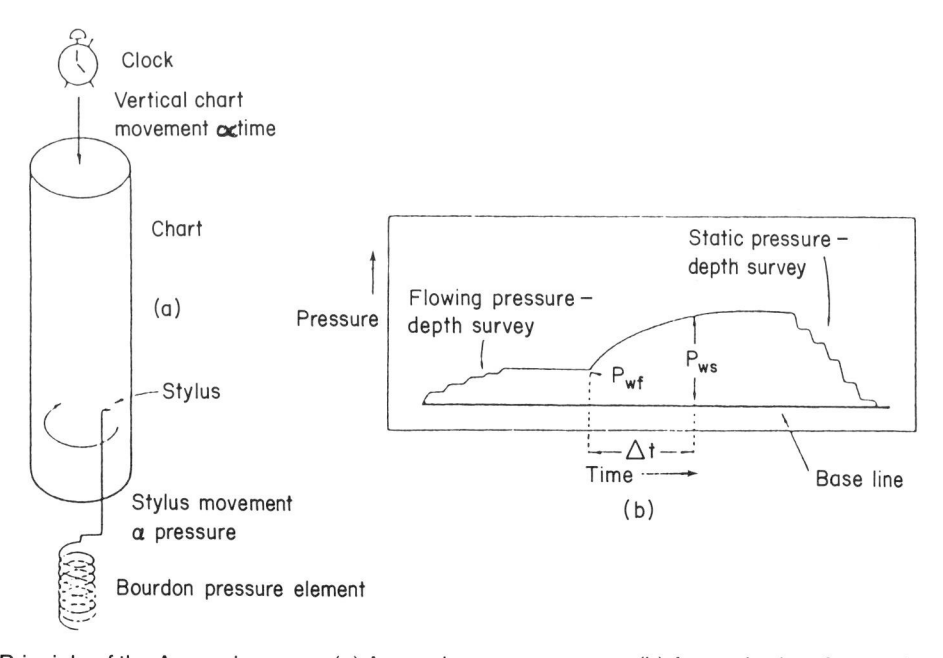

Fig. 9.14 Principle of the Amerada gauge. (a) Amerada pressure gauge, (b) Amerada chart for a typical pressure build-up survey in a producing well.

cant measurement that can be made is the time of arrival of the water cushion at the surface. The difference between this time and instant of valve opening, and the void volume of the test string, can be used to estimate a flow rate for this period.

The end of the water cushion may be more difficult to estimate, because of cutting of the cushion by formation solids. Since there will usually be mud solids, debris, possible formation solids which may plug, erode or damage orifices, this period is not metered, and a robust mass flow meter would be of help in analysing such flow periods. It is, however, possible to use choke pressures to estimate the magnitude of flows, and choke pressures *should* be routinely recorded.

When the well has cleaned up and formation fluids are flowing, attempts are made to stabilize flow – oil being tested through a separator, with gas being vented or flared through a meter and oil being stored or burned off. Figure 9.15 shows test burners. Difficulties with burners, separators, and other surface equipment, with frequent changes of chokes (and so of rate) in attempts to stabilize and control flow, can often result in data going unrecorded for significant periods, materially complicating analysis.

Fig. 9.15 Test burners. (Photo courtesy of BP.)

9.9.4 Wireline testing

1. The formation tester (FT) and formation interval tester (FIT)

These are devices run on electric wireline enabling a small sample of formation fluid, and very limited pressure data, to be obtained from a selected interval.

In operation, a valve is opened and a hydraulic intensifier expands a back-up shoe, forcing the tool and two sealing packers against the formation of casing wall. If there is no flow in open hole, or if the hole is cased, a shaped charge is fired through a sealing packer giving a flow channel from the formation to sample chamber.

Pressures are monitored from the surface, but an Amerada gauge is generally connected to the sampling system for a more accurate measurement.

The sample chamber is small (generally taking a 10 litre sample), and if this fills with formation fluid a pressure build-up is subsequently recorded.

Alternatively, the sample chamber seal valve may be closed and the tool left in place to record a build-up. (This could lead to an overpressured condition in some circumstances.)

Bleeding down the hydraulic pressure releases the tool for pulling. If filtrate invasion is severe, open-hole tests may recover only mud filtrate, but in cased-hole it should generally be possible to recover formation fluid.

The tool may be particularly useful in locating hydrocarbon–water contacts, and the extent of transition zones, but the flow and pressure data are generally of poor quality, and follow-up drill stem or production tests are necessary.

2. The repeat formation tester (vertical pressure logging)

The earlier tools have now been replaced by an open-hole testing tool that measures the vertical pressure distribution in a well, recovers formation fluid samples and, in certain circumstances, measures effective permeability. One such device which has found frequent application is known by its Schlumberger trade name of the *repeat formation tester* or RFT. The device is designed to operate in open-hole sizes between 6″ and 14¾″, in mud pressure ratings up to 20 000 psi and mud temperatures to 350°F. A piston device and an associated packer can be actuated to force a probe through mud cake and allow the flow of small volumes of fluid (about 10 cc) into each of two pretest chambers (Figs. 9.16 and 9.17). Analysis of the flow and

9 RADIAL FLOW ANALYSIS OF WELL PERFORMANCE

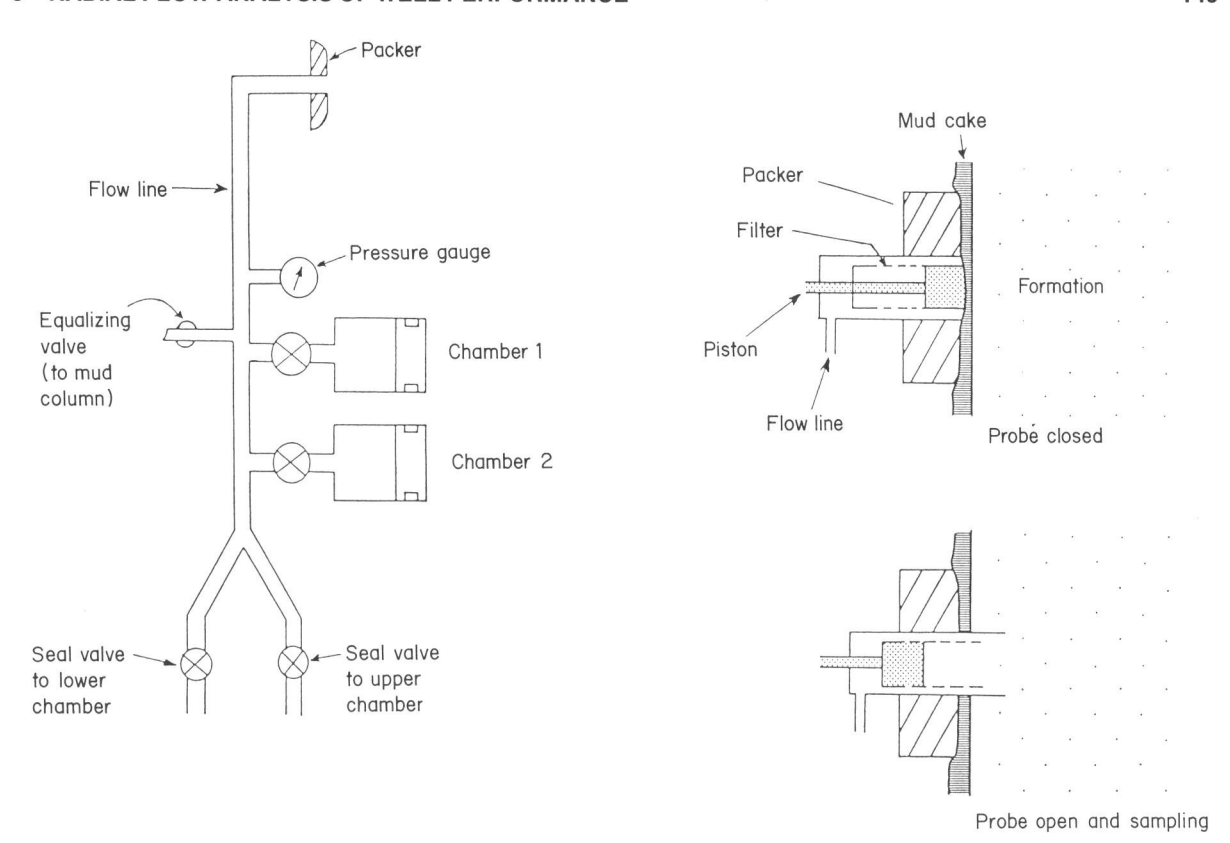

Fig. 9.16 Schematic of the RFT tool.

Fig. 9.17 Photograph of the RFT tool.

build-up curves using spherical and cylindrical flow analysis may lead to an estimate of effective permeability. The built-up pressure is recorded by a strain gauge and can be backed up with a high precision quartz gauge with accuracy typically around 0.5 psi for temperatures $\pm 1°C$ of true. Sample chambers between 1 and 12 gallons can be fitted for saving a fluid sample in extended flow. After each pretest pressure build-up, the probe can be retracted and reset at a different vertical location in the well. In this way any number of pressure data points may be logged in a well. An example of the pretest pressure response in a well is shown in Fig. 9.18. The built-up pressure response at a number of depth points in a well [31] is shown in Fig. 9.19, and the log is compared with the initial pressure gradient of the reservoir. The deviation between current and initial gradients in a producing reservoir can provide a basis for interpretation of reservoir depletion and cross-flow by matching in reservoir simulation[26,30,46].

In many instances the pressure measured by the tool will be the pressure of filtrate which is less than hydrocarbon pressure by the magnitude of capillary threshold pressure. This difference will only be significant in lower permeability formations but

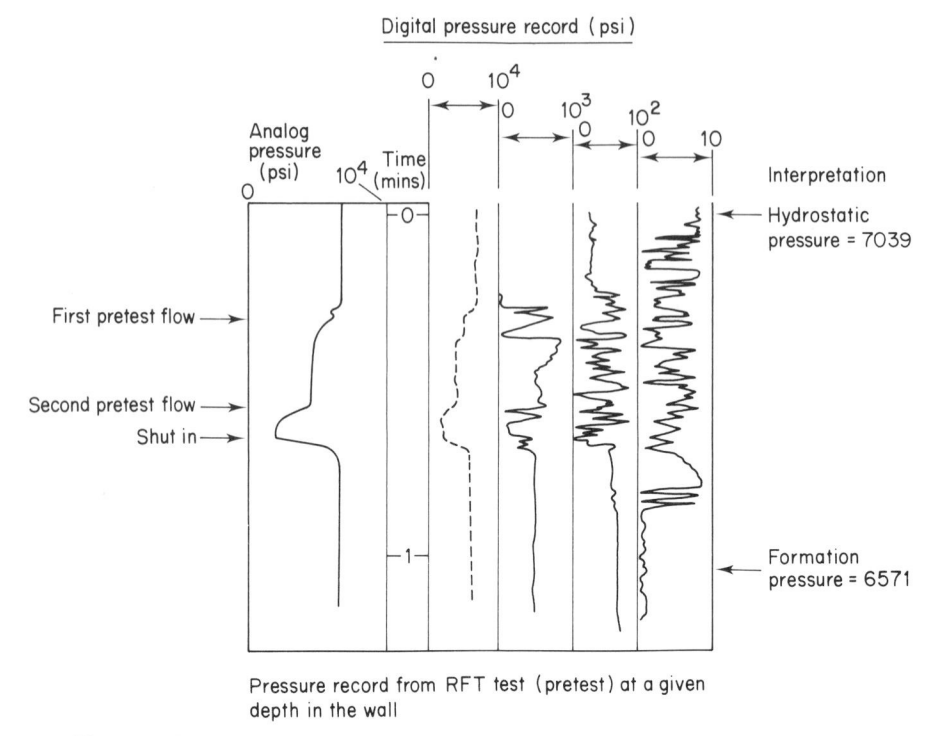

Fig. 9.18 Pressure record from RFT test (pretest) at a given depth in the well.

leads to the conclusion that RFT gradient intersection may represent a hydrocarbon water contact rather than a free water level (see Chapter 6).

9.10 WELL TESTING AND PRESSURE ANALYSIS

9.10.1 Production testing

Production tests are carried out after a well has been completed, with the final casing and liner (if run), flow string and any necessary downhole production equipment (storm chokes, safety valves etc.) installed. The Christmas tree and surface controls will be installed, and testing is limited only by the restraints of production facilities, handling capacities and manpower availability. The production facilities may be temporary (e.g. Rolo tester or other portable tester), or permanent, and the produced hydrocarbon will not generally be flared (although gas associated with produced oil may be burnt off). Under these conditions long drawdown, interference and build-up tests of days' or weeks' duration may be undertaken, enabling very much more significant pressure data to be obtained, and the stabilized flow conditions giving better oil and gas samples than is possible on shorter tests.

Good testing practices involve the monitoring of bottom-hole pressures with a subsurface gauge, and of annulus (when open) and wellhead pressures, and continuous monitoring of oil and gas flow rates.

With prolonged testing, a range of flow rates may be utilized to help establish the productivity index and inflow performance of a well, but this should always be secondary to establishing a valid, essentially constant rate, flow and build-up survey.

It is frequently a practice to flow a well for clean up before beginning a production test proper, but since this period will induce pressure transients it is desirable that some monitoring of this period through orifice meter or burning line pressures should be maintained, unless it is certain that the reservoir returns to an equilibrium state before the production test.

Oil flow rates may be measured by one or more of the following:

orifice meter readings;
positive displacement meter readings;
tank dips.

Gas flow rates will generally be measured by orifice meters.

Separator and tank samples should be monitored regularly for bottom settlings and water.

It is particularly important that all events should be properly recorded – when there is a divided

9 RADIAL FLOW ANALYSIS OF WELL PERFORMANCE

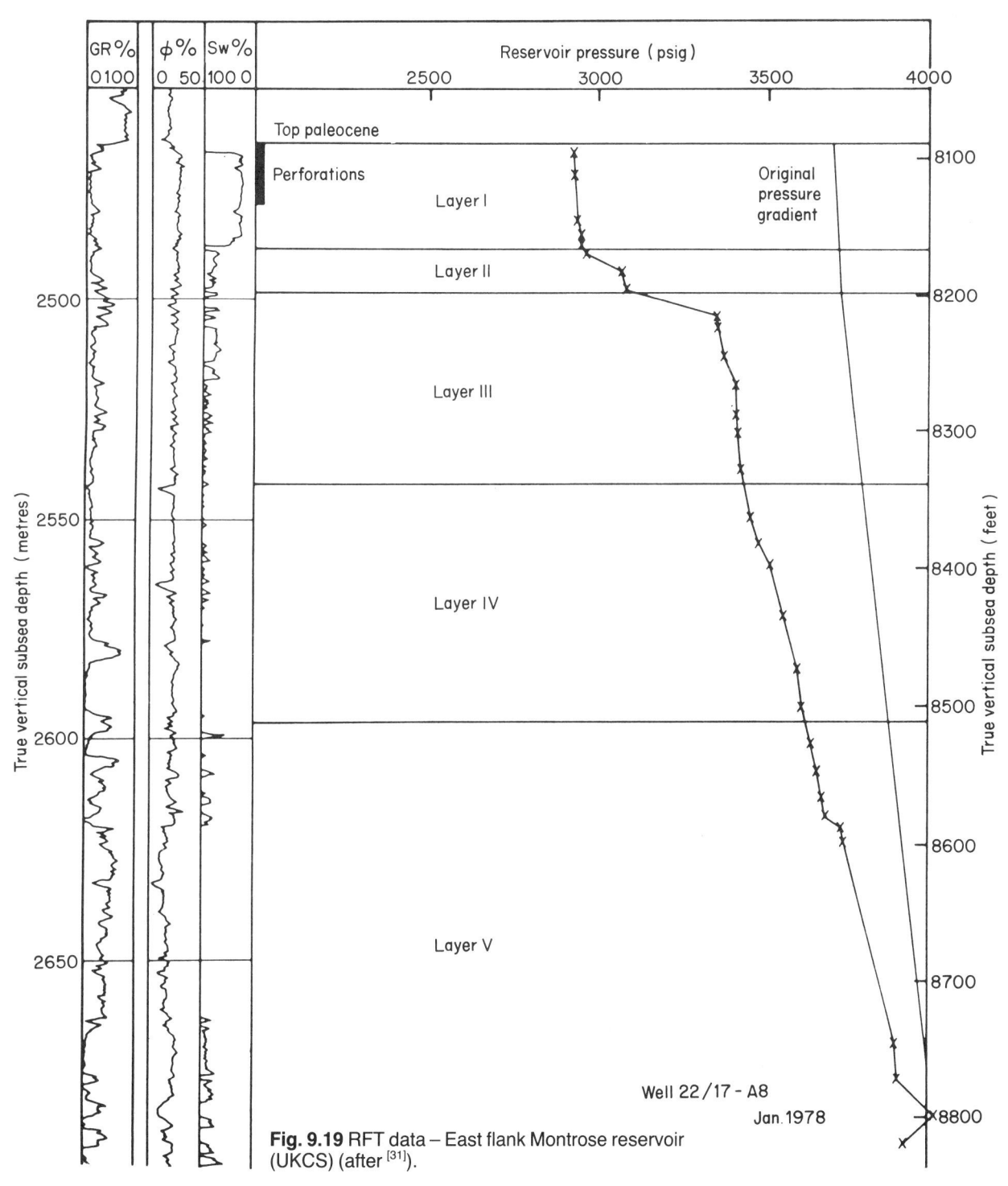

Fig. 9.19 RFT data – East flank Montrose reservoir (UKCS) (after [31]).

responsibility for bottom-hole gauges and surface separator readings, it is sometimes possible for essential data to be omitted from reports. Accurate times, recording of datum levels used (SS, RKB or wellhead tend to be used indiscriminately without being logged), bottom-hole, wellhead, annulus, flow line and separator pressures, choke changes, accurate recording of rates and GORs are essential to accurate interpretation of test data. Interpretation may be difficult enough under ideal conditions; with missing or inaccurate data it may become impossible!

Examples

Example 9.1

Calculate the dimensionless time t_D for the following cases:

(a) $\phi = 0.15$ $r = 10$ cm
      $\mu = 0.3$ cp $t = 10$ s
      $c = 15 \times 10^{-5}$ atm^{-1} $k = 0.1$ D

(b) ϕ, μ, c as above $r = 10$ cm
                        $t = 1000$ s
                        $k = 0.01$ D

(c) ϕ, μ, c as above $r = 100\,000$ cm
                        $t = 10\,000$ s
                        $k = 0.05$ D

Example 9.2

Find the exponential integrals and pressure drops for the following cases:

(a) $\phi = 0.12$ $r = 10$ cm
      $\mu = 0.7$ cp $t = 1$ s
      $c = 10 \times 10^{-5}$ atm^{-1} $k = 0.05$ D
      $h = 2400$ cm $q = 10\,000$ reservoir ccs/s

(b) as above $r = 30\,000$ cm
                 $t = 24$ h

(c) $\phi = 0.12$ $r = 500\,000$ cm
      $\mu = 0.7$ cp $q = 250\,000$ ccs/s
      $c = 10 \times 10^{-6}$ psi^{-1} $t = 365$ days
      $h = 2400$ cm $k = 0.05$ D

Example 9.3

Plot the following drawdown data and estimate the permeability thickness product.

time (h)	0	1.5	3.0	6.0	9.0	12	18	24	48	72
pressure (psi)	5050	4943	4937	4935	4929	4927	4923	4921	4916	4912

Flowing rate was constant at 500 bopd.

Oil viscosity   0.5 cp
Oil formation volume factor   1.7535 RB/STB

If net pay thickness is 60 ft, what is the estimated permeability to oil?

9 RADIAL FLOW ANALYSIS OF WELL PERFORMANCE

Example 9.4

An oil well produces at 500 stb/d for 60 days
Initial reservoir pressure = 5050 psi
Flowing pressure before shut-in = 4728 psi
Pressure build-up data:

Shut-in time (h)	0.25	0.50	1.0	1.5	2.0	3.0	6.0	9.0	18.0	36.0	48
Pressure (psi)	4967	4974	4981	4984	4987	4991	4998	5002	5008	5014	5017

Oil viscosity = 0.7 cp
Oil formation volume factor = 1.454
Estimated net formation thickness = 120 ft
Average porosity = 0.135
Effective fluid compressibility = 17×10^{-6} (psi)$^{-1}$
Effective well radius = 6 in.

Calculate: kh; k; skin factor and completion factor, and pressure drop across skin.

Example 9.5

A well discovers an undersaturated oil reservoir of thickness 50 ft. The static pressure is 1800 psia.
A fluid sample has the following properties:

Oil formation volume factor = 1.365 RB/STB
Effective compressibility of fluid in place = 15×10^{-6} (psi)$^{-1}$

The well was tested at a constant rate of 500 b/d, during which the following pressure record was obtained:

Time (h)	3	6	12	18	24	36	48	72
Pressure (psia)	1438	1429	1420	1415	1412	1407	1403	1398
Time (h)	84	96	120	144	168	192	216	240
Pressure (psia)	1396	1395	1392	1389	1386	1383	1380	1377

Calculate the order of magnitude of the oil in place.

Example 9.6

A test on a gas well gives the following results:

Flow rate MSCF/d	Duration (h)	Bottom hole pressure (psia)
7290	4.5	2506.6
0	4.5	2514.5
16 737	4.5	2489.8
0	4.5	2513.6
25 724	4.5	2467.4
0	4.5	2512.7
35 522	4.5	2435.9

The following build-up was then recorded:

Shut-in time (h)	1	1.5	2	2.5	3	4	5	6
Pressure (psia)	2509.7	2510.7	2511.3	2511.7	2512.1	2512.5	2513.0	2513.2

Over the pressure range considered, viscosity and compressibility factor can be considered constant at $\mu_g = 0.017$ cp $z = 0.856$.

Reservoir temperature	= 138°F
Initial reservoir pressure	= 2515 psia
Well radius	= 0.4 ft
Formation thickness	= 200 ft
Hydrocarbon porosity	= 0.10
Gas gravity, γ_g	= 0.64
r_e	= 5000 ft

(a) Plot the back pressure curve.
    Determine AOF and slope.
(b) Determine permeability and apparent skin factors.
(c) Determine inertial coefficients and inertial pressure drops.

References

[1] Earlougher, R.C.
Advances in Well Test Analysis, Monograph Series 5 SPE of AIME (1977).
[2] Matthews, C.S. and Russell, D.G.
Pressure Build Up and Flow Tests in Wells, Monograph Series 1 SPE of AIME (1967).
[3] Dake, L.P.
Fundamentals of reservoir engineering, *Devel. in Pet. Sci.* **8**, Elsevier (1978).
[4] Cullender, M.H.
The isochronal performance methods of determining the flow characteristics of gas wells, *Trans. AIME* **204** (1955), 137.
[5] Edwards, A.G. and Shryock, S.H.
New generation drillstem testing tools/technology, *Pet. Eng.* (July 1974), 46.
[6] Fetkovich, M.J.
The isochronal testing of oil wells, SPE 4529, *48th Ann. Fall Mtg.* (1973).
[7] Gibson, J.A. and Campbell, A.T.
Calculating the distance to a discontinuity from DST data, SPE 3016, *45th Ann. Fall Mtg.* (1970).
[8] Energy Resources Conservation Board
Guide for the Planning, Conducting and Reporting of Subsurface Pressure Tests, ERCB Report 74–T (Nov. 1974), Calgary, Canada.
[9] Hirasake, G.J.
Pulse tests and other early transient pressure analyses for insitu estimation of vertical permeability, *SPEJ* (Feb. 1974), 75.
[10] Khurana, A.K.
Influence of tidal phenomena on interpretation of pressure build up and pulse tests, *Aust. Pet. Expl. J.* **16**(1), (1976), 99.
[11] Energy Resources Conservation Board
Theory and Practice of Testing Gas Wells, ERCB –75–34 (1975), Calgary, Canada.
[12] Lee, W.J.
Well Testing, Society of Petroleum Engineers, Dallas (1982), Textbook Series.
[13] Matthews, C.S., Brons, F. and Hazebroek, P.
A method for determination of average pressure in a bounded reservoir, *Trans. AIME* **201** (1954), 182.
[14] Miller, C.C., Dyes, A.B. and Hutchinson, C.A.
The estimation of permeability and reservoir pressure from bottom hole pressure build up characteristics, *Trans. AIME* **189** (1950), 91.
[15] Miller, G.B., Seeds, R.W. and Shira, H.W.
A new surface recording down hole pressure gauge, SPE 4125, *47th Ann. Fall Mtg.,* San Antonio (1972).

9 RADIAL FLOW ANALYSIS OF WELL PERFORMANCE

[16] Odeh, A.S., Moreland, E.E. and Schueler, S.
Characterisation of a gas well from one flow test sequence, *J. Pet. Tech.* (Dec. 1975), 1500.
[17] Raghavan, R.
Well test analysis: well producing by solution gas drive, *SPEJ* (Aug. 1976), 196.
[18] Ramey, H.J.
Application of the line source solution to flow in porous media – a review, *Prod. Mon.* (May 1967), 4–7, 25–27.
[19] Ridley, T.P.
The unified analysis of well tests, SPE 5587, *50th Ann. Fall Mtg.* (1975).
[20] Timmerman, E.H. and van Poollen, H.K.
Practical use of drillstem tests, *J. Can. Pet. Tech.* (Apr.–June 1972), 31.
[21] van Everdingen, A.F. and Hurst, W.
The application of the Laplace transformation to flow problems in reservoirs, *Trans. AIME* **186** (1949) 305.
[22] Vogel, J.V.
Inflow performance relationships for solution gas drive wells, *JPT* (Jan. 1968), 83.
[23] Vela, S. and McKinley, R.M.
How areal heterogeneities affect pulse test results, *SPEJ* (June 1970), 181.
[24] Weeks, S.G. and Farris, G.F.
Permagauge – a permanent surface recording downhole pressure monitor through a tube, SPE 5607, *50th Ann. Fall Mtg. SPE* (1975).
[25] Winestock, A.G. and Colpitts, G.P.
Advances in estimating gas well deliverability, *J. Can. Pet. Tech.* (July–Sept. 1965), 111.
[26] Stewart, G. and Wittmann, M.J.
Interpretation of the pressure response of the repeat formation tester, SPE 8362, *54th Ann. Fall Mtg. SPE* (1979).
[27] Pinson, A.E.
Concerning the value of producing time in average pressure determinations from pressure build up analysis, *JPT* (Nov. 1972), 1369.
[28] Peaceman, D.
Interpretation of well block pressure in numerical simulation, *SPEJ* (June 1978), 183.
[29] Schlumberger,
RFT: Essentials of Pressure Test Interpretation, Doc. M–081022 Schlumberger (1981).
[30] Dake, L.P.
Application of the repeat formation tester in vertical and horizontal pulse testing in the Middle Jurassic Brent sands, Paper EUR 270, *Proc. Europ. Off. Pet. Conf.* (1982), 9.
[31] Bishlawi, M. and Moore, R.L.
Montrose field reservoir management, Paper EUR 166, *Proc. Europ. Off. Pet. Conf.* (1980), 205.
[32] Bourdet, D., Ayoub, J.A. and Pirard, Y.
Use of pressure derivative in well test interpretation, SPE Paper 12777, *Proc. Calif. Reg. Mtg.* (1984).
[33] Gringarten, A.C. *et al.*
A comparison between different skin and wellbore storage type curves for early time transient analysis, SPE 8205, *Proc. 54th Ann. Fall Mtg. SPE of AIME* (1979).
[34] Al-Hussainy, R., Ramey, H.J. and Crawford, P.B.
The flow of real gases through porous media, *JPT* (May 1966), 624.
[35] Agarwal, R.G., Al-Hussainy, R. and Ramey, H.J.
An investigation of wellbore storage and skin effect in unsteady liquid flow, *SPEJ* (Sept. 1970), 279.
[36] Streltsova, T.D.
Well pressure behaviour of a naturally fractured reservoir, *SPEJ* (Oct. 1983), 769.
[37] Horner, D.R.
Pressure build up in wells, *Proc. 3rd World Pet. Cong. II* (1951), 503.
[38] Baldwin, D.E.
A Monte Carlo model for pressure transient analysis, SPE 2568, *44th Ann. Fall Mtg. SPE* (1969).
[39] Erhaghi, I. and Woodbury, J.J.
Examples of pitfalls in well test analysis, *JPT* (Feb. 1985), 335.
[40] Ramey, H.J.
Pressure transient testing, *JPT* (July 1982).
[41] McGee, P.R.
Use of a well model to determine permeability layering from selective well tests, *JPT* (Nov. 1980), 2023.
[42] Clark, G. and van Golf-Racht, T.D.
Pressure derivative approach to transient test analysis: a high permeability North Sea reservoir example, SPE 12959, *Proc. Europ. Pet. Conf.,* London (1984), 33.

[43] Wilson, M.R.
Some well test analysis results for a North Sea oil well, SPE 12966, *Proc. Europ. Pet. Conf.* London (1984).
[44] Gringarten, A.C.
Interpretation of transient well test data, In *Developments in Petroleum Eng. –1* (Ed. Dawe and Wilson), Elsevier Applied Science (1985).
[45] Brouse, M.
Proper equipment and techniques ensure better drill stem tests, *World Oil* (May 1983), 87.
[46] Archer, J.S.
Reservoir definition and characterisation for analysis and simulation, Paper PD(1), *Proc. 11th World Pet. Cong.* London (1983).
[47] Diemer, E.
Pressure analysis: the hardware. In: *Developments in Petroleum Engineering,* (Ed. Dawe and Wilson), Elsevier Applied Science, Barking (1985), 89.

Chapter 10
Reservoir Performance Analysis

Recovery of hydrocarbons from a reservoir may make exclusive use of the inherent energy of the *system* (primary recovery); energy may be added to the system in the form of injected fluids (secondary recovery); some of the residual hydrocarbon trapped during *conventional* recovery processes may be mobilized (tertiary or enhanced oil recovery).

10.1 RECOVERY FROM GAS RESERVOIRS

The expansion of gas in the reservoir pore space as pressure declines during production is the most significant mechanism in analysis of gas reservoirs. The compressibility of gas is generally significantly greater than that of the reservoir pore volume, and in the absence of water influx the volumetric material balance reduces to the following expression at reservoir conditions:

Initial gas volume at initial pressure =
  remaining gas volume at lower pressure

Using the terminology:

G = standard condition volume of gas initially in place;
G_p = standard condition volume of cumulative gas produced;
B_g = gas formation volume factor (res.vol/stand.cond.vol.);
i = subscript for initial conditions.

then
$$G(B_{gi}) = (G - G_p)B_g$$

Since the gas formation volume factor represents a ratio between reservoir and standard condition volumes then a simple equation of state can be used in its representation:

$$\frac{(B_{gi})}{B_g} = \frac{V_i}{V} = \frac{z_i n R T_i}{P_i} \cdot \frac{P}{znRT}$$

For an isothermal reservoir $T_i = T$, therefore

$$\frac{(B_{gi})}{B_g} = \frac{z_i \cdot P}{P_i \cdot z}$$

from which

$$\frac{z_i P}{P_i z} = \frac{G - G_p}{G}$$

or

$$\frac{P}{z} = \frac{P_i}{z_i}\left[1 - \frac{G_p}{G}\right]$$

This equation can be arranged in a linear form as shown in Fig. 10.1.

$$\frac{P}{z} = \left(-\frac{P_i}{z_i G}\right)G_p + \frac{P_i}{z_i}$$

A plot of P/z against the cumulative produced gas volume has two significant intercepts, firstly $P/z = P_i/z_i$ at $G_p = 0$ and secondly $G_p = G$ at $P/z = 0$. The application of the P/z against G_p plot in the reservoir analysis of a depletion drive gas reservoir

157

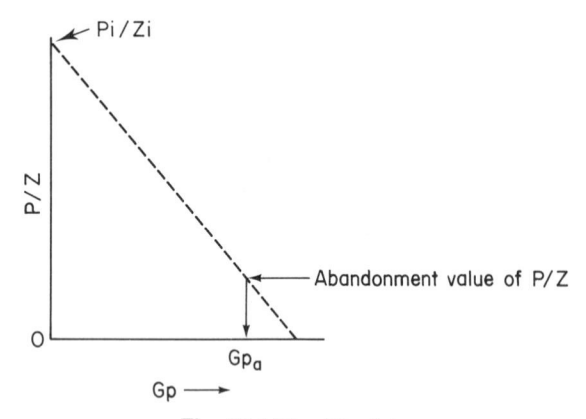

Fig. 10.1 The P/z plot.

without water influx is therefore particularly useful in providing a further estimate of gas in place by extrapolation of early production data. When the value of G indicated by the plot is significantly different from volumetric estimates, then assumptions of reservoir continuity in the field might be questioned. The variation of field data from linearity is a fairly frequent observation and thus may be an indication of water influx (increasing pressure support) or aquifer depletion (decreasing pressure support by fluid transport to another reservoir). Figure 10.2 shows the more usual representation of limited aquifer influx indicated by production data.

In this case the material balance equation must be written as follows, where W_e is the cumulative volume of water influx at reservoir conditions, W_p is the surface condition volume of water produced from wells and B_w is the formation volume factor for water

$$G(B_{gi}) = (G - G_p) B_g + W_e - W_p B_w$$

A linear equation which can be solved by assuming values of W_e to force linearity can be written as follows:

$$\frac{G_p B_g + W_p B_w}{B_g - (B_{gi})} = \left[\frac{1}{B_g - (B_{gi})}\right] W_e + G$$

The relationship between the values of W_e indicated and reservoir pressure at the original gas–water contact can be used to establish the performance classification of the aquifer, i.e. steady state, pseudo steady state, unsteady state. If the production terms $(G_p B_g + W_p B_w)$ are denoted as F and the volume expansion term $(B_g - B_{gi})$ as E_x, the material balance equation becomes

$$F = W_e + G E_x$$

and the linearized equation is, as shown in Fig. 10.3,

$$\frac{F}{E_x} = \frac{W_e}{E_x} + G$$

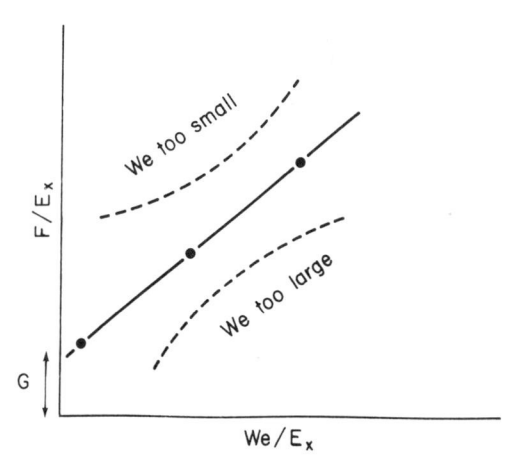

Fig. 10.3 Aquifer performance.

Figure 10.3 shows that the evaluation of W_e is a forcing exercise.

Water influx in a gas reservoir lowers the recovery factor G_p/G by two mechanisms in comparison to normal depletion. Through partial maintenance of reservoir pressure by the influxed water the gas expansion process is arrested. In addition, the water traps gas at relatively high pressures behind the advancing front. The magnitude of trapped gas saturation is likely to be rate dependent, but for many sandstones the sparse literature suggests an

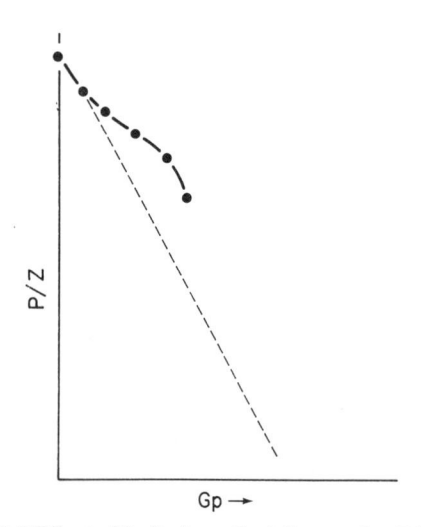

Fig. 10.2 Effect of limited aquifer influx on the P/z plot.

10 RESERVOIR PERFORMANCE ANALYSIS

order of 40% pore volume. A comparison of ultimate recovery factors for natural depletion and water influx can be made.

$$\left.\frac{G_p}{G}\right]_{\substack{\text{depletion,}\\\text{abandonment}}} = 1 - \frac{P_a z_i}{P_i z_a}$$

where a = abandonment pressure conditions. Similarly,

$$\left.\frac{G_p}{G}\right]_{\substack{\text{water influx,}\\\text{abandonment}}} = \left[\frac{1-\bar{S}_{gr}-S_{wi}}{1-S_{wi}}\right]\frac{B_g}{(B_{gi})}$$

where $\bar{S}_{gr}$ is the average residual gas saturation in the reservoir. Such comparisons often yield water influx: depletion recovery factor ratios of about 0.77:1.

10.2 PRIMARY RECOVERY IN OIL RESERVOIRS

Oil can be recovered from the pore spaces of a reservoir rock by expansion or only to the extent that the volume originally occupied by the oil is invaded or occupied in some way. There are several ways in which oil can be produced from a reservoir, and these may be termed mechanisms or *drives*, and where one replacement mechanism is dominant, the reservoir may be said to be operating under a particular *drive*. The analysis of drive mechanisms using a method of material balance follows the general form described by Schilthuis [13].

Possible sources of replacement for produced fluids are:

(a) expansion of undersaturated oil above the bubble-point;
(b) the release of gas from solution in the oil at and below the bubble-point;
(c) invasion of the original oil-bearing reservoir by gas from a free gas cap;
(d) invasion of the original oil-bearing reservoir by water from an adjacent or underlying aquifer.

All replacement processes involve a reduction in pressure in the original oil zone, although pressure drops may be small if gas caps are large, and aquifers large and permeable, and pressures may stabilize at constant or declining reservoir offtake rates under favourable circumstances.

The compressibility of oil, rock and connate water is generally relatively small, so that pressures in undersaturated oil reservoirs will fall rapidly to the bubble-point if there is no aquifer to provide water drive. As a result, these expansion mechanisms are not usually considered separately, and the three principal categories of reservoir drive are:

(1) solution gas drive (or depletion drive) reservoirs;
(2) gas cap expansion drive reservoirs;
(3) water drive reservoirs.

Frequently two or all three mechanisms (together with rock and connate water expansion) may occur simultaneously and result in a *combination drive*.

10.2.1 Solution gas drive: analysis by material balance

If a reservoir at its bubble-point is put on production, the pressure will fall below the bubble-point pressure and gas will come out of solution. Initially this gas may be a disperse discontinuous phase, essentially immobile, until some minimum saturation – the equilibrium, or critical gas, saturation – is attained. The actual order of values of critical saturation are in some doubt, but there is considerable evidence to support the view that values may be very low – of the order of 1% to 7% of the pore volume. Once the critical gas saturation has been established, gas will be mobile and will flow under whatever potential gradients may be established in the reservoir – towards producing wells if the flowing or *viscous* gradient is dominant – segregating vertically if the gravitational gradient is dominant. Segregation will be affected by permeability variations in layers but is known to occur even under apparently unfavourable conditions.

Initially then, a well producing from a closed reservoir will produce at solution GOR. At early times, as pressure declines and gas comes out of solution, but cannot flow to producing wells, the producing GOR will decline. When the critical gas saturation is established and if the potential gradients permit, gas will flow towards producing wells. The permeability to oil will become lower than at initial conditions, and there will be a finite permeability to gas so that the producing gas–oil ratio will rise. As more gas comes out of solution, and gas saturations increase, permeability to gas increases, permeability to oil diminishes and this trend accelerates. Ultimately, as reservoir pressure declines towards abandonment pressure, the change in gas formation volume factor offsets the increasing gas to oil mobility ratio, and the gas–oil ratio trend is reversed, i.e. although the reservoir GOR may continue to increase in terms of standard volumes, the ratio standard cubic ft/stock tank barrel may decline.

In addition to the effect of gas on saturation of,

and permeability to, oil, the loss of gas from solution also increases the viscosity of the oil and decreases the formation volume factor of the oil.

The solution gas drive performance of unconsolidated sand reservoirs and chalk reservoirs (such as those in the Norwegian sector of the North Sea) require consideration and inclusion of pore volume compressibility as it can be of the same order as oil compressibility [15, 22]. In well cemented reservoirs, the pore volume compressibility may be small in comparison to gas and oil compressibility and may be ignored below bubble point.

The analysis of performance of a solution gas drive reservoir can be conducted by use of reservoir condition material balance, or volumetric balance techniques. The methods consider a number of static equilibrium stages of reservoir production during which pressure changes have occurred. At any equilibrium stage a balance is made on the original reservoir content at original pressure and the current reservoir content at current pressure. The pressure reduction between initial and later conditions is accompanied by all or some of the following:

expansion of remaining oil;
liberation and expansion of dissolved gas;
expansion of connate water;
compaction of rock pore volume;
production of a cumulative volume of oil;
production of a cumulative volume of gas;
production of a cumulative volume of water;
injection or influx of a cumulative volume of water;
injection of a cumulative volume of gas.

The nomenclature for describing these processes between initial pressure P_i and some later pressure P is defined as follows:

G_p = cumulative gas produced at surface (standard volumes);
N = stock tank volume of oil initially in place;
NB_{oi} = reservoir condition volume of oil initially in place;
N_p = stock tank volume of cumulative oil produced;
N_pB_o = reservoir condition volume of cumulative oil produced;
R_p = cumulative produced gas–oil ratio (standard volumes) ($= G_p/N_p$);
R_s = solution gas–oil ratio (standard volumes);
c_t = total compressibility $= c_oS_o + c_wS_w + c_gS_g + c_f$;
c_{oe} = effective oil compressibility $= c_t/1-S_w$;
ΔP = pressure change from initial conditions ($= P_i - P$);
W_e' = net water influx in reservoir condition volume units ($= W_e - W_pB_w + W_{inj}B_w$)

For oil production of an undersaturated reservoir from P_i down to bubble-point pressure P_b, the solution gas–oil ratio remains constant in the reservoir. A reservoir condition volumetric balance is thus

$$N_pB_o = N\left(B_o - (B_{oi})\right) + V_p\left(c_wS_{wi} + c_f\right)\Delta P + W_e'$$

The middle term represents the expansion of connate water as fluid pressure is reduced and the compaction of pore volume as grain pressure is increased. For a constant overburden pressure, the decrease in fluid pressure is equal to the increase in grain pressure. The pore volume compressibility c_f is therefore a positive value with respect to fluid pressure reduction. The pore volume is represented by V_p.

Since initial oil saturation S_{oi} can be represented as NB_{oi}/V_p, then

$$V_p = \frac{N(B_{oi})}{S_{oi}} = \frac{N(B_{oi})}{1 - S_{wi}}$$

The equation thus becomes transformed into

$$N_pB_o = N\left(B_o - (B_{oi})\right) + \frac{N(B_{oi})}{1 - S_{wi}}(c_wS_{wi} + c_f)\Delta P + W_e'$$

Since, down to bubble-point pressure $c_o = (B_o - B_{oi})/(B_o)_i\Delta P$ then in terms of the effective oil compressibility c_{oe} this expression becomes

$$N_pB_o = N(B_{oi})\left[c_o + \frac{c_wS_{wi} + c_f}{1 - S_{wi}}\right]\Delta P + W_e'$$

Therefore

$$N_pB_o = N(B_{oi})c_{oe}\Delta P + W_e'$$

The recovery factor for pressures down to bubble-point becomes

$$\left.\frac{N_p}{N}\right]_{P_b<P} = \frac{(B_{oi})c_{oe}\Delta P}{B_o} + \frac{W_e'}{NB_o}$$

As the pressure falls below bubble-point pressure, gas from solution is released and may form free gas saturation in the reservoir or/and be produced. The cumulative gas oil ratio R_p will then become greater than R_{si} and the remaining solution gas–oil ratio R_s in the reservoir will fall below R_{si}.

A reservoir conditions volumetric balance thus

10 RESERVOIR PERFORMANCE ANALYSIS

becomes an equation of the change in reservoir volume, equivalent to total production with the volume change associated with remaining and influxed fluids and adjusted pore volume.

The production of oil and free gas together with any water is

$$N_p B_o + R_p N_p B_g - R_s N_p B_g + W_p B_w$$

The expansion of original oil between pressure P_i and current pressure is

$$NB_o - N(B_{oi})$$

The expansion of liberated solution gas expressed in reservoir volumes at the current pressure is

$$NR_{si} B_g - NR_s B_g$$

The change in hydrocarbon pore volume between pressure P_i and the current pressure is

$$N(B_{oi})\Delta P \left\{ \frac{c_w S_{wi} + c_f}{1 - S_{wi}} \right\}$$

The net influx terms are represented by W_e'.

The balance therefore becomes

$$N_p \left\{ B_o + (R_p - R_s) B_g \right\} = N(B_{oi}) \left\{ \frac{B_o - (B_{oi})}{(B_{oi})} \right.$$
$$\left. + \frac{(R_{si} - R_s) B_g}{(B_{oi})} + \Delta P \frac{(c_w S_{wi} + c_f)}{1 - S_{wi}} \right\} + W_e'$$

In high pore volume compressibility reservoirs such as chalks and unconsolidated sands, the energy contribution of *compaction drive* cannot be ignored even at quite high gas saturations. In other situations, pore volume and connate water compressibility can be small in comparison with gas compressibility and is often ignored in calculations.

Following the nomenclature used by Havlena and Odeh [2], the production and expansion terms can be grouped as follows:

$$F = N_p \left\{ B_o + (R_p - R_s) B_g \right\}$$
$$E_o = \left(B_o - (B_{oi}) \right) + (R_{si} - R_s) B_g$$
$$E_{f,w} = (B_{oi})\Delta P \left\{ \frac{c_w S_{wi} + c_f}{1 - S_{wi}} \right\}$$

and the balance equation can be represented more easily for calculation purposes as

$$F = N(E_o + E_{f,w}) + W_e'$$

A linearization which yields N as an intercept and has unit slope is given by

$$\frac{F}{(E_o + E_{f,w})} = \frac{W_e'}{(E_o + E_{f,w})} + N$$

and is indicated in Fig. 10.4 (a). Values of W_e are chosen to provide the required linearity and will then indicate the aquifer character. Production volumes and *PVT* properties from field data are required for the analysis. In the absence of any influx terms, a plot of F against $(E_o + E_{f,w})$ yields a straight line through the origin with slope N, as shown in Fig. 10.4 (b).

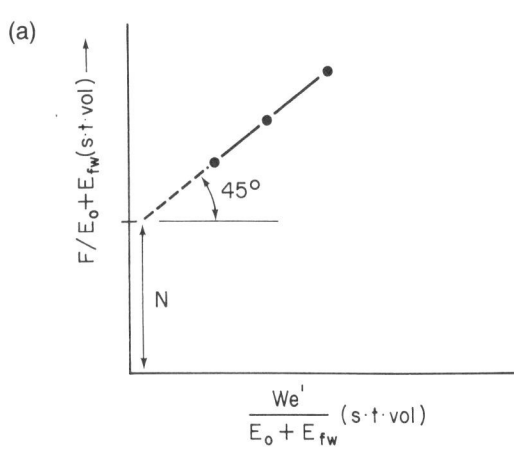

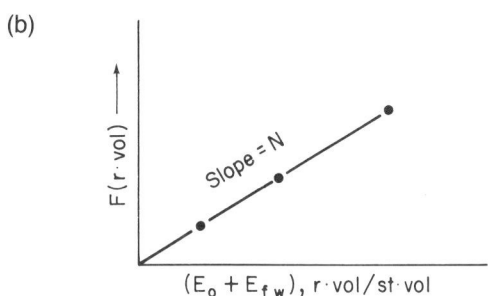

Fig. 10.4 Representation of field data using Havlena and Odeh methods.

The material balance equation for a solution gas drive has also been represented in its static form ignoring pore volume compressibility as follows:

$$N = \frac{N_p (B_o - R_s B_g) + G_p B_g - (W_e')}{\left(B_o - (B_{oi}) \right) + (R_{si} - R_s) B_g}$$

The main expansion terms are in the denominator and by calling them D, Tracy [1] defined the following material balance equation, i.e. $D = (B_o - B_{oi}) + (R_{si} - R_s) B_g$, then:

$$N = N_p F_o + G_p F_g - W_e' F_w$$

where
$$F_o = (B_o - R_s B_g)/D$$
$$F_g = B_g/D$$
$$F_w = 1/D$$

Tracy developed the earlier Tarner [8] method using this formulation to predict recovery performance below bubble-point pressure. Relative permeability data is required which gives k_g/k_o as a function of the average oil saturation $\bar{S}_o$ or liquid saturation $\bar{S}_L$ ($= 1 - \bar{S}_g$) in the reservoir. For the situation where net influx and water production is zero and where pore volume compressibility is insignificant, an analysis in terms of unit stock tank volume of oil in place can be written:

$$1 = N_p F_o + G_p F_g$$

If we use the subscript j to represent the time level in which we are interested, then for a pressure decrement from P_{j-1} to P_j it is necessary to estimate the average producing gas–oil ratio $\bar{R}$ during the decrement:

$$\bar{R} = \frac{R_{j-1} + R_j(k)}{2}$$

This is obtained by an iterative process, so $R_j(k)$ represents the estimated value of R at time j for k guesses from $k = 1$ to n. The correct value of R_j will be obtained by comparing $R(k)$ with a value V/R' calculated as follows.

Between P_{j-1} and P_j the incremental production added to the cumulative production at the time $j-1$ gives the cumulative production at time j, i.e. letting ΔN_p be the cumulative oil production we have

$$(N_p)_j = (N_p)_{j-1} + \Delta N_p$$

and

$$1 = \left((N_p)_{j-1} + \Delta N_p\right)(F_o)_j + \left((G_p)_{j-1} + \bar{R}\,\Delta N_p\right)(F_g)_j$$

from which

$$\Delta N_p = \frac{1 - \left((N_p)_{j-1} (F_o)_j\right) - \left((G_p)_{j-1} (F_g)_j\right)}{(F_o)_j + \bar{R}(F_g)_j}$$

The average oil saturation in the reservoir at time j is given by material balance as follows.

Let initial oil saturation $S_{oi} = 1 - S_{wi}$ in pore volume PV. Therefore

$$PV = \frac{N(B_{oi})}{(1 - S_{wi})}$$

The oil saturation at time j when the cumulative oil production is $(N_p)_j$ is

$$(S_o)_j = \frac{[N - (N_p)_j] (B_o)_j}{PV}$$
$$= \left[1 - \frac{(N_p)_j}{N}\right] \frac{(B_o)_j}{(B_{oi})} (1 - S_{wi})$$

For unit stock tank oil in place this becomes

$$(S_o)_j = \frac{(B_o)_j}{(B_{oi})}\left[1 - \frac{(N_p)_j}{N}\right](1 - S_{wi})$$

The relative permeability ratio k_g/k_o at time j is obtained as a function of S_o or $(S_o + S_{wi})$ as shown in Fig. 10.5. From this, a new estimate $\bar{R}'$ of the average producing gas–oil ratio can be derived and compared to the original estimate $\bar{R}(k)$:

$$\bar{R}' = (R_s)_j + \frac{(B_o)_j}{(B_g)_j} \cdot \frac{(\mu_o)_j}{(\mu_g)_j} \cdot \left[\frac{k_g}{k_o}\right]_j$$

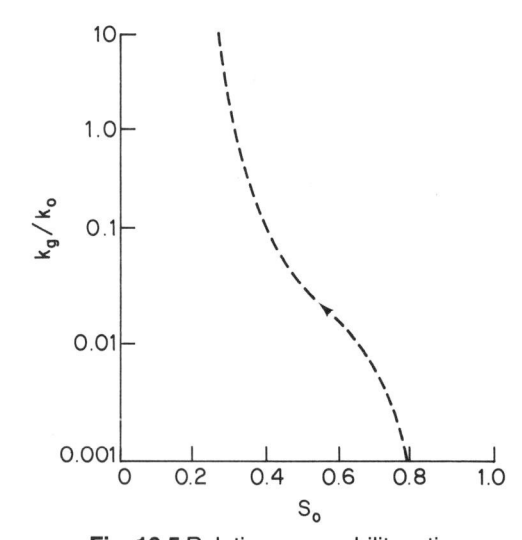

Fig. 10.5 Relative permeability ratio.

A revised estimate of $\bar{R}$ is obtained from $\bar{R} = (\bar{R}(k) + \bar{R}')/2$ and used in a fresh circuit of calculations until a convergence in $\bar{R}$ is obtained (say within 0.1%) for the estimated value of ΔN_p. After convergence

$$(G_p)_j = (G_p)_{j-1} + \bar{R}\,\Delta N_p$$

$$N_p F_o + G_p F_g = 1$$

10 RESERVOIR PERFORMANCE ANALYSIS

and a check is obtained from $N_p F_o + G_p F_g = 1$. The functions F_o and F_g are usually prepared from the PVT data as functions of pressure.

10.2.2 Gas cap expansion drive

The presence of a gas cap at initial reservoir conditions serves to retard the decline in reservoir pressure as oil is produced. The pressure at the original gas–oil contact is by definition the bubble-point pressure since the oil must be saturated. The material balance equation must be formulated by taking into consideration the expansion of the initial gas cap as well as liberation of solution gas from the oil as pressure declines. The efficient recovery of oil will depend on keeping as much gas as possible in the reservoir to act as an expansion energy – completion intervals and location of oil wells are therefore particularly important. The volume of gas at initial reservoir conditions is related to the volume of oil initially in place at reservoir conditions by the ratio term m:

$$\frac{G(B_{gi})}{N(B_{oi})} = m$$

In the formulation a new term must be added to the right-hand side of the solution gas drive equation to represent the gas cap expansion drive process, i.e.

$$\text{Gas cap expansion} = G\left(B_g - (B_{gi})\right)$$

$$= m N(B_{oi})\left[\frac{B_g}{(B_g)_i} - 1\right]$$

If we define E_g as

$$(B_{oi})\left\{\frac{B_g}{(B_{gi})} - 1\right\}$$

then, in Havlena and Odeh formulation

$$F = N\left[E_o + mE_g + (1+m) E_{f,w}\right] + W_e'$$

In this case W_e' can include gas injection $G_{inj} B_g$. It is also expected that the term $E_{f,w}$ can usually be ignored since, in comparison with gas expansion it is small. This may not hold true in chalk reservoirs or other highly compressible unconsolidated sand reservoirs.

In the absence of influx and injection terms the material balance equation can be solved to provide estimates for m and N from production and PVT data by plotting $(F/(E_o + (1+m) E_{f,w}))$ against $(E_g/(E_o + (1+m) E_{f,w}))$ as shown in Fig. 10.6, i.e.

$$\frac{F}{(E_o + (1+m) E_{f,w})} = N + \frac{mN E_g}{(E_o + (1+m) E_{f,w})}$$

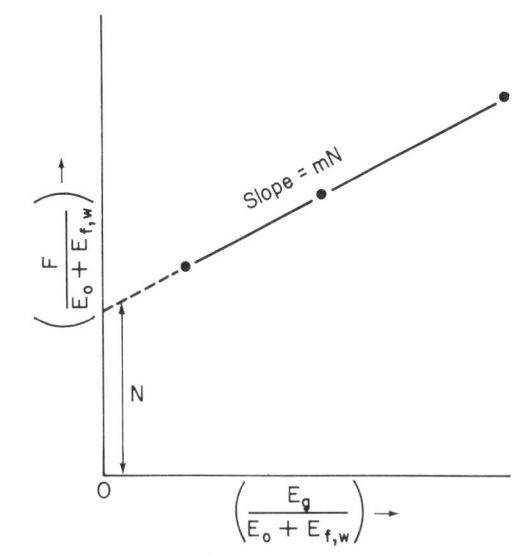

Fig. 10.6 Havlena-Odeh plot in absence of influx and injection terms.

To solve for net influx and injection terms the value of m must be known and, letting $E_T = E_o + E_{f,w} + mE_g$ then a plot of F/E_T against W_e'/E_T gives a line of unit slope and intercept N, as shown in Fig. 10.7, i.e.

$$E_T = E_o + (1+m) E_{f,w} + mE_g$$

$$\frac{F}{E_T} = N + \frac{W_e'}{E_T}$$

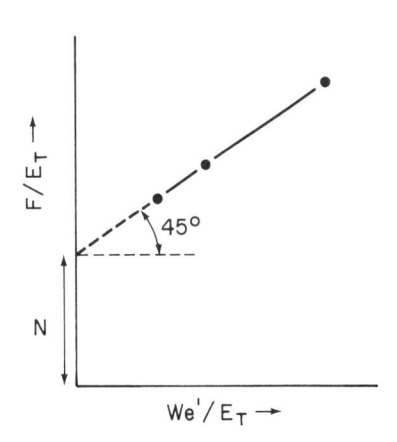

Fig. 10.7 Havlena-Odeh plot using influx and injection terms.

The recovery factor N_p/N evaluated at different pressures during the life of a gas cap drive or solution gas drive recovery process is clearly related to the cumulative produced gas–oil ratio R_p. Since

$$F = NE_T + W_e'$$

therefore for no produced water

$$\left.\frac{N_p}{N}\right]_p = \frac{E_T + W_e'}{B_o + (R_p - R_s)B_g}$$

It is clear from this relationship that low values of R_p lead to higher recovery factors at a given reservoir pressure than high values of R_p. The conclusion may be translated into practice by completing producing wells down flank from primary or secondary gas caps, as shown in Fig. 10.8, to minimize gas saturation around wellbores and also to consider the possibility of gas reinjection to increase W_e'. The expansion energy of gas caps serves oil production best by having gas retained in the reservoir rather than by producing it in large quantities with oil.

10.3 GRAVITY SEGREGATION AND RECOVERY EFFICIENCIES

One mechanism, only briefly referred to, but which has an important role in several aspects of reservoir behaviour is that of gravity segregation – the movement (generally of gas and oil) of phases countercurrent to each other, under the influence of the gravitational potential $g' \Delta \rho$.

Considering the solution gas drive reservoir, the behaviour described earlier assumed essentially that gas saturations build up uniformly throughout the oil zone without any saturation gradients in the vertical direction. (Saturation gradients existing as a result of horizontal pressure gradients, i.e. the pressure drops near the wellbore, will slightly exaggerate this behaviour.)

Under these conditions the expected recovery efficiency will depend on the economic limit for wells and could be as low as 2–3% for low permeability reservoirs with high viscosity and low gas–oil ratio oils, and up to about 15% or so for high permeability reservoirs, normal GOR, low viscosity oils, but will rarely exceed this range. Gravity drainage of oil from a slowly advancing gas cap is reported to give extremely low residual oil saturations – less than 10%.

If the vertical permeability to gas is non-zero, however, there will be a vertical component of gas flow under the gravitational potential, and gas will *segregate* in the reservoir, migrating to structurally high positions, with oil counterflowing downwards. This mechanism has two effects. Firstly, the oil saturation in the lower parts of the reservoir is maintained at a value higher than the average oil saturation – so that permeability to oil is higher, and permeability to gas lower than for the *purely* solution gas drive case. The producing gas–oil ratio is then lower than for solution gas drive alone.

Secondly, the lower producing gas–oil ratio involves smaller gross fluid withdrawals than would otherwise be the case, so that the pressure decline at any given oil cumulative will be smaller, with the usual effects on k_o, B_o and S_o at abandonment.

The segregated gas may form a secondary gas cap, and the later life of a reservoir may then be similar to that of gas cap drive reservoirs.

Under these conditions the recovery efficiencies will be higher – if the economic limit is low, possibly very much higher – and may approach or even exceed the range 20–40% of oil in place.

Gravity drainage plays its greatest role in high contrast dual porosity systems where almost complete segregation can take place in the secondary

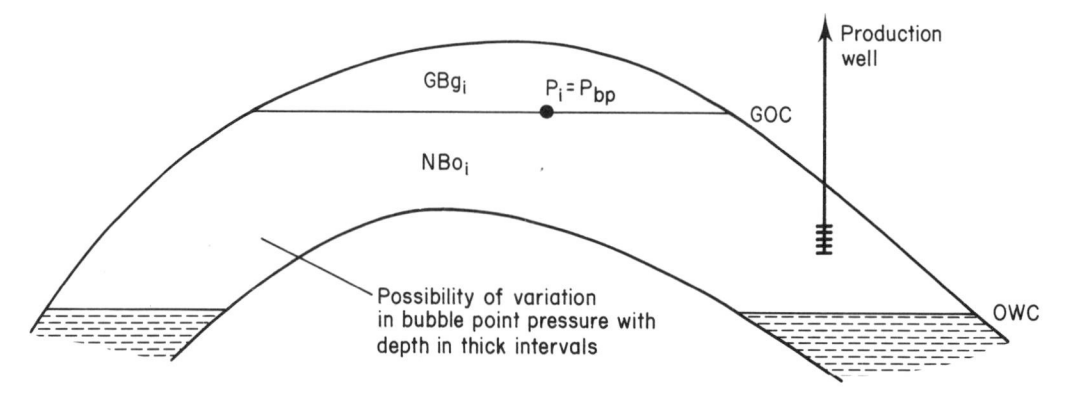

Fig. 10.8 Initial conditions in a reservoir with a gas cap.

10.4 MATERIAL BALANCE FOR RESERVOIRS WITH WATER ENCROACHMENT OR WATER INJECTION

If a reservoir is underlain by, or is continuous with, a large body of water saturated porous rock (an aquifer), then reduction in pressure in the oil zone will cause a reduction in pressure in the aquifer. The total compressibility of an aquifer ($c_w S_w + c_f$) may be relatively large (about $10-6 \times 10^{-6}$ psi^{-1}), and water influx under steady state conditions will obey the rule

$$\Delta V_w = \bar{c} \, V_w \, \Delta P$$

The response time of an aquifer to a change in pressure at the original oil–water contact is of great practical importance. In small aquifers, steady state instantaneous influx may be a good representation but large aquifers tend to behave in an unsteady state manner. When water is required for pressure support in an oil reservoir, the unsteady state response of an aquifer may negate its usefulness and external water injection is frequently used instead. Water injection allows a more immediate replacement of oil zone energy.

The determination of aquifer characteristics is important if water injection is not planned. Two main aquifer geometries, radial and linear, may be considered in analytical analyses, although reservoir simulation can cope with irregular volumes. The properties of an aquifer are rarely known with any confidence since there is usually little well control. The basis for unsteady state aquifer analysis is found in the methods of van Everdingen and Hurst [5], Carter and Tracy [6] and Fetkovitch [7]. They are used to provide estimates of cumulative water encroachment W_e in the total influx term W_e', and can be applied in material balance formulations, such as that due to Havlena and Odeh [2] and discussed in the previous sections.

$$\frac{F}{E_T} = N + \frac{(W_e + W_{inj} B_w + G_{inj} B_g - W_p B_w)}{E_T}$$

where W_{inj} is the cumulative water injection and G_{inj} is the cumulative gas injection both volumes being represented at standard conditions.

The unsteady state aquifer representation of Hurst and van Everdingen requires the discretization of a continuous pressure decline into periods of instantaneous pressure drop. This is achieved using the method of van Everdingen, Timmerman and McMahon [10], where the equivalent instantaneous pressure drop occurring at time zero and subsequent times represented as $\Delta P_o, \Delta P_1, \ldots, \Delta P_j$ are as follows:

$$\Delta P_o = \frac{P_i - P_1}{2}$$

$$\Delta P_j = \frac{P_{j-1} - P_{j+1}}{2}$$

The water influx due to each instantaneous pressure drop is calculated as a time function up to the maximum volume indicated by a steady state instantaneous influx. The total influx into the reservoir is calculated at any time T by superposing the influxes from each pressure drop, which depends on the dimensionless time each has been effective $(T_D - (t_D)_j)$.

For each instantaneous pressure drop from time zero to the end of the nth time step, we can write at some time T

$$W_e(T) = U \sum_{j=0}^{j=n-1} \Delta P_j \cdot W_D \left(T_D - (t_D)_j \right)$$

where:

- T_D is the dimensionless time equivalent to time T;
- $(t_D)_j$ is the dimensionless time at which the instantaneous drop Δp_j commenced;
- t_D is dimensionless time equal in Darcy units to $kt/(\phi \mu \bar{c} r_o^2)$ for radial systems and for linear systems;
- r_o is the radius of the oil zone;
- L is the linear distance from the original OWC to the outer limit of the aquifer;
- U is the aquifer constant equal in Darcy units to $2\pi f \phi h \bar{c} r_o^2$ for radial systems and $wLh\phi\bar{c}$ for linear systems;
- f is the fractional encroachment angle of the radial aquifer = $\theta°/360°$;
- h is the net thickness of the aquifer;
- w is the average width of the linear aquifer;
- $W_D(t_D)$ is the dimensionless cumulative water influx function for a unit pressure drop at the original reservoir OWC at time $t=0$ and easily read from charts or tables of W_D against t_D for different values of re_D. Typical values are shown in Fig. 10.9;
- re_D is the dimensionless radius equal to $r_{aquifer}/r_o$.

In field units where permeability is in mD, time in years, pressure in psi, length in feet, rate in reservoir barrels per day

$U = 1.119 f \phi \, h \, \bar{c} \, r_o^2$ bbl/psi (radial system)

$U = 0.1781 \, w \, L \, h \, \phi \, \bar{c}$ bbl/psi (linear system)

$t_D = 2.309 kt / (\phi \mu \bar{c} r_o^2)$ (radial system, t in years)

$t_D = 2.309 kt / (\phi \mu \bar{c} L^2)$ (linear system, t in years)

If water injection is employed in a reservoir then natural water influx may still occur if the pressure at the original water contact decreases from initial pressure. The combination drive material balance equation which represents a step change from equilibrium at pressure P_i to equilibrium at pressure P can be formulated to show all expansion, production and injection terms as shown below.

In this formulation the subscripts c and s refer in gas terms to conditions in the gas cap and in solution gas. W_{inj} and G_{inj} are cumulative injection volumes at stock tank conditions.

Setting $G(B_g)_i = mN(B_o)_i$; $R_p = G_p/N_p$ and $G_p = (G_p)_c + (G_p)_s$ we can write

$$F = N(E_o + mE_g + (1+m)E_{fw})$$
$$+ (W_e + W_{inj}B_w + G_{inj}B_g - W_pB_w)$$

or

$$F = NE_T + W_e'$$

where: $F = N_p(B_o + (R_p - R_s)B_g)$

$$E_o = (B_o - (B_{oi})) + ((R_{si}) - R_s)B_g$$

$$E_g = (B_{oi})\left[\frac{B_g}{(B_{gi})} - 1\right]$$

$$E_{f,w} = (1+m)(B_{oi})\Delta P \left(\frac{c_w S_{wi} + c_f}{1 - S_{wi}}\right)$$

$$E_T = E_o + mE_g + E_{f,w}$$

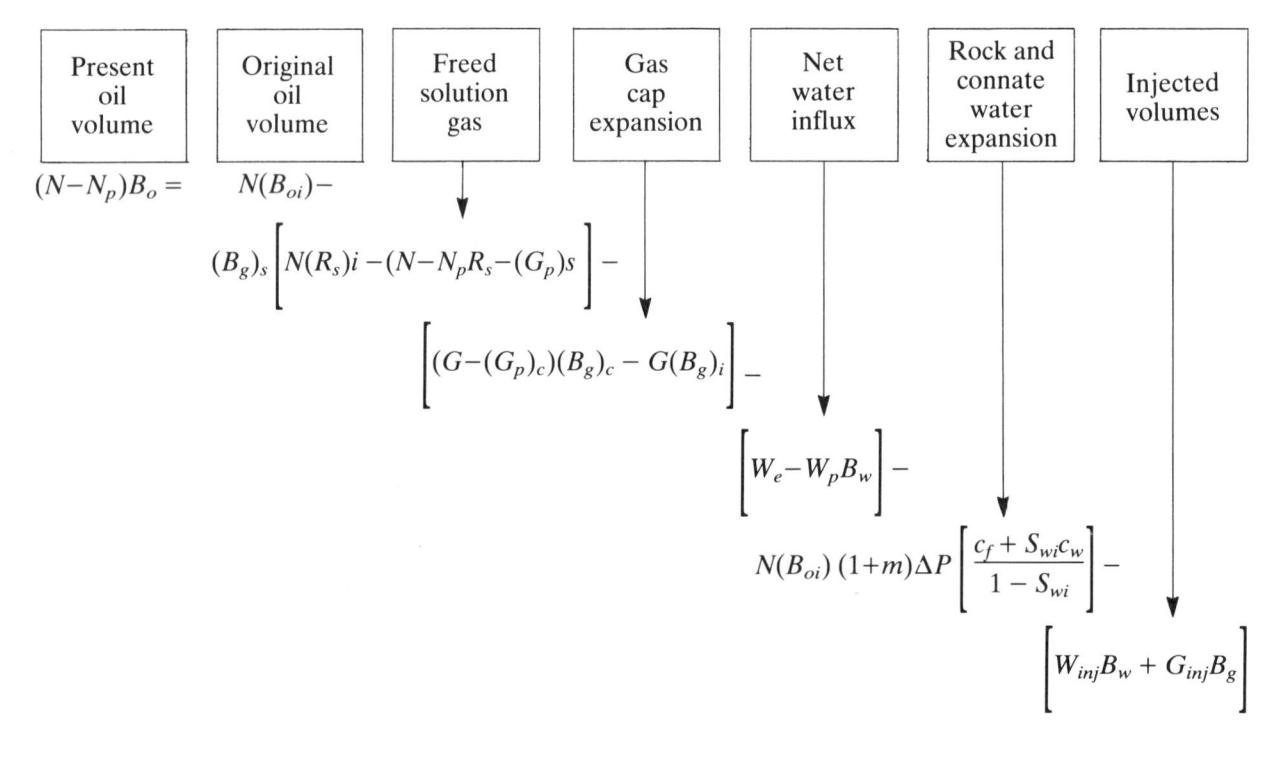

Present oil volume	Original oil volume	Freed solution gas	Gas cap expansion	Net water influx	Rock and connate water expansion	Injected volumes

$(N - N_p)B_o = N(B_{oi}) -$

$(B_g)_s \left[N(R_s)i - (N - N_p R_s - (G_p)s \right] -$

$\left[(G - (G_p)_c)(B_g)_c - G(B_g)_i \right] -$

$\left[W_e - W_p B_w \right] -$

$N(B_{oi})(1+m)\Delta P \left[\dfrac{c_f + S_{wi} c_w}{1 - S_{wi}}\right] -$

$\left[W_{inj} B_w + G_{inj} B_g \right]$

$$N = \frac{N_p(B_o - R_s(B_g)_s) + (G_p)_s(B_g)_s - \left[(G - (G_p)_c)(B_g)_c - G(B_{gi})\right] - \left[W_e - W_p B_w\right] - W_{inj}B_w - G_{inj}B_g}{(B_o - (B_{oi})) + ((R_{si}) - R_s)(B_g)_s + (B_{oi})\Delta P \left[\dfrac{c_f + S_{wi} c_w}{1 - S_{wi}}\right](1+m)}$$

10 RESERVOIR PERFORMANCE ANALYSIS

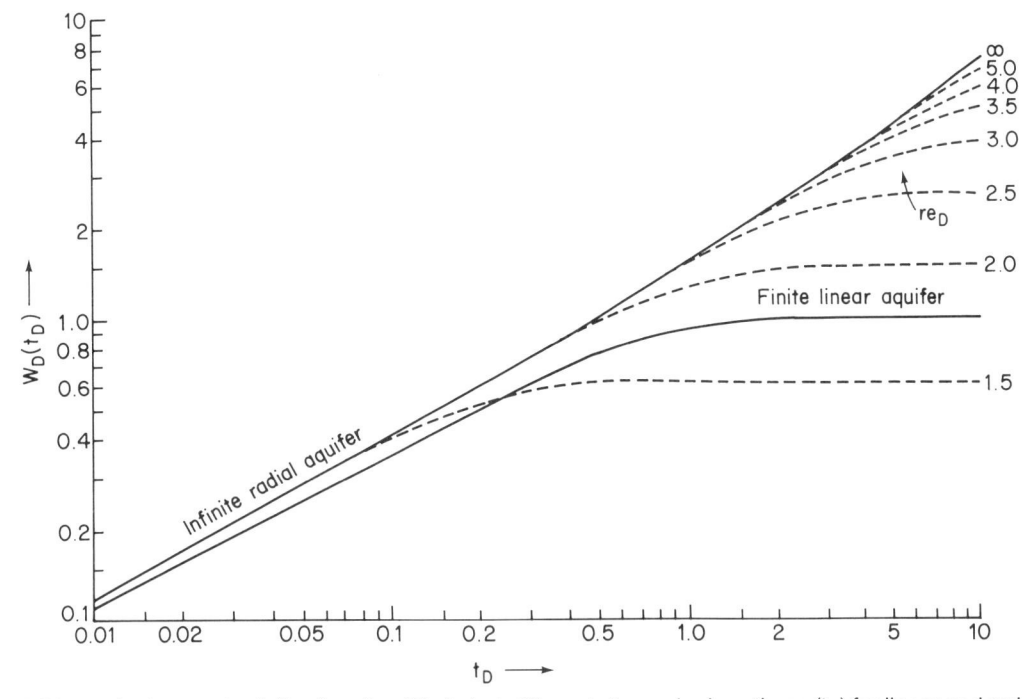

Fig. 10.9 Dimensionless water influx function $W_D(t_D)$ at different dimensionless times (t_D) for linear and radial (r_{e_D} = 1.5 – ∞) constant terminal pressure solutions (after [5]).

The aquifer characteristics may be explored and correlated with an appropriate aquifer model by making use of the linearized material balance formulation and validated production and *PVT* data, as shown in Fig. 10.10.

It is sometimes noted in matching aquifer performance that not all injected water enters the oil zone. This will depend on the transmissibility at the OWC and the pressure gradients established in the aquifer-reservoir system. In general, more than 85% of the injected water is expected to move towards the oil zone rather than repress the aquifer, as shown in Fig. 10.11.

A water drive reservoir may then be particularly rate sensitive, and the reservoir may behave almost as a depletion reservoir for a long period, if offtake rates are very large, or as an almost complete pressure maintained water drive reservoir, if offtake rates are low.

Because of the similarity in oil and water viscosities (for light oils at normal depths), the displacement of oil by water is reasonably efficient, and provided that localized channelling, fingering or coning of water does not occur, water drive will generally represent the most efficient of the natural producing mechanisms for oil reservoirs.

Fig. 10.10 Combination drive, Havlena-Odeh plot.

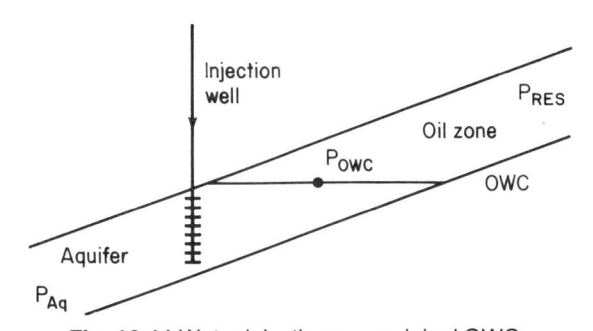

Fig. 10.11 Water injection near original OWC.

As with the gas cap drive reservoirs, a maintained pressure leads to lower viscosities and higher B_o values at any given saturation, reducing the saturation and minimizing the terms S_o/B_o for any given economic limit.

The recovery efficiency of water reservoirs will be governed by an economic limit, the limit in this case being dictated by water handling problems. Provided that water can be controlled reasonably, efficiencies of 30–40% would be expected, and under favourable sweep conditions recovery efficiencies of 50–60% might be calculated. (Ultimately, of course, calculating a recovery efficiency depends on knowing the initial oil in place, and a calculated high recovery factor might simply be the result of underestimating oil in place.)

The first requirement in *maximizing* recovery is to establish the probable natural mechanisms of a reservoir and the extent to which these are likely to be rate sensitive. Obviously, if a reservoir is very much undersaturated, and has a potentially large aquifer, it should be possible to determine a reservoir production rate at which the aquifer response will maintain pressure around the saturation pressure for the whole producing life of the field. With wells located in structurally high positions this would give maximum recovery and maximum *efficiency*. However, if the offtake rate did not generate sufficient income to justify the expenditure on field development it would not be an *economic* rate, and any considerations of maximizing recovery must also involve economic factors.

Where recovery efficiencies seem likely to be low, or where substantially higher offtake rates would be possible if natural mechanisms are augmented, then pressure maintenance or secondary recovery operations may be initiated to improve recovery factors. Normally, this would be either water injection to augment a natural water drive, or gas injection, the former tending to be more efficient because of mobility ratio considerations, but the latter being increasingly important because of gas conservation requirements. Problems with secondary recovery operations are similar to those of the related primary mechanisms – particularly control of gas or water and the prevention of excessive fingering or channeling of injected fluids. This is another aspect of reservoir management in which gravity segregation can play an important (and essentially adverse) part.

10.5 ACCURACY OF THE GROSS MATERIAL BALANCE EQUATION

There are several sources of error in material balance calculations.

10.5.1 Production data

Cumulative oil production is generally measured accurately for royalty and transfer payments. A single reservoir field then presents no difficulty and N_p will be the factor most precisely known. With multireservoir fields the individual reservoirs are not always (rarely!) metered separately, and the allocation of cumulatives to reservoirs depends on intermittent well testing. Nevertheless, this is not a major source of error.

In dealing with past production histories, gas has generally not been measured with any accuracy, and frequently the cumulative produced gas–oil ratio R_p will be subject to very great uncertainty. Reservoirs now being developed should involve less uncertainty since gas is increasingly a marketable product. Water produced is also very uncertain depending upon interpretation of separator, tank farm and well test data taken intermittently, if at all, but the term itself may often be relatively small.

10.5.2 Pressure data

For the simplest calculations, the whole reservoir is assumed to be at some average datum pressure, corresponding to some cumulative production. All pressure dependent values refer to this pressure.

The average pressure must be calculated from a series of essentially transient well tests, taken at varying times and positions, but practically never corresponding to a situation where the whole reservoir is shut in at one time. Regardless of the accuracy of the pressure data itself, the interpretation of average pressure from this data is possibly in error. Reservoirs pressures at datum in given wells may be assigned volumetric or a real regions of influence in the calculation of weighted average reservoir pressures:

i.e. $\bar{P} = \Sigma P_j \phi_j h_j A_j / \Sigma \phi_j h_j A_j$

10.5.3 Pressure dependent oil properties

The values of B and R assigned for any value of reservoir pressure and temperature will depend on either:

(a) the use of generalized correlations relating oil gravity GOR, reservoir temperature and pressure to B_o and R_s. Although generalized correlations are fairly reliable, there is no certainty that they will match the behaviour of

10 RESERVOIR PERFORMANCE ANALYSIS

any given crude system with any particular degree of accuracy; or

(b) a laboratory analysis of a reservoir fluid sample. The problem in this case is the question of whether or not the sample obtained (either as a bottom-hole sample or recombined sample) is truly representative of the reservoir fluid.

All these sources of error contribute to inaccuracy in material balance calculations, but the factor which most dominates the accuracy is the denominator term $(B_t - B_{oi})$. Since this is a difference between two quantities of the same order of magnitude, a small absolute error in either term can lead to a very large error in the function, e.g. if $B_t = 1.3365$ and $B_{oi} = 1.3600$, an error of 5 parts in 1360 (0.4%) can lead to a 100% error in the difference. In view of the uncertainty in pressure dependent values of B_t and R_s, this would appear serious, but provided that a high degree of relative accuracy can be maintained in the values of B_t and B_{oi} (i.e. provided random errors can be reduced), then systematic errors which affect the absolute values, but not the differences, are less important and the error in the difference may not be too important. Even so, material balance calculations at early times of reservoir history are unreliable compared with later calculations when pressure drops are greater and differences in the $(B_t - B_{oi})$ term are greater.

Examples

Example 10.1

Using the data from Problem 4.2, estimate the value of gas in place assuming seismic, log, and geological interpretation have provided the following estimate:

Thickness (average)	500 ft
Area	100 sq miles
Porosity	12%
Water saturation	35%

What is the recoverable gas for an average reservoir abandonment pressure of 500 psia?

Example 10.2

Find the expression for the flowing gas–oil ratio of a well (volume of gas sc/volume of stock tank oil) in a reservoir having a gas saturation in excess of the critical.

Calculate the gas–oil ratio for the following conditions:

μ_o = 0.8 cp B_o = 1.363 rb/stb
μ_g = 0.018 cp B_g = 0.001162 rb/scf
k_o = 1000 mD R_s = 500 scf/stb
k_g = 96 mD

Example 10.3

A reservoir is estimated by a volumetric method to contain approximately 14.5×10^6 barrels of stock tank oil, originally just saturated at the initial reservoir pressure (i.e. no gas cap). The table below gives the properties of the reservoir fluids and production data for the reservoir. There has been no water production.

Pressure (psia)	R_s (scf/stb)	B_o (rb/stb)	B_g (rb/scf)	B_t (rb/stb)	R_p (scf/stb)	N_p (stb)
1850*	690	1.363	0.00124	1.363	—	0
1600	621	1.333	0.00150	1.437	878	1 715 000
1300	535	1.300	0.00190	1.594	996	3 430 000
1000	494	1.258	0.00250	1.748	1100	?

* bubble point
B_g is reservoir barrels/SCF
R_s is SCF/barrel

Calculate the water influx at cumulative oil production of 1.715×10^6 and 3.430×10^6 stb. What would you expect the cumulative production to be at a reservoir pressure of 1000 psi?

Example 10.4

A reservoir may be considered as a right cone with dimensions 750 ft from apex to oil–water contact, and a base radius of 3 miles at the oil–water contact. Estimated values of porosity and water saturation are 0.17 and 0.24 respectively (bed thicker than the column). The oil-water contact is found at 4260 ft subsea, and a fully shut-in pressure of 1919 psi was measured at this depth, the initial average reservoir pressure at reservoir datum being taken as 1850 psi. Given the following *PVT* and production data, what is your estimate of the cumulative production obtainable by the fall in average reservoir pressure to 1000 psi?

Pressure (psi)	R_s (scf/stb)	B_o (rb/stb)	B_g (rb/scf)	B_t (rb/stb)	Oil density (lb/b^3)	N_p (stb)	R_p (scf/stb)	W_p (stb)
1850*	690	1.363	0.00124	1.363	43.4	—	—	—
1600	621	1.333	0.00150	1.437	43.9	3.1×10^8	1100	31×10^6
1300	535	1.300	0.00190	1.594	44.5	5.5×10^8	1350	55×10^6
1000	494	1.258	0.00250	1.748	45.1	—	1800 (est)	63×10^6 (est)

Example 10.5

A reservoir is believed to contain an initial oil in place of 300×10^6 STB and has an initial gas cap of 120.7×10^9 SCF. The initial pressure at the gas–oil contact, which is also a convenient datum, is 5000 psi. The *PVT* properties of the system at reservoir conditions are as follows:

P (psi)	B_o (rb/stb)	R_s (scf/stb)	B_g (rb/scf)
5000	1.3050	500	6.486 E–04
4750	1.2840	450	6.827 E–04
4500	1.2540	390	7.207 E–04
4250	1.2150	325	7.630 E–04

The uniform initial water saturation is 30% and water and pore volume compressibilities are each 3×10^{-6} psi^{-1}. Production started on 1.1.80 and a constant oil production of 60 000 STB/d has been maintained. Water injection started at a constant rate of 70 000 BBL/d on 1.1.81. Assume B_w for injected water is 1.00 RB/BBL. The cumulative gas and reservoir production pressure has been reported as follows:

	Gp	P (psi)
1.1.81	12.045×10^9 SCF	4300
1.1.82	26.280×10^9 SCF	4250

Estimate the cumulative water influx on 1.1.81 and 1.1.82 of the surrounding aquifer.

Example 10.6

An oil reservoir is totally surrounded by a radial aquifer. The radius to the oil–water contact is 9000 ft and the outer radius of the aquifer is 81 000 ft. The aquifer and oil zone net thickness is 200 ft and the aquifer has a porosity of 18%. The water compressibility is 3×10^{-6} psi^{-1} and water viscosity at aquifer conditions of temperature and pressure is 0.4 cp. The pore volume compressibility is 4×10^{-6} psi^{-1}, and the aquifer permeability is 707 mD. The pressures at the original oil-water contact have been determined initially (P_i) and at subsequent yearly intervals (P_1, P_2, P_3) after the start of oil zone production as follows:

Time	P (psia)
Initial	5870
End year 1	5020
End year 2	4310
End year 3	3850

Estimate the aquifer performance in reservoir barrels of influx at the end of each year assuming unsteady state behaviour and using the method of Van Everdingen and Hurst.

References

[1] Tracy, G.W.
A simplified form of the material balance equations, *Trans. AIME* **204** (1955), 243.
[2] Havlena, D. and Odeh, A.S.
The material balance as an equation of a straight line, Pt I *Trans. AIME* **228** (1963), 896, Pt II *Trans. AIME* **231** (1964), 815.
[3] Teeuw, D.
Prediction of formation compaction from laboratory compressibility data, *SPEJ* (Sept. 1971), 263.
[4] Merle, H.A. *et al.*
The Bachaquero study – a composite analysis of the behaviour of a compaction drive – solution gas drive reservoir, *JPT* (Sept. 1976), 1107.
[5] van Everdingen, A.F. and Hurst, W.
The application of the Laplace transformation to flow problems in reservoirs, *Trans. AIME* **186** (1949), 305.
[6] Carter, R.D. and Tracy, C.W.
An improved method for calculating water influx, *Trans. AIME* **219** (196), 415.
[7] Fetkovitch, M.J.
A simplified approach to water influx calculations–finite aquifer systems, *JPT* (July 1971), 814.
[8] Tarner, J.
How different size gas caps and pressure maintenance programs affect the amount of recoverable oil, *Oil Weekly* (12 June 1944), 32.
[9] Coats, K.H., Tek, M.R. and Katz, D.L.
Method for predicting the behaviour of mutually interfering gas reservoirs adjacent to a common aquifer, *Trans. AIME* **216** (1950), 247.
[10] van Everdingen, A.F. Timmerman, E.H. and McMahon, J.J.
Application of the material balance equation to a partial water drive reservoir, *Trans. AIME* **198** (1953), 51.
[11] Stone, H.L. and Gardener, A.O.
Analysis of gas cap or dissolved gas drive reservoirs, *SPEJ* (June 1961), 92.
[12] Wooddy, L.D. and Moscrip, R.
Performance calculations for combination drive reservoirs, *Trans. AIME* **207** (1956), 128.
[13] Schilthuis, R.J.
Active oil and reservoir energy, *Trans. AIME* **118** (1936), 33.
[14] Poston, S.W., Lubojacky, R.W. and Aruna, M.
Maren field – an engineering review, *JPT* (Nov. 1983), 2105.
[15] Blanton, T.L.
Deformation of chalk under confining pressure and pore pressure, Paper EUR 41, *Proc. Europ. Off. Pet. Conf.* (1978), 327.
[16] Newman, G.H.
Pore volume compressibility of consolidated friable and unconsolidated reservoir rocks under hydrostatic loading, *JPT* (Feb. 1973), 129.
[17] Pirson, S.J.
Elements of Oil Reservoir Engineering, McGraw Hill, London (1950).
[18] Dake, L.P.
Fundamentals of reservoir engineering, *Dev. Pet. Sci.* **8** Elsevier, Amsterdam (1978).
[19] Amyx, J.W., Bass, D.M. and Whiting, R.L.
Petroleum Reservoir Engineering, McGraw Hill (1960).
[20] Muskat, M.
Flow of Homogeneous Fluids through Porous Media, IHRDC (1937).

[21] Muskat, M.
Physical Principles of Oil Production, IHRDC (1949).
[22] Newman, G.H.
The effect of water chemistry on the laboratory compression and permeability characteristics of some North Sea chalks, *JPT* (May 1983), 976.
[23] Tehrani, D.H.
An analysis of a volumetric balance equation for calculation of oil in place and water influx, *JPT,* (Sept. 1985), 1664.

Chapter 11
Secondary Recovery and Pressure Maintenance

Secondary recovery techniques involve supplementing the natural energy of a petroleum reservoir by the injection of fluids, normally water or gas. When this is done such that average reservoir pressure is held constant, i.e. reservoir volumetric rate of production is equal to reservoir volumetric rate of fluid replacement, then the process is known as *pressure maintenance*. The level of pressure maintenance in oil production is usually just above bubble-point pressure such that injection costs are minimized. Since production rate is also dependent on reservoir pressure gradients, then the choice of pressure maintenance level will also include rate consideration. In order to provide the capability for natural flow to surface under high water cut, the selection of pressure maintenance level might be determined as shown in Fig. 11.1. When the reservoir condition volumetric rate of fluid replacement is equal to the reservoir condition volumetric rate of production, the technique is known as *complete voidage replacement*. In practice, any fraction of voidage could be replaced if it provides an optimum recovery scheme. Proper design of a secondary recovery scheme is best performed after a period of primary recovery, in order to observe the dynamic response of the reservoir. In offshore field development this is not usually possible and pressure maintenance is implemented early in field life.

The efficiency of secondary recovery and pressure maintenance schemes can be explored by reference to the physical processes occurring. It is assumed that the injected fluid is immiscible with the displaced hydrocarbon. We will consider the effects of fluid properties and saturations, rock characteristics including heterogeneity of permeability, reservoir dip angle, reservoir geometry and lateral continuity, flow rates and well locations. Firstly, however, the principles of fluid displacement in secondary recovery are reviewed.

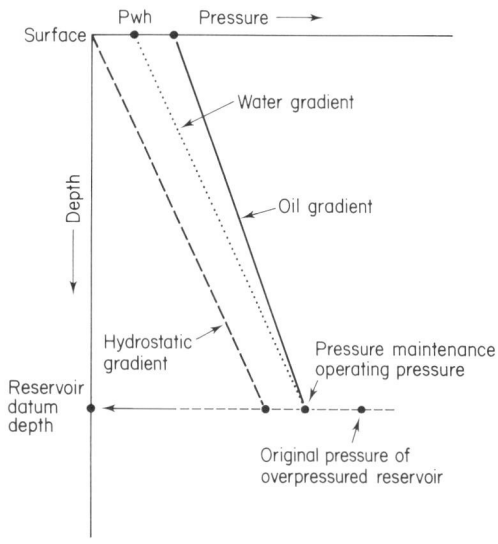

Fig. 11.1 Operating pressure for natural flow in originally overpressured undersaturated oil reservoir under pressure maintenance.

11.1 DISPLACEMENT PRINCIPLES

The displacement of oil by water or gas under immiscible conditions occurs both microscopically and macroscopically in a reservoir. On the microscale we consider the distribution of trapped oil in pores swept by displacing fluid. The distribution of residual oil will depend on competing viscous, capillary and gravitational forces and is particularly influenced by pore size, pore geometry, wettability and displacement rate. The laboratory determination of residual oil saturation in core plugs provides an estimate of the microscopic displacement efficiency, through measurement of an *ultimate* residual oil saturation $S_{or(ULT)}$ for the pore scale recovery process:

$$\text{Recovery factor}_{(ULT)} = \frac{1 - S_{wi} - S_{or(ULT)}}{1 - S_{wi}}$$

The value of $S_{or(ULT)}$ may be less than 10% of pore volume, for low rate gravity drainage of oil below an injected gas cap, to over 40% in a high rate linear water injection in which imbibition between advancing fingers is restricted.

On an interwell scale, the driving force for displacement is represented by a potential gradient (or datum corrected pressure gradient) between an injection well and a producing well. At some economic limiting condition of production rate or fluid cut, the recovery efficiency is represented in terms of an average oil saturation in the reservoir. We may distinguish between an average saturation in the *swept* region of the well pattern $S_{or(swept)}$ and an average residual oil saturation in the whole reservoir including non-contacted or swept regions $\overline{S}_{or}$. This latter oil saturation should be the same as the material balance residual oil saturation.

$$\text{Recovery factor (swept zone)} = \frac{1 - \overline{S}_{wi} - \overline{S}_{or(swept)}}{1 - \overline{S}_{wi}}$$

$$\text{Recovery factor (material balance)} = \frac{1 - \overline{S}_{wi} - \overline{S}_{or}}{1 - \overline{S}_{wi}}$$

For a homogeneous reservoir with constant injection rate and incompressible fluids, the water cut or gas cut will be controlled by the saturation at a producing well. Welge's equations allow a representation of the producing well performance in terms of the average material balance saturation $\overline{S}_w$, the producing well saturation S_2, the fractional oil flow f_{o2} at the producing well, and the gradient of the fractional flow curve at the saturation S_2, i.e.

$$\overline{S}_w = S_2 + f_{o2} \left[\frac{1}{(df/dS)_{S_2}} \right]$$

The fractional flow of displacing phase (water or gas) is given as a saturation dependent expression (since relative permeabilities are considered saturation dependent) as follows:

$$f_D = \frac{1 + 1.127E{-}03 \, \frac{k\, k_{ro} A}{q_T \mu_o} \left[\left(\frac{\partial S_w}{\partial x} \cdot \frac{dP_c}{dS_w} \right) - 0.4335 (\gamma_D - \gamma_o) \sin \alpha \right]}{1 + \left\{ \frac{\mu_w}{k_w} \cdot \frac{k_o}{\mu_o} \right\}}$$

This expression is given in field units where the injection rate q_T is in RB/D, the cross-sectional area A is in ft^3, the absolute permeability k is in mD, the viscosities of oil and water, μ_o and μ_w, are in cp, the densities of oil and water (γ) are in terms of specific gravities relative to water at 60°F, 14.7 psia, the dip angle α is in degrees and by convention is positive for updip flow. The terms dP_c/dS_w and $\partial S_w/\partial x$ are both negative and are normally considered small enough to neglect.

Under semi-steady state conditions, the flow of oil into a wellbore is given by the radial flow equation including mechanical skin as follows:

$$\overline{P} - P_{wf} = \frac{q_o \mu_o}{2\pi k k_{ro} h} \left\{ \ln \frac{r_e}{r_w} - 0.75 + S \right\}$$

In field units, where rate is in RB/D, pressures in psia, length terms in feet, permeability in mD and viscosity in cP, the equation is as follows:

$$\overline{P} - P_{wf} = \frac{141.2 \, q_o \mu_o}{k \cdot k_{ro} h} \left\{ \ln \frac{r_e}{r_w} - 0.75 + S \right\}$$

The term $\{q_o B_o/(\overline{P} - P_{wf})\}$ is known as a productivity index (PI) and has a particular dependence on kh as well as on saturation and pressure influenced terms.

For injection wells an expression known as the injectivity index (II) can be similarly described using the difference between the flowing bottom-hole pressure P_{wf} and the average reservoir pressure $\overline{P}$, in conjunction with the water injection rate Q_i in RB/D:

$$II = \frac{Q_i}{P_{wf} - \overline{P}} = \frac{0.00708 \, (k \cdot k_{rw} \cdot h)}{\mu_w \left[\ln \frac{r_e}{r_w} - 0.75 + S \right]}$$

where $k.k_r$ is the same as the effective fluid permeability k_e.

Well inflow equations are linked to pressure loss calculations in production strings to evaluate producing rates consistent with wellhead choke and separator conditions. The particular influence of $k_e h$ at well locations may influence the total well number

11 SECONDARY RECOVERY AND PRESSURE MAINTENANCE

requirements for a given voidage replacement scheme. In water saturated regions of many reservoirs, diagenetic damage to pore space may have reduced absolute permeability by several orders of magnitude compared with that in the oil zone. In such instances the injectivity of water into an oil zone may prove more attractive than injection into the water at an oil–water contact, even though the relative permeability to water in the presence of residual oil may be less than 30%.

11.2 FACTORS INFLUENCING SECONDARY RECOVERY AND PRESSURE MAINTENANCE SCHEMES

11.2.1 Mobility ratio

The mobility ratio expresses the ratio of the mobility of the displacing phase to the mobility of the displaced phase. If we consider the case of water displacing oil, then

Mobility of water = λ_w

$$= \frac{\text{effective permeability to water}}{\text{viscosity of water}}$$

$$= \frac{k_w}{\mu_w}$$

Similarly,

$$\lambda_o = k_o/\mu_o$$

It is clear that the effective permeabilities of oil and water are saturation and direction dependent. The mobility ratio could therefore be expressed at any saturation condition during a displacement. In general usage, the water mobility is frequently defined at the average water saturation in the water contacted region of a reservoir. For efficient displacement this is often represented as a piston front condition. As shown in Fig. 11.2, a plot of water saturation distribution versus distance (x) from the injection point in a linear system, the frontal saturation S_f occurs at position x_f. The mobility ratio prior to breakthrough at a well (location $x=L$) will use

$$M = \frac{\lambda_w}{\lambda_o} = \left[\frac{k_w}{\mu_w}\right]_{\bar{S}_w} \cdot \left[\frac{\mu_o}{k_o}\right]_{S_{wi}}$$

A particular case of mobility ratio representation known as end-point mobility ratio (M') defines k_w as k_w at S_{or} and k_o as k_o at S_{wi}:

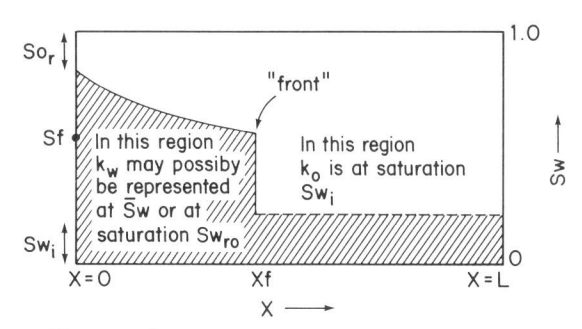

Fig. 11.2 Saturation profile before breakthrough.

$$M' = \frac{k'_w}{\mu_w} \cdot \frac{\mu_o}{k'_o}$$

Figure 11.3 shows results of experimental data from 5 spot patterns for water–oil, gas–oil and miscible displacements. The effect of the volume of displacing fluid injected on total swept area after breakthrough can also be represented as a function of mobility ratio. Using data[10] for a direct line drive, and representing the number of displaceable volumes injected as V_d, Fig. 11.4 has been prepared. In this figure the numerical value of V_d is (volume of injected fluid/$(1-S_{wi}-S_{or})PV$).

11.2.2 Reservoir dip angle

Reservoir dip has an effect on gravity stabilization during immiscible displacement and, by inspection of the fractional flow equation, it can be seen that dip angle influences oil cut. As shown in Fig. 11.5, the magnitude of the effect of dip angle on fractional flow depends on the wetting preference of the rock–fluid system and on the direction of displacement. Compared with the fractional flow curve of a horizontal reservoir with no particular strong wetting preference, a strongly oil wet horizontal reservoir will show a higher water cut at a given saturation, and a strongly water wet one will show a lower water cut at the same saturation. At any given wettability the fractional water flow for updip displacement of oil by water will be lower at a given saturation than for downdip flow. This is a direct result of gravity stabilization of fluids of differing densities. Increasing reservoir dip angle accentuates the stabilization for updip flow and decreases stabilization for downdip water flow. For gas injection, the fractional flow equation indicates that better gravity stabilization will be obtained by updip injection with downdip gas flow at high reservoir dip angles.

The density difference between immiscible fluids

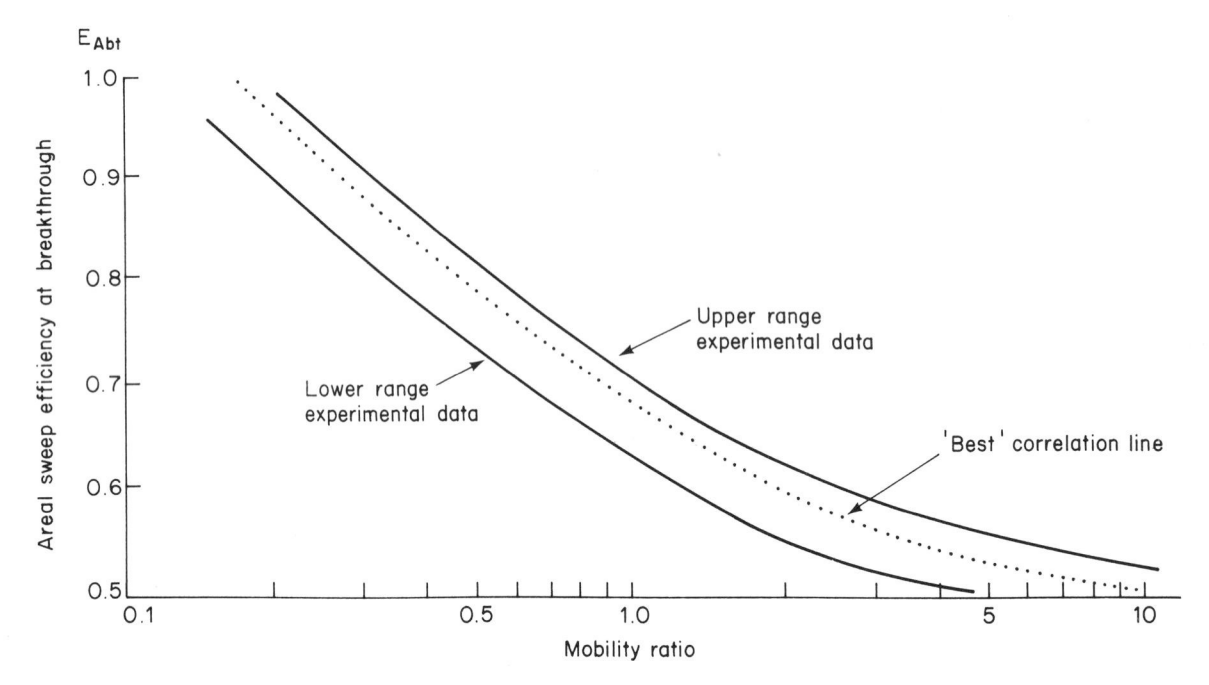

Fig. 11.3 Effect of mobility ratio on areal sweep efficiency at breakthrough: 5 spot pattern (after[3]).

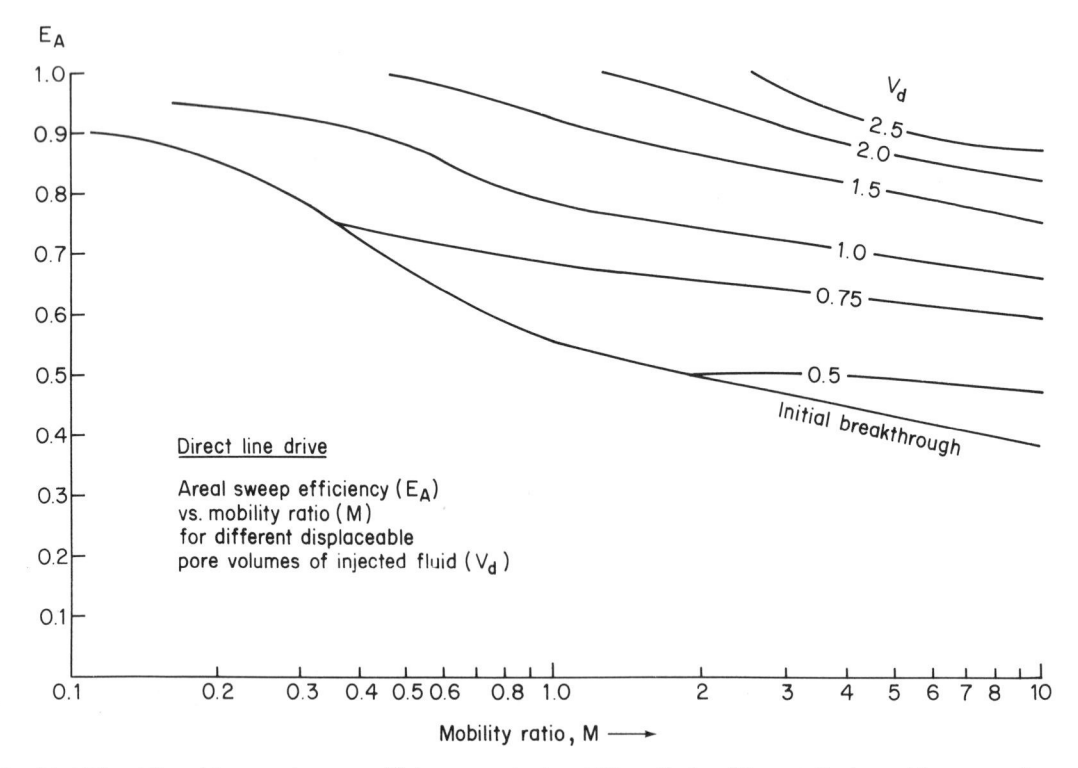

Fig. 11.4 Direct line drive areal sweep efficiency against mobility ratio for different displaceable pore volumes (V_d = $\dfrac{\text{volume of injected fluid}}{\text{PV}(1-S_{wi}-S_{or})}$) of injected fluid. (Data from[10].)

11 SECONDARY RECOVERY AND PRESSURE MAINTENANCE

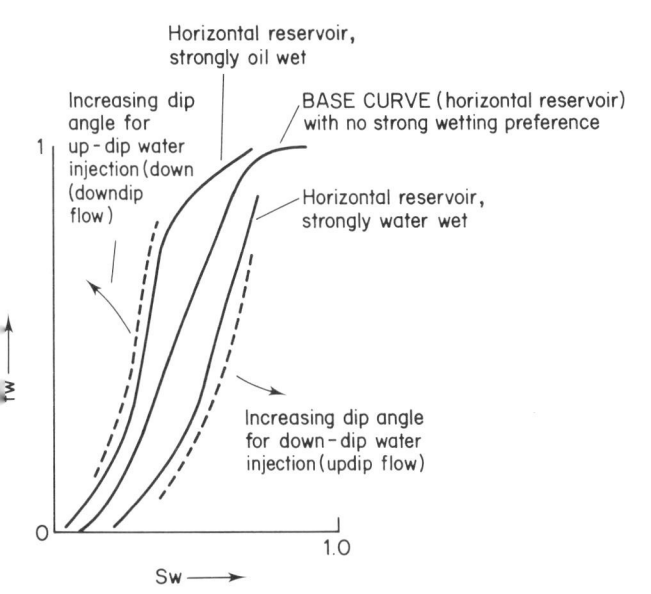

Fig. 11.5 Effect of dip angle and wettability on fractional flow.

may lead to segregation. In steeply dipping reservoirs, the gravity segregation forces may dominate capillary forces, and displacement behaviour is controlled by viscous:gravity force ratios. The mechanism of this process has been presented by Dietz and others [13, 15] and involves the calculation of a critical displacement rate. For rates less than critical, the displacement is stable and underrun or override of displacing fluid through tongue formation should not occur. If the end-point mobility ratio is defined as M', displacing fluid subscripted D, and the dip angle α considered positive for updip flow, then, as indicated in Fig. 11.6, we have:

in Darcy units

$$q_{crit} = \frac{k\,k'_{rD}\,A\,(\rho_D - \rho_o)\,g'\sin\alpha}{1.0133 \times 10^6\,\mu_D\,(M'-1)}$$

in field units

$$q_{crit} = \frac{4.9 \times 10^{-4}\,k\,k'_{rD}\,A\,(\gamma_D - \gamma_o)\sin\alpha}{\mu_D\,(M'-1)}$$

The field units used above are as follows:

q_{crit} = rb/day;
k = mD;
A = ft^2;
γ = res.cond.specific gravity relative to water at standard conditions;
α = degrees;
μ = centipoise.

It should be noted that for an end-point mobility ratio of unity ($M' = 1$) the displacement is unconditionally stable at any rate.

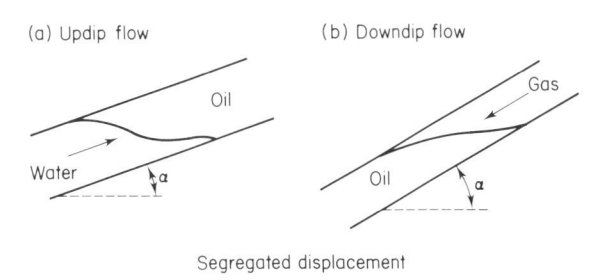

Fig. 11.6 Segregated displacement. (a) Updip (positive) flow, (b) downdip (negative) flow.

11.2.3 Reservoir heterogeneity

Although reservoir heterogeneity can imply many variations in rock properties at different scales of influence, only permeability variation will be considered for the present. So far the representation of reservoir performance by use of a fractional flow equation has assumed a homogeneous reservoir with constant vertical and lateral permeability character. The fractional flow curve can, however, be used in the performance analysis of systems with vertical permeability variation. When there is no cross-flow between layers, then the methods of Stiles[21] and Dykstra and Parsons[23] may be appropriate in generating fractional oil flow performance. When there is pressure communication between vertical *layers* or different rock properties, static or dynamic pseudo-relative permeability functions have to be generated [24, 25, 9]. In Fig. 11.7 a vertical section through a northern North Sea production well has a permeability profile as shown. In reservoir analysis, the connection between vertically adjacent sands is important, as well as the degree of contrast in permeability in a given unit. Where a non-reservoir interval is indicated by core and log in a given well, the geometry of the non-reservoir material becomes significant in assessing whether or not vertical sand connections will occur around the non-reservoir unit at some distance away from the well. The combination of core absolute permeability data, with well test derived effective permeability, and with dynamic RFT response in new wells drilled in producing fields, allows judgement of these possibilities. Figure 11.8 shows RFT data obtained in a Dunlin well in the Brent Sand region of the UKCS North Sea

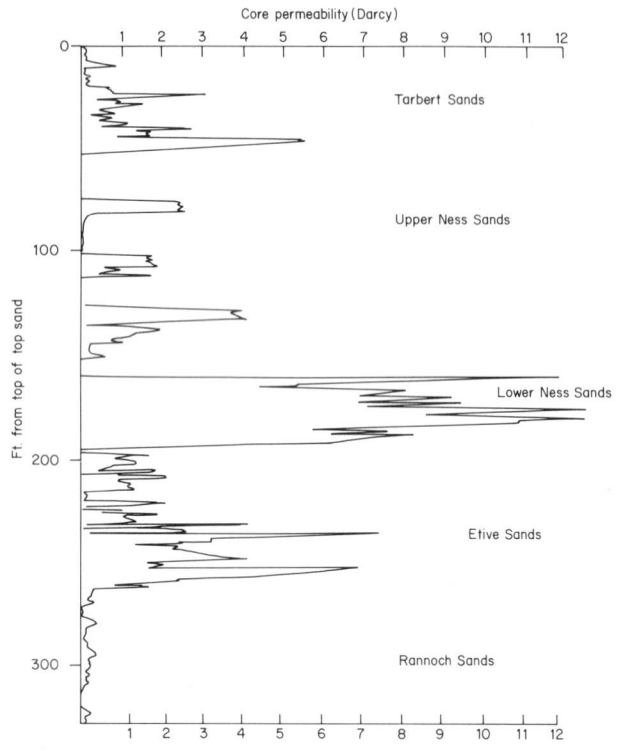

178   PETROLEUM ENGINEERING: PRINCIPLES AND PRACTICE

Fig. 11.7 Permeability distribution through a vertical section of a Brent Sand well in the northern North Sea (UKCS).

compared with detailed sedimentological analysis of nearby core [43]. The degree of vertical connection between sand bodies is suggested from the RFT gradients and may be *matched* by reservoir simulation.

The effect of characteristic sand body types on the flow of injected water and on oil displacement may be anticipated [26] with reference to Figs 11.9 and 11.10. In Fig. 11.9 the depositional environment represents channel sands developed as a lateral accretion surface. Without further significant diagenetic alteration the gamma ray profile and core permeability profile may be as shown. The unit permeability thickness product $(\bar{k}h)$ would be obtained from $\sum k_j h_j$ by subdivision into n sublayers. Water injection into such a sand body would probably result in the combination of gravity forces and viscous forces giving a profile with poor vertical sweep as shown. The sweep will depend particularly on the permeability contrast and the bed thickness.

Figure 11.10 shows the behaviour of a bed with the same $(\bar{k}h)$ product as the channel sand but with the higher permeability sands at the top of the unit. This arrangement frequently results from bar sand deposition. In waterflooding such a unit, the viscous and gravity forces counteract each other and a more efficient sweep may be obtained. Pseudo-relative permeability curves are required if such units are to

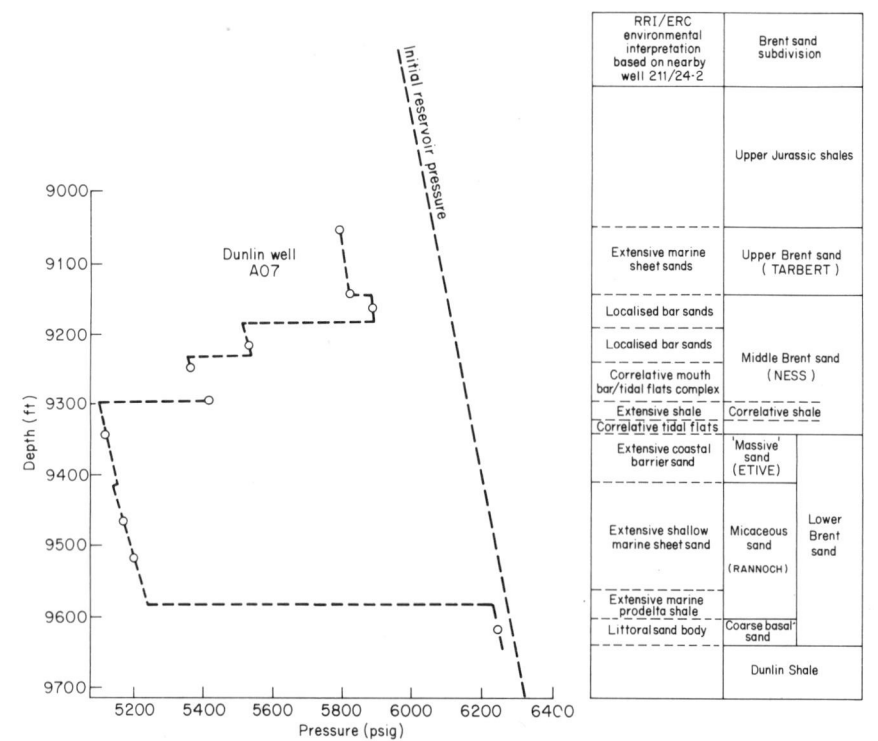

Fig. 11.8 RFT pressure response in a producing reservoir (after [29]).

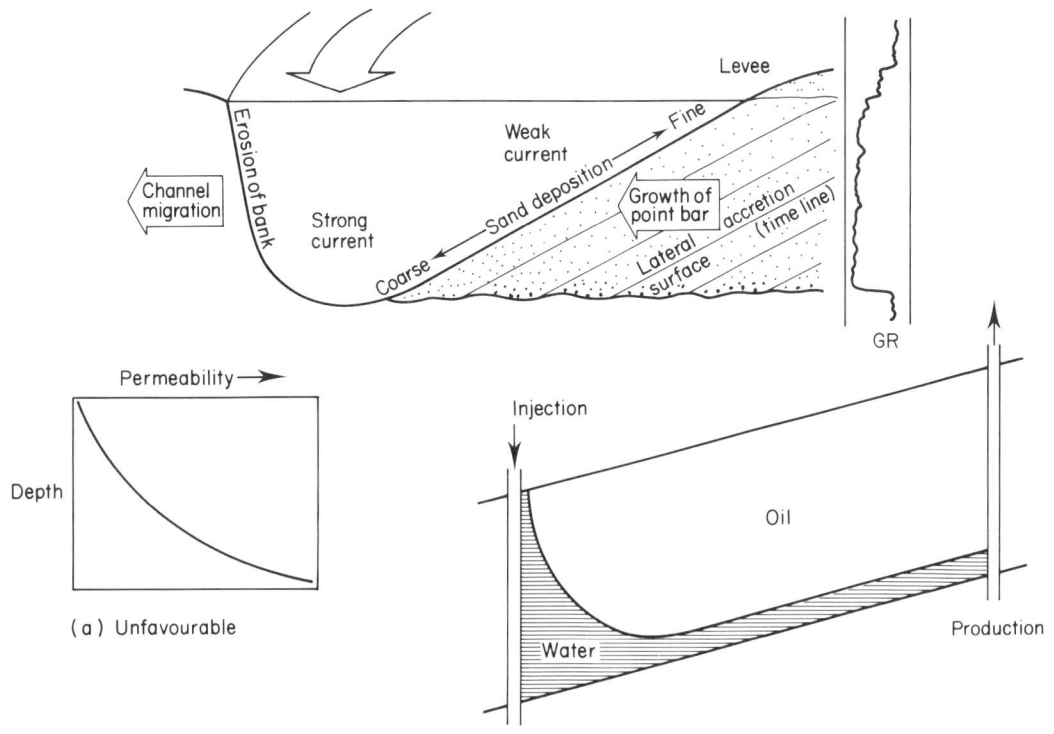

11 SECONDARY RECOVERY AND PRESSURE MAINTENANCE 179

Fig. 11.9 Effect of unfavourable permeability distribution in waterflooding.

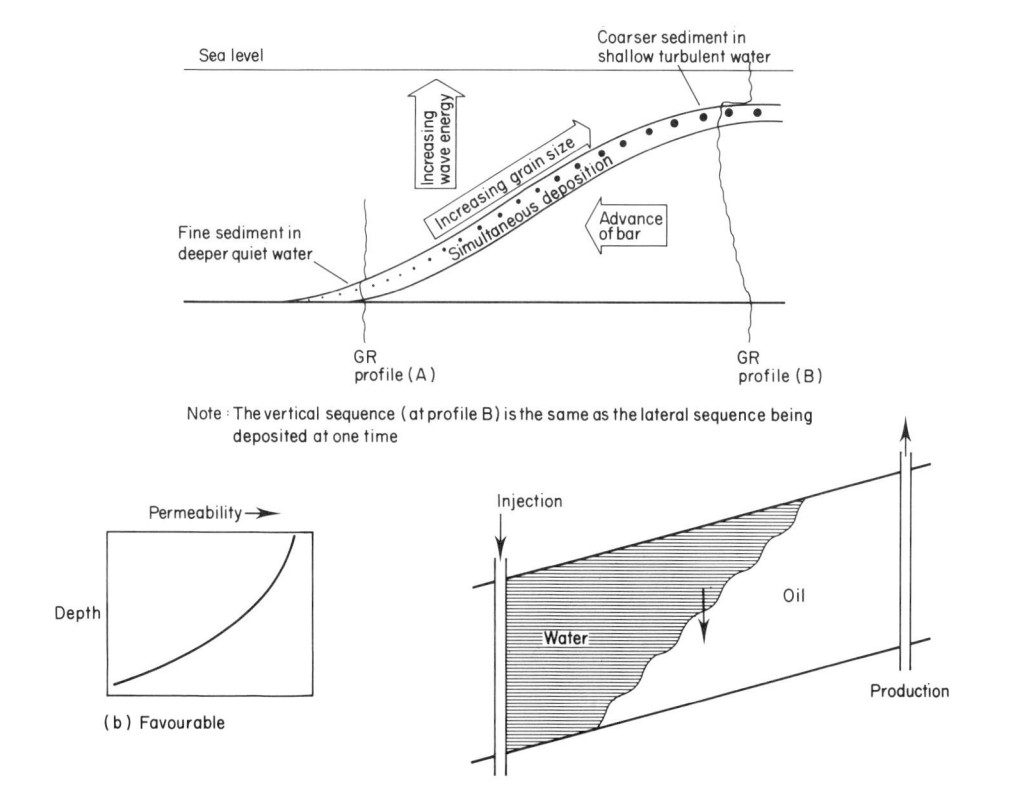

Fig. 11.10 Effect of favourable permeability distribution in waterflooding.

be represented as single layers in numerical calculations.

Even very thin, low permeability layers sandwiched between higher productivity sands can be significant. A thin 1–2 ft very micaceous sand near the top of some North Sea Rannoch Sand reservoirs [43] have had, for example, a major effect on waterflooding. The micaceous sand is not a total seal, but under dynamic conditions prevents vertical movement of injected water. This particular situation has been described by Dake [37]. In a particular example, the consequence of ignoring its presence in waterflood calculations is to predict water breakthrough significantly later than reality.

11.2.4 Reservoir geometry and continuity

Sandbody continuity is largely determined from integration of detailed sedimentological observations with petrophysical and pressure data. The vertical pressure gradients measured in dynamic and static reservoir environments by RFT tools have probably allowed the greatest advances in prediction of continuity. It is fairly obvious that the continuity and directional aspects of sandbodies have significance in the design of water and gas injection facilities. Effective vertical communication dominates the distribution of gas in gas injection schemes, whilst lateral connections and permeability contrast are significant in design of water injection schemes. The areal geometry of a reservoir will influence well spacing and, if offshore, will influence the location and number of platforms required. The position of production and injection wells with respect to faults and original oil water contacts is important. Figure 11.11 shows the effect of a high rate oil producing well in the distortion of an oil–water contact with

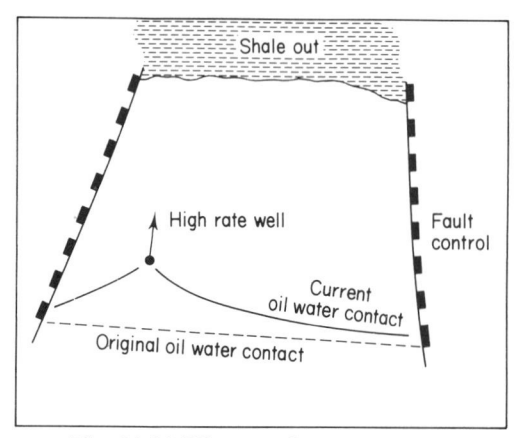

Fig. 11.11 Effect on oil–water contact.

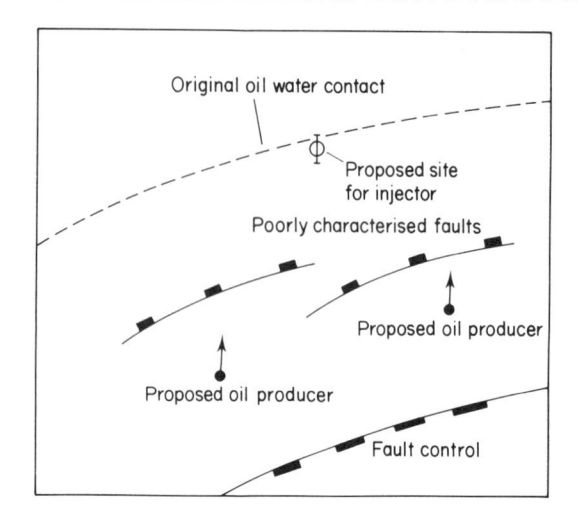

Fig. 11.12 Effect of in-reservoir faults on injection/production well locations.

resulting poor sweep efficiency. Figure 11.12 shows how characterization of partly sealing faults may influence design location of water injection wells. Recognition of *effective* dynamic separation of reservoir beds in complex depositional environments leads to the design of separate injection production facilities for efficient reservoir management. This is illustrated in Fig. 11.13.

11.2.5 Production rate effects

In homogeneous systems there is a lot of evidence to support the contention that rate of oil production, or injection rate, does not affect commercially recoverable oil volume. This assumes that the velocity of oil in a linear geometry is in the order of 0.1–1 m/day and that well spacing ensures high sweep efficiency.

In offshore oilfield production, current economic factors require high initial production rates in order to pay off capital investment. A rule of thumb for a reservoir is to base peak plateau rate on about 10% of the recoverable reserve per annum, or else on about 4% of stock tank oil initially in place (STOIIP) per annum. This is in contrast with traditional onshore operations where peak rates may be around 1–2% of recoverable reserve per annum. Economic factors set the target rates – reservoir characteristics determine whether they are reasonable and prudent.

Reservoir heterogeneity and geometry can lead to lower recoveries than might have been calculated assuming homogeneous properties. The effect of

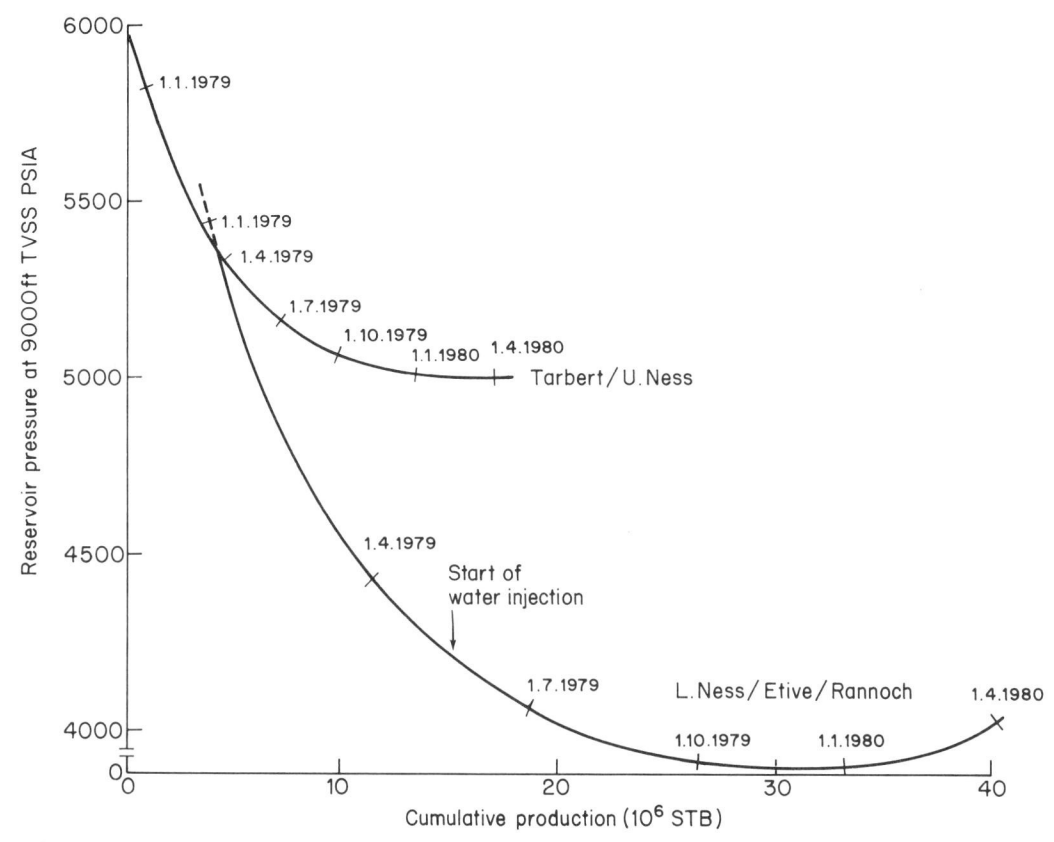

11 SECONDARY RECOVERY AND PRESSURE MAINTENANCE

Fig. 11.13 Pressure history in the UKCS Dunlin reservoir showing dynamic separation of major reservoir units (after [29]).

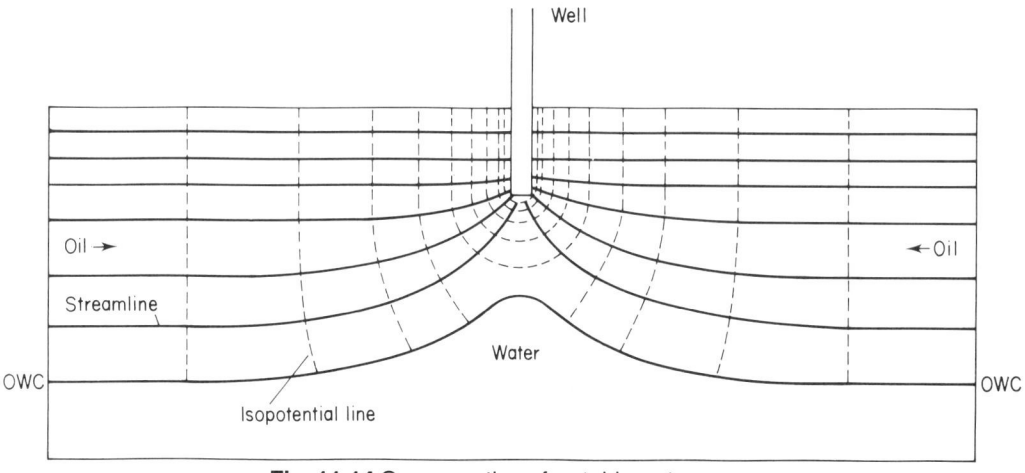

Fig. 11.14 Cross-section of a stable water cone.

reservoir flow rate must be reviewed in the context of competing forces of capillarity, gravity and viscous flow. These operate at both pore scale and interwell scale to differing degrees. Gravity drainage mechanisms require low rate and high dip angle to promote segregation. Reservoirs with high permeability contrast but having vertical pressure communication may require lower rates to promote cross-flow by imbibition from low permeability to higher permeability regions.

In a homogeneous reservoir, analysis of the radial flow behaviour of reservoir fluids moving towards a producing well shows that the rate dependent phenomenon of coning may be important[55]. Coning behaviour has been studied analytically [4,17,18], and in reservoir simulation modes[19,20], and requires average saturation dependent well pseudo-functions to represent well performance in Cartesian grid cell simulators. Figure 11.14 shows the potential and stream line contours around a producing well. The effect of increasing fluid velocity and energy loss in the vicinity of such a well leads to the local distortion of a gas–oil contact or a water–oil contact. The gas and water in the vicinity of the producing wellbore can therefore flow towards the perforations. The relative permeability to oil in the pore spaces around the wellbore decreases as gas and water saturation increase. The local saturations can be significantly different from the bulk average saturations (at distances such as a few hundred metres from the wellbore) as indicated in Fig. 11.15. The prediction of coning behaviour is important since it leads to decisions regarding (a) preferred initial completions, (b) estimation of cone arrival time at a producing well, (c) prediction of fluid production rates after cone arrival and (d) design of preferred well spacing.

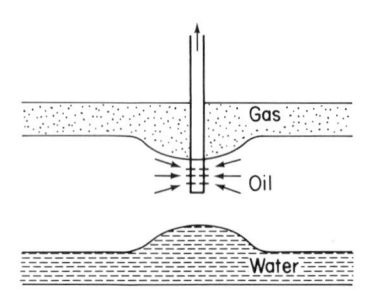

Fig. 11.15 Water and gas cone formation.

The nomenclature of coning recognizes a stable cone as existing in steady state conditions where gravity and viscous forces are balanced. An unstable cone is one that is in the process of advancing or receding. The maximum producing rate at which a stable cone can exist is known as the *critical production rate* for coning – higher rates will cause an advance towards the perforations. At the critical production rate the elevation of the cone from the bulk fluid contact is known as the *critical cone height*. A fairly simplistic representation of the maximum oil potential gradient possible for a stable cone can be written in terms of cone height x and the potential Φ' (or datum corrected pressure). Referring to Fig. 11.16, (a) shows the potential gradient for a stable water cone and (b) shows the gradient at critical rate. Note that at critical rate

$$\frac{d\Phi'}{dr} = \frac{d\Phi'}{dx}$$

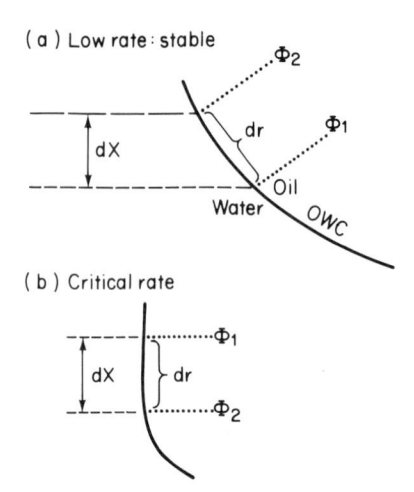

Fig. 11.16 Cone stability. (a) Low rate: stable, (b) critical rate.

From the viscous gravity balance

$$\Delta\Phi' = g' X (\rho_w - \rho_o)$$

Therefore,

$$\left.\frac{d\Phi'}{dr}\right]_{max} = g' (\rho_w - \rho_o)$$

For field units with Φ' in psi, ρ in specific gravity and $g' = 0.433$ psi/ft-g/cc, the viscous gravity balance equation for cone height is

$$x = \frac{\Delta\Phi'}{0.433 (\Delta\rho)}$$

To progress any further in analysis the reader is directed to the works listed in the references at the end of this chapter[2,14,17,18,19,20,48,55].

11.2.6 Well locations and patterns

The development well pattern was established for onshore fields from analysis of areal sweep efficiency. It is more difficult to apply in offshore development and development from pad locations. The main arrangements of wells are shown in the following figures and paragraphs.

(a) *Patterns*
The majority of well patterns defined *historically* pay no attention to gravity effects in dipping reservoirs or to vertical heterogeneity. Areal sweep efficiency at breakthrough and recovery efficiency calculations are often determined by consideration of wells in particular pattern units. The most frequently cited

11 SECONDARY RECOVERY AND PRESSURE MAINTENANCE

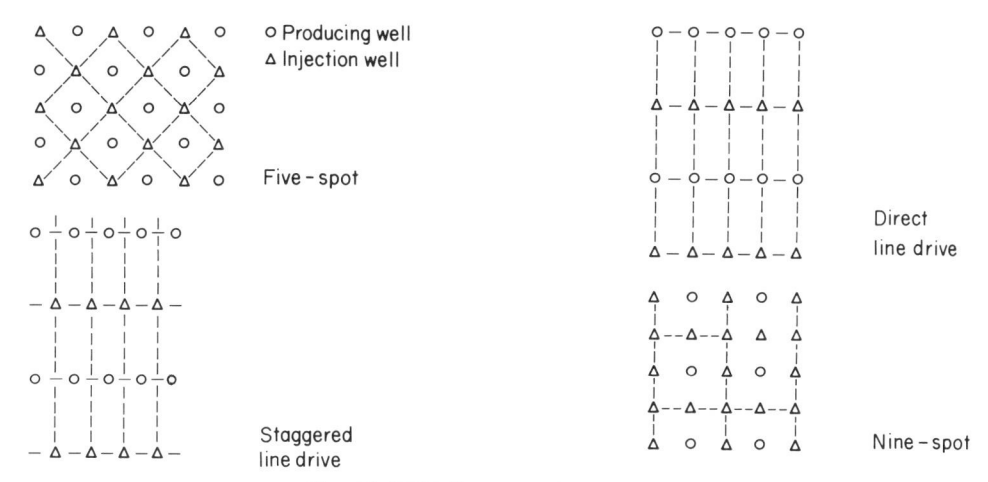

Fig. 11.17 Well patterns for areal sweep.

patterns are shown in Fig. 11.17 as the five spot, nine spot, direct line drive and staggered line drive. Inverted patterns are those with injector locations and producer locations exchanged.

(b) *Well arrangements recognizing structural dip*
Figure 11.18 shows the arrangements frequently encountered for dipping reservoirs. The peripheral water injection scheme is probably the most common. Its success depends on low pressure gradients between producers and injectors so that reservoir energy is restored quickly and pressure maintenance can be employed. The ratio of producers to injectors depends very much on the permeability in the water zone. In some North Sea reservoirs diagenetic damage has reduced water zone permeability by two orders of magnitude in comparison with oil zone permeability.

Crestal injection is usually reserved for gas injectors, although in shallow dips pressure maintenance with water injectors may provide a pressure support that flank wells alone cannot achieve. The conversion of mid-dip producers to injectors after breakthrough may be possible in some reservoirs to minimize interwell gradients. For analytical purposes, well patterns may be analysed as segments of line drives, although reservoir simulation is more frequently employed.

11.3 QUALITY OF INJECTION FLUIDS AND DISPOSAL OF BRINES

The design of secondary recovery and pressure maintenance schemes requires attention to the quality of injection fluids and their compatibility

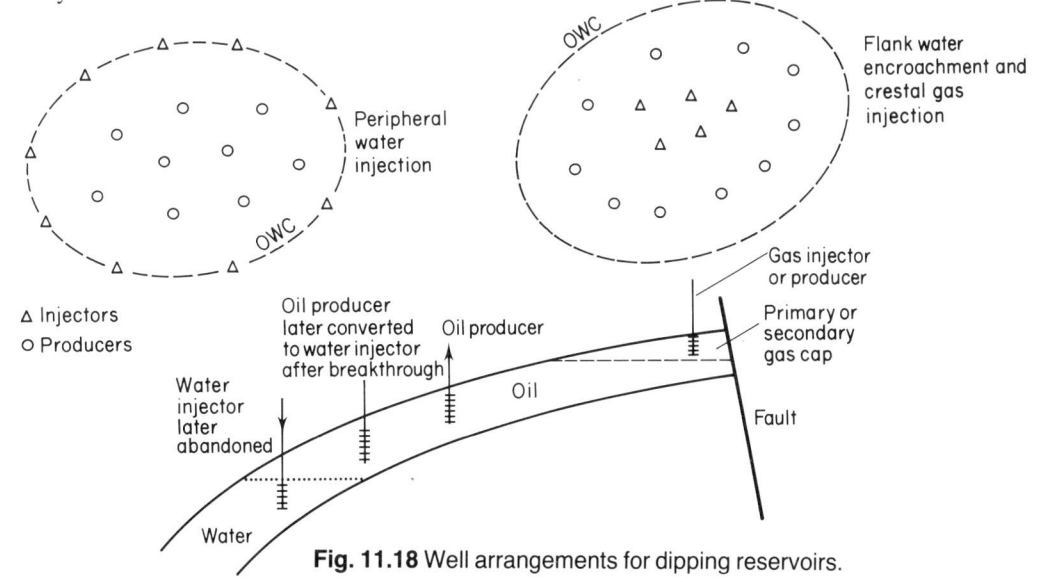

Fig. 11.18 Well arrangements for dipping reservoirs.

184   PETROLEUM ENGINEERING: PRINCIPLES AND PRACTICE

with reservoir fluids, as well as attention to reservoir displacement efficiency. In the same context we may consider the disposal of non-hydrocarbon produced fluids. It is beyond the scope here to deal with these matters in any detail, and the reader is referred to the reference list for more information [1, 31, 34, 39, 42, 51, 54]. In Fig. 11.19 the schematic plan view of the production facilities designed for the Magnus field (Quadrant 211 of UKCS) is shown. The general arrangement shows the platform and subsea wells for oil production by water injection, together with pipeline transport of produced hydrocarbons. Figure 11.20 is a simplified flow sheet indicating the principal components of an offshore production-injection system. Sea water must be filtered and deaerated and biocides added to prevent bacterial growths, as shown in Fig. 11.21. Filtering is designed to a degree which prevents formation plugging with fines. The sea water must be demonstrated to have compatibility with formation water and must not lead to destabilization of clay material in the pore structure. The viscosity of injection water may be significantly higher than formation water as a result of temperature difference, and injectivity calculations are generally made using the higher viscosity.

Produced fluids entering the inlet separator undergo primary separation into oil, water and gas

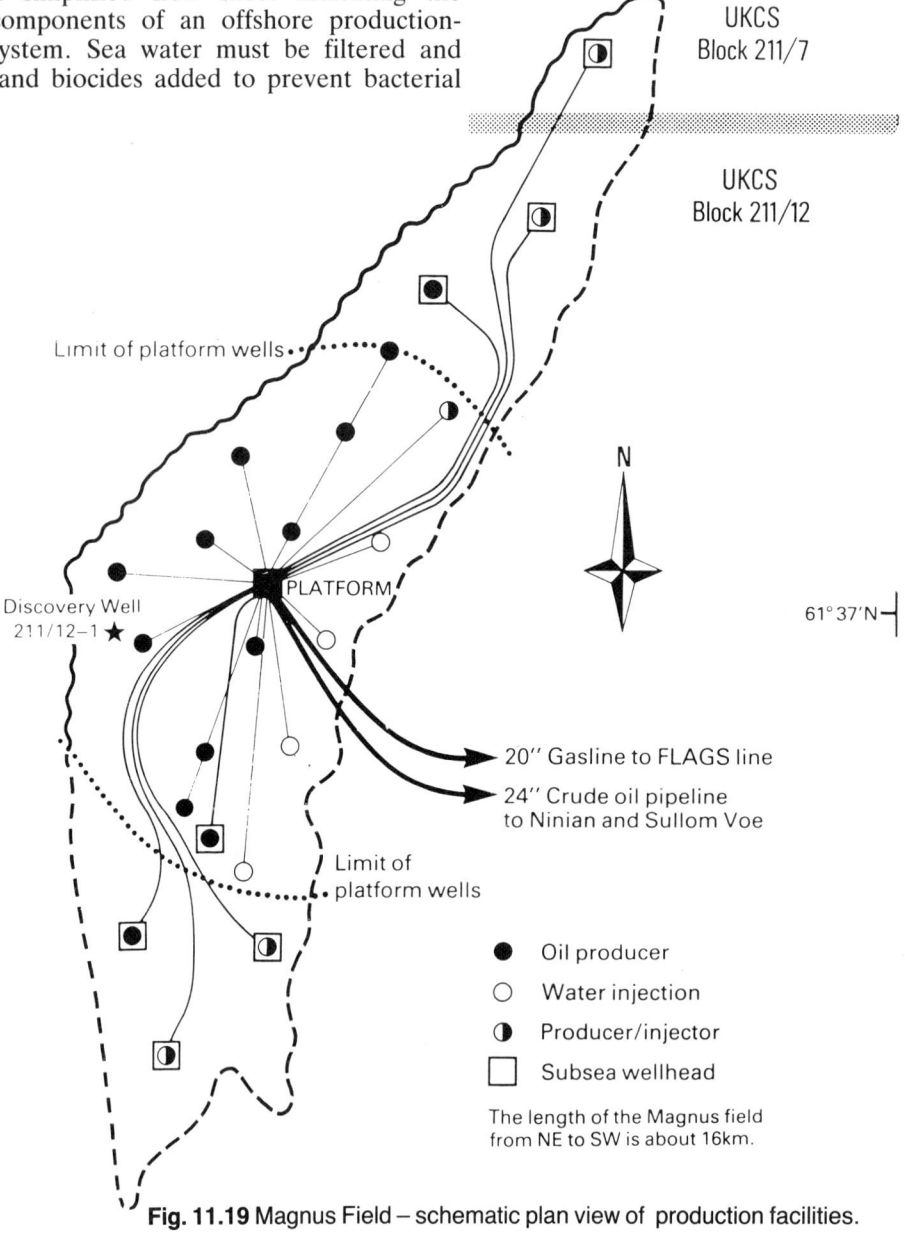

Fig. 11.19 Magnus Field – schematic plan view of production facilities.

11 SECONDARY RECOVERY AND PRESSURE MAINTENANCE

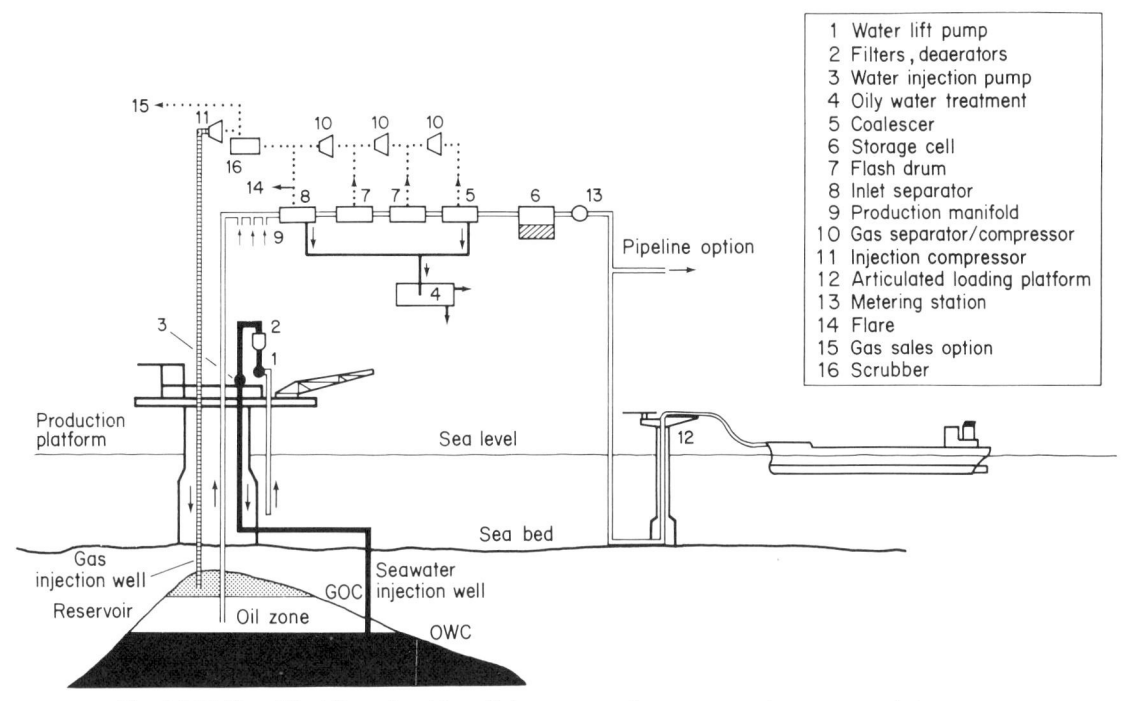

Fig. 11.20 Simplified flow sheet for offshore secondary recovery/pressure maintenance.

1. Water lift pump
2. Filters, deaerators
3. Water injection pump
4. Oily water treatment
5. Coalescer
6. Storage cell
7. Flash drum
8. Inlet separator
9. Production manifold
10. Gas separator/compressor
11. Injection compressor
12. Articulated loading platform
13. Metering station
14. Flare
15. Gas sales option
16. Scrubber

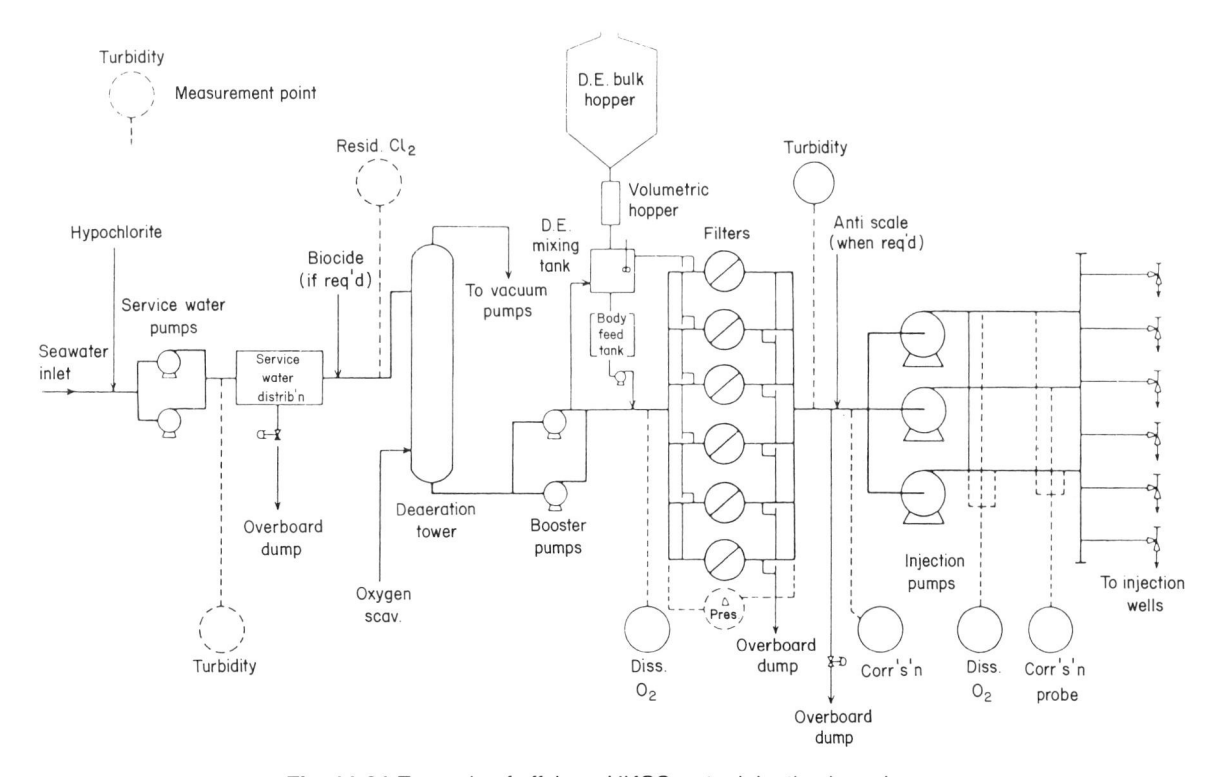

Fig. 11.21 Example of offshore UKCS water injection layout.

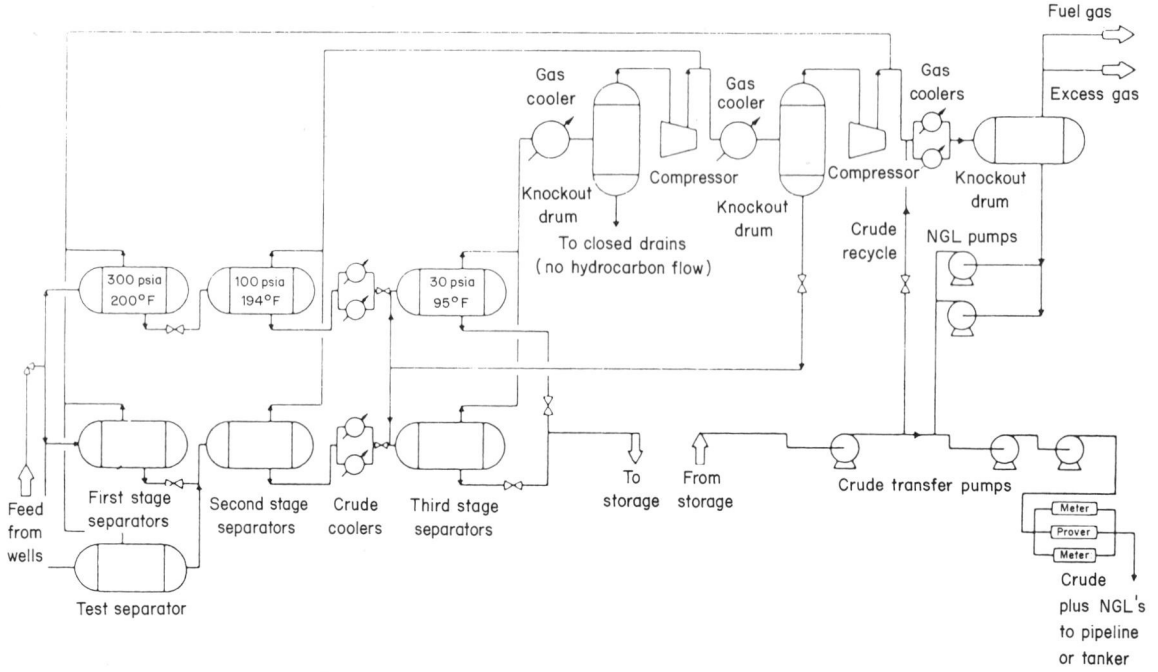

Fig. 11.22 Example of offshore UKCS process layout.

phases, as shown in Fig. 11.22. There is a facility for fluids from this separator to be flared in the event of emergency. All gases separated at decreasing pressures and temperatures in the separator train may undergo liquid knockout, scrubbing and compression for gas sales or reinjection. The products from the final coalescer represent the final separation condition and water passes to the oily water treater (Fig. 11.23) before disposal into the sea, and oil passes perhaps to temporary storage or direct to a pipeline/pump system. For offshore loading of crude oil into tankers, a lower vapour pressure crude is required than for pipeline transportation, and final separator conditions therefore differ. In North Sea operations the tanker loading is often facilitated by use of an articulated loading platform which allows the tankers to take up preferred orientation with respect to winds during loading (Fig. 13.3).

When gas is reinjected into the reservoir, injectivity may be impaired over a period of time by precipitation in the formation of greases from compressor lubricants, and workover treatments may then be required. The gases injected may change *PVT* properties in the gas cap as they will generally be leaner than the original gas. This may have an influence on displacement calculations. When the reinjection gas is a mixture from several reservoirs, the change in *PVT* properties may be even more significant. For reinjection at miscible rather than immiscible pressures with oil, the gas composition also requires careful control.

Examples

Example 11.1

Show the effect on productivity index of oil viscosity for oils of increasing 'viscosity' in the range 0.5 cp, 5 cp, 50 cp, 500 cp, 5000 cp. Assume the reservoir properties are as follows: $r_e = 1500$ ft, $r_w = 0.5$ ft, $s = +4$, $k_{ro} = 0.6$, $h = 100$ ft and $k = 1325$ mD.

Example 11.2

Estimate the ratio in injectivity indices for calculations assuming injection water is colder than formation water ($\mu_w = 0.55$ cp) and calculations assuming it is the same temperature as formation water ($\mu_w = 0.35$ cp) in a particular reservoir.

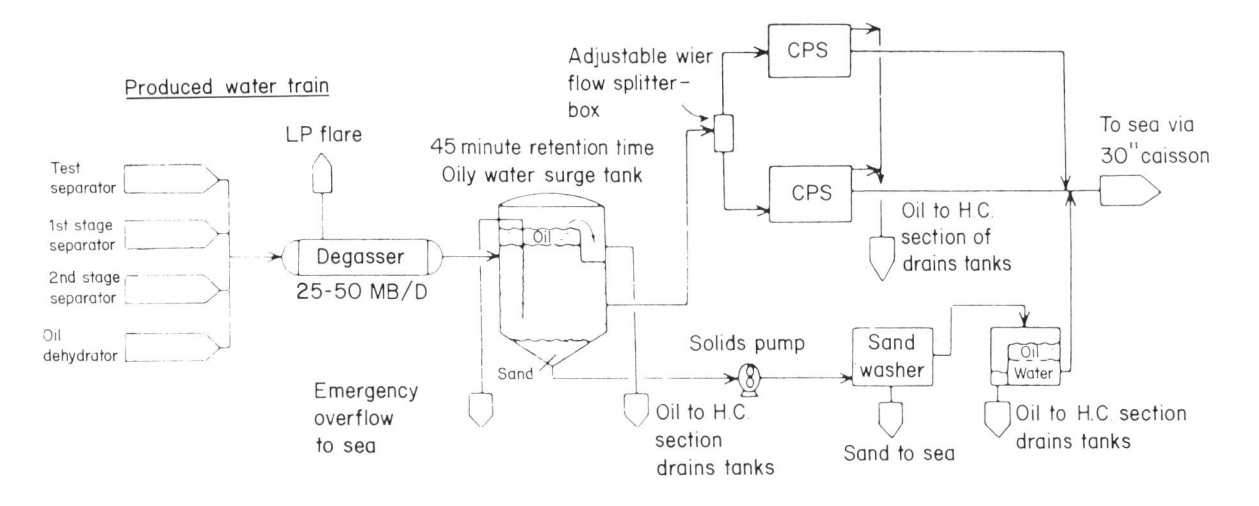

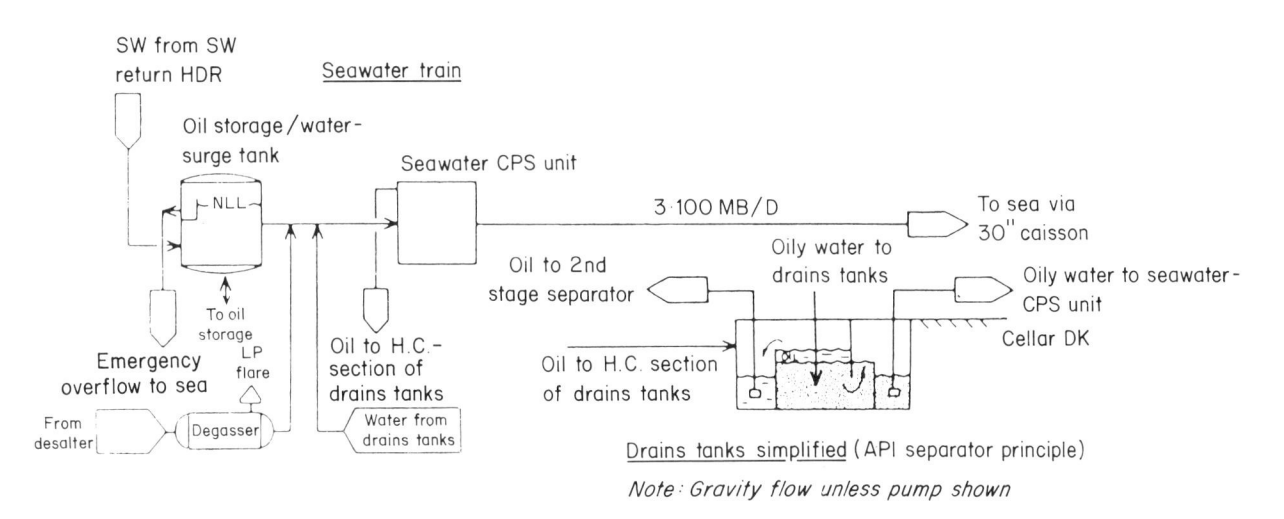

11 SECONDARY RECOVERY AND PRESSURE MAINTENANCE

Fig. 11.23 Maureen field oily water treatment (after [31]).

Example 11.3

A line drive water injection scheme is being operated in a reservoir of length 1 mile between injection and production wells, and of cross-section dimensions 4 miles wide by 98 ft net thickness. The average porosity is 25% and the relative permeability to water in the presence of the residual oil saturation of 30% is 0.4. The relative permeability to oil at the initial water saturation of 30% is 0.85. The oil viscosity at reservoir conditions is 3.4 cp and the water viscosity is 0.4 cp. The water formation volume factor for injected water is 1.005 RB/BBL. Estimate the areal sweep efficiency of the scheme after 10 years if the average daily water injection rate is 53 000 BBL/day.

Example 11.4

An oil well is perforated to within 50 ft of a static water table. At reservoir conditions the specific gravity of the formation water is 1.01 and the specific gravity of the oil is 0.81. For a stable cone to form just below the lowest perforation what is the maximum potential difference (datum corrected pressure differential) that can be allowed during production?

References

[1] Hancock, W.P.
Development of a reliable gas injection operation for the North Sea's largest capacity production platform, *JPT* (Nov. 1983), 1963.

[2] Muskat, M.
The Flow of Homogeneous Fluids through Porous Media, McGraw-Hill Book Co. Inc. (1937).

[3] Craig, F.F.
The Reservoir Engineering Aspects of Water Flooding, SPE Monograph Vol 3 (1971), Soc. Pet. Eng., Dallas, Texas.

[4] Muskat, M.
Physical Principles of Oil Production, McGraw-Hill Book Co. Inc., NY (1950).

[5] Welge, H.
A simplified method for computing oil recovery by gas or water drive, *Trans. AIME* **195** (1952), 91.

[6] Warren, J.E. and Cosgrove, J.J.
Prediction of waterflood behaviour in a stratified system, *SPEJ* (June 1964), 149.

[7] Abernathy, B.F.
Waterflood prediction methods compared to pilot performance in carbonate reservoirs, *JPT* (March 1964), 276.

[8] Calloway, F.H.
Evaluation of waterflood prospects, *JPT* (Oct 1959), 11.

[9] Dake, L.P.
Fundamentals of reservoir engineering, *Dev. Pet. Sci.* **8** (1978). Elsevier, Amsterdam.

[10] Dyes, A.B., Caudle, B.H. and Erickson, R.A.
Oil production after breakthrough – as influenced by mobility ratio, *Trans. AIME* **201** (1954), 81.

[11] Kimber, O.K., Caudle, B.H. and Cooper, H.E.
Areal sweepout behaviour in a ninespot injection pattern, *Trans. AIME* **231** (1964), 199.

[12] Hubbert, M.K.
Darcy's law and the field equations of the flow of underground fluids, *Trans. AIME* **207** (1956), 222.

[13] Dietz, D.N.
A theoretical approach to the problem of encroaching and by-passing edge water, *Proc. Akad. van Wetenschappen,* Amsterdam (1953), V56-B, 83.

[14] Richardson, J.G. and Blackwell, R.J.
Use of simple mathematical models for predicting reservoir behaviour, *JPT* (Sept. 1971), 1145.

[15] Hagoort, J.
Displacement stability of water drives in water wet connate water bearing reservoirs, *SPEJ* (Feb. 1974), 63.

[16] Berruin, N. and Morse, R.
Waterflood performance of heterogeneous systems, *JPT* (July 1979), 829.

[17] Chierici, G.L., Ciucci, G.M. and Pizzi, G.
A systematic study of gas and water coning by potentiometric models, *JPT* (Aug. 1964), 923.

[18] Sobocinski, D.P. and Cornelius, A.J.
A correlation for predicting water coning time, *JPT* (May 1965), 594.

[19] Chappelear, J.E. and Hirasaki, G.J.
A model of oil–water coning for 2D-areal reservoir simulation, *Trans. AIME* **261** (1976), 65.

[20] Letkeman, J.P. and Ridings, R.L.
A numerical coning model, *SPEJ* (Dec. 1970), 418.

[21] Stiles, W.E.
Use of permeability distribution in waterflood calculations, *Trans. AIME* **186** (1949).

[22] Koval, E.J.
A method for predicting the performance of unstable miscible displacement in heterogeneous media, *Trans. AIME* **228** (1963), 145.

[23] Dykstra, H. and Parsons, R.L.
The Prediction of Oil Recovery by Waterflooding, Secondary Recovery of Oil in the United States, API, NY (1950), 160.

[24] Hearn, C.L.
Simulation of stratified waterflooding by pseudo relative permeability curves, *JPT* (July 1971), 805.

[25] Kyte, J.R. and Berry, D.W.
New pseudo functions to control numerical dispersion, SPE (August 1975), 269.

[26] Archer, J.S. and Hancock, N.J.
An appreciation of Middle Brent Sand reservoir features by analogy with Yorkshire coast outcrops, Paper EUR 197, *Proc. Europ. Off. Pet. Conf.* (1980), 501.

[27] Nadir, F.T.
Thistle field development, Paper EUR 165, *Proc. Europ. Off. Pet. Conf.* (1980), 193.
[28] Bishlawi, M. and Moore, R.
Montrose field management, Paper EUR 166, *Proc. Europ. Off. Pet. Conf.* (1980), 205.
[29] van Rijswijk, J. *et al.*
The Dunlin field – review of field development and performance to date, Paper EUR 168, *Proc. Europ. Off. Pet. Conf.* (1980), 217.
[30] Stewart, L.
Piper field – reservoir engineering, Paper EUR 152, *Proc. Europ. Off. Pet. Conf.* (1980), 33.
[31] Nichols, J.H. and Westby, K.A.
Innovative engineering makes Maureen development a reality, Paper EUR 231, *Proc. Europ. Off. Pet. Conf.* (1980), 185.
[32] Tyler, T.N., Metzger, R.R. and Twyford, L.R.
Analysis and treatment of formation damage at Prudhoe Bay, Alaska. *JPT* (June 1985) 1010.
[33] Gesink, J.C.F. *et al.*
Use of gamma ray–emitting tracers and subsequent gamma ray logging in an observation well to determine the preferential flow zones in a reservoir. *JPT* (April, 1985) 711.
[32] Kingston, P.E. and Niko, H.
Development planning of the Brent field, *J. Pet. Tech.* (1975), 1190.
[33] Marcum, B.L., Al-Hussainy, R. *et al.*
Development of the Beryl 'A' field, Paper EUR 97, *Proc. Europ. Off. Pet. Conf.* (1978), 319.
[34] Hillier, G.R., Cobb, R.M. and Dimmock, P.A.
Reservoir development planning for the Forties field, Paper EUR 98, *Proc. Europ. Off. Pet. Conf.* (1978), 325.
[35] Diehl, A.
The development of the Brent field, Paper EUR 108, *Proc. Europ. Off. Pet. Conf.* (1978), 397.
[36] Dufond, R. and Laffont, M.
Frigg, the first giant gas field in the northern North Sea, Paper EUR 110, *Proc. Europ. Off. Pet. Conf.* (1978), 407.
[37] Dake, L.P.
Application of the repeat formation tester in vertical and horizontal pulse testing in the Middle Jurassic Brent sands, Paper EUR 270, *Proc. Europ. Off. Pet. Conf.* (1982), 9.
[38] Tosdevin, M.
Auk Lower Zechstein – drainage mechanism model, Paper EUR 298, *Proc. Europ. Off. Pet. Conf.* (1982), 207.
[39] Hughes, C.T. and Whittingham, K.P.
The selection of scale inhibitors for the Forties field, Paper EUR 313, *Proc. Europ. Off. Pet. Conf.* (1982), 341.
[40] Simmons, P.E.
North Sea offshore compression – future needs, Paper EUR 331, *Proc. Europ. Off. Pet. Conf.* (1982), 511.
[41] Steele, L.E. and Adams, G.E.
A review of the N.North Sea's Beryl field after seven years production, Paper SPE 12960, *Proc. Europ. Pet. Conf.* (1984), 51.
[42] Dawson, A.P. and Murray, M.V.
Magnus subsea wells: design, installation and early operational experience, Paper SPE 12973, *Proc. Europ. Pet. Conf.* (1984), 133.
[43] Robertson Research International/ERC Energy Resource Consultants Ltd
The Brent Sand in the N. Viking Graben, N.Sea – A Sedimentological and Reservoir Engineering Study, RRI/ERC, Llandudno, N. Wales (1980).
[44] Jordan, J.K., McCardell, W.M. and Hocott, C.R.
Effects of rate on oil recovery by waterflooding, *Oil GJ* (13 May 1957), 99.
[45] Deppe, J.C.
Injection rates – the effect of mobility ratio, area swept and pattern, *SPEJ* (June 1961), 81.
[46] Arps, J.J. *et al.*
A statistical study of recovery efficiency, *API Bull* 14D (1967).
[47] Craft, B.C. and Hawkins, M.F.
Applied Petroleum Reservoir Engineering, Prentice Hall Inc., NJ (1959).
[48] Smith, C.R.
Mechanics of Secondary Oil Recovery, Reinhold Pub., NY (1966).
[49] Sandrea, R. and Neilsen, R.
Dynamics of Petroleum Reservoirs under Gas Injection, Gulf Publ., Houston, Texas (1974).
[50] Poettmann, F.H.
Secondary and Tertiary Oil Recovery Processes, Interstate Oil Compact Commission (1974).

[51] Patton, C.C.
Oilfield Water Systems, Campbell Pet Series, Norman, Okl. (1977).
[52] Reznick, A.A., Enick, R.M. and Panvelker, B.
An analytical extension of the Dykestra-Parsons vertical stratification discrete solution to a continuous realtime basis, *SPEJ* **24** (Dec. 1984), 643.
[53] Thomas, W.A.
North Sea development: historic costs and future trends, SPE 12984, *Proc. Europ. Pet. Conf.* (Oct. 1984), 227.
[54] Tinker, G.E.
Design and operating factors that affect waterflood performance in Michigan, *JPT* (Oct. 1983), 1984.
[55] Hurst, W.
Reservoir Engineering and Conformal Mapping of Oil and Gas Fields, Pennwell, Tulsa (1979).

Chapter 12

Improved Hydrocarbon Recovery

12.1 TARGETS

The recovery of light and medium gravity oils by displacement with gas and water, and the recovery of relatively dry gases by expansion, can be considered conventional recovery processes. These techniques lead to gas recovery factors in the range 70–80% and oil recovery factors in the range 20–50%. Any improvement to gas recovery is very limited in terms of design of the recovery mechanism, and efficiency increase mainly involves acceleration of income through well location, well completion and compression choice. Oil recovery projects, on the other hand, appear to offer significant potential for further reducing residual oil saturation [1,2,3,4,6,7,8,11]. In practice, however, the distribution of residual oil in a reservoir is of particular significance; more so than the total oil volume. It can be argued that infill drilling is an effective improved recovery process [83] as is increasing produced water handling facilities. The term residual oil implies nothing absolute about oil saturation and it is process and reservoir property dependent. Distinctions can be made about microscale and macroscale definitions of residual oil saturation, and Table 12.1 shows some of the possible terminology. Measurement of residual oil saturation in the field may be by logging methods, core recovery with fluid control and by material balance. These techniques are discussed in the literature and Fig. 12.1 shows some data on comparative measurements. Near wellbore residual oil measurements may be influenced by *stripping* in the vicinity of an injection well and thus underestimate potentially recoverable oil by improved processes.

In addition to the potential for further recovery from reservoirs containing distributions of residual oil, there must be consideration of techniques to develop reservoirs which might not be developed by conventional processes. In this latter category we may consider higher viscosity heavy oil (API gravity less that 20° API), high pressure gas condensate reservoirs and volatile oil reservoirs, and very low productivity (low permeability-thickness) oil and gas reservoirs. This chapter is designed solely as an introduction to possibilities in improved hydrocarbon recovery and the bibliography and reference list at the end of the chapter should be used as an entry into the more detailed literature.

12.2 THE INFLUENCE OF RECOVERY MECHANISM IN RESIDUAL OIL

In a fully contacted region of a homogeneous oil reservoir undergoing displacement, the competing forces of viscous flow, capillary redistribution and gravity segregation will influence local oil saturation. The pore geometry of the system will particularly influence capillary trapping of oil. The ratio of gravitational to capillary forces is represented by N_B the *Bond number* ($= g'k\,(\rho_w - \rho_o)/(\phi\sigma\cos\theta)$) which has a value around 10^{-6} in normal waterfloods. For systems modelled by core flood experiments, gravity forces have in general been neglected and a term *Capillary number* (N_c) is used to represent the ratio

TABLE 12.1 Residual oil saturation (S_{or} and S_o)

System	Microscale	Macroscale
1–D linear system	S_{or} from core floods in totally bounded system. 100% sweep contact.	$\bar{S}_o$ from material balance. No indication of sweep.
2–D cross-section	Core flood data may be applied in analysis. Need to validate if gravity effects important.	$\bar{S}_o$ from material balance. Cross-flow possibility. Vertical sweep < 100%. Areal sweep assumed 100%.
2–D areal system	Core flood data may require modification in thick sections. Heterogeneity modification.	$\bar{S}_o$ from material balance. S_o in contacted region contrasts with S_o in uncontacted regions. Near wellbore S_o can be different from other regions.
3–D systems	Core flood data only useful in discrete regions. Use of pseudo-functions.	$\bar{S}_o$ from material balance generated from integration throughout system.
All systems – flood out conditions	Usual to provide estimate of S_{or} at less than 1 part oil 1000 in flowing effluent in core flood experiments. Approach to S_{or} depends on wettability, viscosity ratio and core heterogeneity.	Practical. $\bar{S}_o$ limits are attained at economically limiting oil rates or oil cut in produced fluids. These conditions relate to well and gathering centre design capabilities to handle fluids. Could be changed in changed economic climate.

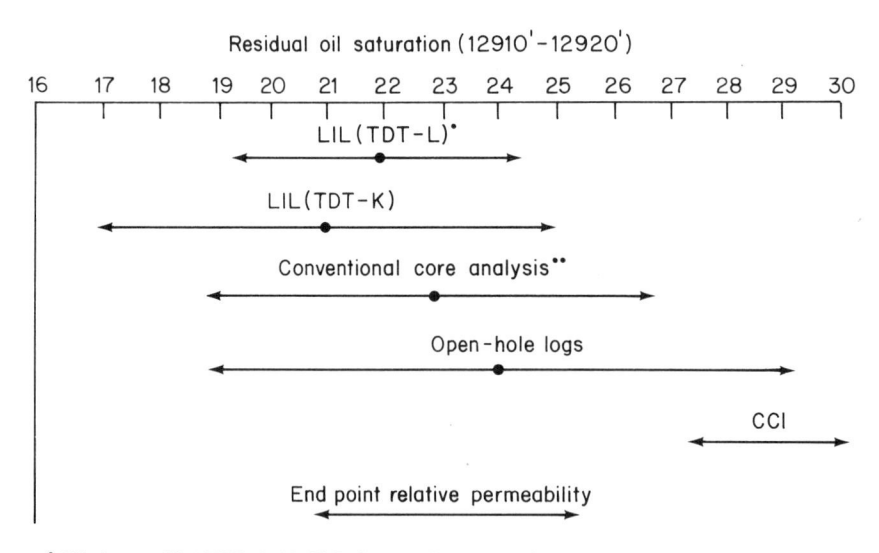

Residual oil saturation (12910'–12920')

```
      16  17  18  19  20  21  22  23  24  25  26  27  28  29  30
                          LIL(TDT-L)*
                       ←————●————→
                      LIL(TDT-K)
              ←——————————●——————————→
                    Conventional core analysis**
                  ←——————————●——————————→
                       Open-hole logs
              ←——————————————●——————————————→
                                              CCI
                                          ←———●———→
                End point relative permeability
                      ←—————●—————→
```

* This is a modified TDT-K with TDT-G electronics source–detector spacing of 60 cm. Readings are taking while stationary.

** Saturation of core plug at 12,918 ft is omitted. The saturation of this plug was more than 2.5 standard deviations from the average.

Fig. 12.1 Residual oil measurements by different techniques in the interval of a single well (after [35]).

between viscous and capillary forces and its effect on oil saturation.

The most frequently used form of the capillary number is that of Moore and Slobod (1956)

$$N_c = \frac{V \mu_D}{\sigma \cos \theta}$$

Taber in a review paper entitled *Research on enhanced oil recovery: past, present and future* [21] has summarized different formulations of the capillary number in the years 1935 to 1979. The general relationship for core plugs and bounded linear flow systems suggests that the form of the relationship N_c against S_{or} is as shown in Fig. 12.2. In this figure S_{or} is the *equilibrium* average oil saturation remaining in the *core plug*, S_{oi} is the initial uniform oil saturation, V is the apparent or superficial velocity of the displacing phase and is equal to the constant volumetric flow rate divided by the cross-section area of the core plug face, μ_D is the Newtonian viscosity of the displacing phase, σ is the interfacial tension between the displacing phase and the oil and θ is the upstream equilibrium contact angle. Units are generally Darcy units with σ in dynes/cm. The interpretation of these data in terms of conventional recovery and enhanced oil recovery potential focuses on an extrapolation and scaling of the laboratory phenomena to field conditions. The flat, essentially constant S_{or} portion of the figure is considered equivalent to conventional field scale displacement mechanisms and covers several orders of magnitude of displacement velocity at normal field interfacial tension and wettability conditions. Through Darcy's law for linear systems, the group $k k_r \Delta \Phi / L$ can be equivalenced to the group $V \mu_D$, where $\Delta \Phi / L$ is the potential gradient, k the absolute permeability and k_r the saturation dependent relative permeability. For practical purposes there is only a limited range of field operation for potential gradients and superficial velocity, and these factors together with effective permeability improvement cannot be changed sufficiently to give much change in S_{or}. At very low displacement rates, which may approach spontaneous imbibition, capillary forces represented by $\sigma \cos \theta$ are dominant and may control S_{or}. The potential for reduction of residual oil saturation is concentrated on modification of the terms, the most significant being σ. In the limit, σ becomes zero as all fluids in a system become miscible. Miscible displacement processes therefore appear very attractive, but in practice their efficiency will be controlled by reservoir heterogeneity, equilibrium approach and economic factors. Low interfacial tension (IFT) systems are another approach to reducing residual oil saturation, and several processes have worked well in laboratory conditions. The reservoir heterogeneity and adsorption of chemical agents inducing the IFT (*surfactants*) tend to lead to adverse economics in field scale applications. The stability of surfactants at reservoir conditions of temperature and salinity is also questionable. The same comments apply to high viscosity polymer fluids which could be added to displacing water to increase the viscosity term in the capillary number. The mechanisms for oil displacement using polymers are complex and involve non-Newtonian flow. Their particular interest at present seems more in the role of diverting agents where they may enter a flooded-out high permeability region and divert flood water into less permeable higher oil saturation regions. The improved oil recovery mechanisms may be considered for conventional oils under the general headings of permeability improvement, miscible processes and chemical processes. Non-conventional systems are then considered as heavy oils and high pressure gas condensates.

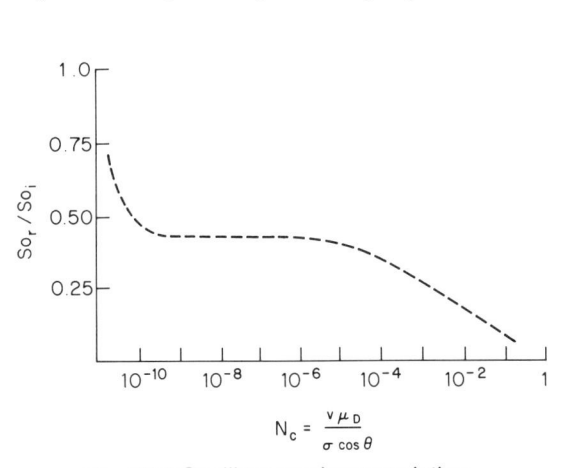

Fig. 12.2 Capillary number correlation.

12.3 PERMEABILITY IMPROVEMENT

This topic focuses attention on the near wellbore region. Improvement can be considered in terms of increased rate of oil production without increase in ultimate volume of recovery (an acceleration project) and in terms of increased total volume of recovery. Well workover activities may be considered as acceleration projects while massive fracture projects may open up otherwise unrecoverable hydrocarbons.

Well productivity improvement can be most easily understood by reference to the semi-steady state radial flow equation for oil flow in an isotropic horizontal system, i.e.

$$\frac{q_o}{\bar{P}-P_{wf}} = \frac{2\pi k k_{ro} h}{\mu_o \left[\ln \frac{r_e}{r_w} - \frac{3}{4} + S\right]}$$

In Darcy units q_o is the reservoir condition volumetric oil flow rate, P the volumetric weighted average reservoir pressure, P_{wf} the flowing bottom-hole pressure at the sand face, k the permeability of the bulk formation, k_{ro} the relative permeability to oil at a saturation in *the vicinity* of the well, h is the average net thickness of the tested interval, μ_o is the oil viscosity at reservoir conditions, r_e is the radial distance from the well to the external boundary of the system, r_w is the radius of the effective wellbore and S is the skin which causes an incremental pressure change in the real system compared with that of a system with bulk formation properties. Hawkins defined the relationship between the radius of the skin zone around a well and the permeability k_s in this zone to the bulk formation permeability as follows (see Chapter 9):

$$S = \left[\ln \frac{r_s}{r_w}\right]\left[\frac{k - k_s}{k_s}\right]$$

It is clear that when $k > k_s$ then S will be positive, and if they are equal S will be zero. The effect of hydraulically fracturing wells or acidizing them may be to make $k_s > k$ and S then becomes negative. The effect on near wellbore pressure is shown on Fig. 12.3 in terms of pressure drops across a skin zone compared with the zero skin case. The magnitude of ΔP_s is influenced by many factors including drilling, drilling fluids, mud cake invasion, borehole rugosity, perforating techniques, completion fluid interaction with formation and formation fluids, saturation changes causing decrease in permeability, injection fluid incompatibility including plugging, precipitation and destabilization of natural formation cements. These factors are largely obvious and receive considerable attention in the literature.

The extension of the simple isotropic radial flow system under semi-steady state conditions to stratified heterogeneous reservoir performance under transient flow control (particularly in low permeability reservoirs) introduces considerable analytical complexity. The analysis of permeability improvement in real heterogeneous reservoirs requires recognition of proper models for perforation, fracturing, acidization and rate dependent sand particle flow. These models are not generally validated and productivity improvement tends to be assessed simplistically on economic criteria, i.e. costs of treatment versus incrementally assigned production increase.

12.4 MISCIBLE DISPLACEMENT MECHANISMS

The displacement of oil by non-aqueous injection of hydrocarbon solvents, lean hydrocarbon gases or high pressure non-hydrocarbon gases such as CO_2, N_2 or flue gases are generally described as miscible floods. The various conditions of pressure, temperature and composition that are required for miscibility (i.e. the elimination of an interface between residual oil and the displacing fluid), whether on first contact or after multiple contacts, are dealt with comprehensively in the literature.

An important factor in most improved oil recovery processes is that of mass transfer between the displaced and displacing phases. In a multicontact system, residual oil behind the displacement front may be stripped of light and intermediate fractions, reducing substantially the residual oil saturation. This mechanism is known as *vaporizing* gas drive. The phase behaviour for miscibility is indicated in Fig. 12.4, where an extended vapour liquid tie line from the two phase region must not pass through the oil composition. It is therefore clear that miscibility between lean gas and oil will occur at pressure P_2 but not at pressure P_1. The minimum miscibility pressure can be defined in such a diagram as the pressure at which the oil composition lies just to the right of the limiting tie line passing through the critical point.

Another mechanism called *condensing* gas drive involves the transfer of intermediate components from the displacing gas to the residual oil and results in a swollen residual oil. The resultant oil is of lower

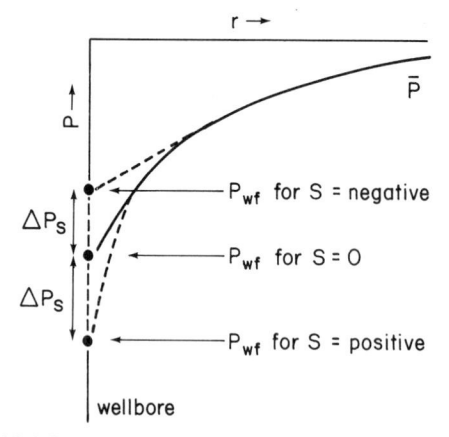

Fig. 12.3 Pressure distribution around producing well.

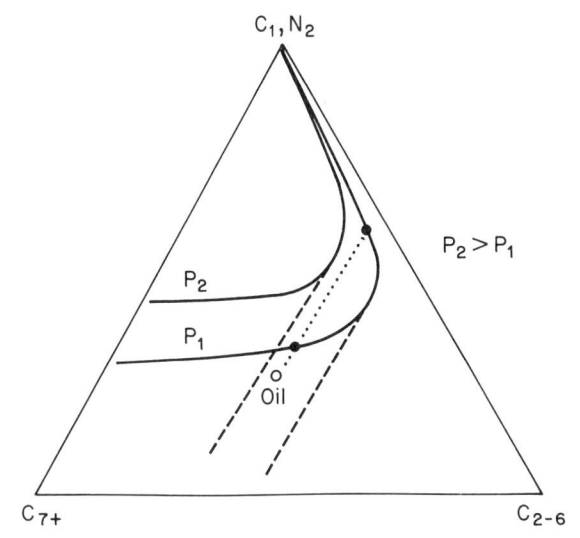

Fig. 12.4 Miscibility in a vaporizing gas process.

viscosity and has an increased oil permeability. These volume effects can be significant even when full miscibility is not attained. In an ideal process, the swelling and mobilization of the disperse, discontinuous residual oil phase leads to the formation of an *oil bank* which may then itself scavenge residual oil as it moves through the formation. Trapping behind the oil bank is prevented by the existence there of miscible conditions, or at least a very large capillary number. Where the formation is contracted by the miscible *solvent* it is expected that oil recovery is complete. Heterogeneity and non-equilibria therefore lead to less than ideal recovery. The oil bank formed by mobilization of previously residual oil may in many instances be of low oil saturation in the presence of the previous displacing fluid (i.e. water). Consequently, the fractional flow of oil may be very low in early time. In implementation of pilot projects in offshore fields, the conclusion is that projects must be started while cash flow from conventional operations exists, so that the characteristics of the scheme can be defined and evaluated.

Several fluids are potentially capable of attaining miscibility with residual oil. The more usual ones together with fluid properties are summarized in Table 12.2 for reservoir conditions of 4000 psia and 200°F.

Experience in the USA has indicated that LPG and NGL can be used in *slugs* in excess of 10% PV, but the risk and expense in using them may not be enough to balance their immediate sale value.

Rich hydrocarbon gases have found use in relatively low pressure environments in condensing gas drive mechanisms.

Lean hydrocarbon has a relatively high miscibility pressure as measured from slim tube experiments. Recent North Sea experience suggests that reinjected gas *cushions* at pressures below expected miscibility undergo significant mass transfer and approach a vaporizing gas drive process.

Carbon dioxide has been shown experimentally to be superior to dry hydrocarbon gas in miscible displacements. Heavy hydrocarbons are volatilized into the gas phase, and under favourable pressure and temperature conditions there is a rapid approach to miscibility.

Nitrogen has a higher miscibility pressure than CO_2 or hydrocarbon fluids but is less effective. Flue gas has been considered in miscible processes as it is predominantly nitrogen, with CO_2, O_2 and SO_2. The presence of additional compounds increases miscibility pressure.

12.5 MISCIBLE FLOOD APPLICATIONS

Displacement stability and the potential for gravity override must be considered in the evaluation of miscible schemes [9,6,5,27,31,32,34,36,77]. Stability problems are likely to be the least with CO_2 which will tend to be gravity stable, although viscous instability may

TABLE 12.2 Properties of miscible fluids

Fluid or solvent	Reservoir condition density		Formation volume factor	Viscosity (cp)
	kg/m³	(lb/ft³)	RB/MSCF	
Light reservoir oil	646	(40)	267	0.2–0.5
Liquefied petroleum gas	450	(25–30)	—	0.08
Rich hydrocarbon gas	300	(19)	0.66	0.02–0.03
Lean hydrocarbon gas	190	(12)	0.80	0.015–0.025
Carbon dioxide	650	(41)	0.50	0.055
Nitrogen	240	(14.6)	0.95	0.026

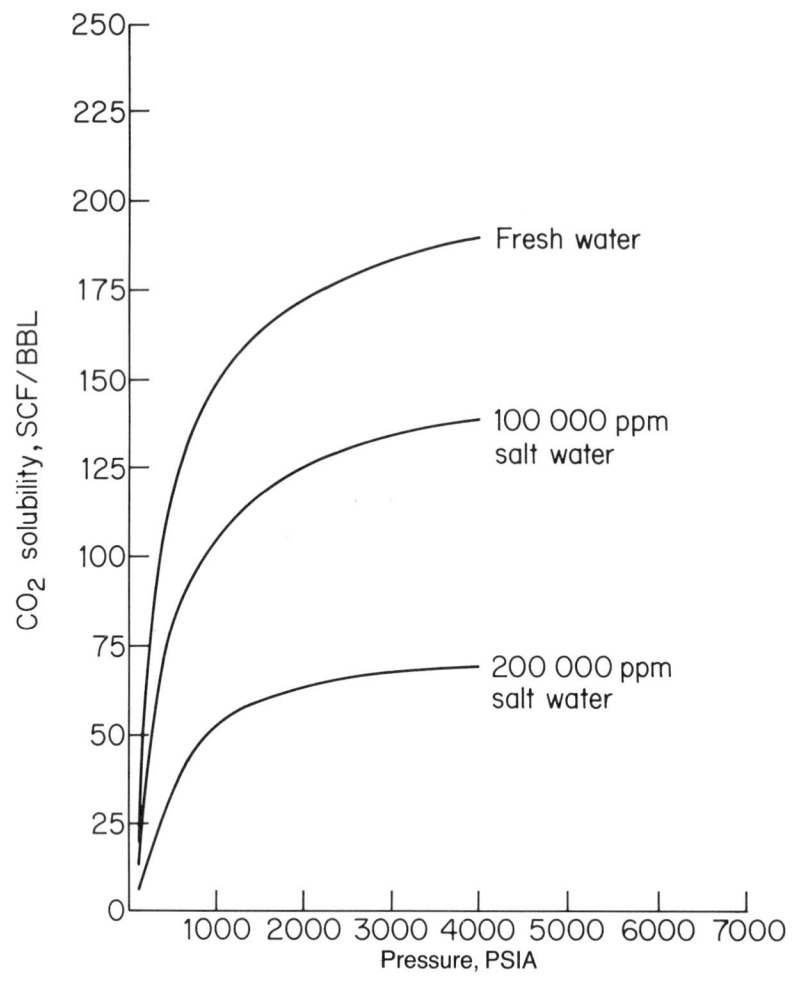

Fig. 12.5 Solubility of CO_2 in water at 100°F (after [64]).

cause concern and will accentuate effects of slug breakdown caused by heterogeneity. Ideal reservoir candidates in a gravity stable miscible process should have steeply dipping beds of good permeability or else be high relief reservoirs with high vertical permeability.

Classic examples of field miscible schemes are the Weeks Island gravity stable CO_2 displacement and the Golden Spike LPG flood. In the North Sea, a degree of miscibility has been achieved with temporary gas reinjection in the Beryl field and into the Statfjord formations of the Brent and Statfjord fields. The fluid choice for miscible displacement projects in North Sea reservoirs is firstly with hydrocarbon gas and secondly consideration of the relative merits of CO_2 and N_2. Offshore generation of CO_2 or N_2 will require additional platform facilities which may render projects uneconomic.

Nearly twice as much CO_2 would be needed as N_2 to occupy one reservoir barrel of pore space. In addition, CO_2 is soluble in formation water, as shown in Fig. 12.5, is more corrosive than N_2 and is more expensive to produce than N_2.

12.6 CHEMICAL FLOOD PROCESSES

Chemical processes for oil displacement are dependent on changes in μ, σ, θ and on the capillary number [12,7,8,15,17,22]. Caustic solutions have been reported to change reservoir rock-fluid wettability and generate *in situ* surfactants, although it appears that results are unpredictable. Wettability change alters both shape and end-points of relative permeability curves and thus leads to improved fractional flow and reduced residual oil saturation in favourable conditions.

12 IMPROVED HYDROCARBON RECOVERY

Polymer applications are centred on thickening of injection water with water soluble polymers and either use as diverting agents or in the consequent improvement in mobility ratio and increased sweep efficiency [20,22]. The polymer solutions are in general non-Newtonian in behaviour.

Biopolymers such as xanthan gums exhibit decreased viscosities at high flow rates and are known as shear thinning fluids. Polyacrylamides tend to have increased viscosity at higher rates and are viscoelastic. Most polymer systems considered for oil displacement are prone to adsorption on reservoir rock surfaces, particularly shaly sites, which decrease their effectiveness. For application in high temperature, high salinity environments, the thermal stability of polymer systems must be demonstrated. Figure 12.6 shows the results of a test at 205°F on 0.15% polymer solutions in 33 000 ppm TDS brine with dissolved oxygen less than 0.02 ppm. The thermal degradation of both polyacrylamide and xanthan solutions was essentially complete in a few hundred days, whereas targets of at least five years might be set. The particular interest in polymer solutions in stratified reservoir systems is in blocking high permeability depleted layers and allowing displacement in the lower permeability thickness layers. The process becomes unattractive when significant cross-flow between layers exists as the polymer solution may become ineffective some distance from the wellbore.

Surfactant processes concentrate on reduction of interfacial tension to increase capillary number and reduce S_{or} in the swept zone. The basic ingredients in a surfactant system are oil, brine and surfactant (plus consurfactants such as alcohols).

Although many real systems will be significantly more complex, a pseudo three-component system can be used to represent phase behaviour at varying compositions. The usual representation (Fig. 12.7) is in the form of equilibrium ternary diagrams with 100% surfactant concentration at the top, 100% brine to the left and 100% oil to the right. Depending on the overall composition, an equilib-

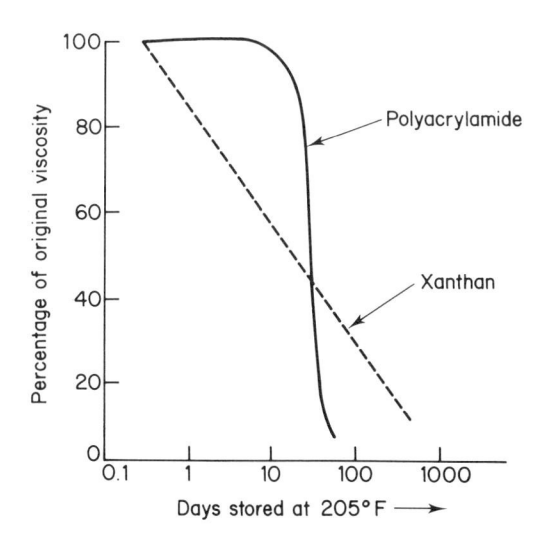

Fig. 12.6 Thermal stability of polymer solutions (after [12])

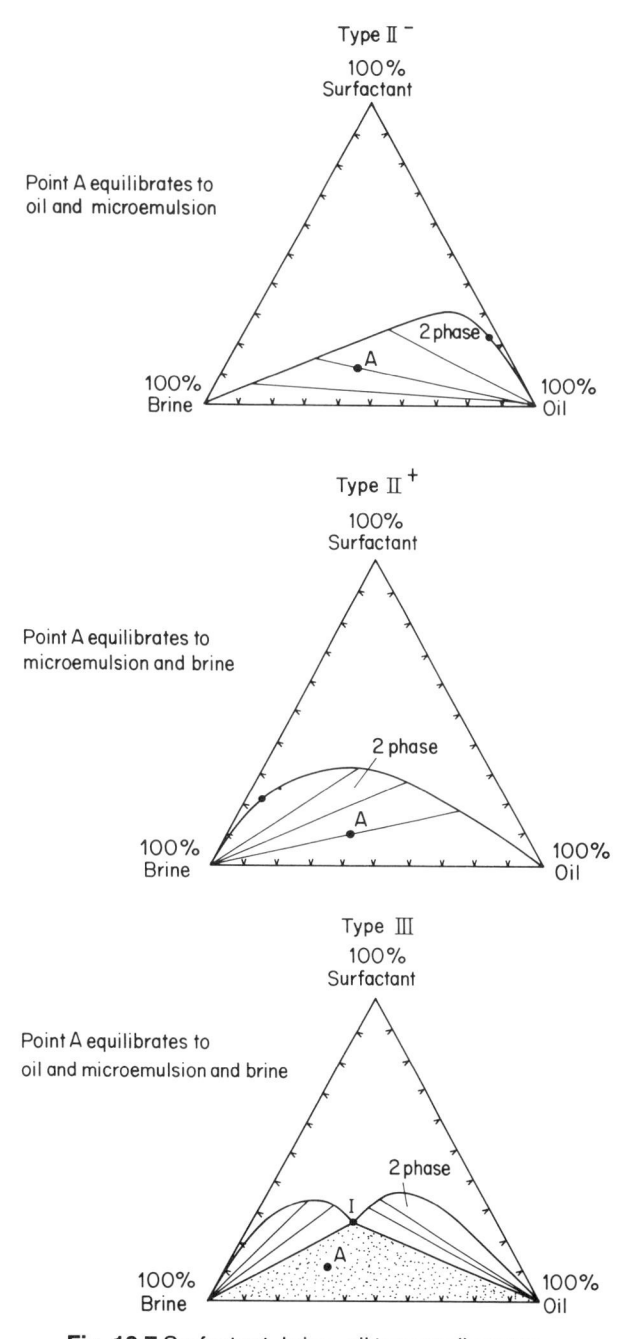

Fig. 12.7 Surfactant–brine–oil ternary diagrams.

TABLE 12.3 Equivalent terminologies

Winsor	Nelson and Pope	Reed and Healy
Type I	Type II$^-$	Lower phase (microemulsion)
Type II	Type II$^+$	Upper phase (microemulsion)
Type III	Type III	Middle phase (microemulsion)

rium can exist in which one, two or three phases are present. The current North Sea interest is in low surfactant concentrations, introduced into a reservoir in a saline aqueous phase. Under such conditions a single-phase system is unlikely with any practical surfactant. In multiphase environments a homogeneous phase containing oil, brine and surfactant, and known as a microemulsion, can form. The microemulsion can be considered in thermodynamic equilibrium with any other phase present in the system. The terminology of surfactant system-phase diagrams has developed through Winsor [8], Nelson and Pope [23] and Reed and Healy [8] and is approximately equivalenced in Table 12.3 Following the Nelson and Pope nomenclature, II indicates a two-phase system and III a three-phase system. As shown in Fig. 12.7, the Type II$^-$ system equilibrates with a lower microemulsion phase and an upper excess oil phase and is characterized with the lines of negative slope in the two-phase region.

The Type II$^+$ system shown in Fig. 12.7 equilibrates with an upper microemulsion phase and a lower excess brine phase and has tie lines with a positive slope in the two-phase region. The Type III system in Fig. 12.7 equilibrates with a middle microemulsion phase, an upper excess oil phase and a lower excess brine phase. Shinoda and others [12,17] have demonstrated that the Type III condition corresponds to a hydrophilic–lipophilic balance in the middle phase which will lead to maximum oil mobilization. A Type III condition during chemical flood displacement may therefore be an important target in flood design. It is, however, inevitable that during the passage of any chemical mixture between an injection location and a production location there will be changes in mixture composition and a continuous change in phase equilibrium. Pope and Nelson [24] have demonstrated that, for practical purposes, flood performance can be calculated as a number of equilibrium steps, as illustrated in Fig. 12.8. There are a number of parameters that, for a given surfactant–oil system, will alter the position of hydrophilic–lipophilic balance. Figure 12.9 shows a plot known as a Shinoda diagram in which the balance variable B is plotted against the fraction of total fluid (oil + brine + microemulsion) that is microemulsion. When B is temperature, the balance

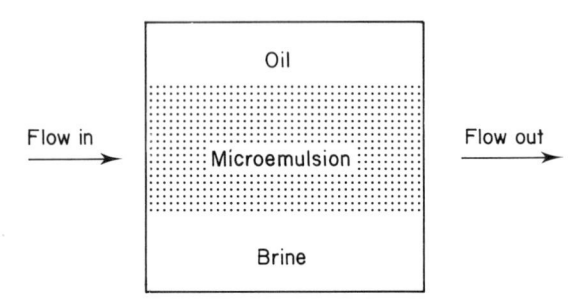

Fig. 12.8 Equilibrium representation of phase distribution in a cell – a three-phase system at some location in the reservoir at some time.

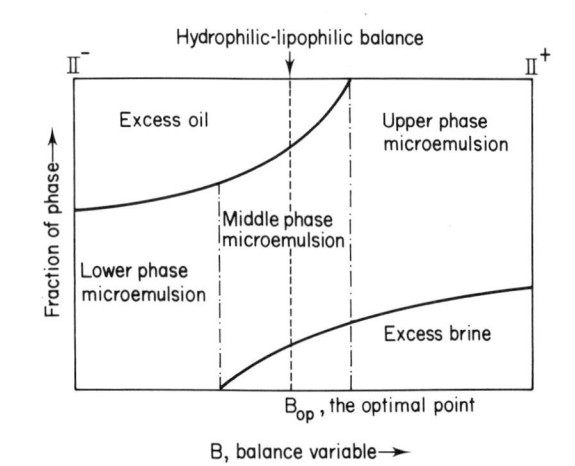

Fig. 12.9 Shinoda diagram (after [12]).

point is called the *phase inversion temperature* [7], PIT; when B is salinity the balance point is known as *optimal salinity*. Surfactants must therefore be formulated for performance at a PIT equivalent to the reservoir temperature. Salinity can sometimes be modified from high to low by preflush, but this introduces the possibility of formation damage in clay sensitive sands.

The main options in surfactant flooding with preselected optimal surfactants are either (i) lower volume, higher concentration surfactant slug, chased by a mobility control polymer, or (ii) lower concentration surfactant in flood water without polymer chase. The onshore North American experience,

12 IMPROVED HYDROCARBON RECOVERY

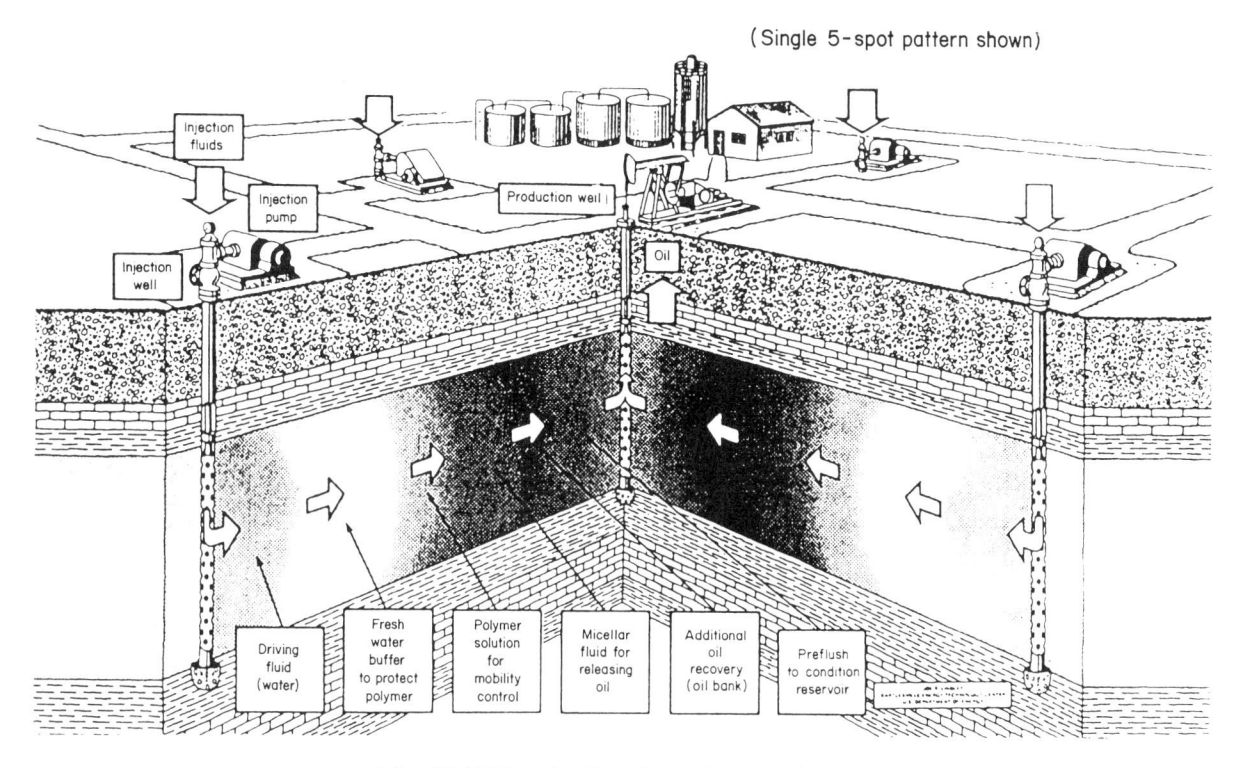

Fig. 12.10 Chemical flooding, micellar polymer.

mainly using petroleum sulphonate surfactants and their derivatives, tends to be of the higher concentration slug type, as shown in Fig. 12.10 [22]. For offshore North Sea reservoirs the interest has focused on the low concentration synthetic surfactant additive to a seawater high volume flood. The circumstances leading to this route have recently been discussed by Grist et al. [15] and centre on the difficulties of offshore handling bulk chemical in volumes needed for higher concentration slugs. A concentration target for the seawater additive system appears to be around 1–2%. The philosophy adopted in the design of a low concentration surfactant flood involves development of a surfactant which exhibits Type III/II$^+$ behaviour and stability at reservoir temperature and salinities and which has an acceptable adsorption character in the specific reservoir. It has been shown that minimum, even though perhaps significant adsorption occurs around the hydrophilic–lipophilic balance condition. Synthetic surfactants, although costly, can be manufactured with relatively narrow molecular weight distributions. A potentially interesting group of oxyalkylated suplhanates with high temperature and high salinity tolerance has been identified by Mattax et al. [12] and has a formulation $C_nO(EO)_xC_3SO_3Na$ where (EO) represents an ethylene oxide group ($-CH_2CH_2O-$). These synthetic surfactant systems are designed to equilibrate in the Type III to Type II$^+$ phase systems. With increasing salinity the phase system moves from Type III towards Type II$^+$ and is analogous to the shift from Type II$^-$ through Type III to Type II$^+$ shown in Fig. 12.11 and reported for petroleum sulphonate systems. In this context, low salinity might be around 6000 ppm TDS and high salinity around 120 000 TDS.

For low surfactant concentration in continuous seawater floods, the majority of effort at present is concentrated on surfactant chemical formulation to meet optimal condition. The adsorption of surfactant on the particular reservoir rock is considered sacrificial in these circumstances and although it can be quantified it probably cannot be changed much. If adsorption is too great to give an economic flood then that surfactant/reservoir system should be rejected. Preliminary surfactant interest can be assessed from ability to mobilize oil from a static residual column and is a precursor to laboratory core floods. In core floods a comparison is made between recovery after waterflooding at reservoir temperature and recovery after surfactant flooding at reservoir temperature and at preferred concentration. Although there is no particular consensus in the

TABLE 12.4 The screening of potential surfactants for use in field operations

---

Define reservoir salinity and surfactant concentration range.
↓
Find surfactants having optimal salinity above and below resident brine salinity and check phase representations.
↓
Blend surfactants at reservoir temperature using equal volumes of reservoir brine and crude oil.
↓
Select formulation so that oil solubility is largest at a given blend and concentration.
↓
Define phase equilibria for salinity scan.
↓
Check adsorption characteristics in flow loop using surfactant composition, heat change and adsorption isotherm techniques.
↓
Initiate short and long core flood tests at reservoir conditions and model results using compositional simulator.
↓
Tune simulator parameters and predict key core flood outcome.
↓
Run experiment for validation.
↓
Use simulator to design pilot flood.
↓
Run pilot flood and retune simulator.
↓
Design field flood or next pilot.

---

literature, it is suggested that tests should eventually be conducted on reservoir zone core lengths up to 2 m. This incidentally indicates that some pre-planning is necessary if preserved core of *in situ* wettability is to be available. The core experiments will show an upper boundary to recovery potential since residual oil saturations will be those of a completely swept zone. The combination of phase equilibrium experiments and core flood experiments will allow calibration of a linear surfactant core flood simulator, of which several versions have been described. A general approach to assessing surfactant reservoir system potential is shown in Table 12.4.

12.7 HEAVY OIL RECOVERY

12.7.1 Characterization of heavy oil

A generalized classification of heavy crude oil considers an association of the following properties:

(a) low API gravity – less than 20°;
(b) high viscosity at normal reservoir conditions ($\mu_o > 20$ cp);
(c) poor reservoir mobility (k_o/μ_o);
(d) dark colour;
(e) sulphur content ($> 3\%$ weight);
(f) metal content (Ni; V; ≈500 ppm);
(g) asphaltene content (up to +50% weight).

Any particular heavy oil may have some of these properties and there is nothing absolute in any classification.

A method proposed by Yen has been used [61] to distinguish the pseudo-ternary composition and origin of heavy oils. The presentation shown in Fig. 12.12 is based on vacuum distillation of the crude oil which results in a volatile component, mainly hydrocarbons and a pot-residue. The residue is mixed with cold *n*-pentane which separates asphaltene from polar compounds. The crude oil composition as plotted on the ternary diagram can be used to distinguish thermally mature oils from weathered and biodegraded heavy oils.

North Sea heavy crude oils are not very well documented. Table 12.5 illustrates some characteristics of heavy crudes in the UK sector. The majority of these oils are not *heavy* in a world wide characterization, but are in offshore reservoirs difficult to develop for many reasons.

Table 12.6 contrasts some selected examples of heavy oils from other parts of the world.

It is clear that some recognition of the characteristics of heavy oil crudes from around the world might be used to guide expected properties from heavy oils to be found in the UKCS. In the USA some 127 ×

12 IMPROVED HYDROCARBON RECOVERY

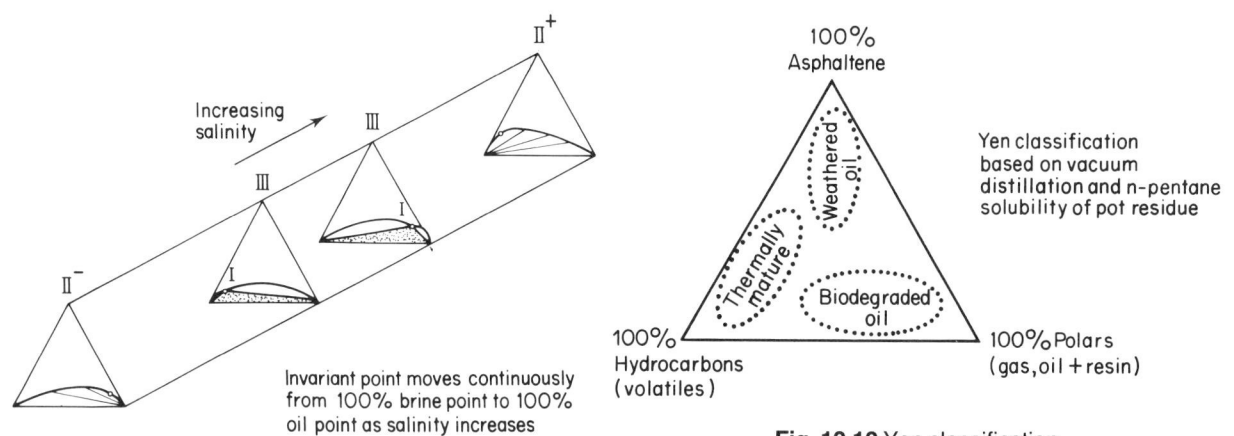

Fig. 12.11 Effect of balance variables on phase distribution. Invasiant point moves continuously from 100% brine point to 100% oil point as salinity increases.

Fig. 12.12 Yen classification.

TABLE 12.5 Characteristics of UKCS heavy crude oils

UKCS quadrant	Gravity (°API)	Reservoir condition Temperature (°F)	Viscosity (cp)	Reservoir
2	20.5–24	120	8–14	Sandstone
3	11–15	115	150–2750	Sandstone
9	15–26	120	(30)	Sandstone
14	25	175	4	Sandstone
16	23	130	5	Sandstone
206	23	135	3.5	Sandstone

TABLE 12.6 Examples of heavy oil reservoirs

Field	Country	Oil gravity	Estimated viscosity cp, (rc)
Gela	Italy	8–13	80–220
Ragusa	Italy	19	25
Duri	Sumatra	20	(25)
Darius	Iran	12–20	—
Harbur	Oman	18–23	—
Karatchok	Syria	19–23	—
Bati Raman	Turkey	12.5	650
Tia Juana	Venezuela	12	3700
Jobo	Venezuela	8	6200
Lloydminster	Canada	10–18	3000
Cold Lake	Canada	10–12	100 000
San Miguel	USA, Texas	12–15	80–100
Kern River	USA, Texas	14	4200
Midway Sunset	USA, California	14	1600

TABLE 12.7 Heavy oil resource distribution in USA

Heavy oil as percentage of all crude (STOIIP)	50%
Heavy oil production as percentage of all crude (annual)	8%
Heavy oil gravity: 20–25°API	48%
less than 20° API	52%
Heavy oil reservoir depths	
less than 500 m	20%
between 500–1000 m	35%
more than 1000 m	45%
Heavy oil reservoir lithology	
limestone reservoirs	3%
sandstone reservoirs more than 3 m thick and shallower than 1000 metres	40%

10^9 STB of heavy crude oil in place has been identified [61]. This figure excludes tar sands and is similar to the volume of medium and light oil historically identified. The US experience in distribution of heavy oil in terms of gravity, reservoirs and depths might also serve as some kind of guide to expectations, and the figures are given in Table 12.7.

These data from a mature exploration area show that heavy oil is widespread geographically and that volumes in place approach that of *conventional* oil. The tables show us nothing of the reservoir sizes and only suggest that production from heavy oil reservoirs is disproportionately small. Recovery factors from heavy oil reservoirs are not a good guide to their potential since they are production process dependent. In the North Sea the heavy oil reservoir potential is linked through economic considerations to reservoir size, geometry, water depth and subsea depth, as well as reservoir rock and fluid properties.

12.7.2 Properties of heavy oil reservoirs

Many of the North Sea examples of heavy oil reservoirs are found in relatively young, friable sandstones of Palaeocene and Eocene age. Reservoir rock and fluid properties are often difficult to obtain as coring can be unsuccessful and the fluids may not flow to surface. Oil viscosity at reservoir conditions may be roughly estimated from dead oil viscosity where the oil is relatively gas free. Figure 12.13 shows the form of relationship between oil viscosity at surface and reservoir temperature conditions for different gravity oils. Cores may be more successfully obtained from friable sands using rubber sleeved or fibre glassed core barrels, typically providing 7 m core lengths. At surface, the recovered core may be quick frozen and packed with solid CO_2 for transportation to a laboratory. The reconstituted samples may not reflect *in situ* reservoir stresses when used in conventional processes for measuring porosity and saturation. Friable and unconsolidated sands can exhibit high pore volume compressibilities in comparison with the often used literature values of Hall around 2×10^6 psi^{-1}. Table 12.8 illustrates some pore volume compressibilities in reservoir rocks having porosities greater than 20%.

The pore volume compressibility is important in correlating core and log data at the same *in situ* stress condition. The interpretation of porosity from

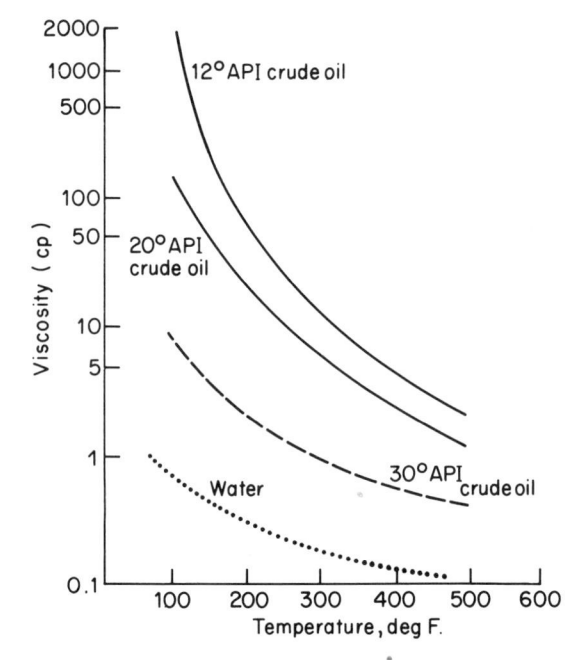

Fig. 12.13 Relationship between kinematic viscosity and temperature

12 IMPROVED HYDROCARBON RECOVERY

TABLE 12.8 Pore volume compressibility of reservoir rocks

Reservoir rock	Surface porosity (Frac)	Pore volume compressibility (psi^{-1})
Oilfield sands (Hall)	0.25	1.75×10^{-6}
Frio sand	0.30	3.5×10^{-6}
Berea sand	0.21	3.4×10^{-6}
Weakly cemented sand	0.30	5.8×10^{-6}
Athabasca sand	0.33	5.0×10^{-6}
Ottawa sand (c–109)	0.345	$10.8 \times 14.5 \times 10^{-6}$
North Sea Chalk	0.39	$> 20 \times 10^{-6}$

formation density logs and their derivatives has been reasonably successful in open-hole conditions. It should, however, be recognized that mud filtrate invasion and oil displacement is likely to be very low, and the density of fluid in the interpreted zone is best represented by that of the heavy oil. Figure 12.14 shows a density log response in an Eocene heavy oil reservoir from the North Sea [58]. Using a mud filtrate density rather than an oil density in the porosity calculation leads to an overestimation by 3 porosity percentage units.

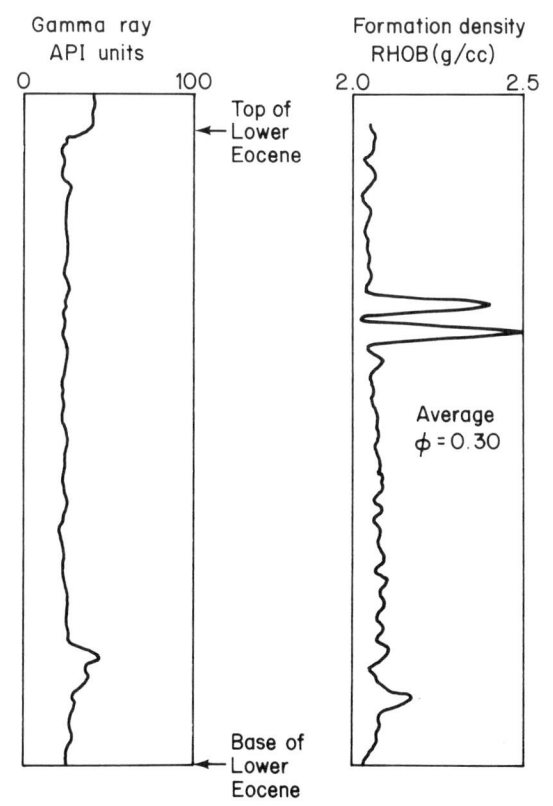

Fig. 12.14 Porosity interpretation from FDC log in North Sea heavy oil well.

12.7.3 Production characteristics of heavy oil reservoirs

Porosity distribution and oil saturation are required for determination of hydrocarbon in place for a heavy oil reservoir, i.e.

$$\text{STOIIP} = \frac{\bar{A}\bar{h}_N(\bar{S}_o\bar{\phi})}{\bar{B}_{oi}}$$

where $\bar{A}$ = area, $\bar{h}_N$ = net thickness, $\bar{S}_o$ = saturation, $\bar{\phi}$ = porosity and $\bar{B}_{oi}$ = initial oil formation volume factor. However, it is flow properties that determine performance. The key parameter in the flow of heavy oil in a reservoir is the mobility, which at some given reservoir pressure (P), temperature (T) and saturation (S) may be written as $[kk_{ro}/\mu_o]_{S,P,T}$ where k is the formation permeability, k_{ro} the relative permeability to oil and μ_o the oil viscosity.

The effect of the term $[kk_{ro}/\mu_o]_{S,P,T}$ on semi-steady state radial flow productivity index is dramatic. In the relationship for PI

$$\text{PI} = \frac{q_o}{\bar{P}-P_{wf}} = \frac{2\pi kk_{ro} h}{\mu_o \left[\log_e \frac{r_e}{r_w} - \frac{3}{4} + S\right]}$$

the particular effect of variation in viscosity is shown in Fig. 12.15.

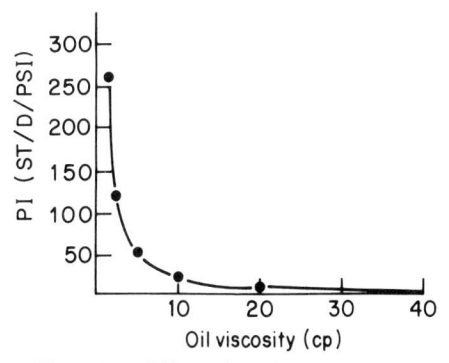

Fig. 12.15 Effect of oil viscosity on PI.

This illustrates one of the reasons why thermal expansion of oil and reduction in viscosity by the addition of thermal energy is attractive for heavy oil systems. Introduction of carbon dioxide in a heavy oil reservoir can also lead to viscosity reduction of the oil in some circumstances. The effect of the ratio between oil viscosity and water viscosity on fractional flow of water in horizontal homogenous reservoirs can also be illustrated as in Fig. 12.16.

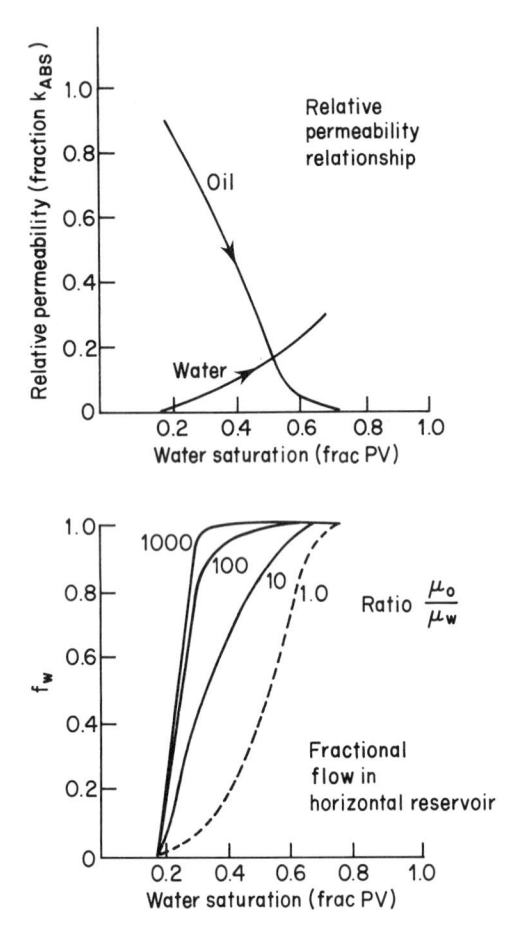

Fig. 12.16 Effect of heavy oil viscosity on fractional flow of water.

Improvement in the productivity of heavy oil reservoirs can be considered through modification of the terms in the following relationship:

$$\text{Productivity} \propto \frac{k \cdot k_{ro}}{\mu_o}$$

Increase in permeability may be achievable through fracturing and acidization of reservoir rock. The reservoir lithology and heterogeneity obviously influences the nature of any improvement in permeability. From a design standpoint, particular consideration must be paid to rock debris flow, proppants and rock strength. The time changing magnitude of improvement as saturation and pressures change is important in the economic assessment of improvement. Uncertainties are attached to the description of induced fractures and acidized permeability and the proper representation of dual porosity/permeability.

Relative permeability improvement relates to wettability change as well as to changes in *irreducible saturations*. The mechanisms of relative permeability change are not well understood and are not capable of being incorporated directly in productivity design. Temperature change, as well as introduction of chemicals such as caustic solutions, may shift relative permeability curves as shown in Fig. 12.17. These changes are observed empirically with particular reservoir rocks and may be partly controlled by pore filling minerals.

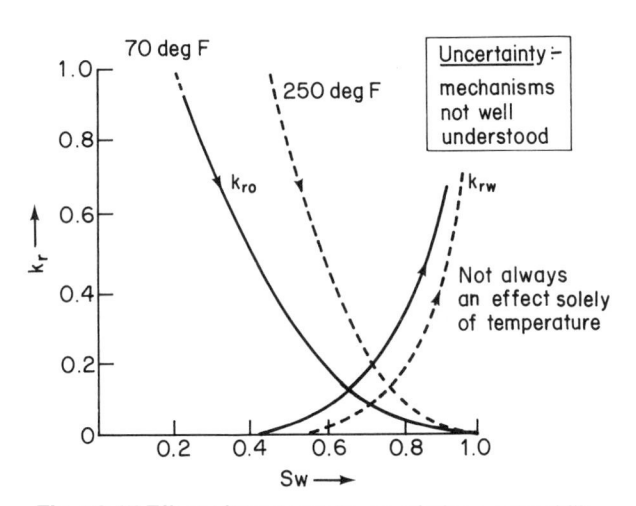

Fig. 12.17 Effect of temperature on relative permeability.

12.8 THERMAL ENERGY

The introduction of thermal energy into a heavy oil reservoir should result in improvement in productivity. The economics of a particular process can be assessed in terms of an energy balance as cost of total energy as compared with value of the product after operating costs and taxes. Thermal injection processes include steam soak, steam drive, hot water drive and a variety of combustion processes from forward to reverse and from air to oxygen with and without water injection (see Figs. 12.18–12.20). In all these processes, consideration must be given to the volume of the heated zone and the mechanisms

12 IMPROVED HYDROCARBON RECOVERY

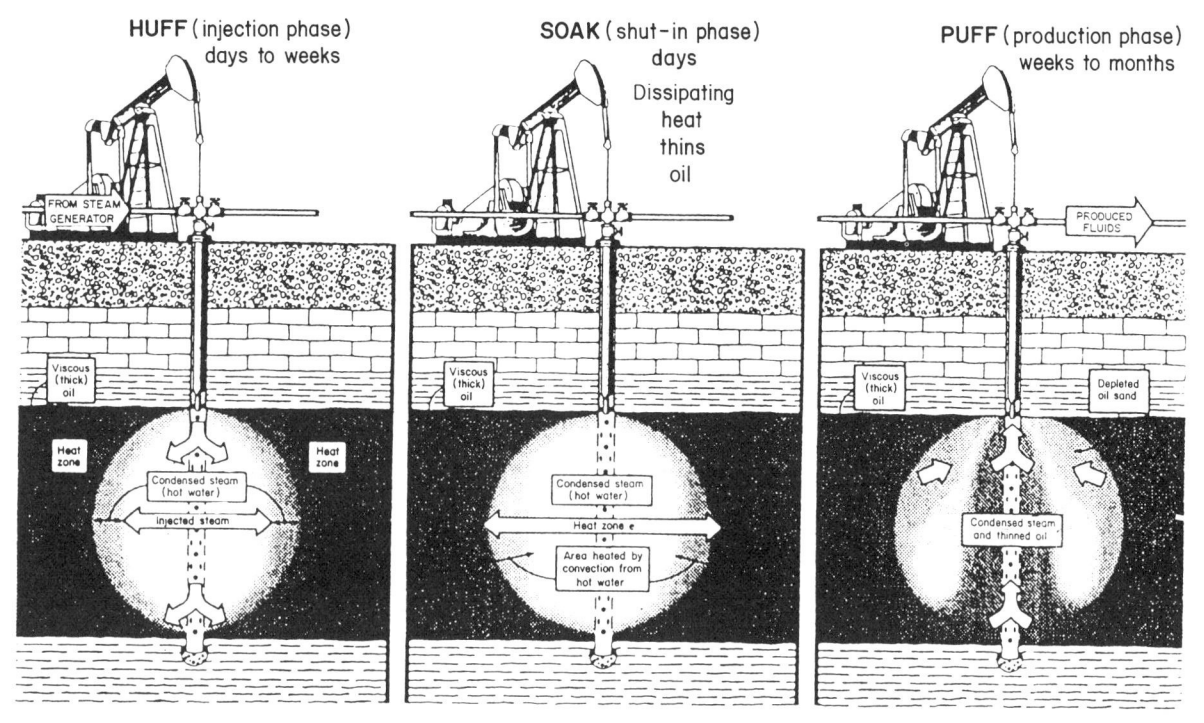

Fig. 12.18 Cyclic steam simulation.

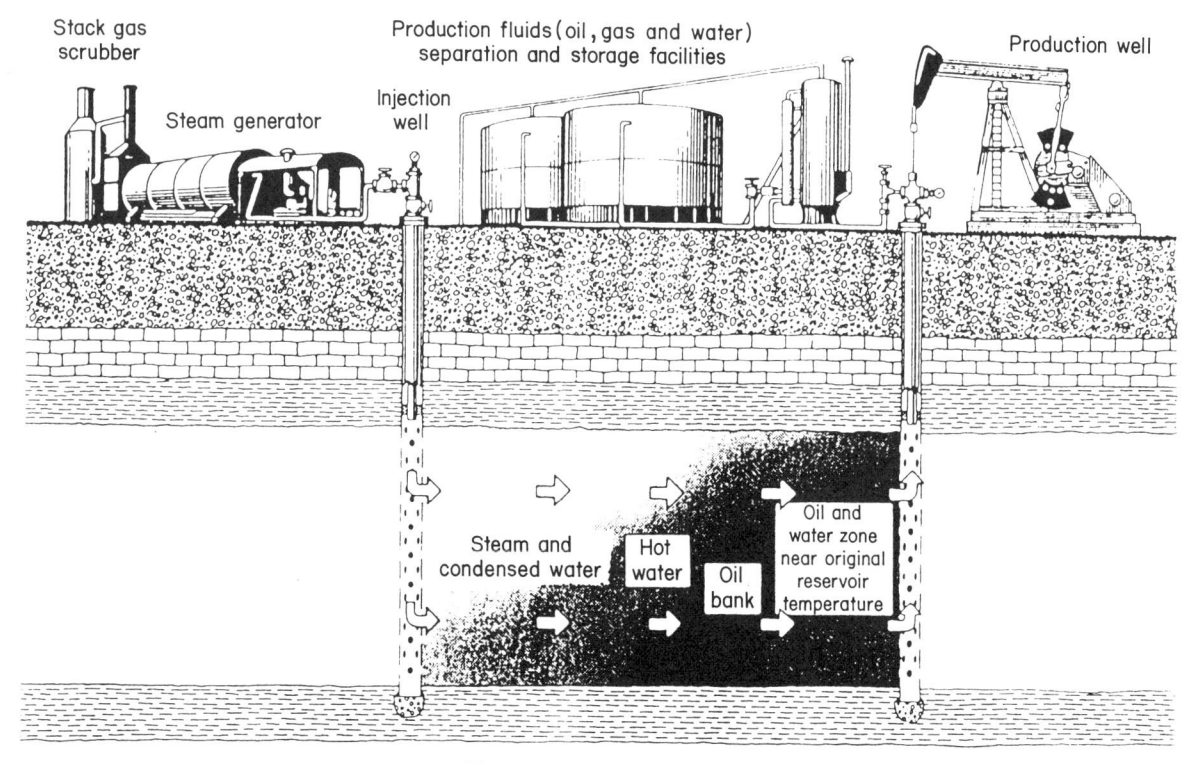

Fig. 12.19 Steam flooding.

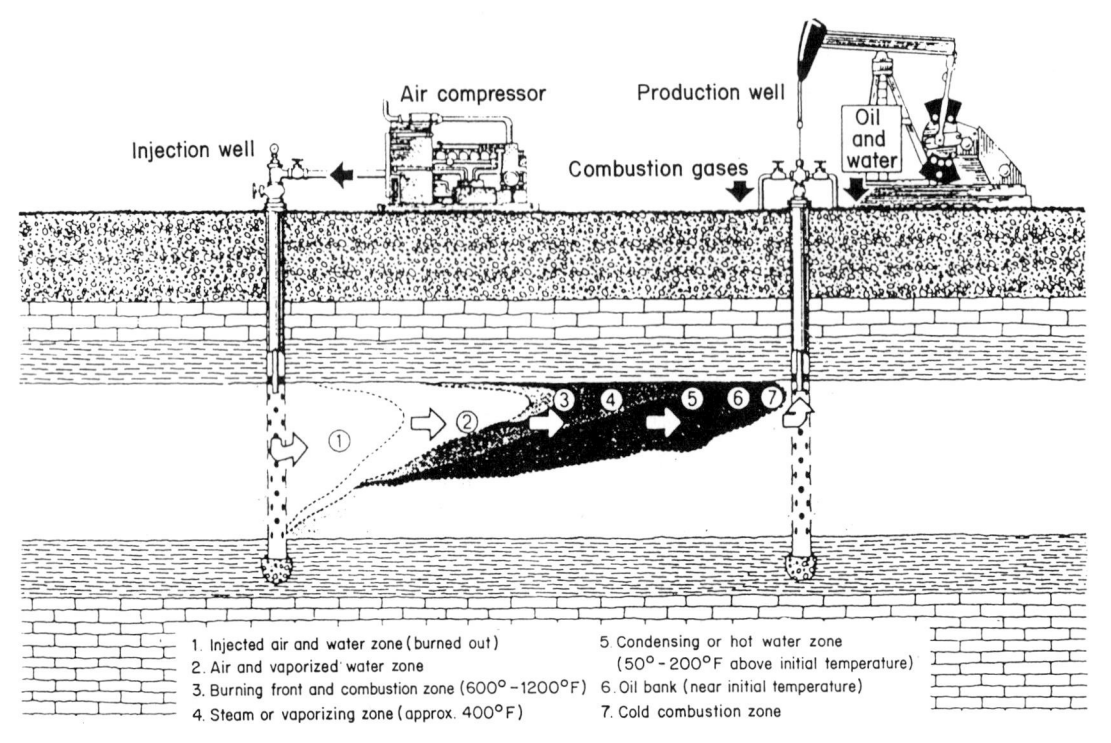

Fig. 12.20 *In-situ* combustion.

1. Injected air and water zone (burned out)
2. Air and vaporized water zone
3. Burning front and combustion zone (600°–1200°F)
4. Steam or vaporizing zone (approx. 400°F)
5. Condensing or hot water zone (50°–200°F above initial temperature)
6. Oil bank (near initial temperature)
7. Cold combustion zone

of heat transfer which result in productivity improvement. For details of the various processes and explanation of mechanisms there is a wealth of current literature [2,3,4,61,62]. The volume of a heated zone in a reservoir, designated V_{rh}, can be represented for consideration of thermal properties in the following form:

$$V_{rh} = \text{Complex function} \left\{ \frac{\begin{bmatrix} \text{Recoverable} \\ \text{latent heat} \end{bmatrix} \begin{bmatrix} \text{Thermal} \\ \text{diffusivity} \end{bmatrix} \begin{bmatrix} \text{Heat} \\ \text{capacity} \end{bmatrix} [\text{Time}]}{\begin{bmatrix} \text{Thermal} \\ \text{conductivity} \end{bmatrix}^2 \begin{bmatrix} \text{Temperature gradient} \\ \text{injector} - \text{producer} \end{bmatrix}} \right\}$$

The heat capacity (C) is applied to reservoir rocks and fluids; thermal conductivity (λ) and thermal diffusivity ($D = \lambda/C\rho$) are applied to the reservoir and the rocks overlying and underlying the reservoir. It is clear also that reservoir structure, heterogeneity and geometry will influence well spacing and heat transfer. In order to solve equations for heated zone volume, consideration of full three-dimensional geometry is important as fluid density differences (i.e. steam, combustion gases, oil, water) play an important role in displacement efficiency. Data requirements include thermal properties of rocks and reservoir fluids together with temperature, pressure and saturation dependent empirical relationships for relative permeability, capillary pressure and *PVT* properties. Some of these data can be obtained quite satisfactorily from correlations and others require specific determination for a given reservoir fluid–rock system. Typical data required in calculations of thermal processes are shown in Table 12.9. The characterization of reservoir transmissibility and continuity is probably a greater uncertainty than any error in fluid and rock properties introduced by use of literature correlations.

In processes typically considered for thermal stimulation of heavy oil reservoirs, gravity override of hot gases, particularly steam, will be important. Steam properties will also be important. Steam properties indicate that superheat at elevated reservoir pressures and temperatures is not proportionately beneficial. For deep reservoirs steam may not condense. The critical point for steam is shown in Fig. 12.21 as 3206.2 psia and 705.4°F. Appreciable latent heat release is not achieved at pressures greater than 2200 psi, equivalent to a depth of some 5000 ft SS in a normally pressured reservoir. Steam override mechanisms have been the subject of laboratory investigations and results support the current interest in application of horizontal well technology.

12 IMPROVED HYDROCARBON RECOVERY

TABLE 12.9 Example data requirements for steamdrive analysis

System	Property	Source	Typical value or units — SI units	Typical value or units — Field units
Oil	Density (21°C)	Measured	912.7 kg/m^3	23.5° API
	Compressibility	Calculated	0.00073 Pa^{-1}	5 × 10^{-6} psi^{-1}
	Coef. thermal expan.	Measured	0.00052 K^{-1}	0.00029°F^{-1}
	Heat capacity	Measured	1.963 kJ/kgK	0.469 Btu/lb°F
	Viscosity	Measured (fn T)	5 mPa at 373 K	5 cp at 212°F
Water and steam	Density	Steam tables	kg/m^3	lb/ft^3
	Compressibility		Pa^{-1}	psi^{-1}
	Heat capacity		kJ/kgK	Btu/lb°F
	Latent heat		kJ/kg	Btu/lb
Reservoir rock	Heat capacity	Correlation (Lit.)	1.7 MJ/m^3 K	25.3 Btu/ft^3°F
	Thermal conductivity	Correlation (Lit.)	238 kJ/m-day-K	38.16 Btu/ft-day°F
	Permeability	Measured	μm^2	mD
	Pore volume compressibility	Measured	Pa^{-1}	psi^{-1}
	Porosity	Measured	fraction	fraction
Overburden and underburden	Heat capacity	Correlation (Lit.)	1.4 MJ/m^3 K	21 Btu/ft^3°F
	Thermal conductivity	Correlation (Lit.)	96.3 kJ/m-day-K	15.5 Btu/ft-day-°F
Oil–water	Relative permeability	Measured	—	—
	Capillary pressure	Measured	Pa	psi
Gas–liquid	Relative permeability	Measured	—	—
	Capillary pressure	Measured	Pa	psi
	Solubility (P, T)	Measured	m^3/m^3	SCF/STB

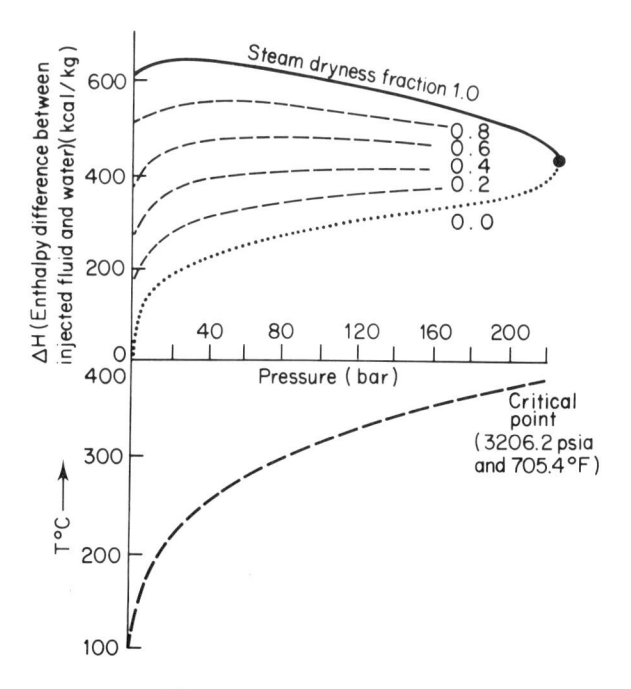

Fig. 12.21 Properties of steam.

12.9 GAS CONDENSATE RESERVOIRS

A pressure–temperature phase envelope of a hydrocarbon mixture known as a gas condensate is shown in Fig. 12.22 for a constant composition system. It can be seen that if the reservoir temperature of the gas phase mixture is above critical and below the

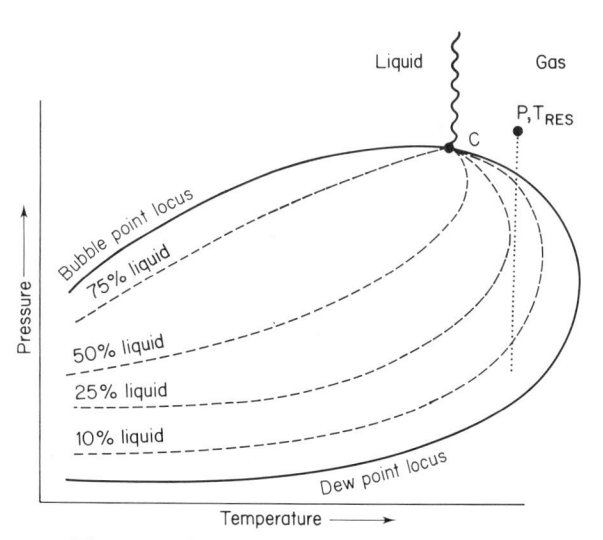

Fig. 12.22 Gas condensate phase diagram.

cricondentherm then, during isothermal pressure reduction, the liquid content of the mixture may increase and then decrease in a phenomenon known as retrograde condensation. This behaviour is characteristic of a gas condensate. In addition the API gravity of resultant stock tank oil is likely to be greater than 45° and producing gas–oil ratios are often in excess of 3000 SCF/STB. The stock tank liquid is often very pale yellow in colour and the Watson characterization factor (K_w) of the C_{7+} fraction is significantly different from black oils. A typical K_w is around 12 for a condensate system and 11.90 for a black oil using the relationship between liquid specific gravity (γ) and molecular weight (M) as follows:

$$K_w = 4.5579 \, M^{(0.15178)} \gamma^{-0.84573}$$

The K_w factor can be used to check if liquid samples are indeed characteristic of the condensate. Sampling gas condensate reservoirs is notoriously difficult because of proximity to critical conditions and retrograde behaviour. Bottom-hole sampling may frequently fail to represent total reservoir fluid, and recombined surface separator samples in these particular conditions are often preferred. Table 12.10 shows an example of a separator gas and liquid composition from a condensate reservoir. Liquid drop out character during isothermal constant volume expansion is shown for three samples with different single stage separator gas–oil ratios in Fig. 12.23.

The magnitude and location in the reservoir of liquid drop out from a condensate reservoir undergoing pressure depletion is vital to the design of a production scheme. It might be predicted from radial flow pressure drop and constant volume depletion data where liquid drop out might occur [66]. The non-equilibrium conditions occurring around a wellbore might invalidate such a calculation. Relative permeability and oil trapping phenomena will play a large part in well productivity. Figure 12.24 shows schematically a liquid saturation and pressure profile in radial flow towards a wellbore of radius r_w.

The mechanisms of liquid drop out and its effect on hydrocarbon recovery and well productivity are not yet fully understood. For more detail the reader is referred to the current literature. Condensate *PVT* properties require matching with equations of state, and validation depends as much as anything on representation of the well stream fluid in sampling. Relative permeability behaviour of condensate systems are also the subject of considerable uncertainty. The numerical modelling of gas condensate reservoir systems is limited by proper representation of flow physics and thermodynamic behaviour in addition to normal reservoir description.

TABLE 12.10 Condensate analysis (after [86])

Oil flow rate	525 BOPD		
Gas flow rate	13475 MSCFD		
Natural gas analysis			Oil analysis
146°F	Separator temperature		157.6°F
664.7 psig	Separator pressure		707.2 psig
2.140 lb/ft³	Flowing density		226.6 lb/BBL
705 580 lb/day	Mass flow		232440 lb/day
Mole %			Mole %
0.59	Nitrogen		0.04
2.09	Carbon dioxide		0.63
83.84	Methane		13.37
7.63	Ethane		4.56
3.01	Propane		4.57
0.49	*Iso*-butane		1.49
0.95	*n*-butane		4.16
0.36	*Iso*-pentane		2.55
0.40	*n*-pentane		3.36
0.52	Hexanes		10.30
0.12	Heptanes +		54.97
Gas gravity = 0.6942			Specific gravity C_{7+} 0.7802
H₂S 10 ppm			Avg. mol. wt. C_{7+} 143.4
			Avg. S content C_{7+} 0.03% wt
			K_w C_{7+} from:
			$4.5579 \, M^{0.15178} \gamma^{-0.84573}$

12 IMPROVED HYDROCARBON RECOVERY

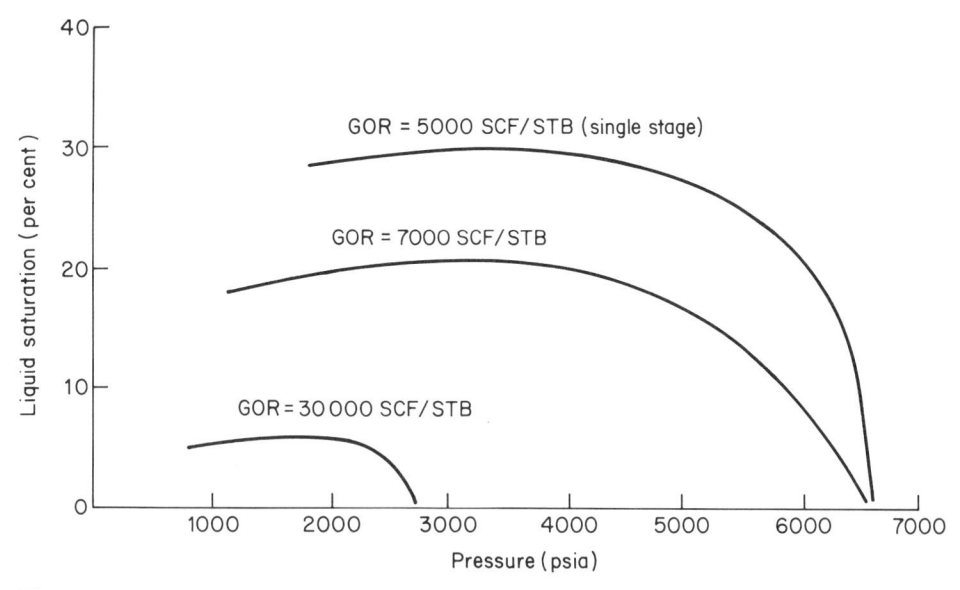

Fig. 12.23 Liquid content during isothermal constant volume expansion of some condensates.

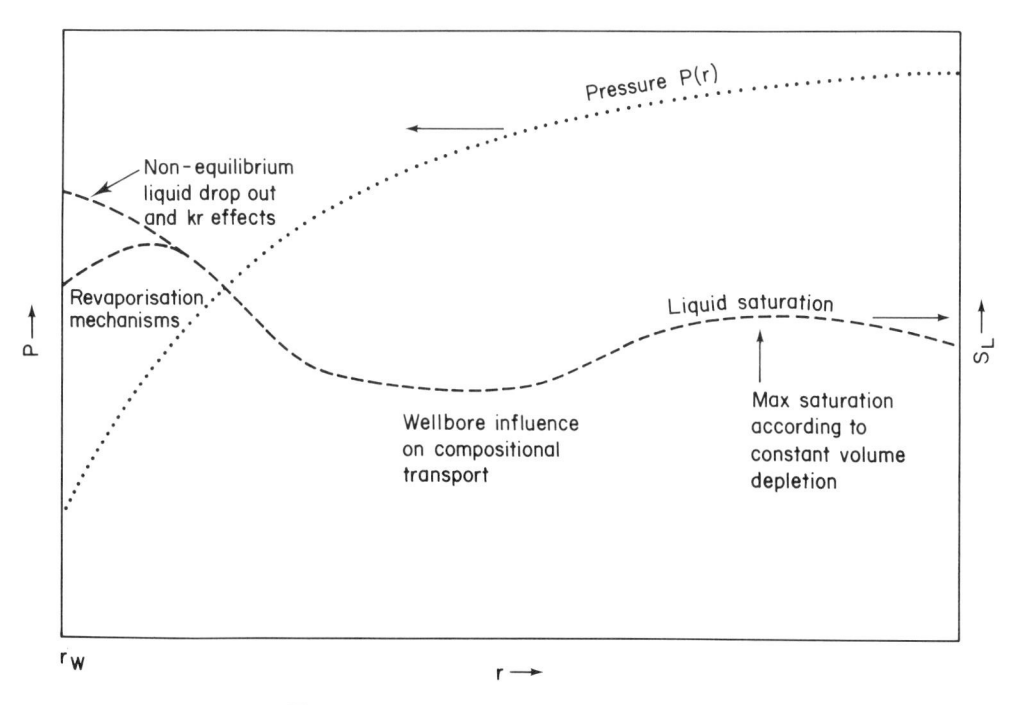

Fig. 12.24 Pressure and saturation profiles.

Production methods for gas condensate reservoirs include [37]:

(a) pressure depletion;
(b) pressure depletion with dry gas recycling;
(c) partial or full pressure maintenance by water drive.

The water drive mechanism may be appropriate for high pressure reservoirs where dry gas compression and reinjection could be costly – this would be particularly true in offshore development of deep reservoirs.

Partial pressure maintenance in gas condensate

reservoirs is conducted with the aim of keeping pressure above dew-point. Gas recycling is a miscible recovery process with mass transfer between advancing dry gas and wet gas in the pore space. After contact of all wet gas and recovery of liquids at surface conditions, the reservoir should contain a single-phase leaner gas which is itself recoverable by pressure depletion or blowdown as if it were a dry gas reservoir. A design drawback to the recovery process is viscous instability which is accentuated by reservoir heterogeneity. Since wet gas might have a significantly greater viscosity at reservoir conditions than dry gas, as well as a greater density, the displacement will then tend to be unstable in terms of Dietz criteria (Chapter 7), and gravity override may occur at quite low rates.

In consideration of processes involving water drive in gas condensate systems, one of the significant uncertainties is trapped gas saturation at high pressure and the effects of pore geometry, capillary number and Bond number on its magnitude. At low rates it is possible that some trapped gas could be recovered during blowdown, although the mechanism would be a complicated three-phase process.

Pressure depletion alone would result in unacceptable recovery factors in most instances, and ultimate recovery can be judged from laboratory simulation of a constant volume depletion using a valid fluid sample. The performance prediction of a gas condensate using the laboratory *PVT* data assumes that liquid saturation remains below some critical value and is thus immobile. The determination of the dry gas and oil in place equivalent to the wet condensate volume can be estimated as follows. From *PVT* laboratory data the following information is available:

R = total gas–oil ratio of the system (scf/stb);
ρ = liquid density (g/cm^3);
γ_g = gas gravity (relative to air = 1);
M_o = molecular weight of liquid (estimate from $M_o = 44.3\rho/[1.03-\rho]$.

Assuming that one pound mole of gas occupies a volume of 379.4 ft^3 at 60°F and 14.7 psia, and that water has a density of 62.4 lb/ft^3 at the same condition, the weight of oil and gas produced at stock tank conditions for each 1 STB liquid is W as follows:

$$W = (5.615 \times 62.4 \times \rho) + \frac{(R\gamma_g)(28.97)}{379.4}$$

For each 1 STB of liquid produced the number of moles (n) produced is therefore

$$n = \frac{R}{379.4} + \frac{(62.4)\rho(5.615)}{M_o}$$

The molecular weight of the reservoir condition fluid $(MW)_{res}$ can therefore be estimated from

$$(MW)_{res} = \frac{W}{n}$$

The reservoir condition gas condensate gravity $(\gamma_g)_{res}$ is therefore

$$(\gamma_g)_{res} = \frac{(MW)_{res}}{28.97}$$

The critical properties of the gas condensate can then be obtained from the pseudo-critical property chart (see Fig. 4.7) or by use of Kay's rule, if reliability is placed in the compositional data. At the pseudo-reduced temperature and pressure in the reservoir the Standing and Katz chart (Fig. 4.7) can be used to derive a value for Z at datum. The gas condensate formation volume factor B_{gi} at initial datum conditions, P (psia) and T (°F), is thus

$$B_{gi} = \frac{0.02829\, z\,(T+460)}{P} \text{ RCF/SCF}$$

The reservoir hydrocarbon pore volume estimated volumetrically is related to the standard condition volume V_{sc} as

$$V_{sc} = \frac{A\, h_n\, \bar{\phi}\, \bar{S}_g}{B_{gi}} \text{ SCF}$$

where A is the effective hydrocarbon reservoir area in ft^3, h_n is the effective hydrocarbon net thickness, $\bar{\phi}$ is the volume weighted average porosity and $\bar{S}_g$ the average saturation of gas condensate in the hydrocarbon region.

At separator conditions the dry gas is that fraction of the total moles of reservoir fluid that are gas, i.e.

$$\text{Dry gas volume} = V_{sc}\left\{\frac{R/379.4}{n}\right\} \text{ SCF}$$

Similarly, the oil content at stock tank conditions is given by

$$N = \frac{V_{sc}}{R}$$

Table 12.11 shows the type of information available from a laboratory constant volume depletion and Table 12.12 indicates that liquid recovery is quite poor. It is this result that focuses interest in dry gas recycling and pressure maintenance/waterdrive projects. Recovery calculations must, however, consid-

12 IMPROVED HYDROCARBON RECOVERY

er the effects of reservoir heterogeneities and constraints imposed by wellbore conditions and processing equipment. Economic considerations are particularly important in evaluation of gas condensate potential.

12.10 VOLATILE OIL RESERVOIRS

Above bubble-point pressure a volatile oil reservoir can be treated as a black oil system. Below bubble-point, oil and gas mobility and reservoir heterogeneity control performance. Compositional approaches to reservoir calculations are used [71] and equilibrium constants (k-values) are required to predict molar relationships between phases. Equations of state may be used to calculate the behaviour of the original fluid composition during production. These equations of state often require tuning of coefficients using *PVT* data.

TABLE 12.11 Constant volume depletion data: example system

Distribution of original fluid between separation stages

Primary separation at 500 psig and 220°F

Gas	912.87 MSCF/MMSCF original fluid
Gas products	C_3: 814 gallons/MMSCF original fluid
	C_4: 489 gallons/MMSCF original fluid
	C_{5+}: 997 gallons/MMSCF original fluid

Second stage separation at 100 psig and 120°F

Liquid	109.14 BBL/MMSCF original fluid
Gas	8.26 MSCF/MMSCF original fluid
Gas products	C_3: 10.6 gallons/MMSCF original fluid
	C_4: 5.1 gallons/MMSCF original fluid
	C_{5+}: 6.3 gallons/MMSCF original fluid

Stock tank fluid 0 psig and 60°F

Liquid	102.64 STB/MMSCF original fluid
Gas	5.47 MSCF/MMSCF original fluid

Total *plant products* in wellstream fluid

C_3 878 gallons/MMSCF original fluid
C_4 564 gallons/MMSCF original fluid
C_{5+} 5416 gallons/MMSCF original fluid

Reservoir fluid initially at 5750 psig

TABLE 12.12 Recovery of fluids by depletion

Pressure (psi)	Original fluid recovery (SCF/SCF)	Stock tank liquid recovery (STB/STB)	Second stage liquid recovery (SSBBL/SSBBL)	Primary separator gas recovery (SCF/SCF)	C_{5+} plant recovery (gall/gall)
5750	0	0	0	0	0
5000	0.0727	0.0445	0.0446	0.0746	0.0502
4000	0.2052	0.0847	0.0851	0.2157	0.1093
3000	0.3720	0.1074	0.1079	0.3958	0.1610
2100	0.5434	0.1195	0.1203	0.5821	0.2051
1300	0.7040	0.1289	0.1288	0.7568	0.2450

Examples

Example 12.1

Using the Stalkup relationships shown in Figure A12.1 (see Appendix II) estimate the breakthrough sweepout efficiency and the dominant flow regime in the following displacements. For five spot patterns assume the Darcy flow velocity can be approximated by $1.25\, q_{inj}/hl$ in field units and the viscous gravity force ratio is given by $2050 \mu UL/(kh\Delta\rho)$. The reservoirs have not been waterflooded previously.

(a) On a five spot pattern with $L = 1500$ ft and thickness $h = 70$ ft a gas is being injected at 4000 RB/d. The reservoir condition densities of gas and oil are 0.4 and 0.8 g/cm^3. The viscosity of the injected gas is 0.02 cp and the oil viscosity is 0.5 cp. Assume that the permeability of the reservoir is isotropic, and can be represented as 130 mD.

(b) On a five spot pattern with $L = 2000$ ft and thickness 50 ft, carbon dioxide is being injected at 1000 RB/d. The density and viscosity of CO_2 at reservoir conditions are taken as 0.64 g/cm^3 and 0.055 cp. The oil density and viscosity at reservoir conditions are 0.75 g/cm^3 and 0.36 cp. Assume that the horizontal permeability is represented as 3 mD and the vertical permeability by 1 mD.

Example 12.2

A simple surfactant system has been discovered. It has only a single- and a two-phase region and can be represented on a ternary diagram, with oil–brine–surfactant being the components.

The following phase equilibrium data have been obtained for the two-phase part of the system. A is in equilibrium with B. The compositions are in weight percent.

A		B	
Surfactant	Oil	Surfactant	Oil
0	2	0	98
9	9	2	97
18	18	4	95
25	32	6	92
27	40	11	84
26	55	17	74

(a) Plot the data and construct the phase envelope. Estimate the composition of the plait (critical) point.
(b) What is the composition and weight fraction for the equilibrium separated phases for 200 g mixture of total composition 4% surfactant, 77% oil?
(c) What weight of surfactant must be added to 100 g of 20% oil in brine mixture to make it just single phase? What is the composition of this final mixture?
(d) What is the composition of the mixture when 150 g of a solution containing 10% oil, 40% surfactant is added to 150 g of a solution 50% oil, 40% surfactant?
(e) What is the composition of the resulting phases when 100 g of a solution composed of 12% surfactant 5% oil is added to 100 g solution composed of 20% surfactant, 77% oil?

Example 12.3

Contrast development by steam stimulation/injection and conventional water injection on a 9 acre five-spot pattern of a reservoir sand containing 150 cp oil at the reservoir temperature of 100°F. The sand is 60 ft thick and has a permeability to oil at reservoir temperature of 1000 mD. Assume the steam is injected at a bottom hole temperature at 380°F with a quality ratio (fsdh) of 0.75. The wellbore radius is 0.5 ft. Other thermal data may be assumed as follows:

Enthalpy of liquid water at T_{res} $(h_w(T_r)) = 69$ Btu/lb m
Enthalpy of liquid water at T_{steam} $(h_w(T_s)) = 355$ Btu/lb m
Latent heat of vaporization at T_{steam} $(L_{vdh}) = 845$ Btu/lb m

12 IMPROVED HYDROCARBON RECOVERY

Heat capacity of reservoir (M_R) = 35 Btu/ft^3–°F
Heat capacity of surrounding rocks (M_s) = 45 Btu/ft^3–°F
Thermal diffusivity of surrounding rocks (α_s) = 0.75 ft^2/Day

In the five spot pattern the side distance, L', is related to the well spacing in acres/pattern, A, by the relationship $L' = 208.71\,(A)^{0.5}$. The steady state flow resistance for injectivity and productivity is considered equal and can be represented by

$$\frac{1}{I} = \frac{1}{J} = \frac{\Delta P}{q_{inj}} = \frac{141.2\,\mu_o\,F_G}{k_o h}$$

where ΔP is the pressure difference at bottom hole flowing conditions between injectors and producers (psi).

q_{inj} is the injection (or production) rate in rb/d
μ_o is the reservoir condition oil viscosity (cp)
h is the reservoir thickness (ft)
F_G is the pattern geometric factor, which for the five spot is $2\,[ln(L'/r_w) - 0.9640]$

For a steam heated injector and a cyclic steam stimulated producer assume that the flow resistance compared to non-steam injection reduces by a factor of 5. The minimum flowing bottom hole pressure permitted in the shallow reservoir is 200 psi and the maximum bottom hole injection pressure is 900 psi. Steam injection should be planned until the steam zone occupies 50% of the pattern volume, for times between 1 and 2.5 years from the start of injection.

Example 12.4

Determine the dry gas volume and liquid volume at standard conditions for a gas condensate reservoir with the following characteristics:

Net thickness = 300 ft
Effective radius = 3 miles
Average porosity = 18%
Average connate water saturation = 25%
Reservoir temperature at datum = 210°F
Reservoir pressure at datum = 4500 psi
Separator liquid gravity = 57.2° API
Separator gas gravity (rel air) = 0.58
Total gas–oil ratio = 5000 *scf/stb*

References

[1] Bond, D.C. (Ed.)
Determination of Residual Oil Saturation, Interstate Oil Compact Commission (1978), Oklahoma.
[2] Poettman, F.H. (Ed.)
Secondary and Tertiary Oil Recovery Processes, Interstate Oil Compact Commission (1974), Oklahoma.
[3] Latil, M.
Enhanced Oil Recovery, IFP Publications, Graham and Trotman, London (1980).
[4] van Poollen, H.K.
Fundamentals of Enhanced Oil Recovery, Penwell Books, Tulsa (1980).
[5] Klins, M.A.
Carbon Dioxide Flooding, IHRDC, Boston. (1984).
[6] Stalkup, F.I.
Miscible Displacement, SPE Monograph No 8, Soc. Pet. Engr., Dallas (1983).
[7] Shah, D.O. (Ed.)
Surface Phenomena in Enhanced Oil Recovery, Plenum Press, NY (1981).
[8] Shah, D.O. and Schechter, R.S. (Eds)
Improved Oil Recovery by Surfactant and Polymer Flooding, Academic Press (1977).
[9] Katz, M.L. and Stalkup, R.I.
Oil recovery by miscible displacement, *Proc. 11th World Pet. Cong.* London (1983), RTD 2 (1).

[10] Thomas, E.C. and Ausburn, B.E.
Determining swept zone residual oil saturation in a slightly consolidated Gulf Coast sandstone reservoir, *JPT* (April 1979), 513.

[11] Bath, P.G.H., van der Burgh, and Ypma, J.G.J.
Enhanced oil recovery in the North Sea, *Proc. 11th World Pet. Cong.* London (1983), ATD 2 (2).

[12] Mattax, C.C., Blackwell, R.J. and Tomich, J.F.
Recent advances in surfactant flooding, *Proc. 11th World Pet. Cong.* London (1983), RTD (3).

[13] Putz, A., Bosio, J., Lambrid, J. and Simandoux, P.
Summary of recent French work on surfactant injection, *Proc. 11th World Pet. Cong.* London (1983), RTD 2 (4).

[14] Kuuskraa, V., Hammershaimb, E.C. and Stosor, G.
The efficiency of enhanced oil recovery techniques: a review of significant field tests, *Proc. 11th World Pet. Cong.* London (1983), RP13.

[15] Grist, D.M., Hill, V.P. and Kirkwood, F.G.
Offshore Enhanced Oil Recovery, IP 84–008 reprinted from *Petroleum Review* (July 1984), 40.

[16] Passmore, M.J. and Archer, J.S.
Thermal properties of reservoir rocks and fluids, In *Developments in Petroleum Engineering–1.* (Ed. Dawe and Wilson), Elsevier, Amsterdam (1985).

[17] Archer, J.S.
Surfactants and polymers – state of the art, *Proc. Conf. on Improved Oil Recovery,* Aberdeen (Nov. 1984), SREA.

[18] Macadam, J.M.
North Sea stimulation logistics and requirements, SPE 12999, *Proc. Europ. Off. Pet. Conf.* London 1984, Soc. Pet. Engr.

[19] Cooper, R.E. and Bolland, J.A.
Effective diversion during matrix acidisation of water injection wells, OTC 4795, *Proc. 16th Ann. OTC,* Houston (1984).

[20] Sorbie, K.S., Robets, L.J. and Foulser, R.W.
Polymer flooding calculations for highly stratified Brent sands in the North Sea, *Proc. 2nd Europ. Symp. on Enhanced Oil Recovery,* Paris (1982).

[21] Taber, J.J.
Research on EOR, past, present and future, In *Surface Phenomena in EOR* (Ed. Shah), Plenum Pub, NY (1981), 13.

[22] Shah, D.O.
Fundamental aspects of surfactant polymer flooding processes, *Proc. 1st Europ. Symp. EOR,* Bournemouth (1981), 1.

[23] Nelson, R.C. and Pope, G.A.
Phase relationships in chemical flooding, *Soc. Pet. Eng. J.* **18** (5), (1978), 325.

[24] Pope, G.A. and Nelson, R.C.
A chemical flooding compositional simulator, *SPEJ* **18** (1979), 339.

[25] National Petroleum Council
Enhanced Oil Recovery, NPC Report (June 1984), Washington.

[26] Coulter, S.R. and Wells, R.D.
The advantage of high proppant concentration in fracture simulation, *JPT* (June 1972), 643.

[27] Doscher, T.M., Oyekan, R. and El-Arabi
The displacement of crude oil by CO_2 and N_2 in gravity stabilised systems, *SPEJ* **24** (1984), 593.

[28] Whitson, C.H.
Effect of C_{7+} properties on equation of state predictions, *SPEJ* **24** (1984), 685.

[29] Bang, H.W. and Caudle, B.H.
Modelling of a miscellar/polymer process, *SPEJ* (1984), 617.

[30] Thomas, C.P., Fleming, P.D. and Winter, W.W.
A ternary, 2 phase mathematical model of oil recovery with surfactant systems, *SPEJ* **24** (1984), 606.

[31] Chase, C.A. and Todd, M.R.
Numerical simulation of CO_2 flood performance, *SPEJ* (1984), 597.

[32] Kantar, K. *et al.*
Design concepts of a heavy-oil recovery process by an immiscible CO_2 application, *JPT* (Feb. 1985), 275.

[33] Fulscher, R.A., Ertekin, T. and Stahl, C.D.
Effect of capillary number and its constituents on two phase relative permeability curves, *JPT* (Feb. 1985), 249.

[34] Reitzel, G.A. and Callow, G.O.
Pool description and performance analysis leads to understanding Golden Spike's miscible flood, *JPT* (July 1977), 867.

[35] Kidwell, C.M. and Guillory, A.J.
A recipe for residual oil saturation determination, *JPT* (Nov. 1980), 1999.

[36] Perry, G.E. and Kidwell, C.M.
Weeks Island 'S' sand reservoir B gravity stable miscible CO_2 displacement, *Proc. 5th Ann. DOE Symp. EOR* (1979).

[37] Wilson, D.C., Prior, E.M. and Fishman, D.M.
An analysis of recovery techniques in large offshore gas condensate fields, EUR 329, *Proc. Europ. Pet. Conf.* London (1982), Soc. Pet. Eng.

[38] Sigmund, P. *et al.*
Laboratory CO_2 floods and their computer simulation, *Proc. 10th World Pet. Cong.* **3** (1979), 243.

[39] Katz, D.L. and Firoozabadi, A.
Predicting phase behaviour of condensate/crude oil systems using methane interaction coefficients, *JPT* (Nov. 1978), 1649.

[40] Rathmell, J., Braun, P.H. and Perkins, T.K.
Reservoir waterflood residual oil saturation from laboratory tests, *JPT* (Feb. 1973), 175.

[41] Schechter, R.S., Lam, C.S. and Wade, W.H.
Mobilisation of residual oil under equilibrium and non equilibrium conditions, SPE 10198, *Proc. 56th Ann. Fall Mtg.* (1981), Soc. Pet. Eng.

[42] Mahers, E.G. and Dawe, R.A.
The role of diffusion and mass transfer phenomena in the mobilisation of oil during miscible displacement, *Proc. Europ. Symp. EOR* (1982).

[43] Wall, C.G., Wheat, M.R., Wright, R.J. and Dawe, R.A.
The adverse effects of heterogeneities on chemical slugs in EOR, *Proc. IEA Workshop on EOR*, Vienna (Aug. 1983).

[44] Aziz, K.
Complete modelling of EOR processes, *Proc. Europ. Symp. EOR*, Bournemouth (1981), 367.

[45] Bristow, B.J.S. de
The use of slim tube displacement experiments in the assessment of miscible gas projects, *Proc. Europ. Symp. EOR*, Bournemouth (1981), 467.

[46] Wilson, D.C., Tan, T.C. and Casinader, P.C.
Control of numerical dispersion in compositional simulation, *Proc. Europ. Symp. EOR*, Bournemouth (1981), 425.

[47] Claridge, E.L.
CO_2 flooding strategy in a communicating layered reservoir, *JPT* (Dec. 1982), 2746.

[48] Koval, E.J.
A method for predicting the performance of unstable miscible displacement in heterogeneous media, *SPEJ* (June 1963), 145.

[49] Metcalfe, R.S.
Effects of impurities on minimum miscibility pressures and minimum enrichment levels for CO_2 and rich gas displacements, *SPEJ* (April 1982), 219.

[50] Jennings, H.Y., Johnson, C.E. and McAuliffe, C.D.
A caustic waterflooding process for heavy oils, *JPT* (Dec. 1974), 1344.

[51] Melrose, J.C. and Brandner, C.F.
Role of capillary forces in determining microscopic displacement efficiency for oil recovery by waterflooding, *J.Can. Pet. Tech.* (Oct. 1974), 54.

[52] Chesters, D.A., Clark, C.J. and Riddiford, F.A.
Downhole steam generation using a pulsed burner, *Proc. 1st Europ. Symp. EOR* Bournemouth (1981), 563.

[53] Lemonnier, P.
Three dimensional numerical simulation of steam injection, *Proc. 1st Europ. Symp. EOR 7Bournemouth (1981), 379.*

[54] Risnes, R., Dalen, V. and Jensen, J.I.
Phase equilibrium calculations in the near critical region, *Proc. 1st Europ. Symp. EOR* Bournemouth (1981), 329.

[55] Sayegh, S.G. and McCafferty, F.G.
Laboratory testing procedures for miscible floods, *Proc. 1st Europ. Symp. EOR* Bournemouth (1981), 285.

[56] Vogel, P. and Pusch, G.
Some aspects of the injectivity of non Newtonian fluids in porous media, *Proc. 1st Europ. symp. EOR* Bournemouth (1981), 179.

[57] Brown, C.E. and Langley, G.O.
The provision of laboratory data for EOR simulation, *Proc. 1st Europ. Symp. EOR Bournemouth* (1981), 81.

[58] Archer, J.S.
Thermal properties of heavy oil rock and fluid systems, *Proc. Conf. Problems Associated with the Production of Heavy Oil,* Oyez Scientific and Technical Services (March 1985).

[59] White, J.L. Goddard, J.E. and Phillips, H.M.
Use of polymers to control water production in oil wells, *JPT* (Feb. 1973), 143.

[60] Bing, P.C., Bowers, B. and Lomas, R.H.
A model for forecasting the economic potential for enhanced oil recovery in Canada, *J. Can. Pet. Tech.* (Nov.–Dec. 1984), 44.

[61] Meyer, R.F. and Steele, C.T.
The future of heavy crude oils and tar sands, *Proc. 1st Int. Conf.* (June 1979), UNITAR (McGraw Hill Inc., NY).

[62] Offeringa, J., Barthel, R. and Weijdema, J.
Keynote paper: thermal recovery methods, *Proc. 1st Europ. Symp. EOR* Bournemouth (Sept. 1981), 527.

[63] Nakorthap, K. and Evans, R.D.
Temperature dependent relative permeability and its effect on oil displacement by thermal methods, SPE 11217, *Proc. 57th Ann. Fall Mtg. SPE* (Sept. 1982).

[64] Okandan, E. (Ed.)
Heavy Crude Oil Recovery, Nato ASI Series, Martinus Nijhoff Pub. (1984), The Hague.

[65] Spivak, A. and Dixon, T.N.
Simulation of gas condensate reservoirs, SPE 4271, *3rd Symp. Numerical Simulation* (1973), Houston.

[66] Dystra, H.
Calculated pressure build up for a low permeability gas condensate well, *JPT* (Nov. 1961), 1131.

[67] Ham, J.D. Brill, J.P. and Eilerts, C.K.
Parameters for computing pressure gradients and the equilibrium saturation of gas condensate fluids flowing in sandstones, *SPE J.* (June 1974), 203.

[68] Saeidi, A. and Handy, L.L.
Flow of condensate and highly volatile oils through porous media, SPE 4891, *Proc. California Reg. Mtg.* (April 1974), Los Angeles.

[69] Fussell, D.D.
Single well performance prediction for gas condensate reservoirs, *JPT* (July 1973), 860.

[70] Eaton, B.A. and Jacoby, R.H.
A new depletion performance correlation for gas condensate reservoir fluid, *JPT* (July 1965), 852.

[71] Ruedelhuber, F.O. and Hinds, R.F.
A compositional material balance method for prediction of recovery from volatile oil depletion drive reservoirs, *JPT* (Jan. 1957), 19.

[72] Firoozabadi, A., Hekim, Y. and Katz, D.L.
Reservoir depletion calculations for gas condensates using extended analyses in the Peng-Robinson equation of state, *Can. J. Chem. Eng.* **56** (Oct. 1978), 610.

[73] Katz, D.L.
Possibility of cycling deep depleted oil reservoirs after compression to a single phase, *Trans. AIME* **195** (1952), 175.

[74] Sprinkle, T.L., Merrick, R.J. and Caudle, B.
Adverse influence of stratification on a gas cycling project, *JPT* (Feb 1971), 191.

[75] Holst, P.H. and Zadick, T.W.
Compositional simulation for effective reservoir management, *JPT* (March 1982), 635.

[76] Fernandes Luque, R., Duns, H. and van der Vlis, A.C.
Research on improved hydrocarbon recovery from chalk deposits, *Proc. Symp. New Technologies for Exploration and Exploitation of Oil and Gas Resources,* Luxembourg (1979), Graham and Trotman.

[77] Perry, G.E.
Weeks Island 'S' sand reservoir B gravity stable miscible CO_2 displacement, SPE/DOE 10695, *Proc. 3rd Jt. Symp. EOR* (April 1982), 309.

[78] Elias, R., Johnstone, J.R. and Krause, J.D.
Steam generation with high TDS feedwater, SPE 8819, *Proc. 1st Jt. SPE/DOE Symp. EOR,* Tulsa (April 1980), 75.

[79] Greaves, M. and Patel, K.M.
Surfactant dispersion in porous media, *Proc. 2nd Europ. Symp. EOR,* Paris (Nov. 1982), 99.

[80] Rovere, A., Sabathier, J.C. and Bossie Codreanu, D.
Modification of a black oil model for simulation of volatile oil reservoirs, *Proc. 2nd Europ. Symp. EOR,* Paris (Nov. 1982), 359.

[81] Monslave, A., Schechter, R.S. and Wade, W.H.
Relative permeabilities of surfactant, steam, water systems, SPE/DOE 12661, *Proc. SPE/DOE Fourth Symp. EOR* (April 1984), 315.

[82] Bragg, J.R. and Shallenberger, L.K.
Insitu determination of residual gas saturation by injection and production of brine, SPE 6047, *Proc. 51st Ann. Fall Mtg.* (1976).

[83] Driscoll, V.J. and Howell, R.G.
Recovery optimization through infill drilling concepts, SPE 4977, *Proc. 49th Ann. Fall Mtg.* (1974).

[84] Kyte, J.R. *et al.*
Mechanism of water flooding in the presence of free gas, *Trans. AIME* **207** (1956), 915.

[85] Weyler, J.R. and Sayre, A.T.
A novel pressure maintenance operation in a large stratigraphic trap, *JPT* (Aug. 1959), 13.

[86] Whitson, C.H.
Practical aspects of characterising petroleum fluids, *Proc. Conf. North Sea Condensate Reservoirs and their Development,* Oyez Sci. Tech. Serv. (May 1983), London.

[87] Prats, M.
Thermal Recovery, SPE Monograph Vol 7, *Soc. Pet. Eng* (1982), Dallas.

[88] Coats, K.H.
Simulation of gas condensate reservoir performance. *JPT* (Oct. 1985), 1870.

[89] Schirmer, R.M. and Eson, R.L.
A direct-fired down hole steam generator – from design to field test *JPT* (Oct 1985), 1903

[90] Nierode, D.E.
Comparison of hydraulic fracture design methods to observed field results *JPT* (Oct. 1985), 1831.

[91] Weiss, W.W. and Baldwin, R.
Planning and implementing a large-scale polymer flood. *JPT* (April 1985), 720.

[92] Kilpatrick, P.K., Scriven, L.E. and Davis, H.T.
Thermodynamic modelling of quaternary systems: oil/brine/surfactant/alcohol. *SPEJ* **25** (1985), 330.

[93] Clancy, J.P. *et. al.*
Analysis of nitrogen injection projects to develop screening guides and offshore design criteria. *JPT* (June 1985), 1097.

Chapter 13

Factors Influencing Production Operations

This chapter is concerned with providing a brief introduction to the principal elements of production systems. The total field of production engineering is very much wider than the subject matter of this chapter and can comprise:

(a) *Reservoir performance*
   Completion intervals, perforations performance, completion practices, completion equipment;
(b) *Well equipment*
   Stress analysis for tubulars and packers, corrosion/erosion considerations, tubular selection, packer selection, wellhead selections, safety valves, wireline services/facilities;
(c) *Well performance analysis*
   Natural flow performance, artificial lift requirements, performance analysis, prediction;
(d) *Stimulation and remedial operations*
   Acidization, fracturing, recompletions redrilling, remedial acceleration. Fig. 13.1 shows a jack-up rig used for a workover at a remote jacket.
(e) *Oil and gas processing*
   Separation, sweetening, dehydration;
(f) *Produced water and injection water treatment.*

13.1 THE PRODUCTION SYSTEM

The total production system is a complex, multiwell, possibly multireservoir, system of producing wells,

Fig. 13.1 Jack-up rig being used to work over a well on a remote jacket platform. (Photo courtesy of BP.)

13 FACTORS INFLUENCING PRODUCTION OPERATIONS

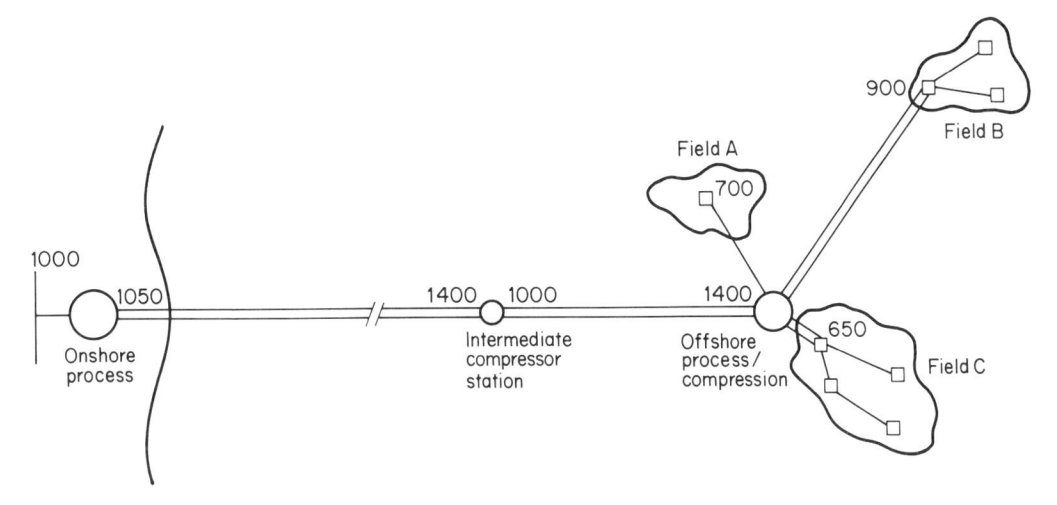

Fig. 13.2 Complex system of reservoirs, pipelines and stations in a gas field development.

flowlines, primary process facilities and delivery lines (Fig. 13.2).

The rates of production required will be checked by a combination of technical factors – especially reservoir and well characteristics – and of economic factors – contract quantities, capital investment and market requirements. One of the primary purposes of production engineering is the evaluation of producing system characteristics and their interactions so that maximum, optimum or design rates of production can be maintained at minimum costs. The analysis of an existing system to identify *bottlenecks* or constraints and to modify such a system for improved performance. Debottlenecking is an important aspect of production engineering.

Offshore alternatives to fixed platform developments are shown in Fig. 13.14. The use of subsea wellheads is more appropriate in deeper water fields, or where wells are required beyond deviation angles of platforms.

At this stage, the principal concern is the design of one representative element of a production system – one producing well. This comprises the associated reservoir volume, the wellbore and flow string, surface control chokes and delivery to the flowline. Interaction with other elements of a complex system is assumed to be defined by constraints on flowline pressure and/or well flow rate. Figure 13.3 shows offshore loading of oil to a tanker, a common alternative to pipeline transportation.

Flow from the reservoir to the flowline through the three elements concerned – the reservoir, the flow string and the choke system – is the system to be considered (Fig. 13.4).

The principal objectives are to determine the initial deliverability of a well under specified conditions. The further considerations of changes in deliverability and prediction of artificial lift or compression requirements belong in a more advanced treatment.

Fig. 13.3 Offshore loading of oil by tanker from a spar. (Photo courtesy of BP.)

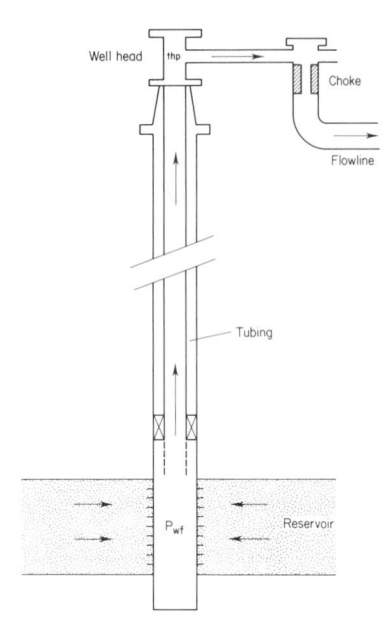

Fig. 13.4 The flow system.

13.2 RESERVOIR BEHAVIOUR IN PRODUCTION ENGINEERING

It is evident that no more oil or gas can be produced from a well than will flow into that well from the reservoir, and any study of well performance must start with an assessment of reservoir/well interactions.

For production engineering purposes a fairly simplified criterion of reservoir behaviour is needed, rather than advanced reservoir engineering, and idealized radial flow equations are modified for this purpose.

13.2.1 Productivity index and well inflow performance

A simple index of well performance is the productivity index (PI) of the well, defined by

$$PI = J = \frac{\text{rate of production}}{\text{drawdown}} = \frac{q}{\bar{P} - P_{wf}}$$

where q = rate of production m³/D or b/D, $\bar{P}$ = reservoir average or static pressure, P_{wf} = flowing bottom-hole pressure at the rate q.

The use of this index implies that it is a constant characteristic of a well, which is by no means true, but it has long been used as a basis for representing well productivity, and as a basis for analysis.

For a constant PI a linear relationship would exist between flow rate q and drawdown $(\bar{P} - P_{wf})$, and at any one instant of time the relationship would be linear with P_{wf}. The flow rate of a well is then determined uniquely by a specified flowing bottom-hole pressure.

In practice, productivity index will vary with flow rate if the range is large and inertial effects arise, will vary with pressure if gas evolves, with effective oil permeability, and with time as oil, gas and water saturations and viscosities change.

Particular care is needed in planning if test rates are artificially restricted to values very much lower than anticipated development well rates, and when straight line extrapolation may be over-optimistic.

The complete relationship between flow rate and drawdown (or flowing bottom-hole pressure) is defined as the inflow performance relationship (IPR).

For an oil well with no inertial effects an equation can be written

$$q = \frac{\text{constant} \cdot kh}{\left[\log_e \frac{r_e}{r_w} - \frac{3}{4} + S\right]} \int_{P_{wf}}^{\bar{P}} \frac{k_{ro}}{\mu_o B_o} dP$$

where μ_o, B_o are functions of pressure, and k_{ro} is a function of saturation.

Above the bubble-point, the product $\mu_o B_o$ is approximately constant and k_{ro} is 1 at zero water saturation, so that the PI is approximately constant for these conditions if inertial effects are absent.

If the pressure near the wellbore drops below the bubble-point, or if inertial effects become significant at high rates, then the PI is not constant, and the IPR becomes curvilinear.

13.2.2 Dimensionless IPRs for oil wells

It has been found that the curvature of the inflow performance relationship is reasonably well fitted by a quadratic equation for a very wide range of reservoir conditions. A general dimensionless quadratic equation can be defined as

$$\frac{q}{q_{max}} = 1 - a\left(\frac{P}{\bar{P}}\right) - (1-a)\left(\frac{P}{\bar{P}}\right)^2$$

where $\bar{P}$ = reservoir average, static pressure, q_{max} = hypothetical rate at zero bottom-hole flowing pressure, p_{wf} = flowing bottom-hole pressure at the rate of q. The factor a was found by Vogel [6] to be 0.2, and the Vogel dimensionless IPR is

$$\frac{q}{q_{max}} = 1 - 0.2\left[\frac{P}{\bar{P}}\right] - 0.8\left[\frac{P}{\bar{P}}\right]^2$$

13 FACTORS INFLUENCING PRODUCTION OPERATIONS

Theoretically this relationship should be used below the bubble-point pressure and a linear relationship above the bubble-point, but as a first approximation to the curvilinear IPR the Vogel relationship will be considered adequate at this stage. Obviously simulation studies could generate IPRs for complex conditions but this is not an objective of production engineering.

13.2.3 IPRs for gas wells

It has been conventional for many years to test gas wells at a series of flow rates and to express the results in the form of a *back pressure* equation:

$$Q = \text{constant} \, (\bar{P}^2 - P_{wf}^2)^n$$

where n is expected normally to have a value between 0.5 and 1.0.

Alternatively, data may be expressed in the form

$$\bar{P}^2 - P_{wf}^2 = AQ + BQ^2$$

In each case it is possible to fit observed data and to generate a complete IPR plot of P_{wf} against rate q.

Having considered the reservoir performance – the deliverability from reservoir to wellbore – the next element of the system is the wellbore itself – the determination of the flow rate that can be sustained from the reservoir depth to the wellhead.

13.3 WELLBORE FLOW

The general assumption will first be made that a specified wellhead pressure should be maintained. Alternative cases will be considered thereafter.

13.3.1 Flow of gases

Although the flow of gas in a wellbore is not strictly single-phase (since gas must be saturated with water at reservoir temperature), it can frequently be approximated as such, and an approximate analytical equation used to describe the pressure drop:flow rate relation in the wellbore. Such an equation is

$$Q_b = \text{constant} \left[\frac{(P_{wf}^2 - e^s P_t^2) d^5 S}{(e^s - 1) \gamma_g T f z L} \right]^{0.5}$$

where Q_b = volume rate of flow of gas at base conditions
P_{wf} = flowing well pressure
P_t = wellhead pressure
d = pipe internal diameter
γ_g = gas specific gravity relative to air
T = mean temperature (absolute)
z = gas deviation factor
f = friction factor (see Fig. 13.5)
L = tubing length
S = dummy variable = constant $\gamma_g H/ZT$
H = vertical depth to tubing shoe

When f is a Fanning friction factor, and Q is in MCSF/d, d in inches, L in feet, P in psi and T in degrees Rankine, then the constant is 0.1.

When field data is available for matching it may be possible to derive a locally valid equation:

$$Q_b^2 = \frac{a \, (P_{wf}^2 - b \, P_t^2) d^5}{L}$$

which can be used for different flow string diameters and drilled depths of wells.

This equation enables a further plot of P_{wf} against rate to be established, as shown in Fig. 13.6.

Tubing flow characteristics can be established for a range of diameters and specified wellhead pressures, and the intersections with the IPR for the well establishes the well deliverability for those conditions.

13.3.2 Flow of oil

The flow of crude oil in a wellbore can be very complex, since under most conditions pressures will fall below the bubble-point and gas will be evolved from solution.

With a very large pressure drop from reservoir to wellhead, the local ratios of gas to oil will change substantially due to gas expansion and to continued increasing evolution of gas. Under these conditions the *patterns* or *regimes* of flow possible in two-phase flow may change continuously, and the calculation of local density and friction loss can be difficult. Since the majority of the head loss in wellbores is the hydrostatic head, it is important to use accurate volumetric data on the gas and oil concerned, but even complex iterative calculation is subject to substantial error in calculating overall pressure drops.

There can be no simple single equation for flow under these conditions, and a simple approach usually involves the use of generalized *pressure traverses* or *lift curves*, unless such traverses have been specifically established for a field.

Figure 13.7 illustrates a typical set of pressure traverses where pressure:relative depth relations for one single flow rate and a series of gas–liquid ratios are illustrated. The use of these can be justified [21,35] because the principal element in the head loss is the hydrostatic term; densities of oil and gas vary only

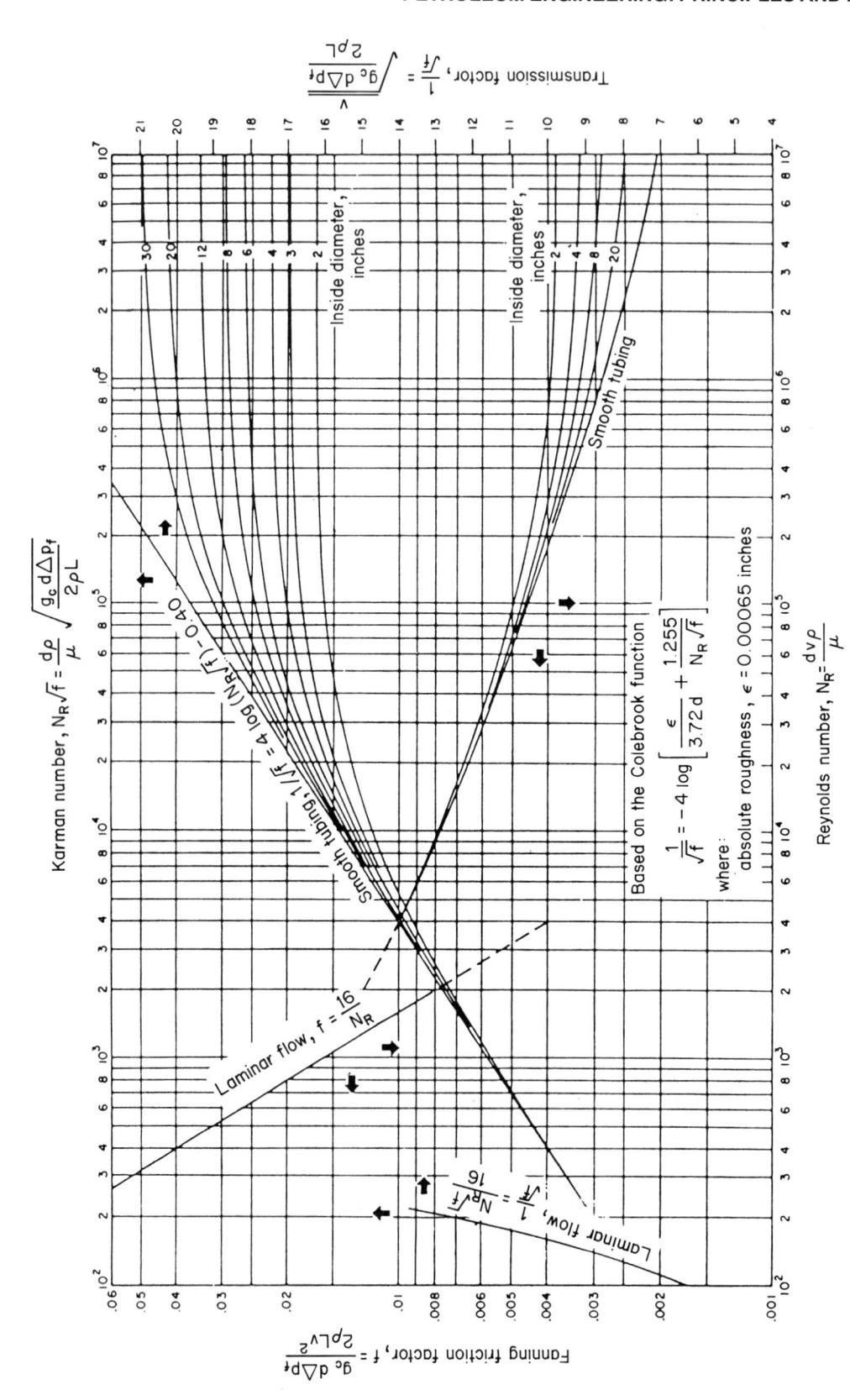

Fig. 13.5 Friction factors in smooth pipe.

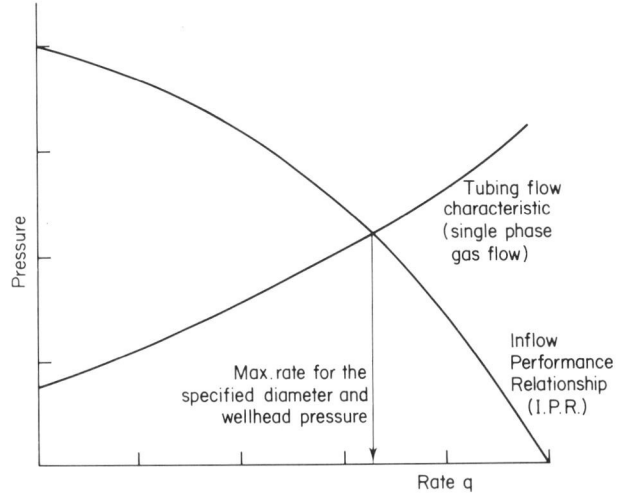

13 FACTORS INFLUENCING PRODUCTION OPERATIONS

Fig. 13.6 Gas well performance.

over small ranges; viscosity is not a factor in the highly turbulent flow involved; other factors are of minor importance. It must be remembered that the depth scale is in no sense an absolute one – only the intervals existing between points of different pressure are represented, not their positions in space.

For example, referring to Fig. 13.7 at a rate of 1500 b/d in a flow string of diameter 4 in., a wellhead pressure of 200 psi is specified for a well with a gas–oil ratio of 200 SCF/b. This gives a relative depth of 2550 ft. The reservoir depth is assumed to be 5000 ft below this, i.e. at a relative depth of 7550 ft (point B). The pressure at this point is found to be 1520 psi at point C. This constitutes one point on the flowing bottom-hole pressure:rate relation. Reference to further pressure traverses gives plots of additional data points, to generate an operating relationship as shown in Fig. 13.8.

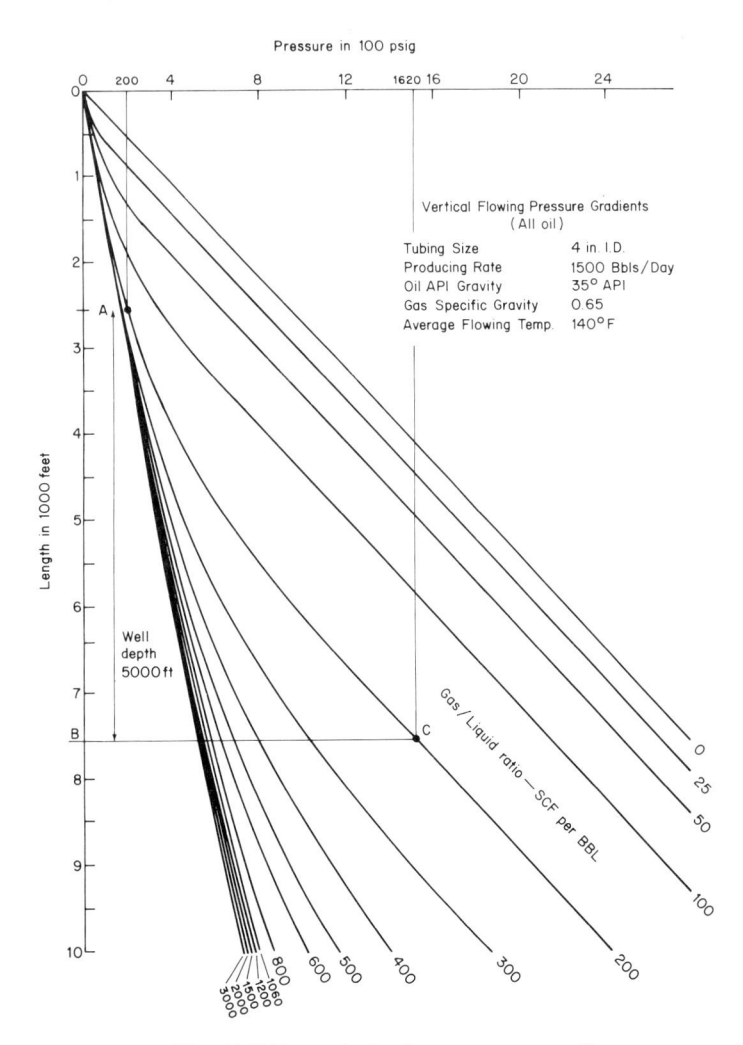

Fig. 13.7 Example flowing pressure gradient.

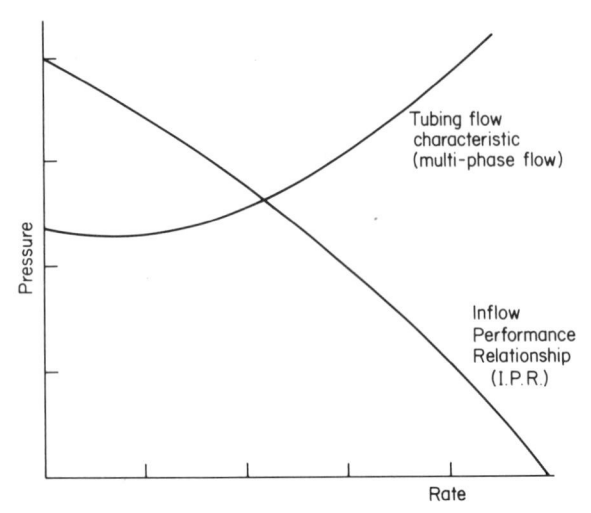

Fig. 13.8 Oil + water (+ gas) well performance.

13.4 FIELD PROCESS FACILITIES

Field processing involves all the mass, heat and momentum transfer processes that are necessary to:

(a) meet sales or delivery specifications of hydrocarbons (whether to pipeline or tanker);
(b) optimize the economic value of hydrocarbons produced;
(c) meet any statutory requirements for the disposal of any part of the production;
(d) meet any specification necessary for fluids for re-injection into the reservoir.

In some cases the objectives are easily met by very simple processing, but in other cases moderately sophisticated processing will be necessary – particularly where delivery is to a tanker, so that a stable, low vapour pressure product must be delivered. In the case of offshore fields, offshore processing may be supplemented, or duplicated, by onshore processing, as in the case of the UK east coast gas process plants for gas, and the Sullom Voe terminal for crude oil handling. Increasingly, also, associated gas streams are processed for liquid recovery before gas is flared, and such liquid recovery will usually be necessary to meet sales gas specifications.

13.5 NATURAL GAS PROCESSING

A sales specification for a natural gas will usually involve:

(a) a water dew-point – the temperature at which water will condense from the gas stream, which must be lower than any temperature likely to occur;
(b) a hydrocarbon dew-point – the temperature at which liquid hydrocarbons will condense from the gas stream, which also must be lower than any temperature likely to occur;
(c) a delivery pressure;
(d) a calorific value (possibly associated with the gas density in an index of burner performance or suitability);
(e) severely restricted maximum values of acid gas content – carbon dioxide and hydrogen sulphide – which are corrosive and, in the latter case, toxic.

All natural gases are produced saturated with water vapour, since they coexist in the reservoir at reservoir temperature with interstitial water in the reservoir. Also, nearly all natural gases contain small proportions of higher molecular weight hydrocarbons which will condense on reduction of temperature. Even the driest gases, of carbonaceous origin, will precipitate a few parts per million (on a volume basis) of liquid hydrocarbons. Very rich condensate streams may produce more than 1000 m^3 liquid condensate per million m^3 of gas produced.

13.5.1 Processing of dry natural gas

The first requirement, at the wellhead, is to take steps to remove water and water vapour before delivery to a pipeline. Water and hydrocarbons can combine together to form crystalline materials known as *hydrates*. These icelike materials are dependent upon both pressure and temperature, but are in general formed only at low temperatures (generally below 70°F). The expansion of gas through valves and fittings can cause such locally low temperatures, even when ambient temperatures are above hydrate formation temperatures.

Consequently, gases must generally be dried, heated or inhibited very near to the wellhead, if wellheads, flowlines and pipelines are not to be subject to the hazard of hydrate blocking. Hydrates are inhibited by the presence of alcohols, and methanol and glycol may be used as inhibitors.

13.5.2 Natural gas dehydration

A typical process flow stream for the offshore processing of natural gas is shown in Fig. 13.9. The well stream is passed to a simple separator (knock out drum) in which free liquid is separated. The (wet) gas is then heated and passed to a drying column. Drying may be effected by:

13 FACTORS INFLUENCING PRODUCTION OPERATIONS

dry desiccant process (e.g. silica gel);
liquid counterflow process (e.g. ethylene glycol).

The schematic of Fig. 13.9 shows a liquid counterflow process. The gas passes in counterflow through a column equipped with a few bubble cap trays, sieve trays or valve trays, and is stripped of water by the glycol. The wet glycol is passed to a regenerator where water is boiled off, and the dry glycol is recirculated to the column.

The dry gas passes to a pipeline. The drying processes also knock out hydrocarbon condensate. This is recovered by simple separation, the light hydrocarbons involved separating very easily from water.

In production from offshore natural gas reservoirs, there is rarely sufficient condensate produced to justify a separate pipeline, and the condensate is usually *spiked* back in to the dry natural gas line for recovery in onshore processing.

13.5.3 Natural gas sweetening

Sweetening is the process of removing acid gases from natural gas. If an offshore field has a significant content of acid gases and a dedicated pipeline to shore, the normal procedure would be to transport the sour gas to shore after dehydration (inhibiting if necessary) and to sweeten the gas on shore. Only if gas from one or more reservoirs, or fields (some sweet, some sour), was to be commingled in a single pipeline system would offshore sweetening normally be conducted. In this case, the offshore processing of a small part of the total gas stream might be more attractive than the onshore processing of a very large gas flow.

Sweetening is accomplished in a manner similar to that of dehydration, i.e. counterflow with a suitable wash agent in a bubble cap tower. For CO_2 removal only, suitable wash agents are:

amine wash;
potassium carbonate wash.

For hydrogen sulphides removal:

amine – ethanolamine or diethanolamine;
patent processes – Sulfinal and Vetrocoke.

Other processes are also available.

13.5.4 Onshore processing

The processing necessary to meet dew-point specifications is normally a moderately cheap refrigeration process. Figure 13.10 shows a typical process flow diagram. The intake gas is first chilled by heat exchange with the cool processed gas, glycol being

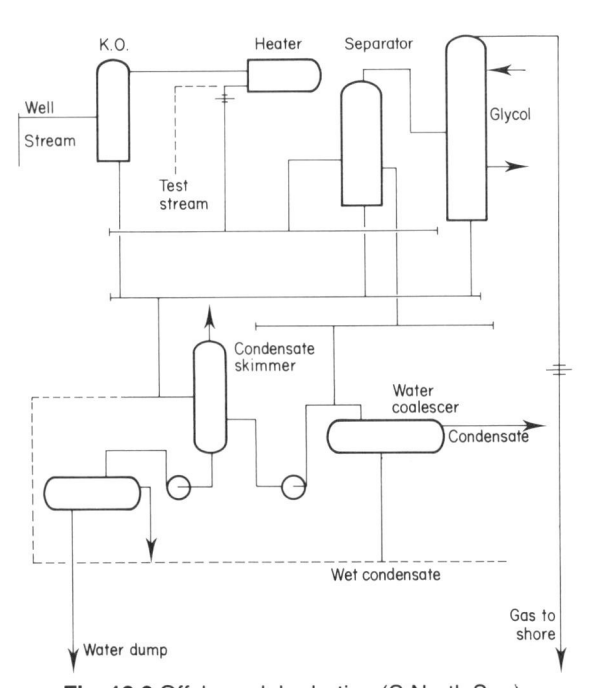

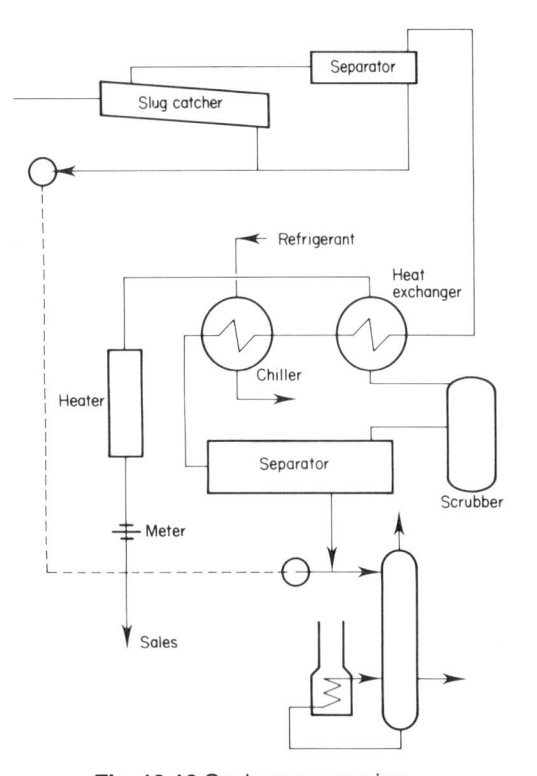

Fig. 13.9 Offshore dehydration (S. North Sea).

Fig. 13.10 Onshore processing.

added to inhibit hydrate formation. The gas then passes to a refrigeration unit (either using freon as a refrigerant, or less efficiently propane from the condensate stream) where it is cooled to −18°C. The resulting gas–liquid mixture is separated, the cool gas being heated first by heat exchange with incoming gas, and then by a fired heater before measurement and transfer. The liquids are boiled off to separate water, condensate and glycol. Liquids from slug catchers and knockouts are blended in, and the resulting stream stabilized to give a stable condensate fraction and a non-specification gas stream. This latter is ideal for fuel.

13.5.5 Calorific value

If no inert gas is present, and dew-point specifications are met, a dry natural gas will meet the usual calorific value/density specifications. If there is a significant content of nitrogen, enrichment may be necessary if the gas is to be put into a national grid. This is done by adding propane (or liquefied petroleum gas) in small quantities. Alternatively, a large industrial user may take a non-specification gas using burners designed for the appropriate calorific value. In the rare cases where the high paraffin content of gas (C_2H_6, C_3H_8) gives an excessively high calorific value, this can be reduced by dilution with nitrogen.

13.5.6 Compression

When compression is needed to meet a delivery specification (as will always be the case with a depletion gas field, and usually with water drive gas fields), the necessary compression is preferably installed as far upstream as is possible (i.e. near the wellheads). This reduces compression power requirements, and only where pressure losses between field and final process plant are small will compression be installed at the process plant.

13.6 CRUDE OIL PROCESSING

The processing of crude oil will have as its objectives:

(a) production of a liquid stream which meets a transport specification. In the case of pipeline transport to an onshore process plant, the vapour pressure specification will not be stringent. However, with common carrier lines, the accurate metering of mass is desirable for equity considerations. For tanker transport, a stringent vapour pressure specification is necessary – crude stabilized at 1 atmosphere and ambient temperature if necessary, together with maximum retention of intermediate hydrocarbons is desirable;

(b) the separation of gas to meet vapour pressure specification, and possibly the processing of the gas to meet a sales specification;

(c) the separation of produced water, and the breaking of any produced emulsions to meet a refinery specification;

(d) the removal of any noxious or toxic materials – especially hydrogen sulphide – from the crude oil prior to delivery;

(e) the removal of salt to meet a refinery specification. This can be a problem when crude oil is produced with small proportions of water and first stage separation occurs at high temperatures. Under these conditions, the produced water can flash to the vapour phase, leaving salt as a residue. Simple freshwater washing is the only necessary process, but this then involves further water separation.

13.6.1 Light oil processing

In this case, gas–oil separation is the major objective, together with maximum recovery of intermediate hydrocarbons. The process adopted will depend primarily on the subsequent use of gas. Where gas surplus to fuel requirements is simply to be flared, gas stream processing will not be justified, as liquids will not be recovered to any significant extent from the gas stream. In this case optimization of the separation process is highly desirable, and a process involving several (3–4) stages of separation with carefully designed separator pressure will be necessary.

Optimization of a separator process can yield a few extra percentage points of stabilized liquid and can make a difference of one or two degrees in the API gravity of the product. With a large crude oil flow this can be highly significant in cash flow terms.

When gas is to be disposed of by a sales outlet, the associated gas will be processed to recover intermediate hydrocarbons and obtain a sales gas specification. In this case, the liquids will be *spiked* back into the crude oil stream and recovered in this way.

13.6.2 Separator design considerations

Separator vessels may be very simple, or may contain several separation elements, depending upon the difficulty of processing. Figure 13.11 shows a separator of moderate complexity. The well stream impinges upon a deflector which effects a crude separation of liquid and gas, the liquid being decelerated and deflected to the lower part of the vessel. A weir retains a high liquid level and, behind this weir, water separates and a level controller maintains an oil–water level within limits. Oil spills over the weir, and gas bubbles can rise and separate upstream and downstream of the weir. The important factor in this phase of separation is the residence time, and separators are usually designed to give a residence time of three to five minutes.

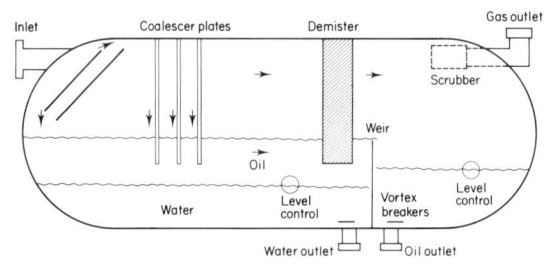

Fig. 13.11 Three-phase separator.

The gas phase may pass through a coalescer in which liquid droplets impinge, coalesce and drip back into the liquid phase. After the coalescer, the gas passes through a demister section (a pad of wire mesh), further to entrap and coalesce entrained liquid droplets.

In the gas region of a separator, gas velocity is the critical design factor, and a rule of thumb expression for maximum gas velocity is

$$V = C \left[\frac{\rho_o - \rho_g}{\rho_g} \right]^{0.5}$$

where V is the critical entrainment velocity (ft/s.), C is the separator coefficient (empirical, about 0.35–0.50 ft/s.), ρ_o, ρ_g are fluid densities. Level controls, level warnings and shutdown systems will keep the separator working within its design limits.

With multiple stages of separation, the pressures and temperatures of each stage of separation are important to the efficiency of separation (Fig. 13.12). Optimum values can be found by laboratory experiment on field samples, or by computation.

13.6.3 Foaming problems

With light gassy crude oils, a separation problem can occur if foams form with flow through the restric-

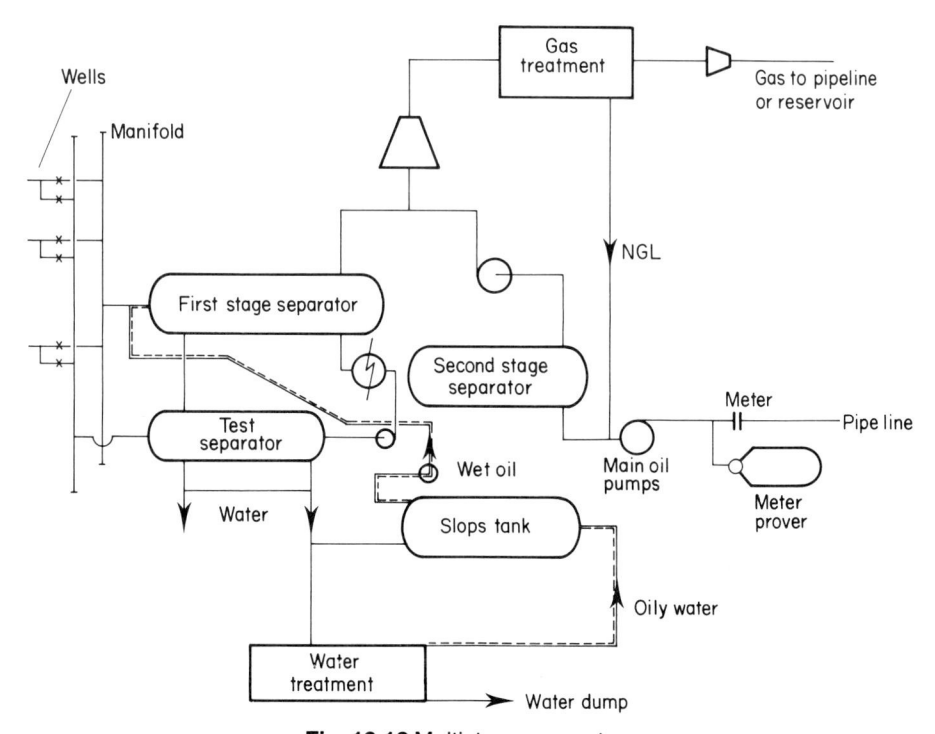

Fig. 13.12 Multistage separator.

tions of a typical separation system. In this case, the residence times necessary for foams to drain effectively and break can be prohibitive and separation highly inefficient. The most effective remedy is chemical foam breaking, the addition of a silicone liquid upstream of the separators being highly effective in promoting foam drainage and breakdown.

13.6.4 Wax problems

Light oils are generally paraffins, and the heaviest components may be paraffin waxes. These waxes may precipitate if the temperature falls below some critical value, and wax build-up in well tubing strings, flowlines and pipelines may occur. Again chemical treatment is effective; chemical inhibitors apparently acting to prevent growth and crystal development of the wax. Alternatively, trace heating or periodic heat treatment and scraping can be used to remove wax after it has built up.

13.7 HEAVY OIL PROCESSING

With heavy oils, gas–oil ratios are usually low, and it is the separation of produced water that is the greater difficulty – the high viscosity of produced crude greatly retarding gravity settling of water. Additionally, heavy crude oils have a greater propensity to emulsion formation than have lighter crudes.

The first procedure adopted in difficult cases is to heat the well stream to reduce viscosity. This may be done in a heater (direct or indirect fired) upstream of the first stage separator, or in a combined heater-treater.

After first gas and water separation, the oil may pass to a heated storage tank where very long residence times may give the necessary separation. When stable emulsions are formed and are a problem which cannot be remedied by heat and settling time, chemical demulsifiers may be used. When used they should be applied as far upstream as is possible – possibly by injection at a downhole pump intake or at least below the wellhead.

Chemical treatment followed by heat treatment will deal with most problems, but in a few cases of very obstinate emulsions, electrostatic precipitation may be necessary as a final last resort.

13.8 PRODUCED WATER TREATMENT

There are stringent specifications for the disposal of water within oilfields – disposal into the North Sea currently requiring a hydrocarbon content of less than 50 ppm. This requires that water from all stages of separation and any oily slops or washings should be cleaned before disposal.

The essential procedures are mainly gravity settling, oily skimmings from a series of settling tanks being recirculated and the final water being treated in a plate or a foam coalescer before disposal. Most systems installed are capable of reducing the hydrocarbon content to less than 30 ppm. Figure 13.13 shows a typical schematic.

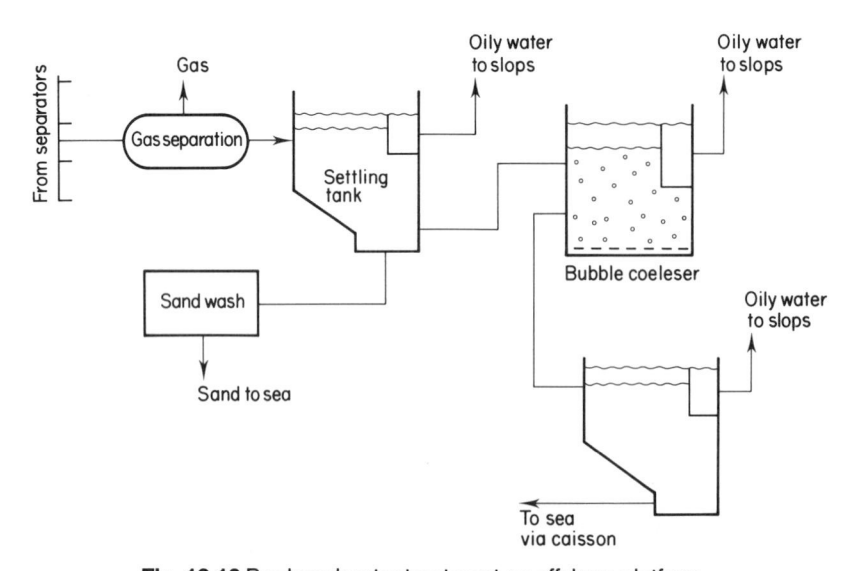

Fig. 13.13 Produced water treatment on offshore platform.

13.9 INJECTION WATER TREATMENT

Water for injection may need to be highly purified before injection into a reservoir. Fine solid materials may plug formations, organic material or bacteria may generate slimes, and oxygen may promote bacterial growth and cause corrosion.

The degree of solids removal necessary is a matter for experiment and experience, and in the North Sea both very fine filtration and virtually no filtration have both been adopted with success. Offshore, the water supply will usually be taken from a level where suspended solids and dissolved oxygen are low – which will be an intermediate depth.

A biocide will be used at the pump intake – sometimes by *in-situ* electrolysis of sea water – followed by a coarse filtration at the surface. De-aeration, either by counterflow in a column or by vacuum de-aeration, will be followed by the addition of any oxygen scavenger (sodium sulphite) and further filtration. This filtration, depending on the degree of filtration needed, may be by means of:

(a) sand and/or anthracite graded beds;
(b) diatomaceous earth/asbestos filter cakes;
(c) polypropylene filter cartridges.

The diatomaceous earth filters are capable of removing practically all solids down to one micron should this be necessary.

The final treatment is likely to be a bactericide addition.

13.10 CRUDE OIL METERING

Metering of crude oil and gas streams is necessary for transfer and sales purposes, and for fiscal purposes.

Metering of crude oil into a common carrier system requires particular care since volumes of mixing of crude oil of different crudes is not the sum of the volumes of the component crude oils. This requires that each contributor with a common pipeline system should maintain a record both of volume delivered, and of the density of the material delivered, and keep records of the overall compositions. Only then can the production from a common carrier system be allocated equitably back to the contributors.

Crude oil streams are metered by turbine meters for the most part, although other types of meter are under study. The turbine meters themselves are calibrated regularly by means of meter provers – a positive displacement device which delivers a measured quantity through the meter. This itself is recalibrated by means of calibrated tanks.

Gas flow rates are usually metered by orifice meter, with density being recorded simultaneously.

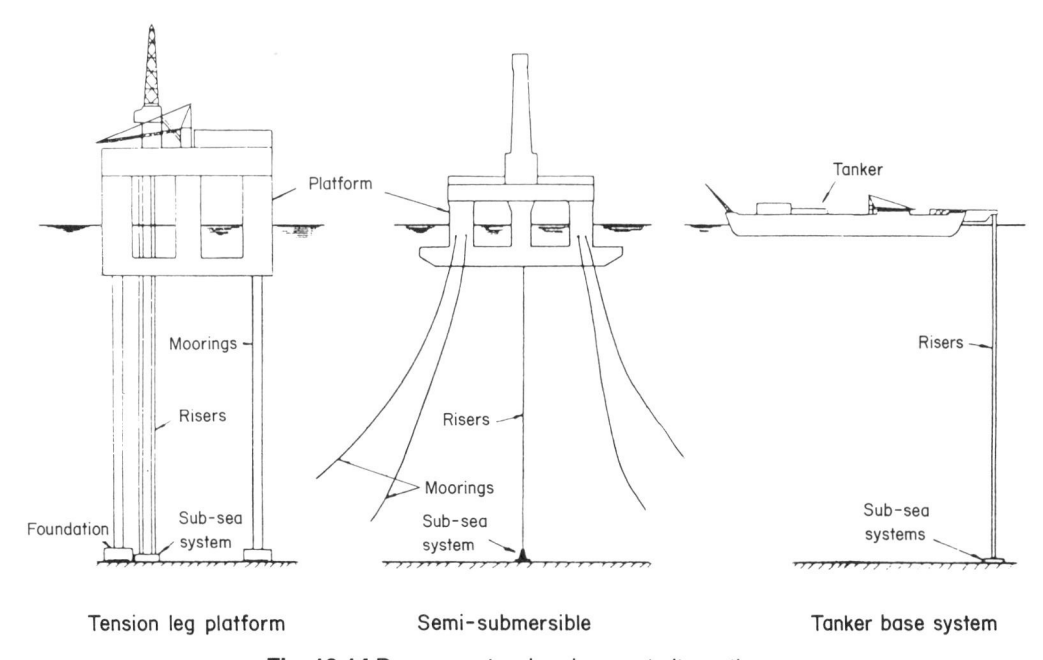

Fig. 13.14 Deeper water development alternatives.

Examples

Example 13.1

(a) A well 6000 ft deep is flowing 2000 b/d at a GOR of 200 SCF/barrel on a 4 in. flow string. If the wellhead pressure is 400 psi, what is the estimated flowing bhp?
(b) A well 5000 ft deep has a flowing bhp of 1200 psi producing 3000 b/d at a GOR of 500 SCF/barrel. What is the estimated wellhead pressure?

Example 13.2

A drill stem test on a well indicates a flowing bottom hole pressure of 1500 psig at a rate of 3315 b/d, with a reservoir static pressure of 2600 psig. The well depth is 5500 ft deep, and the GOR is 200 SCF/barrel. The Vogel IPR relationship is assumed to apply.

If a flowing wellhead pressure of 400 psig is needed on production, evaluate and comment on the performance of a 4 in. flow string for the well.

Example 13.3

Determine the size of a horizontal separator to separate 1000 m^3/day of crude oil from its associated gas (SG = 0.75) at a pressure of 20 bar and a temperature of 40°C. The oil residence time should be 3 min and the oil–gas interface should be half way in the separator volume. The oil has a density of 796 kg/m^3 and a solution gas–oil ratio of 95 m^3/m^3 oil at 0°C and 1 bar. The density of air at 0°C and 1 bar is 1.275 kg/m^3. The design ratio of diameter to length, seam to seam should be between 3 and 4 and the separator must be at least 3 m in length. The maximum gas velocity in the separator is given in m/s by

$$U_{max} = 0.125 \left[\frac{\rho_L - \rho_g}{\rho_g} \right]^{0.5}$$

References

[1] Katz, D.L.
   Overview of phase behaviour in oil and gas production, *JPT* (June 1983), 1205.
[2] Nind, T.E.W.
   Principles of Oil Well Production, McGraw Hill, NY (1981).
[3] Gilbert, W.E.
   Flowing and gas-lift well performance, *API Drill. Prod. Practice* (1954), 126.
[4] Frick, T.C. (ed.)
   Petroleum Production Handbook Vol. 1 (Mathematics and production equipment), SPE (1962).
[5] Vogel, J.V.
   A field test and analytical study of intermittent gas lift, *SPEJ* (Oct. 1974), 502.
[6] Vogel, J.V.
   Inflow performance relationships for solution gas drive wells, *JPT* (Jan. 1968), 83.
[7] Steele, R.D.
   Engineering and economics used to optimize artificial lift methods, *OGJ* (Dec. 1976), 107.
[8] Patton, C.C.
   Oilfield Water Systems, Campbell Pet. Series, Norman (1977).
[9] Boles, B.D.
   Subsea production control (Beryl field), SPE 12971, *Proc. Europ. Pet. Conf.* (Oct. 1984), 117.
[10] Dawson, A.P. and Murray, M.V.
   Magnus subsea wells: design, installation and early operational experience, SPE 12973, *Proc. Europ. Pet. Conf.* (Oct. 1984), 133.
[11] Patel, M.H., Montgomery, J.I. and Worley, M.S.
   A new concept in floating production systems, SPE 12986, *Proc. Europ. Pet. Conf.* (Oct. 1984), 245.
[12] Wray, C.R.
   The fundamental issues in future field development concepts (100 – 250 m waterdepth), SPE 12987, *Proc. Europ. Pet. Conf.* (Oct. 1984), 257.
[13] Ryall, M.L. and Grant, A.A.
   Development of a new high reliability downhole pumping system for large horsepowers, EUR 276, *Proc. Europ. Pet. Conf.* (Oct. 1982), 49.

13 FACTORS INFLUENCING PRODUCTION OPERATIONS

[14] van Staa, R.
Separation of oil from water, EUR 278, *Proc. Europ. Pet. Conf.* (Oct. 1982), 73.

[15] Hankinson, R.W. and Schmidt, T.W.
Phase behaviour and dense phase design concepts for application to the supercritical fluid pipeline system, EUR 330, *Proc. Europ. Pet. Conf.* (Oct. 1982), 505.

[16] Simmons, P.E.
North Sea offshore compression – future needs, EUR 331, *Proc. Europ. Pet. Conf.* (Oct. 1982), 511.

[17] Thambynayagam, R.K.M. and Bristow, P.G.
Design calculations for three phase flow behaviour in wells and flowlines and some problems in their application to the North Sea, EUR 358, *Proc. Europ. Pet. Conf.* (Oct. 1982), 711.

[18] Beggs, H.D. and Brill, J.P.
A study of two phase flow in inclined pipes, *JPT* (May 1973), 607.

[19] Aziz, K., Govier, G.W. and Fogarasi, M.
Pressure drop in wells producing oil and gas, *J. Can. Pet. Tech.* (July – Sept. 1972), 38.

[20] Taitel, Y., Bornea, D. and Dukler, A.E.
Modelling flow pattern transitions for steady upward gas-liquid flow in vertical tubes, *AIChEJ* (May 1980), 345.

[21] Brown, K.E.
The Technology of Artificial Lift Methods, Penwell Books, Tulsa (1980).

[22] Nisbet, R.C.
Efficient operations in a mature oil and gas producing area, SPE 7797, *Proc. Prod. Symp. Tulsa* (Feb. 1979), 15.

[23] Brown, K.E.
Gas Lift Theory and Practice, Petroleum Publishing Corp. (1973).

[24] Wottge, K.E.
Piper field: Surface facilities operating performance and problems experienced, EUR 153, *Proc. Europ. Pet. Conf.* (Oct. 1980), 45 (Vol. 1).

[25] DeMoss, E.E. and Tiemann, W.D.
Gas lift increases high volume production from Claymore field, EUR 189, *Proc. Europ. Pet. Conf.* (Oct. 1980), 429 (Vol. 1).

[26] Ross, D.M., Spencer, J.A. and Stephen, D.D.
Artificial lift by electric submersible pumps in Forties, EUR 190, *Proc. Europ. Pet. Conf.* (Oct. 1980), 437 (Vol. 1).

[27] Wachel, J.C. and Nimitz, W.W.
Assuring the reliability of offshore gas compression systems, EUR 205, *Proc. Europ. Pet. Conf.* (Oct. 1980), 559 (Vol. 1).

[28] Taylor, F.R.
Simulation: a new tool in production operations, EUR 208, *Proc. Europ. Pet. Conf.* (Oct. 1980), 17 (Vol. 2).

[29] Planeix, J.M. and Macduff, T.
The characteristics of shuttle and buffer tankers for offshore fields, EUR 213, *Proc. Europ. Pet. Conf.* (Oct. 1980), 43 (Vol. 2).

[30] Jones, G., Buysse, A.M., Porter, R.H. and Stubbs, J.
Modelling the Brents System production facilities, EUR 228, *Proc. Europ. Pet. Conf.* (Oct. 1980), 159 (Vol. 2).

[31] Nichols, J.H. and Westby, K.A.
Innovative engineering makes Maureen development a reality, EUR 231, *Proc. Europ. Pet. Conf.* (Oct. 1980), 185 (Vol. 2).

[32] Mitchell, R.W. and Finch, T.M.
Water quality aspects of North Sea injection water, EUR 33, *Proc. Europ. Pet. Conf.* (Oct. 1978), 263 (Vol. 1).

[33] McLeod, W.R.
Prediction and control of natural gas hydrates, EUR 116, *Proc. Europ. Pet. Conf.* (Oct. 1980), 449 (Vol. 2).

[34] Allen, T.O. and Roberts, A.P.
Production Operations (Vols 1 and 2), Oil and Gas Cons. Int. (1978).

[35] Aziz, K., Eickmeier, J.R., Fogarasi, M. and Gregory, G.A.
Gradient Curves for Well Analysis and Design, Can. Inst. Min. Met. Monogr. Sp. Vol. 20 (1983).

[36] Arnold, K.E.
Design concepts for offshore produced water treating and disposal systems, *JPT* (Feb. 1983), 276.

[37] Arnold, K. and Stewart, M.
Designing oil and gas producing systems, *World Oil,* (March 1985) 69, (Nov. 1984) 73, (Dec. 1984) 87.

[38] Geertsma, J.
Some rock-mechanical aspects of oil and gas well completions, EUR 38, *Proc. Europ. Pet. Conf.*, London (1978), 301, Vol. 1.

[39] Mukierjee, H. and Brill, J.
Liquid holdup correlations for inclined two phase flow, *JPT* (May 1983), 1003.

[40] Goland, M.
Production Engineering (1986) IHRDC, Boston.
[41] Cox, J.L.
Natural Gas Hydrates: Properties, Occurrence and Recovery. Butterworth (Ann Arbor Science Books), 1983.
[42] Ashkuri, S. and Hill, T.J.
Measurement of multiphase flows in crude oil production systems. *Petroleum Review* (Nov. 1985), 14.
[43] Baker, R.C. and Hayes, E.R.
Multiphase measurement problems and techniques for crude oil production systems. *Petroleum Review* (Nov. 1985) 18.
[44] Jamieson, A.W. *et al.*
Multiphase flow measurements at production platforms. *Petroleum Review* (Nov. 1985), 23.
[45] Brown, K.E. and Lea, J.F.
Nodal systems analysis of oil and gas wells. *JPT* (Oct 1985) 1751.
[46] El-Hattab, M.I.
Scale deposition in surface and subsurface production equipment in the Gulf of Suez. *JPT* (Sept. 1985). 1640.
[47] Beggs, M.D.
Gas Production Operations. OGCI (1984).
[48] Eissler, V.C. and McKee, R.E.
Offshore production operations. *JPT* (April 1985) 583.
[49] Giles, A.J.
Arun Field high pressure gas reinjection facilities. *JPT* (April 1985) 701.

Chapter 14

Concepts in Reservoir Modelling and Application to Development Planning

In this chapter the principles of modelling using reservoir simulators are presented, and emphasis is placed on reservoir description and displacement mechanisms. The application of reservoir models in field development and resource management is illustrated in the rest of the chapter.

14.1 MODELS

In petroleum reservoir development a broad definition of modelling is adopted in which a model is any device by which a predictive understanding of reservoir performance and/or description can be obtained. In this sense it may be physical, conceptual or mathematical.

Physical models include sand packs, cores and core plugs, Hele Shaw models and micromodels. The objective in using physical models is to define physical behaviour, flow patterns, residual saturations and other parameters which may define boundary conditions and perhaps allow scaling to reservoir conditions [6, 7, 9, 10, 11].

Conceptual models provide a basis for exploration of physical processes and are used to guide quantitative estimation. The main types of conceptual models involved in reservoir modelling concern geological models [12, 26, 32, 38, 60, 66, 67, 77, 89]. Depositional and diagenetic history of sediments are presented to account for present day observations of facies character and petrophysical property distribution. In a predictive sense a geological conceptual model is used to guide the values attributed to reservoir properties away from direct well control. An interactive analysis of geological models results from testing their predictions against new well data using fluid flow and vertical pressure gradients. Revised geological models develop from consideration of reservoir performance data as well as from new geological evidence alone [17]. Mathematical models are designed to describe reservoir volumetrics and flow behaviour using the Darcy relationship and conservation of mass, together with empirical parametric relationships. Mathematical models may be simple (tank models, linear or radial one-dimensional (1–D) displacement), or complex (multidimensional, multiphase, multicomponent) flow models. Depending on the definition of the problem and the availability of data, the choice of approach lies with the petroleum engineer [15].

Multidimensional, multiphase reservoir analysis requires definition of a reservoir in discrete regions with given properties and rules of flow. The definition of such regions may be cells or nodes (Fig. 14.1) and leads to the formulation of relationships between saturation and pressure as non-linear differential equations, which can be solved approximately using finite difference [30, 31, 44, 46, 47] or finite element mathematics [88]. The numerical solution of these reservoir equations using high speed computers is known as reservoir simulation modelling. The success of a numerical model depends on two particular conditions, the former of which is more likely to be satisfied.

(1) The ability of the equations to represent

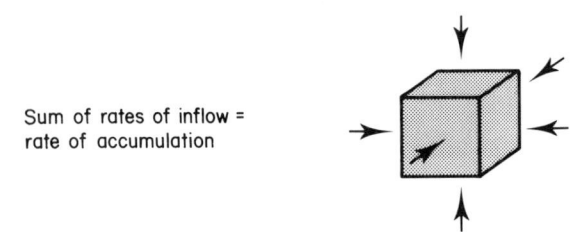

Fig. 14.1 Methodology.

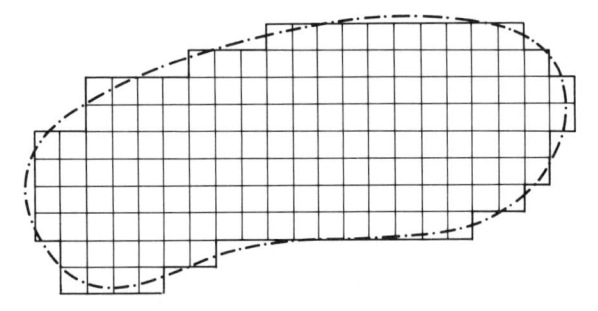

Reservoir split into blocks
Write equations for flow in and out of each block

Sum of rates of inflow = rate of accumulation

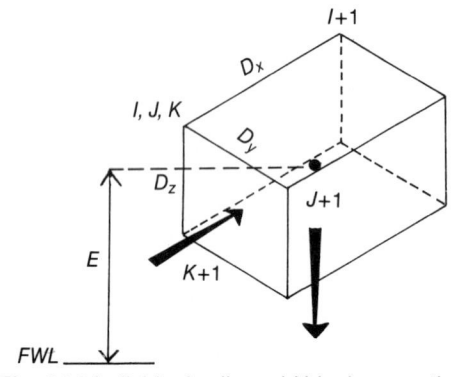

Fig. 14.2 Individual cell or grid block properties.

physics of flow and equilibrium in the reservoir/well system.

(2) the ability of cell or node properties to represent the true three-dimensional (3-D) reservoir description, within a particular cell of dimensions D_x, D_y, D_z and mid-point depth E from a reference datum. The values of cell porosity and directional permeability are defined at the cell centre (Fig. 14.2).

14.2 EQUATIONS OF MULTIPHASE FLOW

These will be illustrated in a linear system for simplicity. An extension to three dimensions simply consists of adding terms in $\partial/\partial y$ and $\partial/\partial z$ and in accounting for gravity effects.

For a conservation term in stock tank units (broadly equivalent to mass, and identical if the API gravity is constant), we can write for the cell illustrated in Fig. 14.3

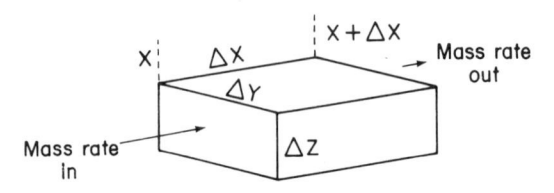

Fig. 14.3 Unit cell.

Mass rate in − mass rate out
= mass rate of accumulation

For the oil phase we have

$$\left\{ \frac{-\Delta Y \Delta Z k_o}{\mu_o B_o} \frac{\partial P_o}{\partial X} \right\}_x - \left\{ \frac{\Delta Y \Delta Z k_o}{\mu_o B_o} \cdot \frac{\partial P_o}{\partial X} \right\}_{x+\Delta_x}$$

$$= \frac{\Delta X \Delta Y \Delta Z}{\Delta t} \left\{ \left(\frac{\phi S_o}{B_o} \right)_{t+\Delta t} - \left(\frac{\phi S_o}{B_o} \right)_t \right\}$$

which in the limit becomes

$$\frac{\partial}{\partial X} \left[\frac{k_o}{\mu_o B_o} \cdot \frac{\partial P_o}{\partial X} \right] = \frac{\partial}{\partial t} \left[\frac{\phi S_o}{B_o} \right]$$

For the water phase we have a similar equation:

$$\frac{\partial}{\partial X} \left[\frac{k_w}{\mu_w B_w} \cdot \frac{\partial P_w}{\partial X} \right] = \frac{\partial}{\partial t} \left[\frac{\phi S_w}{B_w} \right]$$

For the gas phase the equation must include both free gas and gas from solution in the oil (we could at this stage ignore gas dissolved in water). The limit equation thus becomes

$$\frac{\partial}{\partial X} \left[\frac{k_g}{\mu_g B_g} \cdot \frac{\partial P_g}{\partial X} \right] + \frac{\partial}{\partial X} \left[\frac{R_s k_o}{\mu_o B_o} \cdot \frac{\partial P_o}{\partial x} \right] = \frac{\partial}{\partial t} \left[\frac{\phi S_g}{B_g} \right]$$

$$+ \frac{\partial}{\partial t} \left[\frac{\phi R_s S_o}{B_o} \right]$$

The equations presented here show that at any point in space there are at least six unknowns, namely P_o, P_w, P_g, S_o, S_w, S_g. In order to provide a

14 CONCEPTS IN RESERVOIR MODELLING

solution we therefore require three further linking equations defining saturation and capillary pressures of the oil–water and gas–oil systems.

$$S_o + S_w + S_g = 1$$
$$P_{cg} = P_g - P_o$$
$$P_{cw} = P_o - P_w$$

For the three-dimensional system shown in Fig. 14.4 the rate of accumulation at cell is given by

$$(q_o)_{i-N_x-N_y} + (q_o)_{i-N_x} + (q_o)_{i-1} + (q_o)_{i+1}$$
$$+ (q_o)_{i+N_x} + (q_o)_{i+N_x+N_y}$$

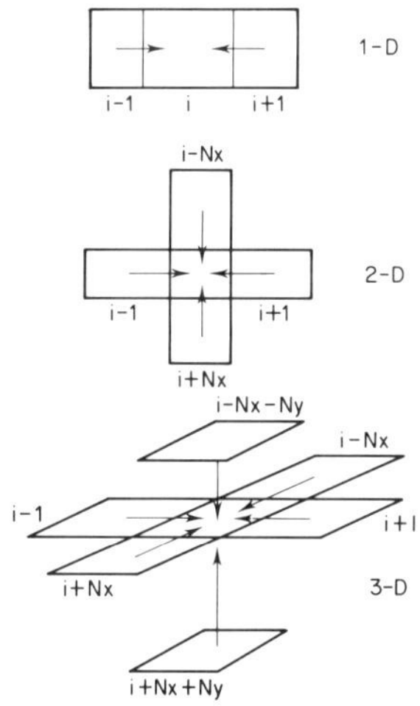

Fig. 14.4 Flow into cell *i* from neighbours in different geometries.

The solution of these equations may be approached by direct solution (Gaussian or matrix decomposition) or by a number of iterative algorithms. For discussion of these techniques the reader is directed to specialist texts [45, 46]. The treatment of error in finite difference and finite element formulations is important in several applications. The effects of cell size and solution time step size are interlinked in the efficiency of solution algorithms. One of the most important tests of reservoir simulation accuracy that can be made concerns numerical dispersion or the *smearing* of a saturation front across several cells. Part of the smearing may result from the definition of an appropriate effective permeability solely at the boundary between two cells undergoing fluid exchange. The use of the effective permeability in the upstream cell only during a time step is widespread. The problem is usually assessed by comparing the results of an analytical Buckley Leverett [2] frontal movement with that predicted by 1-D, 2-phase (2-P) simulation. This should provide an indication of required cell sizes and time steps.

For complex geometry systems there can be no analytical check on results – only comparison with gross simplified analytical estimates and *reasonableness*. For models involving large numbers of cells, phases or components, the direct solution method may involve excessive computer time. Algorithms for several iterative solutions have been published and treat pressure and saturation in all combinations from fully explicit (data known at start time level for the time step) to fully implicit (data known at end time level for the time step). A particularly utilized method of arranging the differential equations in finite difference form results in a solution known as IMPES, meaning IMplicit in Pressure, Explicit in Saturation. Iteration procedures terminate when convergence criteria are satisfied.

14.3 SIMULATOR CLASSIFICATIONS

General classification of reservoir simulators is by dimensionality, phases or composition components, grid arrangement and solution approach, as shown in Table 14.1 and in Fig. 14.5.

Just about all the combinations suggested exist for finite difference simulators. Finite element methods which should be superior in frontal saturation tracking are not common in a 3–D/3-phase mode. The connection of well and operating constraints further serves to delineate different models. Tubing flow is usually considered explicitly in a time step, as at present any implicit treatment is excessive in computing time. The arrangement of a simulator tends to be as shown in Fig. 14.6. The main program directs the calculation and reporting procedures, and the substance of the simulator is contained in subroutines.

14.4 SIMULATOR APPLICATION

As is clear from the number of simulator combinations available, the selection process for a particular task falls to the reservoir engineer. Selection is based on the nature and definitions of the task, the data availability and the economic value of the result [15]. Analytical analyses of a simplified representation of the reservoir and its contents can provide an insight into proper selection of simulator tools and gives a basis for comparison of results. Simulations are frequently conducted to provide information on the sensitivity of ill-defined parameters in reservoir performance prediction. Examples of such sensitivity parameters may well be transmissibility $[kA/L]$, relative permeability, irreducible saturations and

TABLE 14.1 Classification of simulators

Dimensions	Phases/composition	Grid/node arrangement	Options for solution
0–D (material balance)	1–P gas	Cartesian (regular)	Finite element
1–D linear	1–P water	Cartesian (irregular)	Finite difference
1–D vertical	2–P black oil–w	Radial (vertical wells)	– direct soln
	2–P black oil–g	Radial (horizontal wells)	– implicit
			– IMPES
2–D cross-section	3–P bo–w–g	Nodal	Semi implicit
2–D areal	N component		– explicit
3–D sector	N pseudo-components		Conformal mapping
3–D full field	N comp-chemical flood		
	– thermal process		
	– miscible process		
	– inert gas process		
	– volatile oil		
	– gas condensate		

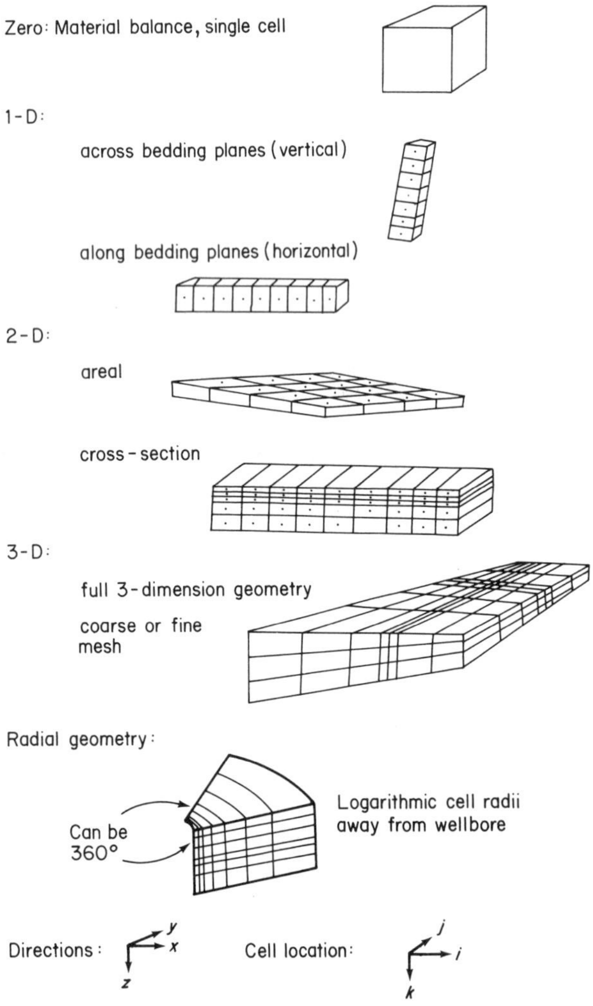

Zero: Material balance, single cell

1-D:
across bedding planes (vertical)
along bedding planes (horizontal)

2-D:
areal
cross-section

3-D:
full 3-dimension geometry
coarse or fine mesh

Radial geometry:
Can be 360°
Logarithmic cell radii away from wellbore

Directions: x, y, z
Cell location: i, j, k

Fig. 14.5 Simulator types.

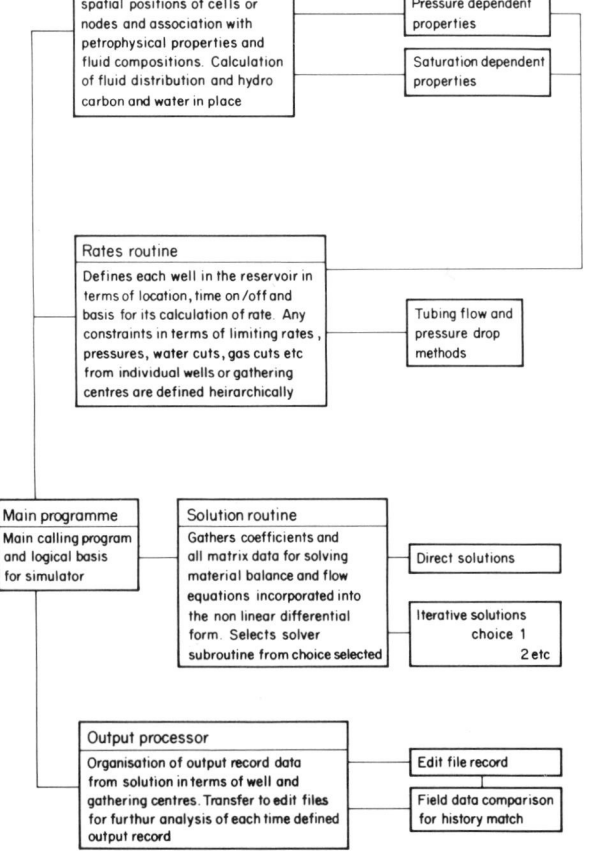

Input processor / HCIIP
Definition of dimensions and spatial positions of cells or nodes and association with petrophysical properties and fluid compositions. Calculation of fluid distribution and hydro carbon and water in place

Pressure dependent properties

Saturation dependent properties

Rates routine
Defines each well in the reservoir in terms of location, time on/off and basis for its calculation of rate. Any constraints in terms of limiting rates, pressures, water cuts, gas cuts etc from individual wells or gathering centres are defined heirarchically

Tubing flow and pressure drop methods

Main programme
Main calling program and logical basis for simulator

Solution routine
Gathers coefficients and all matrix data for solving material balance and flow equations incorporated into the non linear differential form. Selects solver subroutine from choice selected

Direct solutions

Iterative solutions
choice 1
2 etc

Output processor
Organisation of output record data from solution in terms of well and gathering centres. Transfer to edit files for furthur analysis of each time defined output record

Edit file record

Field data comparison for history match

Fig. 14.6 Arrangement of simulator routines.

14 CONCEPTS IN RESERVOIR MODELLING

aquifer character. In addition, preferred well locations and completion intervals can be studied. None of these things necessarily provide a *true* answer, only a comparison with some *base case* set of assumptions.

The reasons why reservoir simulation is so attractive are fairly obvious: since a real reservoir can be produced only once, a series of case studies using a simulator can explore uncertainties in data and resource management options. The validation of a particular simulation in a reservoir cannot be approached until the reservoir has produced for some time (usually several years). At such time a *history match* between reservoir model predicted performance and field observations can lead to improved confidence in future performance predictions. The basis of a history match should include well rates for all fluids, as well as static and dynamic distributions of vertical and lateral pressure gradients. The RFT pressure response of a new well in a producing field provides useful history match data. When miscible or partly miscible processes are being modelled, or condensate/volatile oil reservoirs are modelled, then compositional matches with produced fluids are also needed. Thermal processes and chemical flood matches require even more history match to provide confidence in performance predictions.

In black oil modelling of North Sea reservoirs, the single most important history match parameter is transmissibility, applied to faulted (but not sealing) reservoir intervals, and to vertical restrictions to flow and which result from lithology and facies change in stratified units.

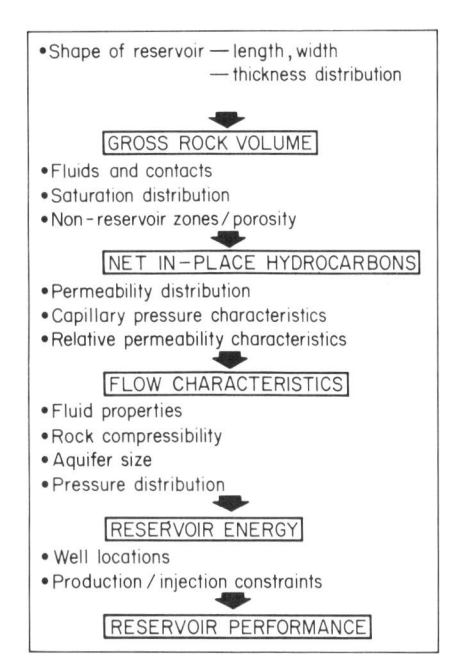

Fig. 14.7 Steps needed to build a reservoir model.

14.5 RESERVOIR DESCRIPTION IN MODELLING

Whether a complex or a simple reservoir model is being applied, a number of steps in analysis and data requirements are common, and are illustrated in Fig. 14.7. The validity of the initialization reservoir model is largely dependent on the geological model and the flow performance is linked to reservoir and production engineering description. Petrophysical

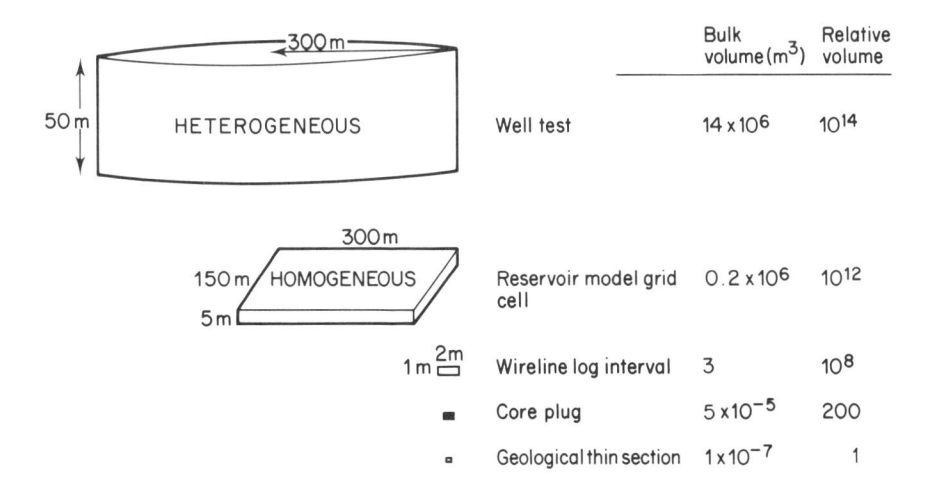

Fig. 14.8 Relative scale of representation.

	Bulk volume (m³)	Relative volume
Well test	14×10^6	10^{14}
Reservoir model grid cell	0.2×10^6	10^{12}
Wireline log interval	3	10^8
Core plug	5×10^{-5}	200
Geological thin section	1×10^{-7}	1

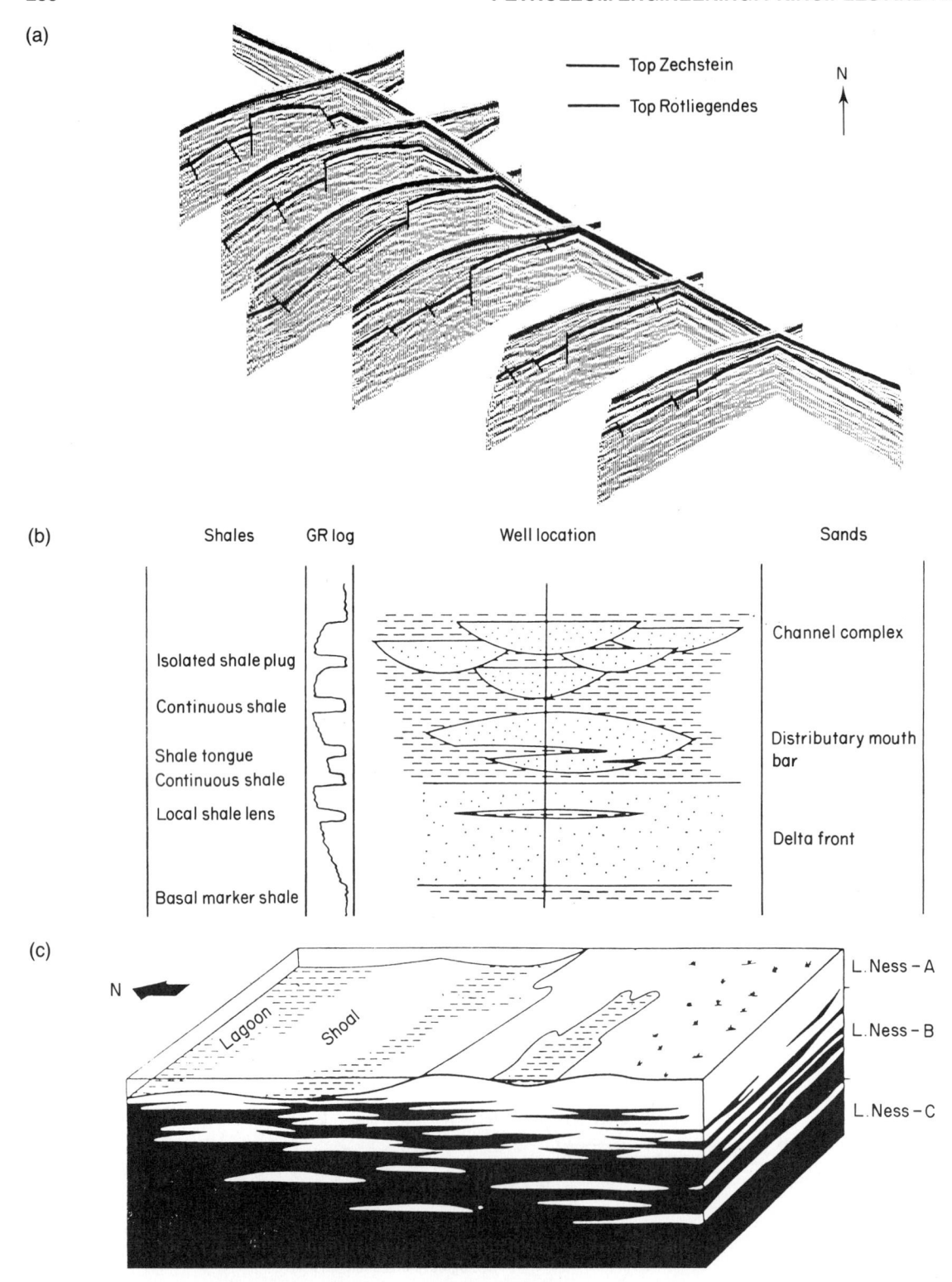

Fig. 14.9 Examples of sandbody continuity. (a) Rotliegendes/Zechstein (North Sea, after[121]); (b) lateral extent of sand and shale bodies (after [12]); (c) conceptual arrangement of sands and shale in Cormorant reservoir (Ness unit) North Sea.

14 CONCEPTS IN RESERVOIR MODELLING

analysis falls between the two since zonation and grid size scale effects must be rationalized and proper attention paid to pore-filling minerals, rock type and saturation representation (Fig. 14.8). The geological model(s) provide the main basis for predicting reservoir description away from direct well control and some discussion of their development and uncertainty is appropriate.

14.5.1 Integration of geological and engineering data

The development of a valid geological model is of necessity an interactive process. The aim is to define vertical and lateral distribution of reservoir and non-reservoir rock in the field (and perhaps in any associated aquifer). This involves recognition of lithology and facies types, correlation between well control and matching with seismic profiles. The particular information available for generating such models tends to be cuttings, cores and log data. Increasingly, development geologists are finding improvements in their geological models result by incorporating pressure analysis data, and some detailed petrophysical interpretation. The particular specialities needed by geologists working with engineers on the generation of the geological framework for reservoir simulation studies are in the fields of sedimentology and palynofacies.

Both these specialities are directed towards an understanding of the sedimentary processes by which a reservoir has formed and the subsequent diagenetic modification of pore space. Core data provides the single most important data base and, through analysis, a palaeogeographical reconstruction of the reservoir may be obtained. Since the vertical record of sediments in a well is related to the lateral processes of sedimentation occurring over a *wide* area at one time, such reconstructions may be made – this adopts the principles defined in Walther's law of facies [83, 84]. The association by analogy of sedimentary processes in ancient systems with observations in modern, active systems leads to an expectation of particular geometries in particular sedimentary forms.

14.5.2 Reservoir geometry and continuity

The typical continuity of reservoir sands and shales is shown in Fig. 14.9. Recognition of appropriate continuity models will allow significant control over recovery of hydrocarbons from reservoirs and influences the type of development scheme employed. In deltaic reservoir models, continuity can vary dramatically according to the depositional environment of individual sandbodies. Deltaic models provide a good example of the influence of conceptual models in reservoir simulation, and will be used here.

The recognition of sandbody type, using core data particularly, is of paramount importance in the development of conceptual geological models and in using them effectively in reservoir simulation. Continuity is usually represented between cells by modification of transmissibility in any dimension or direction (Fig. 14.10).

The extrapolation and correlation of reservoir sands can be severely interrupted by localized faulting subsequent to deposition, which can place reservoir and non-reservoir units against each other. The identification of faults may be apparent from geophysical surveys, from geological hypotheses required to make correlations and from analysis of the pressure versus time behaviour of well tests.

The simplest sub-division of gross deltaic environments is into delta top or delta plain, and delta front. These may then be sub-divided as shown in Fig. 14.11, which can also show characteristic facies

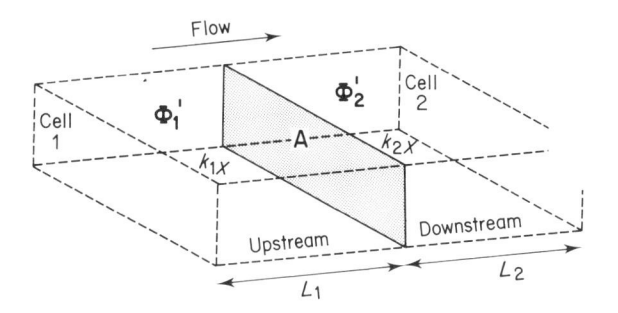

$$T_x (= x\text{-direction transmissibility}) = \frac{2\bar{k}_H \cdot k_r A}{(L_1 + L_2)}$$

Fig. 14.10 Transmissibility in cell models.

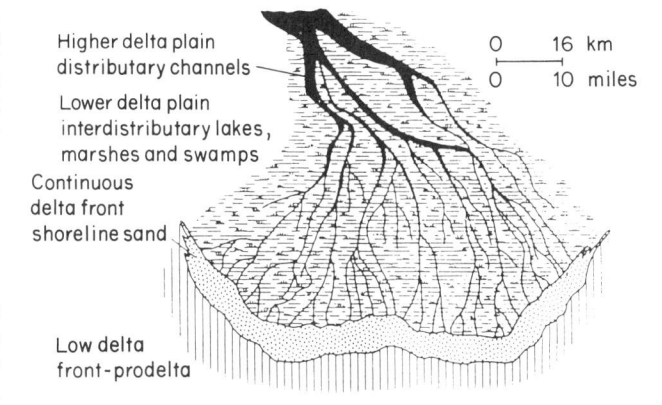

Fig. 14.11 Sub-division of the gross deltaic environment (after [80]).

TABLE 14.2 Delta top characteristics

	Character	Dimensions	Reservoir potential
Crevasse splay	Sand-silty sand interbeds; laminated, current rippled.	Small areal extent: individual sands rarely more than few sq. miles. Few feet thick.	Poor
Natural levee	Sand, silt, clay plant debris. Disturbed bedding.	Linear extent along channel for tens of miles; few feet to tens of feet thick.	Very poor
Marsh/swamp	Silt, clay, coal. Slumps and contorted bedding, wood, plant remains.	May be several square miles (time markers). Thickness from few inches to tens of feet.	None
Channel fill	Trough cross bedding; may show fining upward sequence.	Depends on size of the distributary. Hundreds of feet to tens of miles along length. Few feet to hundreds of feet thick.	Fair to excellent depending on size
Point bar	Well sorted, medium to fine grained sand coarsens downward with increased scale cross bedding.	May be a mile or more wide; several tens of feet thick. Multiple point bars extend many miles.	Excellent

associations for each sub-division. There are complete gradations between the different divisions, and at the upstream end the delta plain passes into a river valley flood plain.

Tables 14.2, 14.3 and 14.4 illustrate reservoir geometry and quality expectations from conceptual models, using deltaic systems as an example.

Most extensive fluvial deposits occupy large areas in lower reaches of rivers and gradually grade into the upper deltaic plain. Sand is deposited in the lower parts of river channels. Overbank sediments (top of point bars, levees, crevasses splays) include much silt and clay.

In fluvial reservoir environments there is a characteristic sequence of sedimentary structures related to increasing current strength. The sedimentary structures can be used to deduce the current as follows:

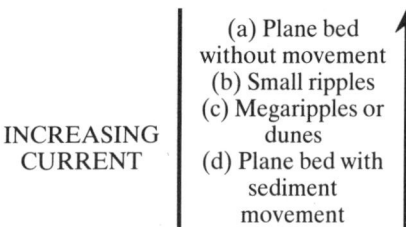

INCREASING CURRENT ↑
(a) Plane bed without movement
(b) Small ripples
(c) Megaripples or dunes
(d) Plane bed with sediment movement
(e) Antidunes
↓ DECREASING CURRENT

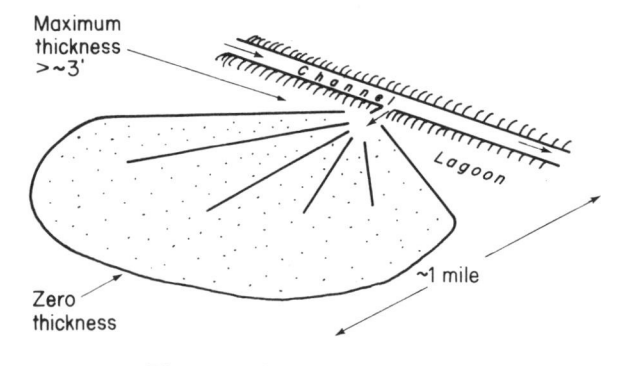

Fig. 14.12 Crevasse splay.

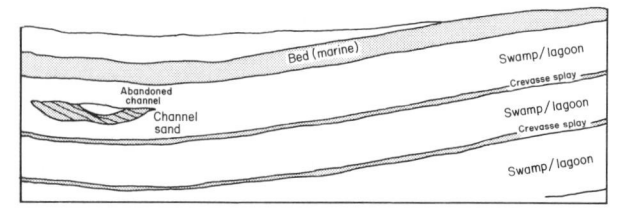

Fig. 14.13 Delta top cross-section.

14 CONCEPTS IN RESERVOIR MODELLING

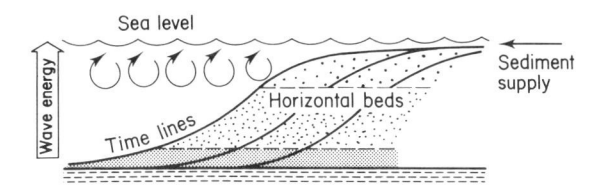

Fig. 14.14 Delta progradation.

TABLE 14.3 Fluvial environments of flood plain and delta top

Single channels	Appearance	Dimensions	Reservoir potential
Braided channel		Typically upstream part of flood plain; thickness up to 50', width ½ mile to 8 miles, length may be hundreds of miles.	Excellent
Meandering channel		Typically downstream part of flood plain, 3–100' thick, multiple sands of meander belt may extend for many miles.	Excellent
Straight channel		Typically deltaic distributaries; 10–60' thick, 100–2000' wide, up to 10 miles long.	Fair to excellent depending on size
Coalesced channels	Vertical stacking (Map: belt) / (Map: continuous sheet) / Isolated stacking (Map: discontinuous sheet)		Reservoir quality depends on size and stacking pattern.

TABLE 14.4 Delta front characteristics

	Character	Dimensions	Reservoir potential
Distributory mouth bar	Well sorted sand grading downwards and outwards into finer grained sediment.	Typically up to 80' thick and 2 miles wide. Length up to 15 miles in cases of progradational growth. May be smaller in shallower water.	Good
Interdistributary bay	Silt and shale with sand lenses and laminae, often burrowed.	May be hundreds of square miles. Thickness from few inches to tens of feet.	Very poor
Barrier beach	Clean well sorted sand at top grading into silty sand at base; parallel laminations and local low-angle cross beds. Possibly cut by tidal channels and forming associated ebb-tide deltas.	Sheet-like cover of area of delta lobe. Tens of feet thick. Possible direction in tidal channel permeability.	Excellent
Shore face	Well sorted sand	Smooth seaward and irregular landward margins. Few miles wide, tens to hundreds of miles long; may prograde and so greatly increase width. Tens of feet thick.	Fair to good
Basal marine sand	Continuous marker shale association	Miles in length and width. Tens to hundreds of feet thick.	None

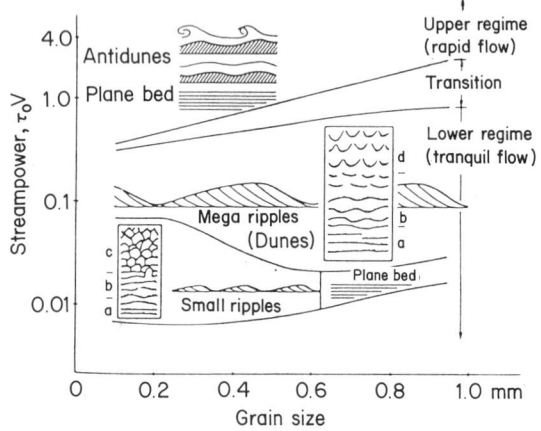

Fig. 14.15 Bedforms in relation to grain size and stream power (after [81, 82]).

Figure 14.15 depicts graphically various bedforms and their relationship to grain size and stream power. Under the influence of flowing water, sediment movement starts in a non-cohesive bed. At low energy conditions, a few grains start rolling and may produce horizontal laminations if sufficient sediment is available and the process continues for a certain length of time.

Observation and correlation from wells can then lead to an expected, consistent model to explain reservoir distribution at a given time or horizon. Such a model is shown in Fig. 14.16 for a deltaic system[17]. In Fig. 14.17 the reservoir character of channel and bar sands are contrasted. The detailed shapes are based on energy for sorting sediments and result from consideration of depositional flow regimes, tidal range and current directions. The resultant models often contain an expectation of vertical and lateral zonation of reservoir properties, which are then related to quantitative estimates of porosity, permeability and saturation. In this exercise the proposed geological zonation must be compared with petrophysical zonation based largely on porosity and permeability (poro-perm). The validity of poro-perm data from core analysis and porosity from log analysis depends very much on recognition of the effects of clay minerals in the pore space and lithological variations from the bulk reservoir properties. Interaction with the geological observations based on X-ray analysis and scanning electron microscope (SEM) studies often helps explain a basis for diagenetic change in pore character, which can be applied as a function of

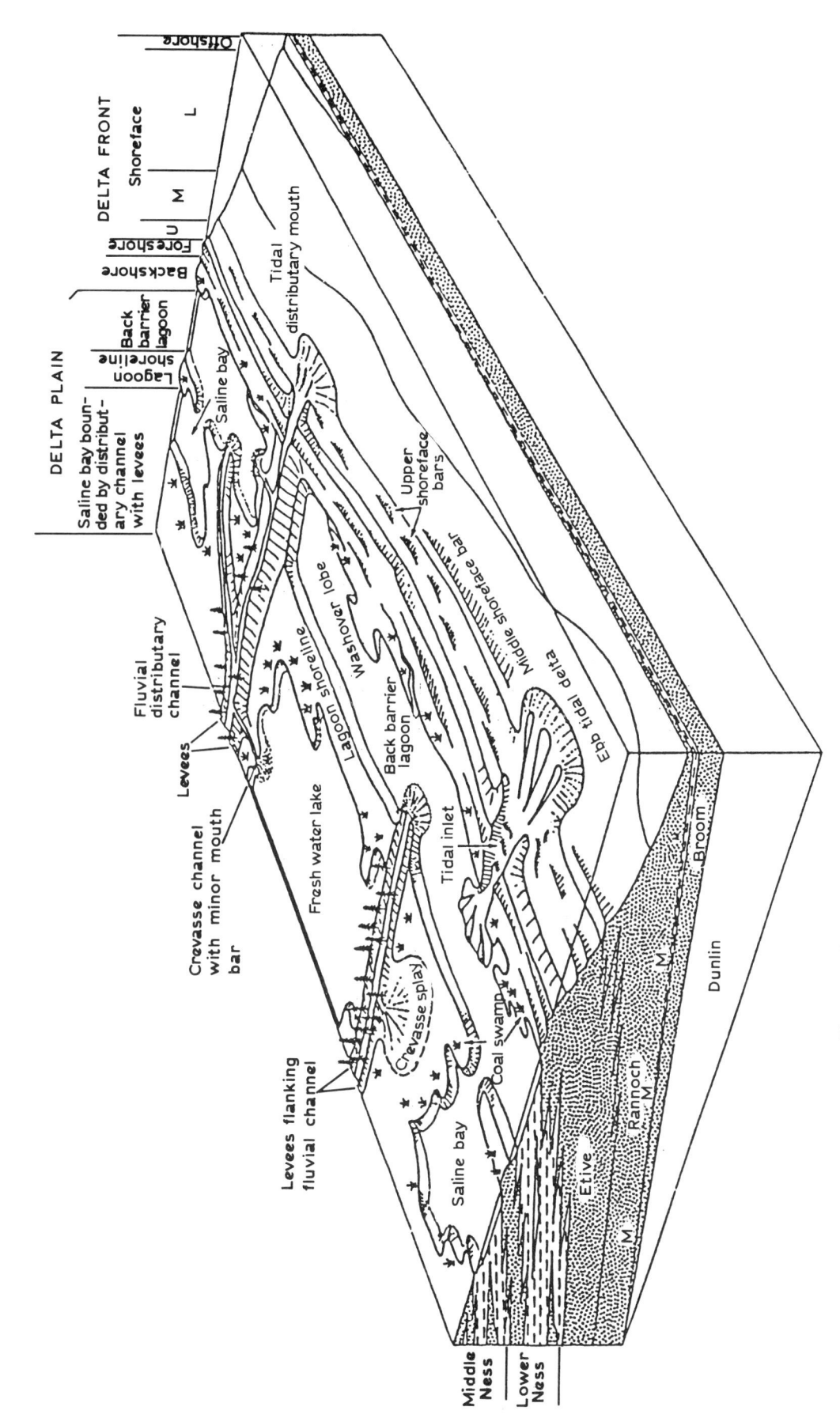

Fig. 14.16 Palaeogeographic representation of complex environments (after [17]).

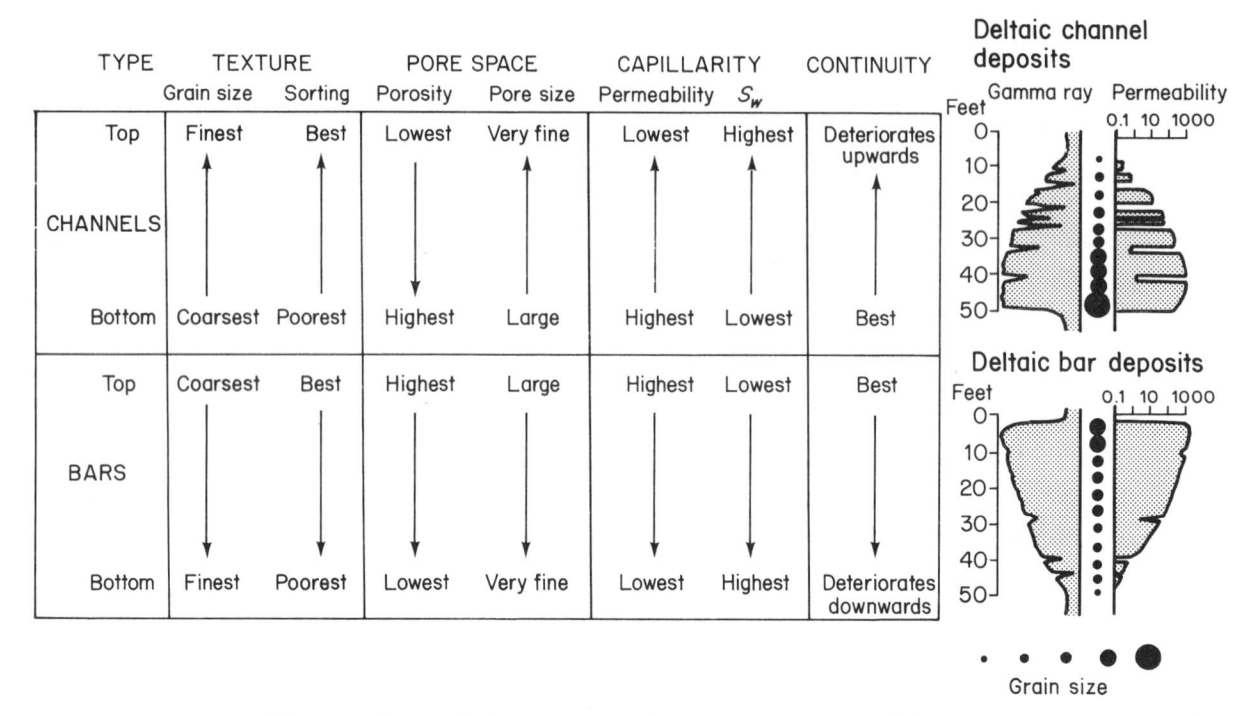

TYPE	TEXTURE		PORE SPACE		CAPILLARITY		CONTINUITY
	Grain size	Sorting	Porosity	Pore size	Permeability	S_w	
CHANNELS Top	Finest	Best	Lowest	Very fine	Lowest	Highest	Deteriorates upwards
CHANNELS Bottom	Coarsest	Poorest	Highest	Large	Highest	Lowest	Best
BARS Top	Coarsest	Best	Highest	Large	Highest	Lowest	Best
BARS Bottom	Finest	Poorest	Lowest	Very fine	Lowest	Highest	Deteriorates downwards

Fig. 14.17 Reservoir characteristics of channel and bar sands (after [76]).

depth, position or saturation in a reservoir model. It should be expected that zones exhibiting diagenetic damage will have different irreducible saturations or relative permeabilities from other zones.

One of the most difficult stages in constructing a reservoir model is compromising scales of observation of geological and petrophysical properties with the scale of model grid cells (Fig. 14.8). For reasons of cost and computing time, the minimization of the number of grid cells used to define a reservoir is often required. The averaging of poro-perm saturation data in volumetric calculation is of less significance than the representation of flow properties by pseudo-functions in stratified reservoir intervals [34] for dynamic reservoir performance calculations. Reservoir zones should represent regions of differing flow properties, primarily dependent on net effective permeability thickness ($k_e h_N$). Figure 14.18 indicates a progression in reservoir description within a region of a cross-section model – any extension to include areal geometry leads to greater complexity in effective transmissibility representation. The upper diagram represents a sandy sequence containing a laterally discontinuous shale and a continuous thin micaceous stratum. The sands are distinguishable from each other by sedimentary facies description but do not have dramatically different permeability contrast in the bedding plane direction. A reservoir simulation model must recognize the role of vertical communication between the sands in controlling saturation distribution and frontal movement under dynamic displacement conditions. A fine grid model recognizes the boundaries between sand units which might control cross-flow (Fig. 14.18 (b)). The boundary between sands 3 and 4 has been removed because permeability thickness contrast was less than a few times, but the boundary between sands 2 and 3 is retained. The micaceous zone is removed from the pore volume of the model but its effect retained as a multiplication factor on the harmonic average vertical permeability calculated between sands 3–4 and 5. A similar vertical transmissibility multiplyer approach is used between sands 1 and 2 to account for the shale wedge – a factor of zero indicates that the shale is sealing and unity indicates that cross-flow is controlled by sand–sand contact. This model has the same fluid content and pore volume as the geological model and is used to study the sensitivity to reservoir description of dynamic saturation distributions.

A coarse grid model of this region (Fig. 14.18 (c)) can be developed to reduce computer time in field performance prediction studies. The model again contains the same total fluid and pore volume as the geological model. Pseudo-relative permeability and capillary pressure functions of pore volume weighted

14 CONCEPTS IN RESERVOIR MODELLING

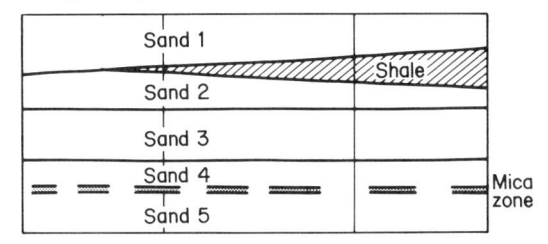

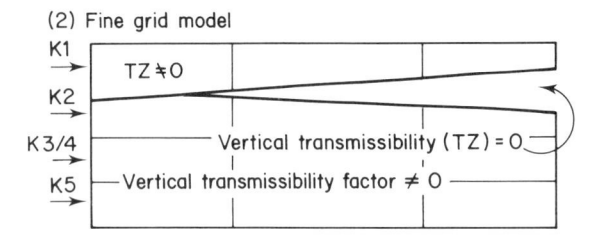

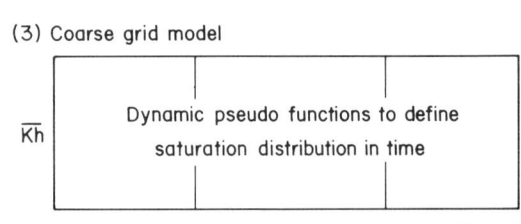

Fig. 14.18 Zonation in cross-section modelling. (a) Geological representation, (b) fine grid model, (c) coarse grid model.

average saturation, generated from results of the fine grid model, are used in the coarse grid model to demonstrate the same displacement behaviour. The method of Kyte and Berry [34] is most frequently adopted for this purpose. Well functions may also be used to represent partial penetration and local radial flow coning character [33]. The characterization of a well productivity index in reservoir models requires modification from analytical forms since in numerical models there is no concept of saturation pressure gradient within a grid cell. Potential gradients exist as step changes between cells and serve to move fluids across intercell boundaries. Peaceman [35] has shown that a semi-steady state productivity index in equidimensional (D_x) grid cells can be represented as follows:

$$PI = \frac{q_o}{\bar{P} - P_{wf}} = \frac{\text{constant} \cdot k_o h}{\mu_o B_o \left[\log_e \frac{0.2(\Delta x)}{r_w} - \frac{(0.2\Delta x)^2}{2 r_e^2} \right]}$$

where $r_w = r_w e^{-s}$. Subsequent work has shown that for cells of different sizes a good approximation over a range of values can be obtained from $\Delta x = (\Delta x_1 \cdot \Delta x_2)^{0.5}$

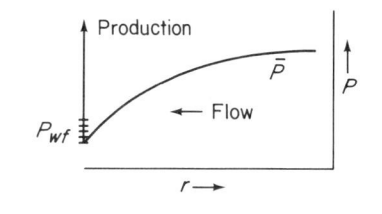

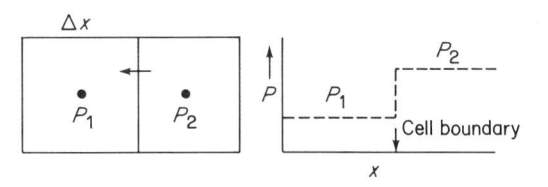

Fig. 14.19 Contrast in pressure representation in analytical and simulator calculations.

The problems of cross-flow and inflow to wellbores from stratified systems, together with application of directional relative permeabilities and the representation of the flow of fluids across faults and partially displaced layers, has been discussed in the context of reservoir simulation by Smith [56].

14.5.3 Uncertainty in reservoir model description

The minimization of uncertainty in reservoir simulation is time dependent and occurs as more reservoir simulation performance prediction is confirmed by historical field measurement. The *history* matching process is not, however, unique since several variables could be modified to obtain a match. Modifications are required that are reasonable and can be defended on both geological and engineering grounds. Increasingly the history match procedure involves multidisciplinary teams of reservoir engineers, petrophysicists and development geologists, together with geophysicists as necessary. Figure 4.20 shows a cross-section and areal grid representation in the Statfjord field, of a Middle Jurassic, Brent reservoir used in early development planning and controlled with data from seven exploration wells. A three-dimensional model on this grid base would contain some 11400 cells. In the areal model, well control exists in 7 cells of the total 760, about 1%. The conceptual model of the field provides the basis for inferring the properties of the other 99%.

In an individual stratum, the mapping of reservoir characteristics, such as areal extent, thickness, porosity, net:gross variation and permeability, involves use of conceptual models, the validity of which emerge during reservoir production. As has been shown by Archer [17], reservoir mapping and

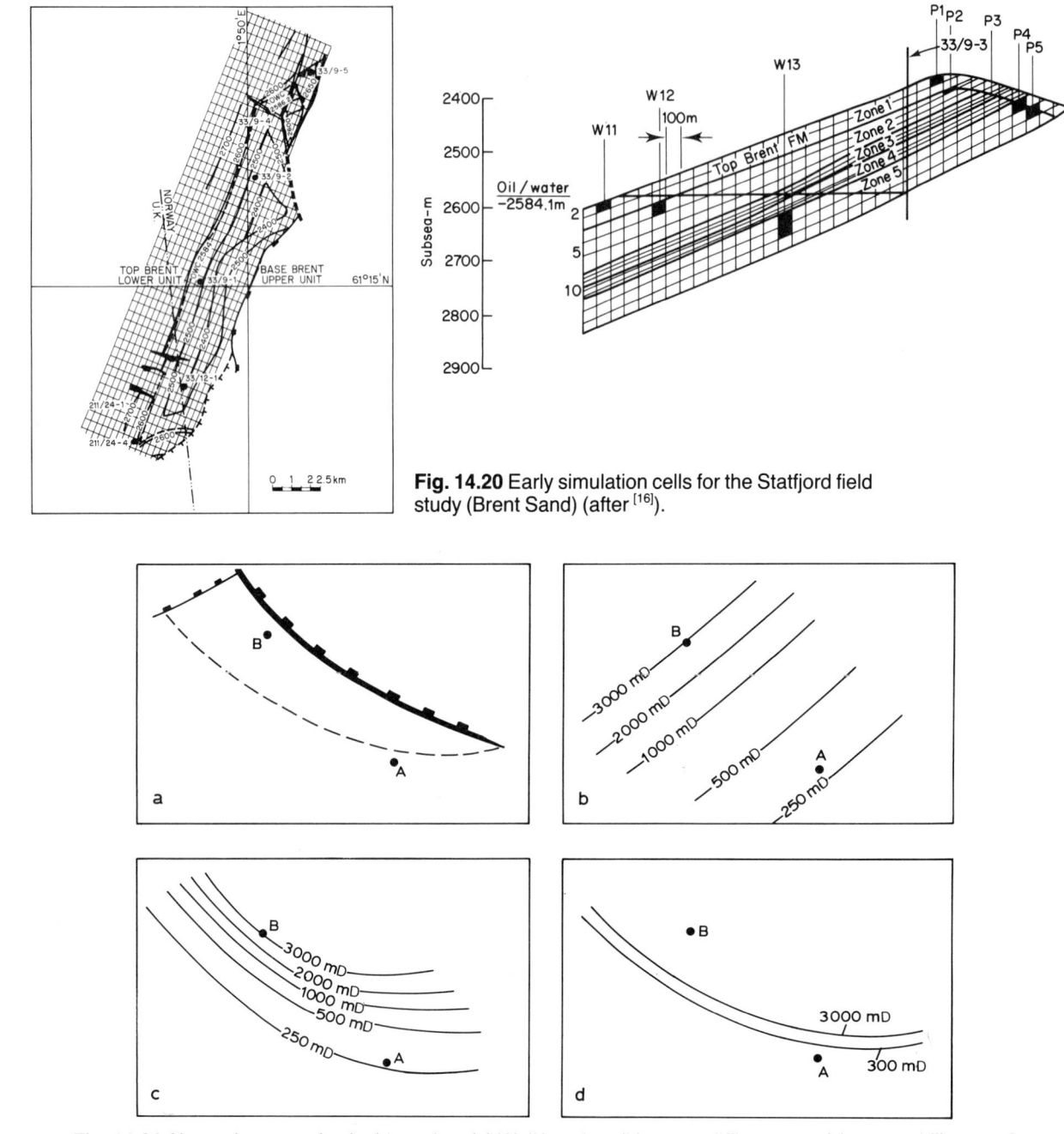

246   PETROLEUM ENGINEERING: PRINCIPLES AND PRACTICE

Fig. 14.20 Early simulation cells for the Statfjord field study (Brent Sand) (after [16]).

Fig. 14.21 Alternative petrophysical mapping. (a) Well location, (b) permeability map 1, (c) permeability map 2, (d) permeability map 3.

cross-section interpretation can be varied even with a given control data set (Figs 14.21 and 14.22). Development planning calls for flexibility in design so that early key wells can be used to help differentiate between model possibilities.

The greatest uncertainties in black oil reservoir modelling tend to be in appropriate zonation, inter-zone transmissibility and in saturation dependent relative permeability terms. These uncertainties can be explored in terms of their impact on proposed development by sensitivity studies, the results of which may point to the need for key well

14 CONCEPTS IN RESERVOIR MODELLING

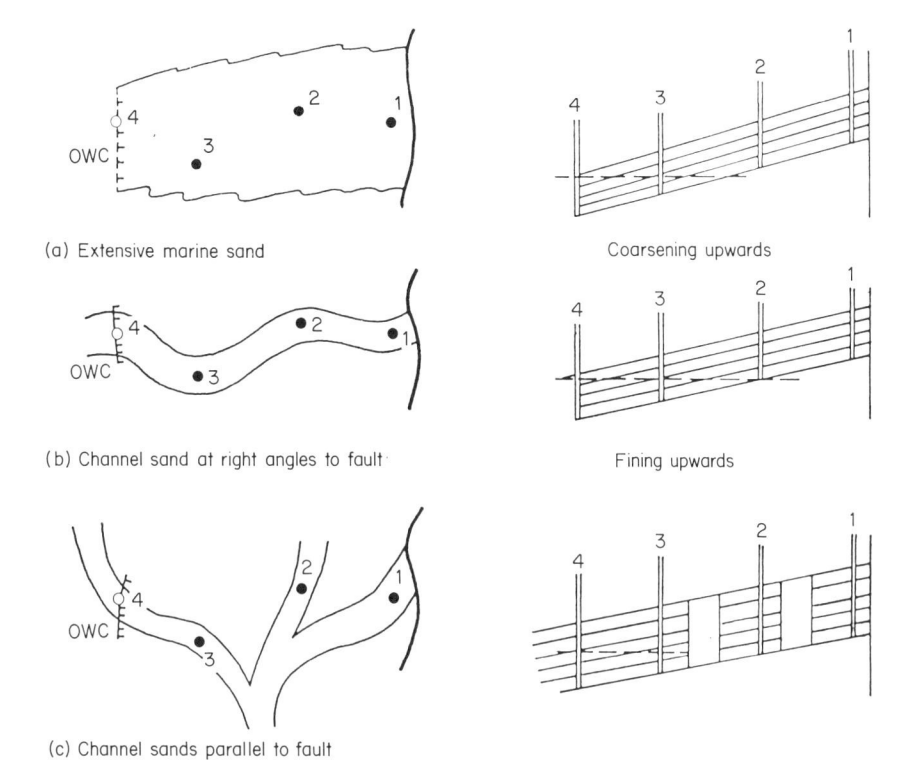

Fig. 14.22 Alternative sand models giving different performance predictions. (a) Extensive marine sand, (b) channel sand at right angles to fault, (c) channel sands parallel to fault.

data. History matching measured performance (pressure distribution and producing fluid ratios) with reservoir simulation is the only way to validate a model. Examples, which are not unique, are shown in Figs 14.23 and 14.24.

In gas condensate and volatile oil reservoirs the greatest uncertainties in addition to those mentioned for black oils are in valid fluid properties as functions of pressure and temperature. Sampling in these reservoirs at bottom-hole conditions is generally unreliable and in these particular circumstances recombined surface samples may be preferred.

In heavy oil reservoirs fluid sampling is also difficult and the reservoir fluid may not flow – composition may sometimes then be obtained from extracting core. The interpretation of viscosity at proposed reservoir development conditions becomes a particular uncertainty.

Schemes for enhanced oil recovery which involve miscible processes and chemical processes often have great uncertainty attached to the modelling of physical mechanisms for displacement. This is particularly true when multicontact or partial miscible processes are considered and for adsorption and microemulsion formation in surfactant processes. In many of these processes there is significant numerical dispersion which makes displacement front tracking difficult and which may cause uncertainty in performance predictions.

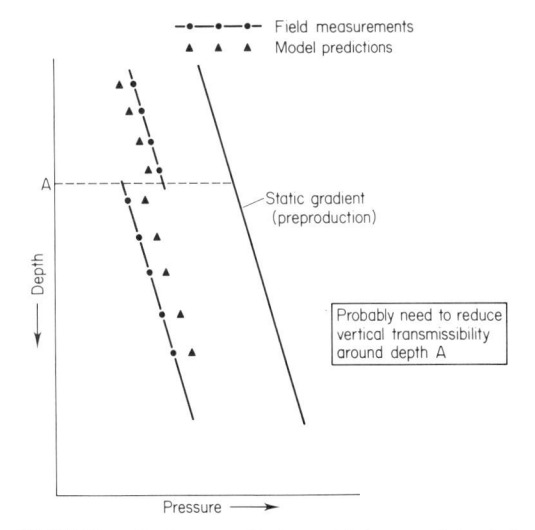

Fig. 14.23 Results from preliminary history match at given well. (Could vary: vertical permeability; horizontal permeability; relative permeabilities; net pay volume.)

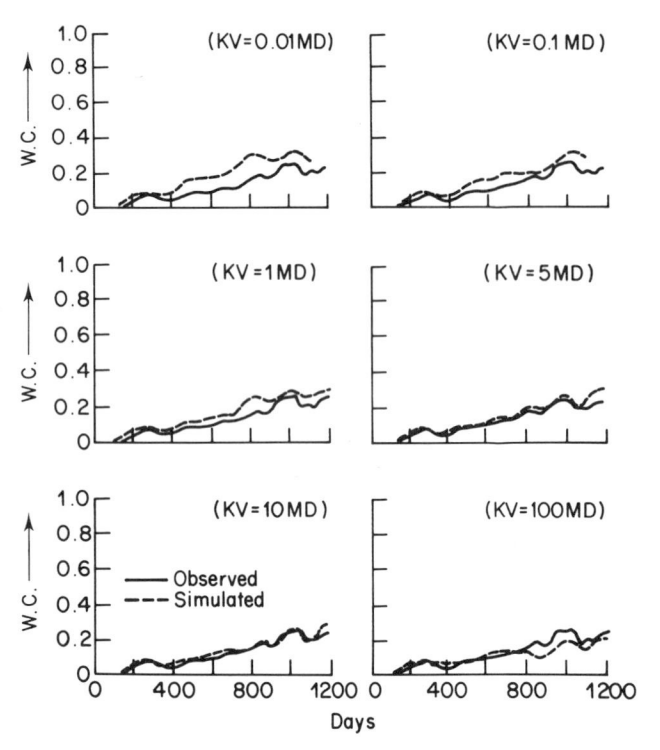

Fig. 14.24 History match of water cut development in Etive/Rannoch sand system by assumption of vertical permeability between sands (after [118]).

14.6 APPLICATION OF RESERVOIR MODELS IN FIELD DEVELOPMENT

The decision base for reservoir development is both technical and economic[120]. Onshore development should proceed stepwise and is often unconstrained by development well locations. It should in general be cheaper than any offshore project of comparable reserves. In this part of the chapter we shall concentrate on offshore field development as uncertainty in reservoir characteristics is more significant.

14.6.1 Application Sequence

Perhaps the sequence of field development considerations follows these steps:

(1) Exploration drilling location chosen on basis of potential structure, mature source rocks, reservoir rocks, trap and migration path. Well drilled – discovery, tested and indication of commercial productivity index PI.
(2) Appraisal wells to delineate structure and establish fluid contacts. Core to define sedimentology and provide basis for reservoir recovery mechanisms. Fluid samples for *PVT* properties. Well tests/core log data rationalization. Pressure regime and aquifer contribution assessed.
(3) Development of preliminary geological model. Petrophysical data used to define porosity and saturation vertical and lateral distribution. Reservoir engineering data added for volumetric and dynamic analyses. Preliminary estimation of recovery factors for potential recovery processes. Analytical methods used to define stability of displacement. Assessment of vertical and lateral heterogeneity.
(4) Preliminary economic analysis based on notional cost estimates and value of products. Assume a peak production rate for oil of, for instance, 10% of recoverable reserve per annum and a plateau duration such that at least 30–40% of the recoverable reserve are recovered at peak rate, and decline is based on 10–20% per annum (depends strongly on heterogeneity). Estimate well requirements based on semi-steady state completion PI. For gas wells consider plateau rate as fraction of reserve per annum according to typical contracts, i.e. for notional 20-year life of reservoirs containing more than 1 TCF, assume peak rate ACQ (annual contract quantity) is 0.05 × recoverable reserve per year until 60% of the recoverable reserve has been produced. Facility requirements should be designed for day rate offtakes where SDC is the seller's delivery capacity – a seasonal factor having a maximum value about 1.7.
(5) Develop more detail in geological model and consider mapping and correlation options. Define basis for net pay and rationalize geological and petrophysical definitions in zonation. Complete petrophysical analysis on standardized basis. Define uncertainties and represent net thickness, porosity, saturation and field area (limits) in probability distributions for each zone.
(6) Represent stratified reservoir character in permeability contrast distributions. Check zonation using capillary pressure character and irreducible saturations. Develop relative permeability data for zones and regions in the field. Develop transmissibility modification factor maps for all interlayers.
(7) Represent *PVT* data regionally if appropriate.
(8) Run Monte Carlo type volumetric analyses. Define hydrocarbon in place level for simulation models.
(9) Select cross-sections of the reservoir (along dip

14 CONCEPTS IN RESERVOIR MODELLING

axes) to ensure gravity effects on flow represented properly. Represent layer nature transmissibility and reservoir permeability as sensitivity parameters. Examine vertical sweep efficiency in analytical and reservoir simulation calculations for different recovery mechanisms and well locations/completion intervals. Determine character of pseudo-functions for use in coarser grid models. Check in 1–D mode.

(10) Run radial simulation models to evaluate coning potential or to calculate well functions in terms of saturation in some defined region.

(11) Run three-dimensional sector models to study both areal and vertical sweep efficiency and sensitivity of options to reservoir description uncertainties.

(12) Extend to coarser grid full reservoir model in three dimensions for the recovery mechanism selected. Study effect of well locations, rates, completion and recompletion intervals and production/operation constraints. Production profiles and facility implications used in economic analyses are evaluated.

(13) Refine field plan and consider effects of tubing flow constraints, pumps and separators and pass results to project management with recommendations. Define preferred well development sequence and design early data collection program. Plan for model updates and history matching. Represent recoverable reserves as probabilistic distribution. The use of a number of early development wells and a decline in field pressure may allow development of refined correlations, such as shown in Fig. 14.25.

14.6.2 Recent Field Studies

The petroleum engineering literature contains many examples of field studies using reservoir models (see reference list). A particularly constructive example is the Stiles and Bobeck [55] account of the Fulmer pre-development simulation study in Blocks 30/16 and 30/11-b in the UKCS North Sea. The Fulmer

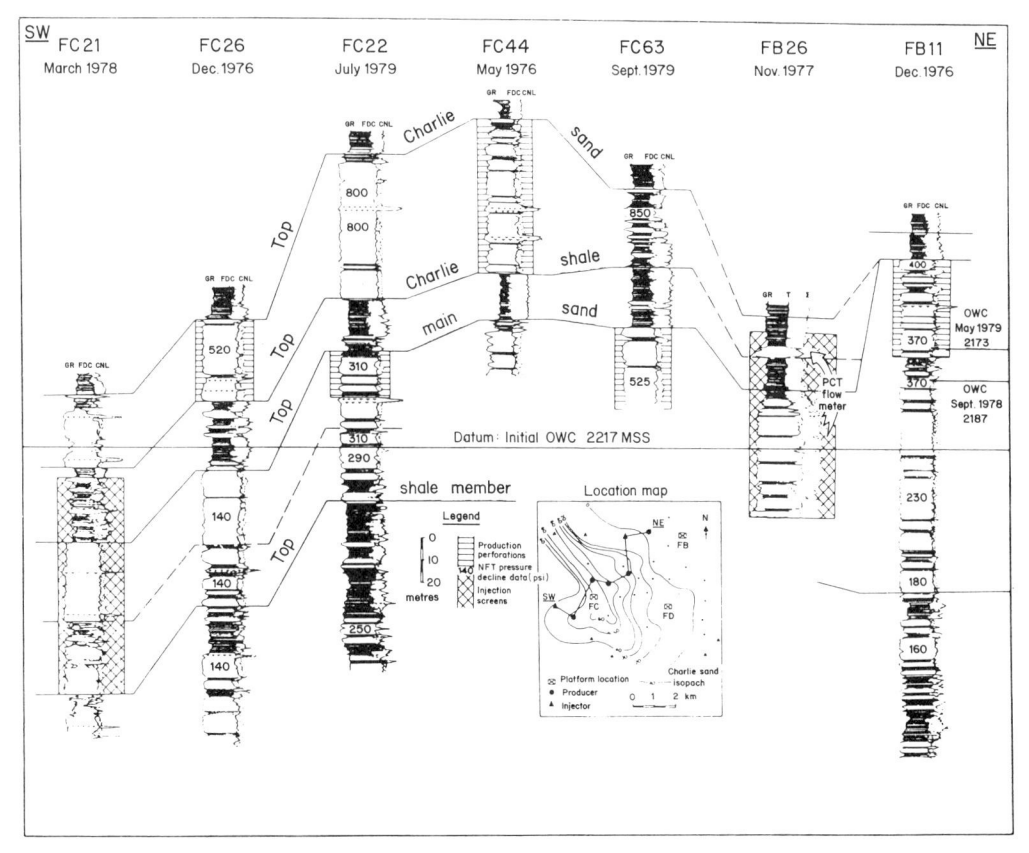

Fig. 14.25 Reservoir zonation in the Forties reservoir showing use of geological and petrophysical correlation enhanced with RFT and production logging data. (After [122].)

field is a stratified Upper Jurassic shallow marine sandstone in a faulted anticlinal structure, and contains 41° API undersaturated oil. The datum depth and pressure is 10 000 ft SS and 5700 psi. The reported value of oil in place is 824×10^6 STB. The field is situated in the central sector of the North Sea about 170 miles from Aberdeen and water depths are some 275 ft. The development plan preferred for the reservoir employs flank water injection and temporary gas storage in the crestal region. Reservoir studies were therefore directed at both gas and water displacement of oil and coning potential.

Cross-section and single well studies were used in addition to full field studies. An early production system was evaluated and installed. This employed four production wells drilled through a subsea template prior to installation of the main platform. The template well provided early information on which to improve the geological and reservoir engineering models. Figure 14.26 shows the structure map with west and north cross-sections indicated, and Fig. 14.27 shows the general geological cross-section. Figure 14.28 shows the grid pattern for a gas coning model. Fig. 14.29 shows the north flank cross-section model and Fig. 14.30 the west flank models.

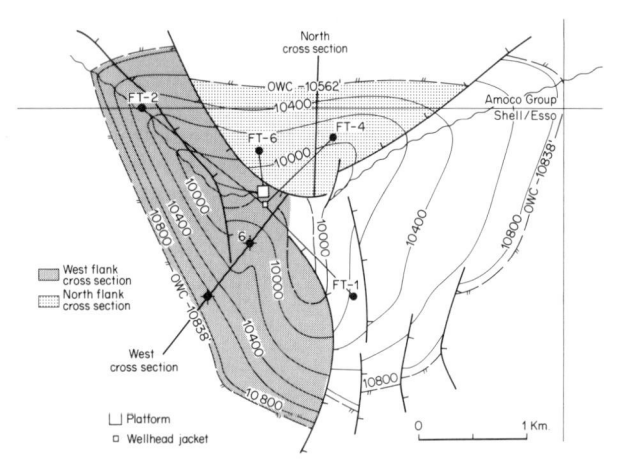

Fig. 14.26 Fulmar structure map showing cross-section locations. (After [55].)

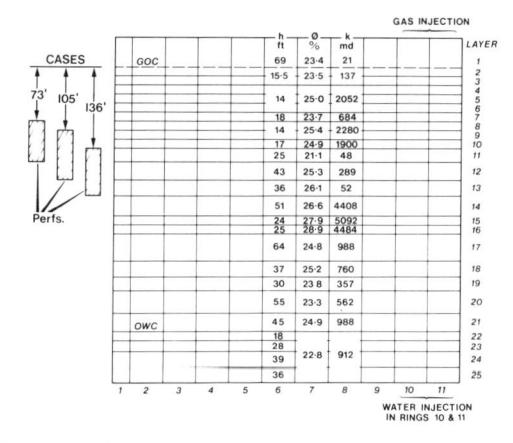

Fig. 14.28 Fulmar gas coning radial model description.

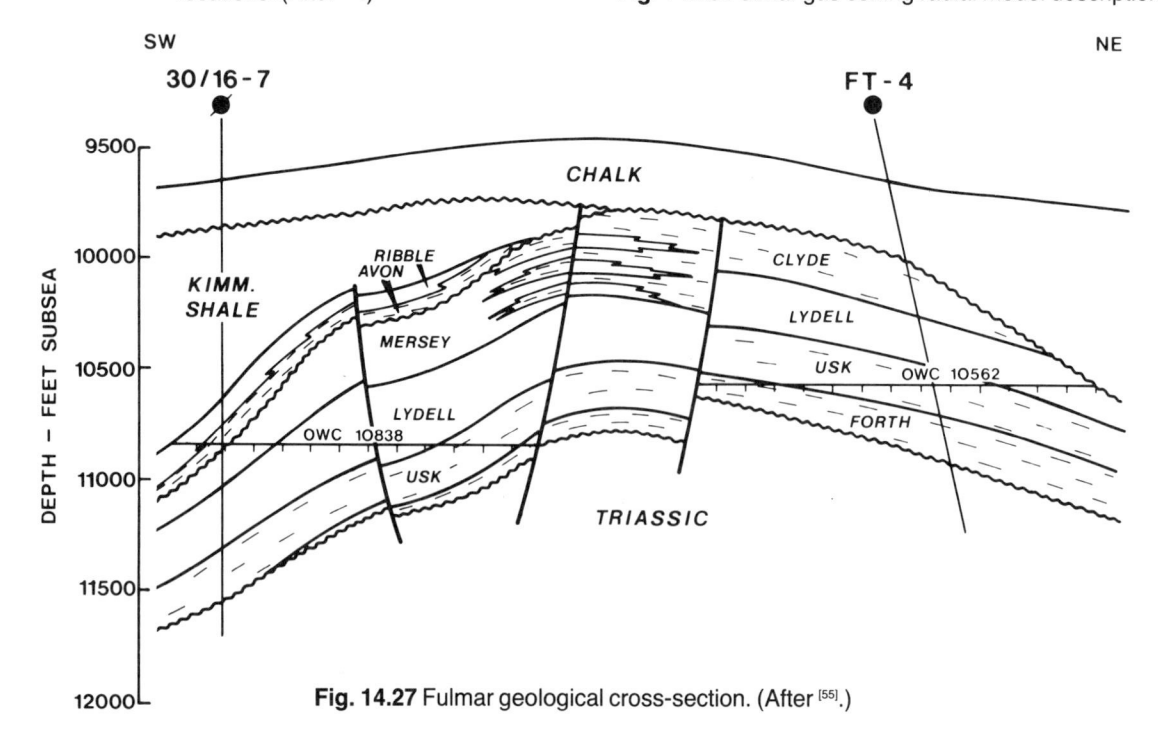

Fig. 14.27 Fulmar geological cross-section. (After [55].)

14 CONCEPTS IN RESERVOIR MODELLING

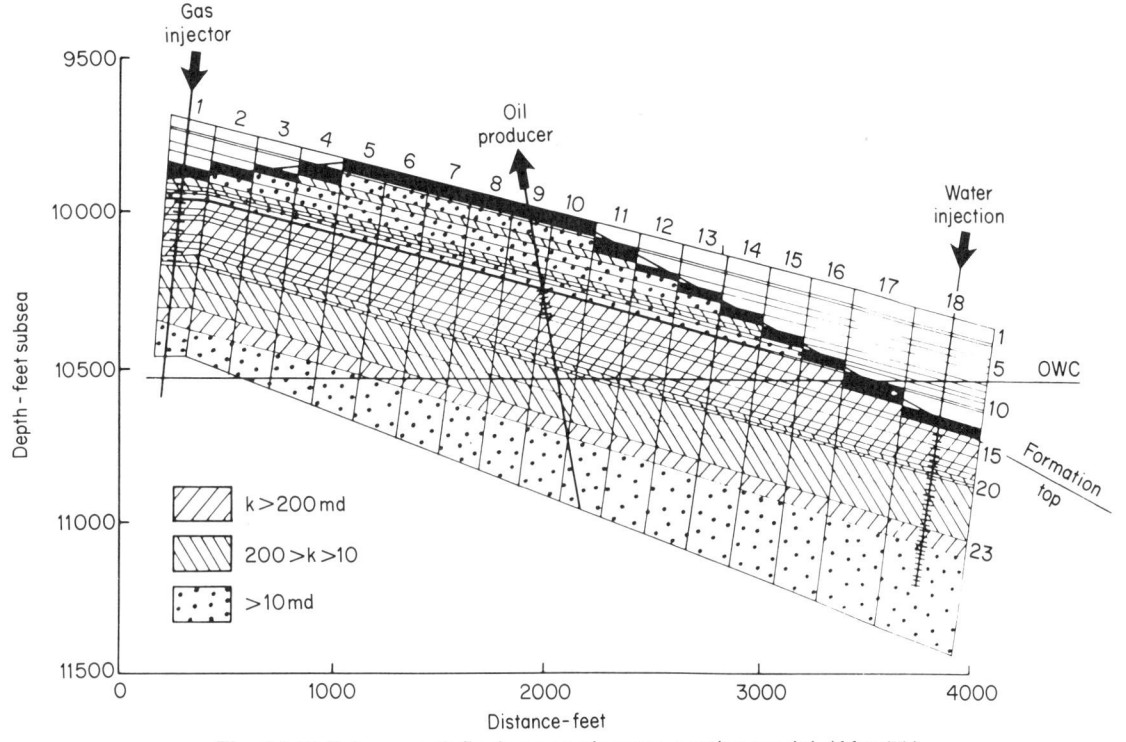

Fig. 14.29 Fulmar north flank reservoir cross-section model. (After [55].)

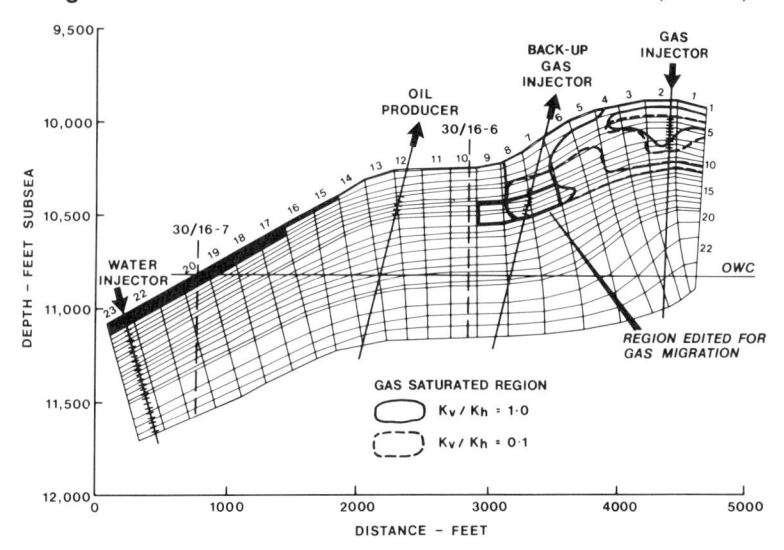

Fig. 14.30 Gas migration path predicted in west flank cross-section model. (After [55].)

The simulation and reservoir engineering study recognised that gravity forces would play an important role at planned reservoir withdrawal rates and that gas override and water underrun might be reduced. The apparent lack of restriction to vertical flow however gave concern about coning and resulted in recommendations about completion locations for wells. It also showed that long term gas storage in the reservoir crest was not a good plan but that gas could be injected temporarily into the oil column where it should rapidly migrate upwards. The Petroleum Engineering literature contains numerous examples of reservoir simulation studies at all stages of field exploitation and the reader is referred to the reference list for further case studies.

References

[1] Craig, F.F.
Reservoir Engineering Aspects of Waterflooding, SPE Monograph, Vol. 3 (1971).

[2] Buckley, S.E. and Leverett, M.C.
Mechanism of fluid displacements in sand, *Trans. AIME* **146** (1942), 107.

[3] Rapoport, L.A. and Leas, W.J.
Properties of linear waterfloods, *Trans. AIME* **189** (1953), 139.

[4] Terwilliger, P.L. *et al.*
An experimental and theoretical investigation of gravity drainage, *Trans. AIME* **192** (1951), 285.

[5] Levine, J.S.
Displacement experiments in a consolidated porous system, *Trans. AIME* **201** (1954), 55.

[6] Johnson, E.F., Bossler, D.P. and Naumann, V.O.
Calculation of relative permeability from displacement experiments, *Trans. AIME* **216** (1959), 370.

[7] Croes, G.A. and Schwarz, N.
Dimensionally scaled experiments and theories on the water drive process, *Trans. AIME* **204** (1955), 35.

[8] Engelberts, W.L. and Klinkenberg, L.J.
Laboratory experiments on displacement of oil by water from rocks of granular materials, *Proc. 3rd World Pet. Cong. II* (1951), 544.

[9] van Meurs, P.
The use of transport three dimensional models for studying the mechanisms of flow processes in oil reservoirs, *Trans. AIME* **210** (1957), 295.

[10] Egbogah, E.O. and Dawe, R.A.
Microvisual studies of size distribution of oil droplets in porous media, *Bull. Can. Pet. Geol.* **28** (June 1980), 200.

[11] Bonnet, J. and Lenormand, R.
Constructing micromodels for the study of multiphase flow in porous media, *Rev. d. IFP* **42** (1977), 477.

[12] Archer, J.S. and Hancock, N.J.
An appreciation of Middle Jurassic Brent sand reservoir features by analogy with Yorkshire coast outcrops, EUR 197, *Proc. Europ. Pet. Conf.* (1980), 501, Vol. 1.

[13] Archer, J.S. and Wilson, D.C.
Reservoir simulation in the development of North Sea oil fields, *The Chemical Engineer* (July 1978), 565.

[14] Archer, J.S. and Hurst, A.H.
The role of clay mineralogy on reservoir description in petroleum engineering, *Proc. Clay Mineralogy Conf.,* Cambridge (1985).

[15] Archer, J.S., Keith, D.R. and Letkeman, J.P.
Application of reservoir simulation models in the development of North Sea reservoirs, SPE 5285, *Proc. Europ. Symp. SPE* (1975).

[16] Archer, J.S. and Wong, S.W.
Interpretation of laboratory waterflood data by use of a reservoir simulator, *SPEJ* (Dec. 1973), 343.

[17] Archer, J.S.
Reservoir definition and characterisation for analysis and simulation, *Proc. 11th World Pet. Cong.,* London (1983), PD 6 (1).

[18] Havlena, D.
Interpretation, averaging and use of the basic geological engineering data, *J. Can. Pet. Tech.* **6** (1967), 236.

[19] Bishop, K.A., Breit, V.S., Oreen, D.W. and McElhiney, J.C.
The application of sensitivity analysis to reservoir simulation, SPE 6102, *Proc. 51st Ann. Fall Mtg.* (1976).

[20] Killough, J.E.
Reservoir simulation with history dependent saturation functions, *Trans. AIME* **261** (1976), 37.

[21] Slater, G.E. and Durrer, E.J.
Adjustment of reservoir simulation models to match field performance, *Trans. AIME* **251** (1971), 295.

[22] Thomas, L.K., Lumpkin, W.B. and Reheis, G.M.
Reservoir simulation of variable bubble point problems, *SPEJ* (1976), 10.

[23] Toronyi, R.M. and Ali, S.M.F.
Determining interblock transmissibility in reservoir simulators, *JPT* (1974), 77.

[24] Wilson, D.C., Tan, T.C. and Casinader, P.C.
Control of numerical dispersion in compositional simulation, *Proc. 1st Europ. Symp. EOR,* Bournemouth (1981), 425.

[25] Mrosovsky, I., Wong, J.Y. and Lampe, H.W.
Construction of a large field simulator on a vector computer, *JPT* (Dec. 1980), 2253.

[26] Wadman, D.H., Lamprecht, D.E. and Mrosovsky, I.
Joint geologic/engineering analysis of the Sadlerachit reservoir, Prudhoe Bay field, *JPT* (July 1979), 933.

[27] Ramey, H.J.
Commentary on the terms 'transmissibility' and 'storage', *JPT* (March 1975), 294.

[28] Nobles, M.A.
Using Computers to Solve Reservoir Engineering Problems, Gulf Pub. (1984).

[29] van Rosenberg, D.U.
Methods for Numerical Solution of Partial Differential Equations, Farrar (1969).

[30] Critchlow, H.
Modern Reservoir Engineering : a Simulation Approach, Prentice Hall (1977).

[31] Society of Petroleum Engineers
Proceedings of the SPE Symposiums of Reservoir Simulation, (1973, 1976, 1982, 1985).

[32] RRI/ERC
The Brent Sand in the N. Viking Graben, N. Sea – a Sedimentological and Reservoir Engineering Study, Robertson Research Intl/ERC Energy Resource Consultants Ltd., Llandudno, N. Wales (1980).

[33] Chappelear, J.E. and Hirasaki, G.J.
A model of oil water coning for 2D areal reservoir simulation, *SPEJ* (April 1976), 65.

[34] Kyte, J.R. and Berry, D.W.
New pseudo functions to control numerical dispersion, *SPEJ* (Aug. 1975), 269.

[35] Peaceman, D.W.
Interpretation of wellblock pressure in numerical simulation, *SPEJ* (June 1978), 183.

[36] Hirasaki, G.J. and O'Dell, P.M.
Representation of reservoir geometry for numerical simulation, *Trans. SPE* **249** (1970), 393.

[37] Robertson, G.E. and Woo, P.T.
Grid orientation effects and use of orthogonal curvilinear coordinates in reservoir simulation, SPE 6100, *Proc. 51st Ann. Fall Mtg. SPE* (1976).

[38] Rose, W.
A note on the role played by sediment bedding in causing permeability anisotropy, *JPT* (Feb. 1983), 330.

[39] Poston, S.W., Lubojacky, R.W. and Aruna, M.
Merren field – an engineering review, *JPT* (Nov. 1983), 2105.

[40] Clutterbuck, P.R. and Dance, J.E.
The use of simulations in decision making for the Kuparuk River field development, *JPT* (Oct. 1983), 1893.

[41] Odeh, A.S.
Comparison of solutions to a three dimensional black oil reservoir simulation problem, *SPEJ* (Jan. 1981), 13.

[42] West, W.J., Garvin, W.W. and Sheldon, J.W.
Solution of the equations of unsteady state two phase flow in oil reservoirs, *Trans. AIME* **201** (1954), 217.

[43] Ertekin, T.
Principles of numerical simulation of oil reservoirs – an overview, In *Heavy Crude Oil Recovery* (ed. Okandan), Nato ASI Series, Martinus Nijhoff, The Hague (1984), 379.

[44] Aziz, K. and Settari, A.
Petroleum Reservoir Simulation, Applied Science Publishers, Barking (1979).

[45] Aziz, K.
Numerical methods for three dimensional reservoir models, *J. Can. Pet. Tech.* (Jan., Feb. 1968), 7.

[46] Peaceman, D.W.
Fundamentals of Numerical Reservoir Simulation, Elsevier (1977).

[47] Thomas, G.W.
Principles of hydrocarbon reservoir simulation, *IHRDC* (1982).

[48] Diehl, A.
The development of the Brent field – a complex of projects, EUR 108, *Proc. Europ. Pet. Eng. Conf.,* London (1978), 397, Vol. II.

[49] Bath, P.G.
The Brent field : a reservoir engineering review, EUR 164, *Proc. Europ. Pet. Conf.,* London (1980), 179, Vol. I.

[50] Nadir, F.T.
Thistle field development, EUR 165, *Proc. Europ. Pet. Conf.,* London (1980), 193, Vol. I.

[51] Bishlawi, M. and Moore, R.L.
Montrose field reservoir management, EUR 166, *Proc. Europ. Pet. Conf.,* London (1980), 205, Vol. I.

[52] van Rijswijk *et al.*
The Dunlin field, EUR 168, *Proc. Europ. Pet. Conf.,* London (1980), 217, Vol. I.

[53] Poggiagliomi, E.
Detailed reservoir delineation by interactive seismic stratigraphic extrapolation, EUR 198, *Proc. Europ. Pet. Conf.*, London (1980), 513, Vol. I.
[54] Johnson, R., Riches, H. and Ahmed, H.
Application of the vertical seismic profile to the Piper field, EUR 274, *Proc. Europ. Pet. Conf.*, London (1982), 39.
[55] Stiles, J.H. and Bobek, J.E.
Fulmar pre-development simulation study, EUR 282, *Proc. Europ. Pet. Conf.*, London (1982), 99.
[56] Smith, C.A.
Methods for modelling the performance of layered reservoir systems, EUR 284, *Proc. Europ. Pet. Conf.*, London (1982), 123.
[57] Thambynayagam, R.K.M.
Analytical solutions for PBU and FO analysis of water injection tests of partially penetrating wells: non-unit mobility, SPE 12965, *Proc. Europ. Pet. Conf.*, London (1984), 83.
[58] Aron, D., Ashbourne, T.J. and Oloketuyi, D.O.
The secondary recovery project, Ogharefe field, Nigeria, *JPT* (April 1984), 671.
[59] Gwinn, V.E.
Deduction of flow regime from bedding character in conglomerates and sandstones, *J. Sed. Petrol.* (1964), 656.
[60] Hails, J.R. and Hoyt, J.H.
The significance and limitations of statistical parameters for distinguishing ancient and modern sedimentary environments of the lower Georgia coastal plain, *J. Sed. Petrol.* (1969), 559.
[61] Houser, J.F. and Neasham, J.W.
Bed continuity and permeability variations of recent deltaic sediments, *Bull. AAPG* (1976), 681.
[62] Stearns, D.W. and Friedman, M.
Reservoirs in fractured rock, *AAPG Memoir* **16** (1972), 82.
[63] Harpole, K.J.
Improved reservoir characterisation – a key to future reservoir management, *JPT* (Nov. 1980), 2009.
[64] Selley, R.C.
Subsurface environmental analysis of N. Sea sediments, *Bull. AAPG* (1976), 184.
[65] Visher, G.C.
Use of vertical profile in environmental reconstruction, *Bull. AAPG* (1965), 41.
[66] Nagetegaal, P.J.
Relationship of facies and reservoir quality in Rotliegendes desert sandstones, SNS region, *J. Pet. Geol.* **2** (1979), 145.
[67] Weber, K.J.
Influence of common sedimentary structures on fluid flow in reservoir models, *JPT* (March 1982), 665.
[68] Weber, K.J. *et al.*
Simulation of water injection in a barrier-bar type, oil rim reservoir in Nigeria, *JPT* (Nov. 1978), 1555.
[69] Zeito, G.A.
Interbedding of shale breaks and reservoir heterogeneities, *JPT* (Oct. 1965), 1223.
[70] Verrien, J.P., Courand, G. and Montadert, L.
Applications of production geology methods to reservoir characteristics analysis from outcrop observations, *Proc. 7th World Pet. Cong.*, Mexico II (1967), 425, Elsevier.
[71] Polasek, T.L. and Hutchinson, C.A.
Characterisation of non-uniformities within a sandstone reservoir from a fluid mechanics standpoint, *Proc. 7th World Pet. Cong.*, Mexcio II (1967), 397, Elsevier.
[72] Prats, M.
The influence of oriented arrays of thin impermeable shale lenses or of highly conductive natural fractures on apparent permeability anisotropy, *JPT* (Oct. 1972), 1219.
[73] Richardson, J.C., Harris, D.G., Rossen, R.H. and van Hee, Q.
The effect of small discontinuous shales on oil recovery, *JPT* (Nov. 1978), 1531.
[74] Huppler, J.D.
Numerical investigation of effects of core heterogeneity on waterflood relative permeability, *SPEJ* (Dec. 1970), 381.
[75] Chapman, R.E.
Petroleum Geology – a Concise Study, Elsevier (1973).
[76] Sneider, R.M., Tinker, C.N. and Meckel, L.D.
Deltaic environment reservoir types and their characteristics, *JPT* (Nov. 1978), 1538.
[77] Harris, D.G. and Hewitt, C.H.
Synergism in reservoir management – the geologic perspective, *JPT* (July 1977), 761.

[78] Pryor, W.A. and Fulton, K.
Geometry of reservoir type sand bodies and comparison with ancient reservoir analogs, SPE 7045, *Proc. Symp. on Improved Oil Recovery,* Tulsa (April 1978), 81.

[79] Ruzyla, K. and Friedman, G.M.
Geological heterogeneities important to future enhanced recovery in carbonate reservoirs, SPE/DOE 9802, *Proc. 2nd Jt. Symp. EOR,* Tulsa (April 1981), 403.

[80] Fisher, W.L., Brown, L.F., Scott, A.J. and McGowen, J.H.
Delta systems in the exploration for oil and gas, *Bur. Econ. Geol.* (1969), Univ. Tex. Austin.

[81] Simons, D.B., Richardson, E.V. and Nordin, C.F.
Sedimentary structures generated by flow in alluvial channels, In *Primary Sedimentary Structures and their Hydrodynamic Interpretation* (Middleton, G.V., ed.), Soc. Econ. Pet. Min. Spr. Pub. 12 (1965), 34.

[82] Allen, J.R.L.
Current Ripples, their Relations to Patterns of Water and Sediment Motion, North Holland Pub. Co., Amsterdam (1968).

[83] Reading, H.G. (ed.)
Sedimentary Environments and Facies, Blackwell Scientific, Oxford (1978).

[84] Reineck, H.E. and Singh, I.B.
Depositional Sedimentary Environment, Springer Verlag, Berlin (1973).

[85] Chauvin, A.L. *et al.*
Development planning for the Statfjord field using 3–D and areal reservoir simulation, SPE 8384, *Proc. Ann. Fall Mtg.* (1979).

[86] Utseth, R.H. and Macdonald, R.C.
Numerical simulation of gas injection in oil reservoirs, SPE 10118, *Proc. Ann. Fall Mtg.* (1981).

[87] Addington, D.V.
An approach to gas coning correlations for a large grid cell reservoir simulation, *JPT* (Nov. 1981), 2267.

[88] Darlow, B.L., Ewing, R.E. and Wheeler, M.F.
Mixed finite element method for miscible displacement problems in porous media, *SPEJ* (Aug. 1984), 391.

[89] Haldorsen, H.H. and Lake, L.W.
A new approach to shale management in field-scale models, *SPEJ* (Aug. 1984), 447.

[90] Craig, F.F., Willcox, P.J., Ballard, J.R. and Nation, W.R.
Optimised recovery through continuing interdisciplinary cooperation, *JPT* (July 1977), 755.

[91] Le Blanc, R.J.
Distribution and continuity of sandstone reservoirs, *JPT* (July 1977), Pt 1, 776, Pt 2, 793.

[92] Jardine, D., Andrews, D.P., Wishart, J.W. and Young, J.W.
Distribution and continuity of carbonate reservoirs, *JPT* (July 1977), 873.

[93] Harris, D.G.
The role of geology in reservoir simulation studies, *JPT* (May 1975), 625.

[94] Groult, J., Reiss, L.H. and Montadort, L.
Reservoir inhomogeneities deduced from outcrop observations and production logging, *JPT* (July 1966), 883.

[95] Campbell, C.V.
Reservoir geometry of a fluvial sheet sandstone, *Bull. AAPG* (1976), 1009.

[96] Davies, D.K., Ethridge, F.G. and Berg, R.R.
Recognition of barrier environments, *Bull. AAPG* (1971), 550.

[97] Barwis, J.H. and Makurath, J.H.
Recognition of ancient tidal inlet sequences, *Sedimentology* 25 (1978), 61.

[98] Budding, M.C. and Inglin, H.F.
A reservoir geological model of the Brent sands in Southern Cormorant, In *Petroleum Geology of the Continental Shelf N.W. Europe* (eds Illing and Hobson), Inst. Pet. (1981), 326.

[99] Craig, F.F.
Effect of reservoir description on performance predictions, *JPT* (Oct. 1970), 1239.

[100] Yusun, J., Dingzeng, L. and Changyan, L.
Development of Daqing oil field by waterflooding, *JPT* (Feb. 1985), 269.

[101] Simlote, V.N., Ebanks, W.J., Eslinger, E.V. and Harpole, K.J.
Synergistic evaluation of a complex conglomerate reservoir for EOR, Barrancas Formation, Argentina, *JPT* (Feb. 1985), 269.

[102] Hutchinson, C.A., Dodge, C.F. and Polasek, T.L.
Identification, classification and prediction of reservoir inhomogeneities affecting production operations, *JPT* (March 1961), 223.

[103] Treiber, L.E., Archer, D.L. and Owens, W.W.
Laboratory evaluation of the wettability of fifty oil producing reservoirs, *SPEJ* (Dec. 1972), 531.

[104] Kumar, N. and Sanders, J.E.
Inlet sequence: a vertical succession of sedimentary structures and textures created by the lateral migration of tidal inlets, *Sedimentology* **21** (1974), 491.

[105] Friend, P.F., Slater, M.J. and Williams, R.C.
Vertical and lateral building of river sandstone bodies, *J. Geol. Soc. London* **136** (1979), 39.

[106] Gilreath, J.A. and Stephens, R.W.
Interpretation of log response in deltaic sediments, *Finding and Exploring Ancient Deltas in the Subsurface*, AAPG Pub. (1975).

[107] Delaney, R.P. and Tsang, P.B.
Computer reservoir continuity study at Judy Creek, *J. Can. Pet. Tech.* (Jan.–Feb. 1982), 38.

[108] Raymer, L.L. and Burgess, K.A.
The role of well logs in reservoir modelling, SPE 9342, *Proc. 55th Ann. Fall Mtg.* (1980).

[109] Gretener, P.E.
Geology, geophysics and engineering: a case for synergism, *J. Can. Pet. Tech.* (May–June 1984), 55.

[110] Richardson, J.G. and Stone, H.L.
A quarter century of progress in the application of reservoir engineering, *JPT* (Dec. 1973), 1371.

[111] Denison, C. and Fowler, R.M.
Palynological identification of facies in a deltaic environment, *Proc. Nor. Pet. Conf. on Sedim. of N. Sea Res. Rocks,* Geilo, Norway (May 1980).

[112] Bain, J.S., Nordberg, M.O. and Hamilton, T.M.
3D seismic applications in the interpretation of Dunlin field, UK N.Sea, SPE 9310, *Proc. 55th Ann. Fall Mtg. SPE* (1980).

[113] Stewart, G., Wittmann, M.J. and Lefevre, D.
Well performance analysis: a synergetic approach to dynamic reservoir description, SPE 10209, *Proc. 56th Ann. Fall Mtg. SPE* (1981).

[114] Hallett, D.
Refinement of the geological model of the Thistle field, In *Petroleum Geology of the Continental Shelf of N.W. Europe* (eds Illing and Hobson), Inst. Pet. London (1981), 315.

[115] Hancock, N.J.
Diagenetic modelling in the Middle Jurassic Brent sand of the N. North Sea, EUR 92, *Proc. Europ. Pet. Conf.,* London (1978), 275, Vol. II.

[116] McMichael, C.L.
Use of reservoir simulation models in the development planning of the Statfjord field, EUR 89, *Proc. Europ. Pet. Conf.,* London (1978), 243, Vol. II.

[117] Hillier, G.R., Cobb, R.M. and Dimmock, P.A.
Reservoir development planning for the Forties field, EUR 98, *Proc. Europ. Pet. Conf.,* London (1978), 325, Vol. II.

[118] Dake, L.P.
Application of the RFT in vertical and horizontal pulse testing in the Middle Jurassic Brent sands, EUR 270, *Proc. Europ. Pet. Conf.,* London (1982), 9.

[119] Coats, K.H.
Simulation of gas condensate reservoir performance. *JPT* (Oct. (1985), 1870.

[120] Archer, J.S.
Some aspects of reservoir description for reservoir modelling, In *North Sea Oil and Gas Reservoirs,* Proc. Intl. Seminar, Trondheim (December 1985), Eds. Butler, A.T. and Kleppe, J., Graham & Trotman, London 1986.

[121] Gray, I.
Viking gas field. In *Petroleum and the Continental Shelf of N.W. Europe* (Ed. Woodward), Elsevior Applied Science, London (1975) 241.

[122] Carman, G.S. and Young, R.
Reservoir geology of the Forties oilfield, In *Petroleum Geology of the Continental Shelf of N.W. Europe.* (Eds. Illing and Hobson), Heyden, London (1981) 371.

Appendix I
SPE Nomenclature and Units*

Standard letter symbols for reservoir engineering and electric logging have been defined by the AIME (Society of Petroleum Engineers). Some non-standard terms, subscripts and nomenclature are still in use and may be encountered.

No effective standardization or metrication of units has yet occurred, and the industry uses American mixed units to a large extent, although some metric units mixed with American still may be encountered. An application of the SI metric system is found in the *Journal of Petroleum Engineerinng* (1985) in the issues for August (p.1415) and October p.1801.

UNITS

Volume
acre-foot for large volumes
barrel
cubic ft
cubic metre

Liquid volume
barrel = 5.615 cubic ft
cubic metre = (35.31) ft^3
(Unless otherwise specified, an oil volume will be tank oil measured at 1 atmosphere and 60°F.)

Gas volume
cubic foot)   measured at 1 atmosphere
cubic metre) and 60°F
MCF = thousands of cubic feet
MMCF = millions of cubic feet
(The billion is the American billion = 10^9;
the trillion is the American trillion = 10^{12}.)

Pressure
pounds force per square in (psi)
atmosphere
bar

Temperature
degrees Fahrenheit °F
degrees Rankine °R = 460 + °F
degrees Kelvin K

Length
pipelines – miles, feet, kilometres
well depths – feet or metres

Diameters
tubular diameters generally inches or centimetres
feet/metres

Viscosity
centipoise

Density
lb mass per cubic foot
kg mass per cubic metre
g per cubic centimetre

---

* Reprinted from *Journal of Petroleum Technology*, 1984, pp. 2278–2323 by permission. © SPE–AIME, 1984.

Specific gravity
liquids relative to water (62.4 lb/ft^3)
gases relative to air (0.0765 lb/ft^3)
API scale for tank oil

Gas–oil ratio
standard cubic feet of gas per stock tank barrel of oil
cubic metres of gas (s.c.) per cubic metre tank oil

Flow rate
liquids – barrel per day (b/d)
  cubic metres per day (m^3/d)
gases – standard cubic ft per day SCF/d, MCF/d
  and MSCFD/d, MMSCFD
  cubic metres per day (m^3/d) MSCFD/d

Oil densities
API gravity

$$°API = \frac{141.5}{(SG)_{oil}} - 131.5$$

SG = specific gravity of water = 1.0

Recommendation for metrication and appropriate conversion factors for units are given:

Recommended units: conversions

Quantity	SI unit	Industry unit	SPE preferred unit	Conversion factor (industry → preferred)
Length	m	mile	km	1.609344
		metre	m	1.0
		foot	m	0.3048
		inch	mm	25.4
Area	m^2	sq. mile	km^2	2.589988
		acre	km^2	4.046873 × 10^3
		sq. ft	m^2	0.0920304
		sq. inch	mm^2	6.4516 × 10^2
Volume	m^3	m^3	m^3	1.0
		acre foot	m^3	1.233482 × 10^3
		barrel	m^3	1.589873 × 10^{-1}
		ft^3	m^3	2.831685 × 10^{-2}
		US gallon	m^3	3.785412 × 10^{-3}
Capacity/length	m^3/m	barrels/ft	m^3/m	5.216119 × 10^{-1}
		ft^3/ft	m^3/m	9.02903404 × 10^{-2}
		US gall./ft	m^3/m	1.241933 × 10^{-2}
Mass	kg	lb mass	kg	4.535924 × 10^{-1}
		short ton	Mg	0.9071847
Temperature gradient	K/m	$°$F/ft	K/m	1.822689
Pressure	Pa	atmosphere	kPa	1.013250 × 10^2
		bar	kPa	1.0 × 10^2
		kgf/sq. cm	kPa	9.806650 × 10^1
		lbf/sq. in.	kPa	6.894757
		dyne/sq. cm	Pa	1 × 10^{-1}
Pressure gradient	Pa/m	lbf/sq. in./ft	kPa/m	2.262059 × 10^1
Density	kg/m^3	lbm/ft^3	kg/m^3	1.601846 × 10^{-1}
		lbm/US gall.	kg/m^3	1.198264 × 10^2
Volume rate	m^3/s	b/d	m^3/d	1.589873 × 10^{-1}
		US gall./min	m^3/hr	0.2271247
Viscosity	Pa.s	cP	Pa.s	1.0 × 10^{-3}
Permeability	m^2	Darcy	μm^2	9.869233 × 10^{-1}
		milliDarcy	μm^2	9.869233 × 10^{-4}

SPE SYMBOLS STANDARD
Preface
Objectives

The primary objectives of the 1984 Symbols Standards are to combine prior standards and supplements into one publication so as to provide (1) consistency of usage and maximum ease of understanding of mathematical equations for the readers of technical papers, and (2) to codify symbols lists, rules and guides for the writers of technical papers.

Structure of lists

The 1984 Symbol Standards are a consolidation of the 1956 Standard and all later supplements. Some of the cross-grouping and obsolete quantities have been eliminated. The complete symbols list is given in four different forms as follows:

A. Symbols alphabetized by physical quantity,

B. Subscripts alphabetized by physical quantity,

C. Symbols alphabetized by symbols,

D. Subscripts alphabetized by symbols.

The names or labels for the quantities are for identification only and are not intended as definitions. Defining equations are given in a few cases where further identifications may be needed. For the present, the specification of units and conditions of measurement is left to the user.

For convenience in dimensional checking of equations, a column has been included giving the dimensions of each quantity in terms of mass, length, time, temperature and electrical charge (m, L, t, T, q). The term *various* also appears in this column for several symbols. This terminology permits maximum flexibility for quantities that may require different dimensions in different problems. Examples are symbols: (1) m for slope of a line (two variables of any dimensions can be related); (2) C for concentration (dimensions might be m/L^3, dimensionless or other); (3) F (factor) when it represents ratio (dimensions might be L^3/m, m, dimensionless or other). This flexibility in dimensions permits desirable shortening of the symbols list.

Additional standard symbols

The extraordinary growth in all phases of petroleum and computer technology has necessitated the adoption of additional standard symbols, since the original standards were published in 1956 following five years of intensive development. Additions resulted from requests from members and from editorial reviews of the numerous papers submitted to SPE for publication.

Principles of symbols selection

Once the original reservoir Symbols Standard was established in 1956, the principles employed in the selection of additional symbols have been as follows:

A. (1) *Use single letters only* for the main letter symbols. This is the universal practice of the American National Standards Institute (ANSI), the International Organization for Standardization (ISO) and the International Union of Pure and Applied Physics (IUPAP) in more than 20 formal Standards adopted by them for letter symbols employed in mathematical equations. (2) Make available single and multiple subscripts to the main letter symbols to the extent necessary for clarity.

  Multiple letters such as abbreviations are prohibited for use as the main symbol (kernel) for a quantity. A few exceptions are some traditional mathematical symbols such as log, ln and lim. Thus quantities that are sometimes represented by abbreviations in textual material, tables or graphs are required in the SPE Symbols Standards to have single-letter kernels. Examples are: gas–oil ratio (GOR), bottom-hole pressure (BHP), spontaneous potential (SP), static SP (SSP), which, respectively, have the following SPE Standard symbols: R_{pbh}, E_{SP}, E_{SSP}.

B. Adopt the letter symbols of original or prior author usage, where not in conflict with principles C and D below.

C. Adopt letter symbols consistent or parallel with the existing SPE Standard, minimizing conflicts with that Standard.

D. Where pertinent, adopt the symbols already standardized by such authorities as ANSI, ISO, or IUPAP (see A); minimize conflicts with these Standards.

E. Limit the list principally to basic quantities, avoiding symbols and subscripts for combinations, reciprocals, special conditions, etc.

F. Use initial letters of materials, phase, processes, etc., for symbols and subscripts, as being suggestive and easily remembered.

G. Choose symbols that can be readily handwritten, typed, and printed.

Principles of letter symbol standardization

A. **Requirements for Published Quantity.**
Each published letter symbol should be:

1. *Standard, where possible.* In the use of published symbols, authors of technical works (including textbooks) are urged to adopt the symbols in this and other current standard lists and to conform to the principles stated here. An author should give a table of the symbols used and their respective interpretations, or else refer to a standard list as a source for symbols used but not explained. For work in a specialized or developing field, an author may need symbols in addition to those already contained in standard lists. In such a case the author should be careful to select simple suggestive symbols that avoid conflict in the given field and in other closely related special fields. Except in this situation, the author should not introduce new symbols or depart from currently accepted notation.

2. *Clear in reference.* One should not assign to a given symbol different meanings in such a manner as to make its interpretation in a given context ambiguous. Conflicts must be avoided. Often a listed alternative symbol or a modifying subscript is available and should be adopted. Except in brief reports, any symbol not familiar to the reading public should have its meaning defined in the text. The units should be indicated whenever necessary.

3. *Easily identified.* Because of the many numerals, letters and signs that are similar in appearance, a writer should be careful in calling for separate symbols that in published form might be confused by the reader. For example, many letters in the Greek alphabet (lower case and capital) are practically indistinguishable from English letters; the zero is easily mistaken for a capital O.

4. *Economical in publication.* One should try to keep at a minimum the cost of publishing symbols. In particular: (1) Notations which call for handsetting of movable type should be rejected in favour of forms adapted to modern mechanical methods of composition. (2) No one work should use a great variety of types and special characters. (3) Handwriting of inserted symbols, in copy largely typewritten and to be reproduced in facsimile, should not be excessive. (4) Often a complicated expression appears as a component part of a complex mathematical formula – for example, as an exponent of a given base. Instead, one may introduce locally, a single non-conflicting letter to stand for such a complicated component. An explanatory definition should then appear in the immediate context.

B. **Secondary symbols.** Subscripts and superscripts are widely used and for a variety of conventional purposes. For example, a subscript may indicate: (1) the place of a term in a sequence or matrix; (2) a designated state, point, part, or time, or system of units; (3) the constancy of one independent physical quantity among others on which a given quantity depends for its value; (4) a variable with respect to which the given quantity is a derivative. Likewise, for example, a superscript may indicate: (1) the exponent for a power, (2) a distinguishing label, (3) a unit, or (4) a tensor index. The intended sense must be clear in each case. Several subscripts or superscripts sometimes separated by commas may be attached to a single letter. A symbol with a superscript such as prime (') or second ("), or a tensor index, should be enclosed in parentheses, braces or brackets before an exponent is attached. So far as logical clarity permits, one should avoid attaching subscripts and superscripts to subscripts and superscripts. Abbreviations, themselves standardized, may appear among subscripts. A conventional sign, or abbreviation, indicating the adopted unit may be attached to a letter symbol, or corresponding numeral. Reference marks, such as numbers in distinctive type, may be attached to words and abbreviations, but not to letter symbols.

C. **Multiple subscript-position order.** The wide variety and complexity of subject matter covered in the petroleum literature make it impossible to avoid use of multiple subscripts with many symbols. To make such usage less confusing, the following guides were employed for the order of appearance of the individual letters in multiple subscripts in the symbols list. Use of the same rules is recommended when it becomes necessary to establish a multiple subscript notation that has not been included in this list.

1. When the subscript *r* for 'relative' is used, it should appear first in subscript order. Examples: K_{ro}; K_{rg}.

2. When the subscript *i* for 'injection' or

SPE NOMENCLATURE AND UNITS

'injected' or 'irreducible' is used, it should appear first in subscript order (but after r for 'relative'). Examples: B_{ig}, formation volume factor of injected gas; c_{ig}, compressibility of injected gas.

3. Except for Cases 1 and 2 above (and symbols K_h and L_v), phase, composition and system subscripts should generally appear first in subscript order. Examples: B_{gi}, initial or original gas formation volume factor; B_{oi}, initial or original oil formation volume factor; $C_{O,i}$, initial or original oxygen concentration; B_{ti}, initial or original total system formation volume factor; ρ_{sE}, density of solid particles making up experimental pack; also F_{aF}, G_{Lp}, G_{wgp}, G_{Fi}.

4. Abbreviation subscripts (such as 'ext', 'lim', 'max', 'min'), when applied to a symbol already subscripted, should appear last in subscript order and require that the basic symbol and its initial subscript(s) be first enclosed in parentheses. Examples: $(i_a)_{max}$, $(S_{hr})_{min}$.

5. Except for Case 4 above, numerical subscripts should appear last in subscript order. Examples: q_{oD3}, dimensionless oil production rate during time period 3; p_{R2}, reservoir pressure at time 2; $(i_{a1})_{max}$, maximum air injection rate during time period 1.

6. Except for Cases 4 and 5 above, subscript D for 'dimensionless' should usually appear last in subscript order. Examples: p_{tD}; q_{oD}; $(q_{oD3})_{max}$.

7. Except for Cases 4, 5 and 6 above, the following subscripts should usually appear last in subscript order: regions such as bank, burned, depleted, front, swept, unburned (b, b, d, f, s, u); separation, differential and flash (d, f); individual component identification (i orQl other). Examples: E_{Db}; R_{sf}, n_{pf}.

D. **Typography.** Letter symbols for physical quantities, and other subscripts and superscripts, whether upper case, lower case, or in small capitals, when appearing as light-face letters of the English alphabet, are printed in italic (sloping) type. Arabic numerals, and letters or other alphabets used in mathematical expressions, are normally printed in vertical type. When a special alphabet is required, boldface type is to be preferred to German, Gothic, or script type. In material to be reproduced in facsimile, from copy largely typewritten, letters that would be boldface in print may be indicated to be such by special underscoring, while the few distinct letters used from other alphabets, if carefully made, should be self-explanatory. It is important to select a type face that has italic forms, and clearly distinguished upper case, lower case and small capitals. Only type faces with serifs are recommended.

E. **Remarks.** Quantity symbols may be used in mathematical expressions in any way consistent with good mathematical usage. The product of two quantities is indicated by writing ab. The quotient may be indicated by writing

$$\frac{a}{b}, a/b \text{ or } ab^{-1}$$

If more than one solidus is used in any algebraic term, parentheses must be inserted to remove any ambiguity. Thus, one may write $(a/b)/c$, or a/bc, but not $a/b/c$.

F. **Special notes.** Observe the following:

1. When the mobilities involved are on opposite sides of an interface, the mobility ratio will be defined as the ratio of the displacing phase mobility to the displaced phase mobility, or the ratio of the upstream mobility to the downstream mobility.
2. Abbreviated chemical formulas are used as subscripts for paraffin hydrocarbons: C_1 for methane, C_2 for ethane, C_3 for propane . . . C_n for C_nH_{2n+2}.
3. Complete chemical formulas are used as subscripts for other materials: CO_2 for carbon dioxide, CO for carbon monoxide, O_2 for oxygen, N_2 for nitrogen, etc.
4. The letter R is retained for electrical resistivity in well logging usage. The symbol ρ is to be used in all other cases and is that preferred by ASA.
5. The letter C is retained for conductivity in well logging usage. The symbol σ is to be used in all other cases and is that preferred by ASA.
6. Dimensions: L = length, m = mass, q = electrical charge, t = time, and T = temperature.
7. Dimensionless numbers are criteria for geometric, kinematic and dynamic similarity between two systems. They are derived by one of three procedures used in methods of similarity: integral, differential, or dimensional. Examples of dimensionless numbers are Reynolds number (N_{Re}) and Prandtl number (N_{Pr}). For a discussion of methods

of similarity and dimensionless numbers, see "Methods of Similarity", by R.E. Schilson, *J. Pet. Tech.* (August, 1964) 877.

8. The quantity x can be modified to indicate an average or mean value by an overbar, $\bar{x}$.

Principles of computer symbol standardization

A. **Symbol Structure.** The computer symbols are structured from four possible parts representing respectively arithmetic mode, mathematical operators, basic quantities and subscripts, exclusive of time and space designations. Each of these parts has a defined number of characters and, when all are used in a single symbol, the total length may be ten characters. Example ten-character notations are:

XDELPRSTQQ, XDELCMPPRD

When any of the four parts are not used, the remaining characters are to be right- or left-justified to form a string of characters without blank positions.

In practice, the combined notations will not usually exceed six characters. In those cases where the complete computer symbol does exceed six characters, and the computer language being used will not allow more than six, a shortened notation must be employed. The part of the notation representing the basic mathematical quantity (letter) symbol should be retained and the other parts of the notation shortened. Shortened symbols are no longer standard, and therefore must be defined in the text or appendix as is appropriate.

1. The first part of the notation consists of one character position to define the arithmetic mode of the complete computer symbol. It is suggested that X be used for floating point variables and I for integers. This notation position should be used only if absolutely necessary, the preferred approach being the use of a declaration within the program.

2. The second part of the notation (operator field) consists of three characters and is used for mathematical operators. The notation should suggest the operation.

3. The third part of the notation (quantity symbol field) consisting of three characters, is used to represent the basic mathematical quantity (letter) symbol. The three letter notation mnemonically denotes the quantity name as closely as possible. This part of the computer notation is thus of the nature of an abbreviation. All three character positions must be employed.

Fixed characters are utilized in this part of the notation when heat quantities, indexes and exponents are being assigned computer symbols. When a heat quantity is denoted, H appears in the first character position, as exemplified by thermal conductivity HCN. Indexes such as resistivity index are denoted by X in the third character position. Exponents are characterized by XP in the second and third positions, such as porosity exponent MXP.

4. The fourth part of the notation (subscript field) is used to represent the subscripts of the mathematical letter symbol and normally consists of one of the three character positions. Computer symbol subscripts are normally designated by using the mathematical letter subscripts of the SPE Symbols Standard.

Though usually not required, more characters may be used when necessary for designation of multiple mathematical letter subscripts. For example, dimensionless average reservoir pressure would be denoted by PRSAVQ.

The computer subscript designation is placed immediately to the right of the quantity symbol field with no intervening space.

Dimensionless numbers are denoted by Q in the last required subscript position. Average, maximum, minimum, extrapolated or limiting values of a quantity are denoted respectively by AV, MX, MN, XT, of LM in the first two subscript positions; additional subscripting occurs immediately to the right of these defined notations. Other than in these cases, the order of subscripting should follow the rules given in the 'Multiple Subscripts – Position Order'.

5. No binding rule is made for the notation of space and time subscripts, since the method of subscripting is often dictated by the characteristics of a particular computer. However, the vital importance of these subscripts makes it necessary to establish a standard and require an author to define any deviations. The system outlined below should be used when the subscripts are not implied by an array location or an index specified by the program logic.

The following sketch indicates the coordinate system used to denote special position in multi-dimensional arrays.

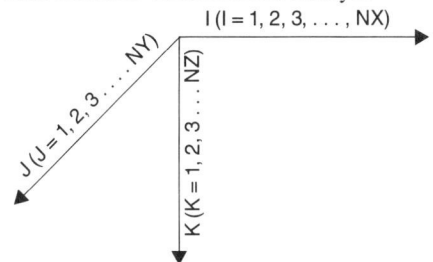

```
         I (I = 1, 2, 3, ..., NX)
J(J = 1, 2, 3, ..., NY)  ↗
         ↓ K(K = 1, 2, 3, ..., NZ)
```

This convention was adopted so that the page position of printed output obtained in a normal I, J, K sequence would correspond to position as viewed on maps as normally used in petroleum engineering. Similarly, I, K or J, K sequences would correspond to cross-sections as normally used.

The space and time subscripts are constructed by placing a letter code (I, J, K, T) before the following symbols:

Machine Symbol	Definition
P2	present location plus 2
P3H	present location plus 3/2
P1	present location plus 1
P1H	present location plus 1/2
M1H	present location minus 1/2
M1	present location minus 1
M3H	present location minus 3/2
M2	present location minus 2

Hence, the subscript for the present time t would be T, and that for subscript $t-2$ would be TM2.

If an array contains information corresponding to points halfway between the normally indexed points, then the convention is to shift the plus-direction elements to the node being indexed.

In the following example, the permeability at the $i-\frac{1}{2}$ point would be referenced as PRMIP1H(I − 1), and that for the $i+\frac{1}{2}$ point would be referenced as PRMIP1H(I). See sketch below.

```
   i−1      i−½       i        i+½
────o────────|────────o────────|────→
   I−1  PRMIP1H(I−1)  I    PRMIP1H(I)
```

B. Units. Each complete computer symbol represents a mathematical letter symbol and its associated subscripts. The mathematical letter symbol in turn designates a physical quantity. Neither the complete computer symbol nor the mathematical letter symbol implies any specific units of measure. Authors are urged to familiarize themselves with the SI System of units and use them as much as practical. The choice of units (*Trans. AIME* **263** (1977) 1685) and their designation is, however, left to the author.

C. Restriction to computer programs. Use of the computer symbols is restricted to the description of programming for computers. As a consequence, the computer symbols *must not be used* in works of portions of papers where programming is not discussed or as abbreviations in text or graphical material.

D. Character set. The computer symbols must be constructed from the 26 English letters and 10 Arabic numerical characters. Each complete computer symbol must begin with a letter and not a numeral.

The computer symbols are always represented by vertical type in printed text. English capital letters and Arabic numerals are used in hand or typewritten material.

E. Nonstandard symbols. The rules for establishing the computer symbols contained in this standard are such that quantities not covered can, in most instances, be given a notation that is compatible with it. Such additional computer symbols are, by definition, nonstandard.

Duplication of computer symbols for quantities that can occur simultaneously in an equation or computer program must be avoided. Elimination of a duplication may lead to a computer symbol that is at variance with the standard; i.e., a notation that is nonstandard.

When nonstandard computer symbols occur in a technical work, they should be clearly defined in the text or appendix, as is appropriate, and in the program.

F. Special notes. No computer symbols have been defined here for numerical quantities, functions, and arithmetic, relational, or logical operators. When employed in programs, their usage should be fully explained by comments in the program text. Some of these special cases are noted below:

1. No computer symbols to designate common or natural logarithms have been established. Rather, these functions should be designated by the notations compatible with the computer system being employed. The notation used should be defined in the paper.
2. The computer symbol for dimensionless numbers in general (unnamed dimensionless numbers) is NUMQ. Named dimen-

sionless numbers have the mnemonic title designation in the field representing the quantity and a Q in the last subscript position employed. Thus, Reynolds number is designated as REYQ. Similarly, Prandtl number could be designated as PRDQ, Grashof number as GRSQ, Graetz number as GRTQ. Any dimensionless number not contained in this standard should be defined in the paper.

3. No computer subscript notations corresponding to these mathematical letter subscripts are established. See section G.
4. No mathematical letter subscripts correspond to these computer subscripts. See section G.

G. Permissible format changes. In preparing the computer symbols it became necessary to modify the format of certain of the basic letter symbols, subscripts or symbol-subscript combinations. These changes are in accord with the General Principles of Computer Symbol Standardization. They do not imply that changes in the form of the economics, well logging and formation evaluation, reservoir engineering, or natural gas engineering letter symbols as contained elsewhere in this SPE Standard are authorized. Rather these changes are shown as a matter of record to prevent confusion and to present examples of permissible format changes in the computer symbols that may be followed when it becomes necessary to construct a computer notation not included in the list.

1. Basic symbolic subscripts of SPE Letter Symbols Standard represented by different designation in Computer Symbols Subscript List. (Only changes in the basic subscripts are shown. Combination subscripts that contain these items are also changed accordingly.)

SPE letter subscript	Computer symbol	Subscript title
c	CP	capillary
D	Q	dimensionless quantity
Dm	QM	dimensionless quantity at condition m
E	EX	experiment
ext	XT	extrapolated
F	FU	fuel
lim	LM	limiting value
m	FU	fuel (mass of)
max	MX	maximum
min	MN	minimum
$\bar{p}$	PAV	mean or average pressure
pr	PRD	pseudo-reduced
r	RD	reduced
tD	TQ	dimensionless time

2. Quantities represented by single symbol in SPE Letter Symbols Standard but by symbol-subscript combination in Computer Symbols List.

SPE letter symbol	Computer symbol	Quantity title
G	GASTI	total inital gas in place in reservoir
L	MOLL	moles of liquid phase
N	NUMQ	dimensionless number in general
N	OILTI	initial oil in place in reservoir
u	VELV	volumetric velocity (flow rate or flux, per unit area)
V	MOLV	moles of vapour phase
W	WTRTI	initial water in place in reservoir
x	MFRL	mole fraction of component in liquid phase
y	MFRV	mole fraction of component in vapour phase
z	MFRM	mole fraction of component in mixture

3. Quantities represented by symbol-subscript combination in SPE Letter Symbols Standard but by a Computer Symbol Notation only.

SPE letter symbol-subscript combination	Computer symbol	Quantity title
k_h	HCN	thermal conductivity

4. Symbol-subscript combinations of SPE Letter Symbols Standard represented by Computer Symbol-Subscript Notation wherein subscript notations are not the same.

SPE NOMENCLATURE AND UNITS

SPE letter symbol–subscript combination	Computer symbol	Quantity title
G_L	NGLTI	initial condensate liquids in place in reservoir
G_{Lp}	NGLP	cumulative condensate liquids produced
N_{Re}	REYQ	Reynolds number (dimensionless number)
R_{sw}	GWRS	gas solubility in water

5. Subscripts of SPE Letter Symbols Standard not assigned Computer Subscript Notations as a result of actions noted in 4.

SPE letter subscript	Subscript title
L_p	liquid produced, cumulative (usually with condensate, G_{Lp})
Re	Reynolds (used with Reynolds number only, N_{Re})
sw	solution in water (usually with gas solubility in water, R_{sw})

6. Letter operator–symbol combination of SPE Letter Symbols Standard represented by Computer Symbol Notation only.

SPE letter symbol	Computer symbol	quantity Title
t	TAC	interval transit time

Distinctions between, and descriptions of, abbreviations, computer symbols, dimensions, letter symbols, reserve symbols, unit abbreviations and units

Confusion often arises as to the proper distinctions between abbreviations, computer symbols, dimensions, letter symbols, reserve symbols, unit abbreviations and units used in science and engineering. The Society of Petroleum Engineers has adhered to the following descriptions:

A. **Abbreviations** – (for use in textual matter, tables, figures, and oral discussions) – an abbreviation is a letter or group of letters that may be used in place of the full name of a quantity, unit, or other entity. *Abbreviations are not acceptable in mathematical equations.* SPE provides a list of preferred abbreviations in its 'Style Guide' for authors.

B. **Computer Symbols** – (for use in computer programs) – a computer symbol is a letter or group of letters and numerals used to represent a specific physical or mathematical quantity in the writing and execution of computer programs. One computer symbol may be employed to represent a group of quantities, properly defined. *Computer symbols are not acceptable as substitutes for letter symbols* in the required mathematical (equational) developments leading up to computer programs. At the present time, all SPE computer symbols employ capital letters and numerals.

C. **Dimensions** – dimensions identify the physical nature of or the general components making up a specific physical quantity; SPE employs the five basic dimensions of mass, length, time, temperature, and electrical charge (m, L, t, T, q).*

D. **Letter symbols** – (for use in mathematical equations) – a letter symbol is a *single* letter, modified when appropriate by one or more subscripts or superscripts, used to represent a specific physical or mathematical quantity in a mathematical equation. A single letter may be employed to represent a group of quantities, properly defined. The same letter symbol should be used consistently for the same generic quantity, or special values, being indicated by subscripts or superscripts.

E. **Reserve symbols** – a reserve symbol is a single letter, modified when appropriate by one or more subscripts or superscripts, which can be used as an alternate when two quantities (occurring in some specialized works) have the same standard letter symbol. These conflicts may result from use of standard SPE symbols or subscript designations that are the same for two different quantities, or use of SPE symbols that conflict with firmly established, commonly used notations and signs from the fields of mathematics, physics, and chemistry.

To avoid conflicting designations in these cases, use of reserve symbols, reserve subscripts, and reserve symbol–reserve subscript combinations is permitted, *but only in cases of symbols conflict.* Author preference for the reserve symbols and subscripts does not justify their use.

In making the choice as to which of two quantities should be given a reserve designation,

---

* Electrical charge is current times time, ISO uses: Mass (M), Length (L), Time (T), Temperature (θ), Electric current (I), Amount of substance (N) and Luminous intensity (J).

an attempt should be made to retain the standard SPE symbol for the quantity appearing more frequently in the paper; otherwise, the standard SPE symbol should be retained for the more basic item (temperature, pressure, porosity, permeability, etc.).

Once a reserve designation for a quantity is employed, it must be used consistently throughout a paper. Use of an unsubscripted reserve symbol for a quantity requires use of the same reserve symbol designation when subscripting is required. Reversion to the standard SPE symbol or subscript is not permitted within a paper. For larger works, such as books, consistency within a chapter or section must be maintained.

The symbol nomenclature, which is a required part of each work, must contain each reserve notation that is used together with its definition.

F. **Unit Abbreviations** – a unit abbreviation is a letter or group of letters (for example, cm for centimeter), or in a few cases a special sign, that may be used in place of the name of a unit. The International Organization for Standardization (ISO) and many other national and international bodies concerned with standardization emphasize the special character of these designations and rigidly prescribe the manner in which the unit abbreviations shall be developed and treated.

G. **Units** – units express the system of measurement used to quantify a specific physical quantity. In SPE usage, units have 'abbreviations' but do not have 'letter symbols'. Up to this time, SPE has not standardized a general system of units, nor units for individual quantities; it has signified willingness, however, to join in a future national effort to convert from the English to a metric system of units.

SPE's practices showing the above distinctions are illustrated in the table of example quantities. Authors can materially aid themselves, editors, and readers by keeping the distinctions in mind when preparing papers for SPE review. Manuscripts submitted to SPE are subject to review on these aspects before being accepted for publication.

Examples

Quantity	Abbrev. for text, tables, figures, oral use	Letter symbol for mathematical equations	Reserve symbol used only in case of symbols conflict	Computer symbol for programs	Dimensions	Unit abbrev. and units*
gas–oil ratio, producing	GOR	R	none	GOR	none	cu ft/BBL
gas–oil ratio, solution, initial	initial solution GOR	R_{si}	none	GORSI	none	cu ft/BBL
productivity index	PI	J	j	PDX	$L^4 t/m$	b/d/psi
productivity index, specific	SPI	J_s	j_s	PDXS	$L^3 t/m$	b/d/psi/ft

* Examples only; SPE has not standardized units.

Contrasting symbol usage

SPE and certain American Standards Association, American National Standards Institute and International Organization for Standardization symbols lists do not use the same letter symbols to represent identical quantities. The variations in notations result from the application of the SPE guides in choosing symbols as detailed herein, the lack of agreement between various ASA standards, the ASA's policy of allowing several symbols to represent the same quantity in any list and the large number of quantities assigned symbols by the SPE. It is to be emphasized that the symbols contained in the SPE list are standard for use in petroleum engineering, but the symbols of other disciplines as sanctioned by the American Standards Association should be used when working outside the area of petroleum production. These ASA symbol standards are published by the American Society of Mechanical Engineers, United Engineering Center, 345 East 47th Street, New York, NY 10017.

The Society Board of Directors has approved the SPE 1984 Symbols Standards, and recommends them to the membership and to the industry. All authors must include Nomenclatures in any manuscript submitted to SPE for publication.

Acknowledgement

The work done in sorting and combining the various standard lists by Schlumberger Well Services Engineering personnel in Houston, Texas and Schlumberger-Doll Research Center personnel in Ridgefield, Connecticut is gratefully acknowledged.

SPE NOMENCLATURE AND UNITS

A. Symbols alphabetized by physical quantity

Letter symbol	Reserve SPE letter symbol	Computer letter symbol	Quantity	Dimensions
w	z	ARR	Arrhenius reaction rate velocity constant	L^3/m
k	K	PRM	absolute permeability (fluid flow)	L^2
g		GRV	acceleration of gravity	L/t^2
Z_a		MPDA	acoustic impedance	$m/L^2 t$
v	V, u	VAC	acoustic velocity	L/t
a		ACT	activity	
F_{aF}		FACAFU	air/fuel ratio	various
i_a		INJA	air injection rate	L^3/t
a	F_a	AIR	air requirement	various
a_E	F_{aE}	AIREX	air requirement, unit, in laboratory experimental run, volumes of air per unit mass of pack	L^3/m
a_R	F_{aR}	AIRR	air requirement, unit, in reservoir, volumes of air per unit bulk volume of reservoir rock	
μ_a	η_a	VISA	air viscosity	m/Lt
m_k		AMAK	amortization (annual write-off of unamortized investment at end of year k)	M
A		AMP	amplitude	various
A_c		AMPC	amplitude, compressional wave	various
A_r		AMPR	amplitude, relative	various
A_s		AMPS	amplitude, shear wave	various
α	β, γ	ANG	angle	
θ	β, γ	ANG	angle	
Θ	α_d	ANGD	angle of dip	
θ_c	Γ_c, γ_c	ANGC	angle, contact	
ω			angular frequency	$1/t$
K_{am}	M_{am}	COEANI	anisotropy coefficient	
I_R		INCK	annual operating cash income, over year k	M
G_{an}	f_{Gan}	GMFAN	annulus geometrical factor (muliplier or fraction)	
t_a script t	Δt	TACA	apparent interval transit time	t/L
C_a	σ_a	ECNA	apparent conductivity	tq^2/mL^3
ρ_a	D_a	DENA	apparent density	m/L^3
r_{wa}	R_{wa}	RADWA	apparent or effective wellbore radius (includes effects of well damage or stimulation)	L
ϕ_a	f_a, ε_a	PORA	apparent porosity	
R_a	ρ_a, r_a	RESA	apparent resistivity	$mL^3 tq^2$
R_z	ρ_z, r_z	RESZ	apparent resistivity of the conductive fluids in an invaded zone (due to fingering)	$mL^3 tq^2$
$\approx$		APPR	approximately equal to or is approximated by (usually with functions)	
A	S	ARA	area	L^2
E_A	η_A, e_A	EFFA	areal efficiency (used in describing results of model studies only); area swept in a model divided by total model reservoir area (see E_P)	

Letter symbol	Reserve SPE letter symbol	Computer letter symbol	Quantity	Dimensions
$\sim$		ASYM	asymptotically equal to	
p_a	P_a	PRSA	atmospheric pressure	m/Lt^2
Z		ANM	atomic number	
A		AWT	atomic weight (atomic mass, relative)	m
α	M_α	COEA	attenuation coefficient	$1/L$
$\bar{q}$	$\bar{Q}$	RTEAV	average flow rate or production rate	L^3/t
$-$		AV	average or mean (overbar)	
$\bar{p}$	$\bar{P}$	PRSAV	average pressure	m/Lt^2
$\bar{p}_R$	$\bar{P}$	PRSAVR	average reservoir pressure	m/Lt^2
ϕ	β_d	DAZ	azimuth of dip	
μ	M	RAZ	azimuth of reference on sonde	
n		NGW	backpressure curve exponent, gas well	
C		CGW	backpressure curve (gas well), coefficient of	$L^{3-2n}t^{4n}/m^{2n}$
n		NGW	backpressure curve (gas well), exponent of	
$\log_a$			base a, logarithm	
β	γ	BRGR	bearing, relative	
h	d,e	THK	bed thickness, individual	L
p_{wf}	P_{wf}	PRSWF	bottomhole flowing pressure	m/Lt^2
p_{bh}	P_{BH}	PRSBH	bottomhole pressure	m/Lt^2
p_{wf}	P_{wf}	PRSWF	bottomhole pressure flowing	m/Lt^2
p_{iwf}	P_{iwf}	PRSIWF	bottomhole flowing pressure, injection well	m/Lt^2
P_{Iws}	P_{iws}	PRSIWS	bottomhole static pressure, injection well	m/Lt^2
p_{ws}	P_{ws}	PRSWS	bottomhole pressure at any time after shut-in	m/Lt^2
p_w	P_w	PRSW	bottomhole pressure, general	m/Lt^2
p_{ws}	P_{ws}	PRSWS	bottomhole pressure, static	m/Lt^2
p_{ww}	P_{ww}	PRSWW	bottomhole (well) pressure in water phase	m/Lt^2
T_{bh}	θ_{BH}	TEMBH	bottomhole temperature	T
b	w	WTH	breadth, width, or (primarily in fracturing) thickness	L
p_e	P_e	PRSE	boundary pressure, external	m/Lt^2
r_e	R_e	RADE	boundary radius, external	L
B_{gb}	F_{gb}	FVFGB	bubble-point formation volume factor, gas	
B_{ob}	F_{ob}	FVFOB	bubble-point formation volume factor, oil	
p_b	p_s, P_s, P_b	PRSB	bubble-point (saturation) pressure	m/Lt^2
b_{gb}	f_{gb}, F_{gb}	RVFGB	bubble-point reciprocal gas formation volume factor at bubble-point conditions	
V_{bp}	v_{bp}	VOLBP	bubble-point pressure, volume at	L^3
R_{sb}	F_{gsb}	GORSB	bubble-point solution gas–oil ratio	
Δt_{ws}	$\Delta \tau_{ws}$	DELTIMWS	buildup time; shut-in time (time after well is shut in) (pressure buildup, shut-in time)	t
p_b	D_b	DENB	bulk density	m/L^3
K	K_b	BKM	bulk modulus	m/Lt^2
V_b	v_b	VOLB	bulk volume	L^3
V_{bE}	v_{bt}	VOLBEX	bulk volume of pack burned in experimental tube run	L^3
f_V	f_{Vb}, V_{bt}	FRCVB	bulk (total) volume, fraction of	
V_{Rb}	v_{Rb}	VOLRB	burned reservoir rock, volume of	L^3
v_b	v_b, u_b	VELB	burning-zone advance rate (velocity of)	L/t

SPE NOMENCLATURE AND UNITS

Letter symbol	Reserve SPE letter symbol	Computer letter symbol	Quantity	Dimensions
C		ECQ	capacitance	q^2t^2/mL^2
Q_V	Z_V	CEXV	capacity, cation exchange, per unit pore volume	
Q_{Vt}	Z_{Vt}	CEXUT	capacity, cation exchange, per unit pore volume, total	
P_c	P_C, p_C	PRSCP	capillary pressure	m/Lt^2
C_i		INVI	capital investment, initial	M
C_k		INVK	capital investment, subsequent, in year k	M
C	C_t	INVT	capital investments, summation of all	M
P_{PV}		CFLPV	cash flow, discounted	M
P		CFL	cash flow, undiscounted	M
I_k		INCK	cash income, annual operating, over year k	M
I		INC	cash income, operating	M
I_a		INCA	cash income, operating, after taxes	M
I_b		INCB	cash income, operating, before taxes	M
p_{cf}	P_{ct}	PRSCF	casing pressure, flowing	m/Lt^2
p_{cs}	P_{cs}	RSCS	casing pressure, static	m/Lt^2
Q_V	Z_V	CEXV	cation exchange capacity per unit pore volume	
Q_{Vt}	Z_{Vt}	CEXUT	cation exchange capacity per unit pore volume, total	
m		MXP	cementation (porosity) exponent (in an empirical relation between F_R and ϕ)	
Q	q	CHG	charge (current times time)	q
K_{ani}	M_{ani}	COEANI	coefficient, anisotropy	
α	M_α	COEA	coefficient, attenuation	I/L
h	h_h, h_T	HTCC	coefficient, convective heat transfer	m/t^3T
D	μ, δ	DFN	coefficient, diffusion	L^2/t
K_c	M_c, K_{ec}	COEC	coefficient, electrochemical	mL^2/t^2q
K_R	$M_R a, C$	COER	coefficient, formation resistivity factor ($F_R\phi^m$)	
K		KSP	coefficient in the equation of the electro-chemical component of the SP (spontaneous electromotive force)	mL^2/t^2q
C		CGW	coefficient of gas-well backpressure curve	$L^{3-2n}t^{4n}/m^{2n}$
U	U_T, U_θ	HTCU	coefficient heat transfer, over-all	m/t^3T
I	I_T, I_θ	HTCI	coefficient, heat transfer, radiation	m/t^3T
β	b	HEC	coefficient, thermal cubic expansion	I/T
K	M	COE	coefficient or multiplier	various
S_L	p_L, s_L	SATL	combined total liquid saturation	
log			common logarithm, base 10	
n_{pj}	N_{pj}	MOLPJ	component j, cumulative moles produced	
n_j	N_j	MOLJ	component j, moles of	
x		MFRL	component, mole fraction of, in liquid phase	
z		MFRM	component, mole fraction of, in mixture	
y		MFRV	component, mole fraction of, in vapour phase	
C	n_C	NMBC	components, number of	
E_c	Φ_c	EMFC	component of the SP, electrochemical	mL^2/t^2q
E_k	Φ_k	EMFK	component of the SP, electrokinetic	mL^2/t^2q
c	k, κ	CMP	compressibility	Lt^2/m
z	Z	ZED	compressibility factor (gas deviation factor, $z=PV/nRT$)	

Letter symbol	Reserve SPE letter symbol	Computer letter symbol	Quantity	Dimensions
$z_{\bar{p}}$	$Z_{\bar{p}}$	ZEDPAV	compressibility factor or deviation factor for gas, at mean pressure	
c_f	k_f, κ_f	CMPF	compressibility, formation or rock	Lt^2/m
c_g	k_g, κ_g	CMPG	compressibility, gas	Lt^2/m
c_o	k_o, κ_o	CMPO	compressibility, oil	Lt^2/m
C_{pr}	k_{pr}, κ_{pr}	CMPPRD	compressibility, pseudo-reduced	
C_w	k_w, κ_w	CMPW	compressibility, water	Lt^2/m
A_c		AMPC	compressional wave amplitude	various
C	c, n	CNC	concentration	various
C_{C1}	c_{C1}	CNCC1	concentration, methane (concentration of other paraffin hydrocarbons would be indicated similarly, C_{C2}, C_{C3}, etc.)	various
C_{O_2}	c_{O_2}	CNCO2	concentration, oxygen (concentration of other elements or compounds would be indicated similarly, C_{CO_2}, C_{N_2}, etc.)	various
C_m	c_m, n_m	CNCFU	concentration, unit fuel (see symbol m)	various
G_L	g_L	NGLTI	condensate liquids in place in reservoir, initial	L^3
G_{Lp}	g_{Lp}	NGLP	condensate liquids produced, cumulative	L^3
C_L	C_L, n_L	CNTL	condensate or natural gas liquids content	various
σ	γ	SIG	conductivity (other than logging)	various
C	σ	ECN	conductivity (electrical logging)	tq^2/mL^3
C_a	σ_a	ECNA	conductivity, apparent	tq^2/mL^3
C_{fD}		CNDFQ	conductivity, fracture, dimensionless	
k_h	λ	HCN	conductivity, thermal (always with additional phase or system subscripts)	mL/t^3T
w	z	ARR	constant, Arrhenius reaction rate velocity constant	L^3/m
λ	C	LAM	constant, decay (I/τ_d)	I/t
ε		DIC	constant, dielectric	q^2t^2/mL^3
γ			constant, Euler's = 0.5772	
D_c		DSCC	constant-income discount factor	
h		HPC	constant, hyperbolic decline $$q = q_i / \left[1 + \frac{a_i t}{h}\right]^h$$	
R		RRR	constant, universal gas (per mole)	mL^2/t^2T
C		WDC	constant, water-drive	L^4t^2/m
C_L		WDCL	constant, water-drive, linear aquifer	L^4t^2/m
m	F_F	FCM	consumption, fuel	various
m_E	F_{FE}	FCMEX	consumption of fuel in experimental tube run	m/L^3
m_{tg}	F_{FEg}	FCMEXG	consumption of fuel in experimental tube run (mass of fuel per mole of produced gas)	m
m_R	F_{FR}	FCMR	consumption of fuel in reservoir	m/L^3
Θ_c	Γ, γ_c	ANGC	contact angle	
C_L	c_L, n_L	CNTL	content, condensate or natural gas liquids	various
C_{wg}	c_{wg}, n_{wg}	CNTWG	content, wet-gas	various
h	h_h, h_T	HTCC	convective heat transfer coefficient	m/t^3T

SPE NOMENCLATURE AND UNITS 271

Letter symbol	Reserve SPE letter symbol	Computer letter symbol	Quantity	Dimensions
g_c		GRVC	conversion factor in Newton's second law of Motion	
B	C	COR	correction term or correction factor (either additive or multiplicative)	
N	n, C	NMB	count rate (general)	$1/t$
N_N	N_n, C_N	NEUN	count rate, neutron	$1/t$
N_{GR}	N_γ, C_G	NGR	count rate, gamma ray	$1/t$
S_{gc}	ρ_{gc}, s_{gc}	SATGC	critical gas saturation	
p_c	P_c	PRSC	critical pressure	m/Lt^2
T_c	θ_c	TEMC	critical temperature	T
S_{wc}	ρ_{wc}, s_{wc}	SATWC	critical water saturation	
A	S	ARA	Cross-section (area)	L^2
Σ	S	XSTMAC	cross-section, macroscopic	I/L
σ		XSTMIC	cross-section, microscopic	I/L
σ	s	XNL	cross-section of a nucleus, microscopic	L^2
β	b	HEC	cubic expansion coefficient, thermal	I/T
G_{Lp}	g_{Lp}	NGLP	cumulative condensate liquids produced	L^3
G_{Fp}	g_{Fp}	GASFP	cumulative free gas produced	L^3
G_e	g_e	GASE	cumulative gas influx (encroachment)	L^3
G_i	g_i	GASI	cumulative gas injected	L^3
R_p	F_{gp}, F_{gop}	GORP	cumulative gas–oil ratio	
G_p	g_p	GASP	cumulative gas produced	L^3
n_{pj}	N_{pj}	MOLPJ	cumulative moles of component j produced	
N_e	n_e	OILE	cumulative oil influx (encroachment)	L^3
N_p	n_p	OILP	cumulative oil produced	L^3
Q_p		FLUP	cumulative produced fluids (where N_p and W_p are not applicable)	
W_e	w_e	WTRE	cumulative water influx (encroachment)	L^3
W_i	w_i	WTRI	cumulative water injected	L^3
F_{wop}		FACWOP	cumulative water–oil ratio	mL/t^2
W_p	w_p	WTRP	cumulative water produced	L^3
G_{wgp}	g_{wgp}	GASWGP	cumulative wet gas produced	L^3
∇x			curl	
I	i script i, i	CUR	current, electric	q/t
r_s	R_s	RADS	damage or stimulation radius of well (skin)	L
F_s	F_d	DMRS	damage ratio or condition ratio (conditions relative to formation conditions unaffected by well operations)	
Z	D, h	ZEL	datum, elevation referred to	L
λ	C	LAM	decay constant ($1/\tau_d$)	$1/t$
τ_d	t_d	TIMD	decay time (mean life) ($1/\lambda$)	t
t_{dN}		TIMDN	decay time, neutron (neutron mean life)	t
h		HPC	decline constant, hyperbolic [from equation $q = q_i / \left[1 + \dfrac{a_i t}{h} \right]^h$	
d		DECE	decline factor, effective	
a		DEC	decline factor, nominal	
δ	Δ	DCR	decrement	various
F		DGF	degrees of freedom	

Letter symbol	Reserve SPE letter symbol	Computer letter symbol	Quantity	Dimensions
∇		DEL	del (gradient operator)	
t_d	τ_d	TIMDY	delay time	t
D		DLV	deliverability (gas well)	L^3/t
ρ	D	DEN	density	m/L^3
ρ_a	D_a	DENA	density, apparent	m/L^3
ρ_b	D_b	DENB	density, bulk	m/L^3
ρ_f	D_f	DENF	density, fluid	m/L^3
ρ_{xo}	D_{xo}	DENXO	density, flushed zone	m/L^3
n	N	NMB	density (indicating 'number per unit volume')	$1/L^3$
ρ_F	D_F	DENFU	density, fuel	m/L^3
ρ_g	D_g	DENG	density, gas	m/L^3
ρ_{ma}	D_{ma}	DENMA	density, matrix (solids, grain)	m/L^3
n_N		NMBN	density (number) of neutrons	$1/L^3$
$\bar{\rho}_L$	$\bar{D}_L$	DENAVL	density of produced liquid, weight-weighted avg.	m/L^3
ρ_{sE}	D_{sE}	DENSEX	density of solid particles making up experimental pack	m/L^3
ρ_o	D_o	DENO	density, oil	m/L^3
γ	s, F_s	SPG	density, relative (specific gravity)	
ρ_t	D_t	DENT	density, true	m/L^3
ρ_w	D_w	DENW	density, water	m/L^3
D_E		EDE	depletion	
N_R	N_F	FUDR	deposition rate of fuel	$m/L^3 t$
D_P		EDP	depreciation	
D	y, H	DPH	depth	L
δ	r_s	SKD	depth, skin (logging)	L
z	Z	ZED	deviation factor (compressibility factor) for gas ($z = pV/nRT$)	
$z_{\bar{p}}$	$Z_{\bar{p}}$	ZEDPAV	deviation factor (compressibility factor) for gas, at mean pressure	
δ		ANGH	deviation, hole (drift angle)	
p_d	P_d	PRSD	dew-point pressure	m/Lt^2
d	D	DIA	diameter	L
d_h	d_H, D_h	DIAH	diameter, hole	L
d_i	d_I, D_i	DIAI	diameter, invaded zone (electrically equivalent)	L
$\bar{d}_p$	$\bar{D}_p$	DIAAVP	diameter, mean particle	L
ε		DIC	dielectric constant	$q^2 t^2/mL^3$
Δ		DEL	difference or difference operator, finite ($\Delta x = x_2 - x_1$ or $x - x_2$)	
D	μ, δ	DFN	diffusion coefficient	L^2/t
η		DFS	diffusivity, hydraulic ($k/\varphi c\mu$ or $\lambda/\varphi c$)	L^2/t
Q_{LtD}	Q_{LtD} script l	ENCLTQQ	dimensionless fluid influx function, linear aquifer	
Q_{tD}		ENCTQQ	dimensionless fluid influx function at dimensionless time t_D	
C_{fD}		CNDFQ	dimensionless fracture conductivity	
q_{gD}	Q_{gD}	RTEGQ	dimensionless gas production rate	
N		NUMQ	dimensionless number, in general (always with identifying subscripts) (Example: Reynolds number, N_{Re})	
q_{oD}	Q_{oD}	RTEOQ	dimensionless oil production rate	

SPE NOMENCLATURE AND UNITS

Letter symbol	Reserve SPE letter symbol	Computer letter symbol	Quantity	Dimensions
V_{pD}	v_{pD}	VOLPQ	dimensionless pore volume	
p_D	P_D	PRSQ	dimensionless pressure	
p_{tD}	P_{tD}	PRSTQQ	dimensionless pressure function at dimensionless time t_D	
q_D	Q_D	RTEQ	dimensionless production rate	
x_D			dimensionless quantity proportional to x	
r_D	R_D	RADQ	dimensionless radius	
t_D	τ_D	TIMQ	dimensionless time	
t_{Dm}	τ_{Dm}	TIMMQ	dimensionless time at condition m	
q_{wD}	Q_{wD}	RTEWQ	dimensionless water production rate	
Θ	α_d	ANGD	dip, angle of	
Θ_a	α_{da}	ANGDA	dip, apparent angle of	
Φ_a	β_{da}	DAZA	dip, apparent azimuth of	
Φ	β_d	DAZ	dip, azimuth of	
D_c		DSCC	discount factor, constant-income	
D		DSC	discount factor, general	
D_{sp}		DSCSP	discount factor, single-payment $[1/(1+i)k;$ or $e^{-jk}, j = \ln(1+i)]$	
D_{spc}		DSCSPC	discount factor, single-payment (constant annual rate) $[e^{-jk}(e^j - 1)/j]$	
i		RTED	discount rate	
P_{PV}		CFLPV	discounted cash flow	
K	d	DSP	dispersion coefficient	L^2/t
ψ		DSM	dispersion modulus (dispersion factor)	
s	L	DIS	displacement	L
E_{Db}	η_{Db}, e_{Db}	EFFDB	displacement efficiency from burned portion of *in situ* combustion pattern	
E_{Du}	η_{Du}, e_{Du}	EFFDU	displacement efficiency from unburned portion of *in situ* combustion pattern	
E_D	η_D, e_D	EFFD	displacement efficiency: volume of hydrocarbons (oil or gas) displaced from individual pores or small groups of pores divided by the volume of hydrocarbons in the same pores just prior to displacement	
δ	F_d	DPR	displacement ratio	
δ_{ob}	F_{dob}	DPROB	displacement ratio, oil from burned volume, volume per unit volume of burned reservoir rock	
δ_{ou}	F_{dou}	DPROU	displacement ratio, oil from unburned volume, volume per unit volume of unburned reservoir rock	
δ_{wb}	F_{dwb}	DPRWB	displacement ratio, water from burned volume, volume per unit volume of burned reservoir rock	
d	L_d, L_2	DUW	distance between adjacent rows of injection and production wells	L
a	L_a, L_1	DLW	distance between like wells (injection or production) in a row	L
L	s, l script l	LTH	distance, length, or length of path	L
Δr	ΔR	DELRAD	distance, radial (increment along radius)	L

Letter symbol	Reserve SPE letter symbol	Computer letter symbol	Quantity	Dimensions
∇			divergence	
r_d	R_d	RADD	drainage radius	L
Δt_{wf}	$\Delta \tau_{wf}$	DELTIMWF	drawdown time (time after well is opened to production) (pressure drawdown)	t
δ		ANGH	drift angle, hole (deviation)	
i_r		RORI	earning power or rate of return (internal, true, or discounted cash flow)	
s	S, σ	SKN	effect, skin	
d		DECE	effective decline factor	
r_{wa}	R_{wa}	RADWA	effective or apparent wellbore radius (includes effects of well damage or stimulation)	L
k_g	K_g	PRMG	effective permeability to gas	L^2
k_o	K_o	PRMO	effective permeability to oil	L^2
k_w	K_w	PRMW	effective permeability to water	L^2
ϕ_e	f_e, ε_e	PORE	effective porosity	
E	η, e	EFF	efficiency	
E_A	η_A, e_A	EFFA	efficiency, areal (used in describing results of model studies only): area swept in a model divided by total model reservoir area (see E_P)	
E_{Db}	η_{Db}, e_{Db}	EFFDB	efficiency, displacement, from burned portion of *in situ* combustion pattern	
E_{Du}	η_{Du}, e_{Du}	EFFDU	efficiency, displacement, from unburned portion of *in situ* combustion pattern	
E_D	η_D, e_D	EFFD	efficiency, displacement: volume of hydrocarbons (oil or gas) displaced from individual pores or small groups of pores divided by the volume of hydrocarbon in the same pores just prior to displacement	
E_I	η_I, e_I	EFFI	efficiency, invasion (vertical): hydrocarbon pore space invaded (affected, contacted) by the injection fluid or heat front divided by the hydrocarbon pore space enclosed in all layers behind the injected-fluid or heat front	
E_R	η_R, e_R	EFFR	efficiency, over-all reservoir recovery: volume of hydrocarbons recovered divided by volume of hydrocarbons in place at start of project ($E_R = E_P E_I E_D = E_V E_D$)	
E_p	η_P, e_P	EFFP	efficiency, pattern sweep (developed from areal efficiency by proper weighting for variations in net pay thickness, porosity and hydrocarbon saturation): hydrocarbon pore space enclosed behind the injected-fluid or heat front divided by total hydrocarbon pore space of the reservoir or project	
E_{Vb}	η_{Vb}, e_{Vb}	EFFVB	efficiency, volumetric, for burned portion only, *in situ* combustion pattern	
E_V	η_V, e_V	EFFV	efficiency, volumetric: product of pattern sweep and invasion efficiencies	
E	Y	ELMY	elasticity, modulus of (Young's modulus)	m/Lt^2

SPE NOMENCLATURE AND UNITS

Letter symbol	Reserve SPE letter symbol	Computer letter symbol	Quantity	Dimensions
I	i script i,i	CUR	electric current	q/t
Z_e	Z_E, η	MPDE	electric impedance	mL^2/tq^2
ρ	R	RHO	electrical resistivity (other than logging)	mL^3tq^2
R	ρ, r	RES	electrical resistivity (electrical logging)	mL^3tq^2
τ_e		TORE	electrical tortuosity	
d_i	d_I, D_i	DIAI	electrically equivalent diameter of the invaded zone	L
K_c	M_c, K_{ec}	COEC	electrochemical coefficient	mL^2/t^2q
E_c	Φ_c	EMFC	electrochemical component of the SP	mL^2/t^2q
E_k	Φ_k	EMFK	electrokinetic component of the SP	mL^2/t^2q
E	V	EMF	electromotive force	mL^2/t^2q
Z	D, h	ZEL	elevation referred to datum	L
G_e	g_e	GASE	encroachment or influx, gas, cumulative	L^3
ΔG_e	Δg_e	DELGASE	encroachment or influx, gas during an interval	L^3
N_e	n_e	OILE	encroachment or influx, oil, cumulative	L^3
ΔN_e	Δn_e	DELOILE	encroachment or influx, oil, during an interval	L^3
e	i	ENC	encroachment or influx rate	L^3/t
e_g	i_g	ENCG	encroachment or influx rate, gas	L3/t
e_o	i_o	ENCO	encroachment or influx rate, oil	L^3/t
e_w	i_w	ENCW	encroachment or influx rate, water	L3/t
W_e	w_e	WTRE	encroachment or influx, water, cumulative	L^3
ΔW_e	Δw_e	DELWTRE	encroachment or influx, water, during an interval	L^3
E	U	ENG	energy	mL^2/t^2
H	I	HEN	enthalpy (always with phase or system subscripts)	mL^2/t^2
H_s	I_s	HENS	enthalpy (net) of steam or enthalpy above reservoir temperature	mL^2/t^2
h	i	HENS	enthalpy, specific	L^2/t^2
s	σ	HERS	entropy, specific	L^2/t^2T
S	σ_t	HER	entropy, total	mL^2/t^2T
$\geqslant$		GE	equal to or larger than	
$\leqslant$		LE	equal to or smaller than	
K	k, F_{eq}	EQR	equilibrium ratio (y/x)	
d_i	d_I, D_i	DIAI	equivalent diameter (electrical) of the invaded zone	L
t_p	τ_p	TIMP	equivalent time well was on production prior to shut-in (pseudo-time)	t
R_{we}		RWE	equivalent water resistivity	mL^3tq^2
erf		ERF	error function	
$erfc$		ERFC	error function, complementary	
E_n			Euler number	
γ			Euler's constant = 0.5772	
β	b	HEC	expansion coefficient, thermal cubic	1/T
ϕ_E	f_E, ε_E	POREX	experimental pack porosity	
n		NGW	exponent of back-pressure curve, gas well	
m		MXP	exponent, porosity (cementation) (in an empirical relation between F_R and ϕ)	
n		SXP	exponent, saturation	
e^z	$\exp^z$	EXP	exponential function	

Letter symbol	Reserve SPE letter symbol	Computer letter symbol	Quantity	Dimensions
$-Ei(-x)$			exponential integral $\int_{x}^{\infty} \frac{e^{-t}}{t} dt$, x positive	
$Ei(x)$			exponential integral, modified $\lim_{\varepsilon \to 0} \left[\int_{-\infty}^{-\varepsilon} \frac{e^t}{t} dt + \int_{\varepsilon}^{\infty} \frac{e^t}{t} dt \right]$, x positive	
p_e	P_e	PRSE	external boundary pressure	m/Lt^2
r_e	R_e	RADE	external boundary radius	L
p_{ext}	P_{ext}	PRSXT	extrapolated pressure	m/Lt^2
z	Z	ZED	factor, compressibility (gas deviation factor $z = PV/nRT$)	
D		DSC	factor, discount	
d		DECE	factor, effective decline	
a		DEC	factor, nominal decline	
g_c		GRVC	factor, conversion, in Newton's second law of Motion	
F_R		FACHR	factor, formation resistivity, equals R_o/R_w (a numerical subscript to f indicates the value of R_w)	
f		FACF	factor, friction	
G	f_G	GMF	factor, geometrical (multiplier) (electrical logging)	
G_{an}	f_{Gan}	GMFAN	factor, geometrical (multiplier) annulus (electrical logging)	
G_i	f_{Gi}	GMFI	factor, geometrical (multiplier) invaded zone (electrical logging)	
G_p	f_{Gp}	GMFP	factor, geometrical (multiplier) pseudo (electrical logging)	
G_{xo}	f_{Gxo}	GMFXO	factor, geometrical (multiplier) flushed zone (electrical logging)	
G_m	f_{Gm}	GMFM	factor, geometrical (multiplier) mud (electrical logging)	
G_t	f_{Gt}	GMFT	factor, geometrical (multiplier) true (non-invaded zone) (electrical logging)	
F		FAC	factor in general, including ratios (always with identifying subscripts)	various
F_B		FACB	factor, turbulence	
w	m	MRT	flow rate, mass	m/t
Q	q, Φ	HRT	flow rate, heat	mL^2/t^3
u	ψ	VELV	flow rate or flux, per unit area (volumetric velocity)	L/t
q	Q	RTE	flow rate or production rate	L^3/t
$q_{\bar{p}}$	$Q_{\bar{p}}$	RTEPAV	flow rate or production rate at mean pressure	L^3/t
$\bar{q}$	$\bar{Q}$	RTEAV	flow rate or production rate, average	L^3/t
p_{iwf}	P_{iwf}	PRSIWF	flowing bottom-hole pressure, injection well	m/Lt^2
p_{wf}	P_{wf}	PRSWF	flowing pressure, bottom-hole	m/Lt^2
p_{cf}	P_{cf}	PRSCF	flowing pressure, casing	m/Lt^2
p_{tf}	P_{tf}	PRSTF	flowing pressure, tubing	m/Lt^2

SPE NOMENCLATURE AND UNITS

Letter symbol	Reserve SPE letter symbol	Computer letter symbol	Quantity	Dimensions
Δt_{wf}	$\Delta \tau_{wf}$	DELTIMWF	flowing time after well is opened to production (pressure drawdown)	t
F	f	FLU	fluid (generalized)	various
v_f	V_f, u_f	VACF	fluid interval velocity	L/t
Z	D, h	ZEL	fluid head or height or elevation referred to a datum	L
t_f script t	Δt_f	TACF	fluid interval transit time	t/L
ρ_f	D_f	DENF	fluid density	m/L^3
Q_{tD}		ENCTQQ	fluid influx function, dimensionless, at dimensionless time t_D	
Q_{LtD}	Q_{ltD} script l	ENCLTQQ	fluid influx function, linear aquifer, dimensionless	
Q_p	Q_{ltD} script l	FLUP	fluids, cumulative produced (where N_p and W_p are not applicable)	
ρ_{xo}	D_{xo}	DENXO	flushed-zone density	m/L^3
R_{xo}	ρ_{xo}, r_{xo}	RESXO	flushed-zone resistivity (that part of the invaded zone closest to the wall of the hole, where flushing has been maximum)	mL^{3}tq^2
G_w	f_{Gxo}	GMFXO	flushed-zone geometrical factor (fraction or multiplier)	
u	ψ	FLX	flux	various
u	ψ	VELV	flux or flow rate, per unit area (volumetric velocity)	L/t
F	Q	FCE	force, mechanical	mL/t^2
E	V	EMF	force, electromotive (voltage)	mL^2/t^{2}q
ϕ_R	f, ε_R	PORR	formation or reservoir porosity	
c_f	k_f, κ_f	CMPF	formation or rock compressibility	Lt^2/m
F_R		FACHR	formation resistivity factor – equals R_0/R_w (a numerical subscript to F indicates the value R_w)	
K_R	M_R, a, C	COER	formation resistivity factor coefficient ($F_R \phi^m$)	
R_t	ρ_t, r_t	REST	formation resistivity, true	mL^{3}tq^2
R_0	ρ_0, r_0	RESZR	formation resistivity when 100% saturated with water of resistivity R_w	mL^{3}tq^2
T_f	θ_f	TEMF	formation temperature	T
B_{gb}	F_{gb}	FVFGB	formation volume factor at bubble-point conditions, gas	
B_{ob}	F_{ob}	FVFOB	formation volume factor at bubble-point conditions, oil	
B_g	F_g	FVFG	formation volume factor, gas	
B_o	F_o	FVFO	formation volume factor, oil	
B_t	F_t	FVFT	formation volume factor, total (two-phase)	
B	F	FVF	formation volume factor volume at reservoir conditions divided by volume at standard conditions	
B_w	F_w	FVFW	formation volume factor, water	
f	F	FRC	fraction (such as the fraction of a flow stream consisting of a particular phase)	
f_g	F_g	FRCG	fraction gas	

PETROLEUM ENGINEERING: PRINCIPLES AND PRACTICE

Letter symbol	Reserve SPE letter symbol	Computer letter symbol	Quantity	Dimensions
f_L	F_L, f_t script l	FRCL	fraction liquid	
f_V	f_{vb}, V_{bf}	FRCVB	fraction of bulk (total) volume	
$f_{\phi sh}$	ϕ_{igfsh}	FIGSH	fraction of intergranular space ('porosity') occupied by all shales	
$f_{\phi w}$	ϕ_{igfw}	FIGW	fraction of intergranular space ('porosity') occupied by water	
$f_{\phi shd}$	ϕ_{imfshd}	FIMSHD	fraction of intermatrix space ('porosity') occupied by nonstructural dispersed shale	
C_{fD}		CNDFQ	fracture conductivity, dimensionless	
L_f	x_f	LTHFH	fracture half-length (specify 'in the direction of' when using x_f)	L
I_f	i_f, I_F, i_F	FRX	fracture index	
G	F	GFE	free energy (Gibbs function)	mL^2/t^2
I_{Ff}	i_{Ff}	FFX	free fluid index	
R_F	F_{gF}, F_{goF}	GORF	free gas–oil ratio, producing (free-gas volume/oil volume)	
G_{Fp}	g_{Fp}	GASFP	free gas produced, cumulative	L^3
G_{Fi}	g_{Fi}	GASFI	free-gas volume, initial reservoir ($=mNB_{oi}$)	L^3
R_F	F_{gf}, F_{goF}	GORF	free producing gas–oil ratio (free-gas volume/oil volume)	
f	ν	FQN	frequency	$1/t$
f		FACF	friction factor	
p_f	P_f	PRSF	front or interface pressure	m/Lt^2
C_m	c_m, n_m	CNCFU	fuel concentration, unit (see symbol m)	various
m	F_F	FCM	fuel consumption	various
m_E	F_{FE}	FCMEX	fuel consumption in experimental tube run	m/L^3
m_{Eg}	F_{FEg}	FCMEXG	fuel consumption in experimental tube run (mass of fuel per mole of produced gas)	m
m_R	F_{FR}	FCMR	fuel consumption in reservoir	m/L^3
ρ_F	D_F	DENFU	fuel density	m/L^3
N_R	N_F	FUDR	fuel deposition rate	$m/L^3 t$
f		FUG	fugacity	m/Lt^2
N_{GR}	N_γ, C_G	NGR	gamma ray count rate	$1/t$
γ		GRY	gamma ray [usually with identifying subscript(s)]	various
G	g	GAS	gas (any gas, including air) always with identifying subscripts	various
S_{og}	ρ_{og}, s_{og}	SATOG	gas-cap interstitial-oil saturation	
S_{wg}	ρ_{wg}, s_{wg}	SATWG	gas-cap interstitial-water saturation	
c_g	k_g, κ_g	CMPG	gas compressibility	Lt^2/m
z	Z	ZED	gas compressibility factor (deviation factor) ($z = PV/nRT$)	
R		RRR	gas constant, universal (per mole)	mL^2/t^2T
ρ_g	D_g	DENG	gas density	m/L^3
$z_{\bar{p}}$	$Z_{\bar{p}}$	ZEDPAV	gas deviation factor (compressibility factor) at mean pressure	
z	Z	ZED	gas deviation factor (compressibility factor, $z = PV/nRT$) (deviation factor)	
k_g	K_g	PRMG	gas, effective permeability to	L^2
B_g	F_g	FVFG	gas formation volume factor	

SPE NOMENCLATURE AND UNITS

Letter symbol	Reserve SPE letter symbol	Computer letter symbol	Quantity	Dimensions
B_{gb}	F_{gb}	FVFGB	gas formation volume factor at bubble-point conditions	
f_g	F_g	FRCG	gas fraction	
G	g	GASTI	gas in place in reservoir, total initial	L^3
G_e	g_e	GASE	gas influx (encroachment), cumulative	L^3
ΔG_e	Δg_e	DELGASE	gas influx (encroachment) during an interval	L^3
e_g	i_g	ENCG	gas influx (encroachment) rate	L^3/t
G_i	g_i	GASI	gas injected, cumulative	L^3
ΔG_i	Δg_i	DELGASI	gas injected during an interval	L^3
i_g		INJG	gas injection rate	L^3/t
C_L	c_L, n_L	CNTL	gas liquids, natural, or condensate content	various
λ_g		MOBG	gas mobility	$L^3 t/m$
f_g	F_g	FRCG	gas, fraction	
f_g	F_g	MFRTV	gas mole fraction $\left[\dfrac{V}{L+V}\right]$	
k_g/k_o	K_g/K_o	PRMGO	gas–oil permeability ratio	
R_p	F_{gp}, F_{gop}	GORP	gas–oil ratio, cumulative	
R_F	F_{gF}, F_{goF}	GORF	gas–oil ratio, free producing (free-gas volume/oil volume)	
R	F_g, F_{go}	GOR	gas–oil ratio, producing	
R_{sb}	F_{gsb}	GORSB	gas–oil ratio, solution at bubble-point conditions	
R_s	F_{gs}, F_{gos}	GORS	gas–oil ratio, solution (gas solubility in oil)	
R_{si}	F_{gsi}	GORSI	gas–oil ratio, solution, initial	
G_p	g_p	GASP	gas produced, cumulative	L^3
ΔG_p	Δg_p	DELGASP	gas produced during an interval	L^3
G_{pE}	g_{pE}	GASPEX	gas produced from experimental tube run	L^3
q_g	Q_g	RTEG	gas production rate	L^3/t
q_{gD}	Q_{gD}	RTEGQ	gas production rate, dimensionless	
b_g	f_g, F_g	RVFG	gas reciprocal formation volume factor	
b_{gb}	f_{gb}, F_{gb}	RVFGB	gas reciprocal formation volume factor at bubble-point conditions	
G_{pa}	g_{pa}	GASPUL	gas recovery, ultimate	L^3
k_{rg}	K_{rg}	PRMRG	gas, relative permeability to	
S_g	ρ_g, s_g	SATG	gas saturation	
S_{gc}	ρ_{gc}, s_{gc}	SATGC	gas saturation, critical	
S_{gr}	ρ_{gr}, s_{gr}	SATGR	gas saturation, residual	
R_s	F_{gs}, F_{gos}	GORS	gas solubility in oil (solution gas–oil ratio)	
R_{sw}		GWRS	gas solubility in water	
γ_g	s_g, F_{gs}	SPGG	gas specific gravity	
μ_g	η_g	VISG	gas viscosity	m/Lt
μ_{ga}	η_{ga}	VISGA	gas viscosity at 1 atm	m/Lt
C		CGW	gas-well back-pressure curve, coefficient of	$L^{3-2n} t^{4n}/m^{2n}$
n		NGW	gas-well back-pressure curve, exponent of	
D		DLV	gas-well deliverability	L^3/t
G_{wgp}	g_{wgp}	GASWGP	gas, wet, produced, cumulative	L^3
h	d, e	THK	general and individual bed thickness	L
N		NUMQ	general dimensionless number (always with identifying subscripts)	

Letter symbol	Reserve SPE letter symbol	Computer letter symbol	Quantity	Dimensions
G	f_G	GMF	geometrical factor (multiplier) (electrical logging)	
G_{an}	f_{Gan}	GMFAN	geometrical factor (multiplier), annulus (electrical logging)	
G_{xo}	f_{Gxo}	GMFXO	geometrical factor (multiplier), flushed zone (electrical logging)	
G_i	f_{Gi}	GMFI	geometrical factor (multiplier), invaded zoned (electrical logging)	
G_m	f_{Gm}	GMFM	geometrical factor (multiplier), mud (electrical logging)	
G_t	f_{Gt}	GMFT	geometrical factor, (multiplier), true (non-invaded zone) (electrical logging)	
G_p	f_{Gp}	GMFP	geometrical factor (multiplier), pseudo (electrical logging)	
G_t	f_{Gt}	GMFT	geometrical factor (multiplier), true (electrical logging)	
g	γ	GRD	gradient	various
g_G	g_g	GRDGT	gradient, geothermal	T
∇			gradient operator	
g_T	g_h	GRDT	gradient, temperature	T
ρ_{ma}	D_{ma}	DENMA	grain (matrix, solids) density	m/L^3
g		GRV	gravity, acceleration of	L/t^2
γ	s, F_s	SPG	gravity, specific, relative density	
γ_g	s_g, F_{gs}	SPGG	gravity, specific, gas	
γ_o	s_o, F_{os}	SPGO	gravity, specific, oil	
γ_w	s_w, F_{ws}	SPGW	gravity, specific, water	
h_t	d_t, e_t	THKT	gross (total) pay thickness	L
V_u	R_u	GRRU	gross revenue ('value') per unit produced	M/L^3
V	R, V_t, R_t	GRRT	gross revenue ('value'), total	M
$t_{1/2}$		TIMH	half life	t
Q	q, Φ	HRT	heat flow rate	mL^2/t^3
L_v	λ_v	HLTV	heat of vaporization, latent	L^2/t^2
α	a, η_h	HTD	heat or thermal diffusivity	L^2/t
C	c	HSP	heat, specific (always with phase or system subscripts)	L^2/t^2T
h	h_h, h_T	HTCC	heat transfer coefficient, convective	m/t^3T
U	U_T, U_θ	HTCU	heat transfer coefficient, over-all	m/t^3T
I	I_T, I_θ	HTCI	heat transfer coefficient, radiation	m/t^3T
Z	D, h	ZEL	height, or fluid head or elevation referred to a datum	L
h	d, e	ZHT	height (other than elevation)	L
A	S_H	HWF	Helmholtz function (work function)	mL^2/t^2
y	f	HOL	hold-up (fraction of the pipe volume filled by a given fluid: y_o is oil hold-up, y_w is water hold-up $\Sigma$of all hold-ups at a given level is one)	
δ		ANGH	hole deviation, drift angle	
d_h	d_H, D_h	DIAH	hole diameter	L
η		DFS	hydraulic diffusivity ($k/\phi c \mu$ or $\lambda \phi c$)	L^2/t

SPE NOMENCLATURE AND UNITS

Letter symbol	Reserve SPE letter symbol	Computer letter symbol	Quantity	Dimensions
r_H	R_H	RADHL	hydraulic radius	L
T_H		TORHL	hydraulic tortuosity	
ϕ_h	f_h, ε_h	PORH	hydrocarbon-filled porosity, fraction or percent of rock bulk volume occupied by hydrocarbons	
I_R	i_R	RSXH	hydrocarbon resistivity index R_t/R_0	
S_{hr}	ρ_{hr}, s_{hr}	SATHR	hydrocarbon saturation, residual	
I_H	i_H	HYX	hydrogen index	
h		HPC	hyperbolic decline constant (from equation) $q = q_i / \left[1 + \dfrac{a_i t}{h}\right]^h$	
$g(z)$ script I			imaginary part of complex number z	
Z		MPD	impedance	various
Z_a		MPDA	impedance, acoustic	m/L²t
Z_e	Z_E, η	MPDE	impedance, electric	mL²/tq²
I	ι	_X	index (use subscripts as needed)	
I_f	i_f, I_F, i_F	FRX	index, fracture	
I_{Ff}	i_{Ff}	FFX	index, free fluid	
I_H	i_H	HYX	index, hydrogen	
I	i	IJX	index, injectivity	L⁴t/m
n	μ	RFX	index of refraction	
I_ϕ	i_ϕ	PRX	index, porosity	
$I_{\phi 1}$	$i_{\phi 1}$	PRXPR	index, primary porosity	
J	j	PDX	index, productivity	L⁴t/m
I_R	i_R	RXSH	index, (hydrocarbon) resistivity R_t/R_0	
$I_{\phi 2}$	$i_{\phi 2}$	PRXSE	index, secondary porosity	
I_{shGR}	i_{shGR}	SHXGR	index, shaliness gamma-ray $(\gamma_{\log} - \gamma_{cn})/(\gamma_{sh} - \gamma_{cn})$	
I_s	i_s	IJXS	index, specific injectivity	L³t/m
J_s	j_s	PDXS	index, specific productivity	L³t/m
h	d, e	THK	individual bed thickness	L
G_e	g_e	GASE	influx (encroachment), cumulative, gas	L³
N_e	n_e	OILE	influx (encroachment), cumulative, oil	L³
W_e	w_e	WTRE	influx (encroachment), cumulative, water	L³
ΔG_e	Δg_e	DELGASE	influx (encroachment) during an interval, gas	L³
ΔN_e	Δn_e	DELOILE	influx (encroachment) during an interval, oil	L³
ΔW_e	Δw_e	DELWTRE	influx (encroachment) during an interval, water	L³
Q_{LtD}	Q_{ltD} script l	ENCLTQQ	influx function, fluid, linear aquifer, dimensionless	
Q_{tD}	Q_{ltD} script l	ENCTQQ	influx function, fluid, dimensionless (at dimensionless time t_D)	
e	i	ENC	influx (encroachment) rate	L³/t
e_g	i_g	ENCG	influx (encroachment) rate, gas	L³/t
e_o	i_o	ENCO	influx (encroachment) rate, oil	L³/t
e_w	i_w	ENCW	influx (encroachment) rate, water	L³/t
G_L	g_L	NGLTI	initial condensate liquids in place in reservoir	L³
C_i		INVI	initial capital investment	M

Letter symbol	Reserve SPE letter symbol	Computer letter symbol	Quantity	Dimensions
N	n	OILTI	initial oil in place in reservoir	L^3
p_i	P_i	PRSI	initial pressure	m/Lt^2
G_{Fi}	g_{Fi}	GASFI	initial reservoir free-gas volume ($=mNB_{oi}$) ($=GB_{gi}$)	L^3
R_{si}	F_{gsi}	GORSI	initial solution gas–oil ratio	L/t^2
W	w	WTRTI	initial water in place in reservoir	L^3
S_{wi}	ρ_{wi}, s_{wi}	SATWI	initial water saturation	
G_i	g_i	GASI	injected gas, cumulative	L^3
ΔG_i	Δg_i	DELGASI	injected gas during an interval	L^3
W_i	w_i	WTRI	injected water, cumulative	L^3
ΔW_i	Δw_i	DELWTRI	injected water during an interval	L^3
i		INJ	injection rate	L^3/t
i_a		INJA	injection rate, air	L^3/t
i_g		INJG	injection rate, gas	L^3/t
i_w		INJW	injection rate, water	L^3/t
p_{iwf}	P_{iwf}	PRSIWF	injection well bottom-hole pressure, flowing	m/Lt^2
p_{iws}	P_{iws}	PRSIWS	injection well bottom-hole pressure, static	m/Lt^2
I	i	IJX	injectivity index	$L^4 t/m$
I_s	i_s	IJXS	injectivity index, specific	$L^3 t/m$
G_L	g_L	NGLTI	in-place condensate liquids in reservoir, initial	L^3
G	g	GASTI	in-place gas in reservoir, total initial	L^3
N	n	OILTI	in-place oil in reservoir, initial	L^3
W	w	WTRTI	in-place water in reservoir, initial	L^3
F_{wo}		FACWO	instantaneous producing water–oil ratio	
b	Y	ICP	intercept	various
i		IRCE	interest rate, effective compound (usually annual)	
i_M		IRPE	interest rate, effective, per period	
j	r	IRA	interest rate, nominal annual	
p_f	P_f	PRSF	interface or front pressure	M/Lt^2
σ	y, γ	SFT	interfacial, surface tension	m/t^2
ϕ_{ig}	f_{ig}, ε_{ig}	PORIG	intergranular 'porosity' (space) $(V_b - V_{gr})/V_b$	
$-Ei(-x)$			integral, exponential $\int_x^\infty \frac{e^{-t}}{t} dt$, x positive	
$Ei(x)$			integral, exponential, modified $\lim_{\varepsilon \to 0} \left[\int_{-\infty}^{-\varepsilon} \frac{e^t}{t} dt + \int_\varepsilon^\infty \frac{e^t}{t} dt \right]$, x positive	
$f_{\phi sh}$	ϕ_{igfsh}	FIGSH	intergranular space (porosity), fraction occupied by all shales	
$f_{\phi w}$	ϕ_{igfw}	FIGW	intergranular space (porosity), fraction occupied by water	
$f_{\phi shd}$	ϕ_{imfshd}	FIMSHD	intermatrix space (porosity), fraction occupied by non-structural dispersed shale	
ϕ_{im}	f_{im}, ε_{im}	PORIM	intermatrix 'porosity' (space) $(V_b - V_{ma})/V_b$	
U	E_i	INE	internal energy	mL^2/t^2
S_{og}	ρ_{og}, s_{og}	SATOG	interstitial-oil saturation in gas cap	

SPE NOMENCLATURE AND UNITS

Letter symbol	Reserve SPE letter symbol	Computer letter symbol	Quantity	Dimensions
S_{wg}	p_{wg}, S_{wg}	SATWG	interstitial-water saturation in gas cap	
S_{wo}	S_{wb}	SATWO	interstitial-water saturation in oil band	
t script t	Δ_t	TAC	interval transit time	t/L
t_a script t	Δt_a	TACA	interval transit time, apparent	t/L
M	$m_{\theta D}$	SAD	interval transit time-density slope (absolute value)	tL^2/m
t_f script t	Δt_f	TACF	interval transit time, fluid	t/L
t_{ma} script t	Δt_{ma}	TACMA	interval transit time, matrix	t/L
t_{sh} script t	Δt_{sh}	TACSH	interval transit time, shale	t/L
d_i	d_I, D_i	DIAI	invaded zone diameter, electrically equivalent	L
G_i	f_{Gi}	GMFI	invaded zone geometrical factor (multiplier) (electrical logging)	
R_i	ρ_i, r_i	RESI	invaded zone resistivity	mL^3tq^2
E_I	η_I, e_I	EFFI	invasion (vertical) efficiency: hydrocarbon pore space invaded (affected, contacted) by the injected-fluid or heat front divided by the hydrocarbon pore space enclosed in all layers behind the injected-fluid or heat front	
S_{iw}	ρ_{iw}, S_{iw}	SATIW	irreducible water saturation	
v	N	VSK	kinematic viscosity	L^2/t
E_k		ENGK	kinetic energy	mL^2/t^2
$\mathcal{L}(y)$ script L			Laplace transform of y $\int_0^\infty y(t) e^{-st} dt$	
s			Laplace transform variable	
∇			Laplacian operator	
$>$		GT	larger than	
L_v	λ_v	HLTV	latent heat of vaporization	L^2/t^2
L	s, ℓ script l	LTH	length, path length, or distance	L
$\bar{\tau}$	t	TIMAV	lifetime, average (mean life)	t
lim		LM	limit	
C_L		WDCL	linear aquifer water-drive constant	L^4t^2/m
f_L	$F_L, f,$ script l	FRCL	liquid fraction	
f_L	$F_L, f,$ script l	MFRTL	liquid mole fraction $L/(L+V)$	
x		MFRL	liquid phase, mole fraction of component in	
L	n_L	MOLL	liquid phase, moles of	
S_L	ρ_L, s_L	SATL	liquid saturation, combined total	
G_L	g_L	NGLTI	liquids, condensate, in place in reservoir, initial	
G_{Lp}	g_{Lp}	NGLP	liquids, condensate, produced cumulative	L^3
$\log_a$			logarithm, base a	
$\log$			logarithm, common, base 10	
Ln			logarithm, natural, base e	
Σ	S	XSTMAC	macroscopic cross section	$1/L$
σ	s	XNL	macroscopic cross section of a nucleus	L^2
μ	m	PRMM	magnetic permeability	mL/q^2
k	κ	SUSM	magnetic susceptibility	mL/q^2
M	I	MAG	magnetization	m/qt
M_f		MAGF	magnetization, fraction	

Letter symbol	Reserve SPE letter symbol	Computer letter symbol	Quantity	Dimensions
m		MAS	mass	m
w	m	MRT	mass flow rate	m/t
t_{ma} script	Δt_{ma}	TACMA	matrix interval transit time	t/L
ρ_{ma}	D_{ma}	DENMA	matrix (solids, grain) density	m/L^3
V_{ma}	v_{ma}	VOLMA	matrix (framework) volume (volume of all formation solids except dispersed clay or shale)	L^3
$\bar{\tau}$	$\bar{t}$	TIMAV	mean life (average lifetime)	t
$\bar{\tau}$	$\bar{t}$	TIMD	mean life (decay time) (1/τ)	t
$\bar{p}$	$\bar{P}$	PRSAV	mean or average pressure	m/Lt^2
—		AV	mean or average (overbar)	
μ		MEN	mean value of a random variable	
$\bar{d}$	$\bar{D}_p$	DIAAVP	mean particle diameter	L
$\bar{x}$		MENES	mean value of a random variable, x, estimated	
F	Q	FCE	mechanical force	mL/t^2
C_{c_1}	c_{C_i}	CNCCI	methane concentration (concentration of various other paraffin hydrocarbons would be indicated similarly C_{c_2}, C_{c_3}, etc.)	
σ		XSTMIC	microscopic cross section	L^2
z		MFRM	mixture, mole fraction of component	
λ		MOB	mobility (k/μ)	L^{3}t/m
λ_g		MOBG	mobility, gas	L^{3}t/m
λ_o		MOBO	mobility, oil	L^{3}t/m
M	F_λ	MBR	mobility ratio, general ($\lambda_{displacing}/\lambda_{displaced}$)	
$M_{\bar{S}}$	M_{Dd}, M_{su}	MBRSAV	mobility ratio, diffuse-front approximation $[(\lambda_D + \lambda_d)_{swept}/(\lambda_d)_{unswept}]$; D signifies displacing; d signifies displaced; mobilities are evaluated at average saturation conditions behind and ahead of front	
M	F_λ	MBR	mobility ratio, sharp-front approximation (λ_D/λ_d)	
M_t	$F_{\lambda,t}$	MBRT	mobility ratio, total, $[(\lambda_t)_{swept}/(\lambda_t)_{unswept}]$; 'swept' and 'unswept' refer to invaded and uninvaded regions behind and ahead of leading edge of displacement front	
λ_t	Λ	MOBT	mobility, total, of all fluids in a particular region of the reservoir, e.g., $(\lambda_o + \lambda_g + \lambda_w)$	L^{3}t/m
λ_w		MOBW	mobility, water	L^{3}t/m
K	K_b	BKM	modulus, bulk	m/Lt^2
ψ		DSM	modulus, dispersion, (dispersion factor)	
G	E_s	ELMS	modulus, shear	m/Lt^2
E	Y	ELMY	modulus of elasticity (Young's modulus)	m/Lt^2
V_M	v_m	VOLM	molal volume (volume per mole)	L^2
f_g	F_g	MFRTV	mole fraction gas V/(L + V)	
f_L	F_L, f_l script l	MFRTL	mole fraction liquid L/(L + V)	
x		MFRL	mole fraction of a component in liquid phase	
z		MFRM	mole fraction of a component in mixture	

SPE NOMENCLATURE AND UNITS

Letter symbol	Reserve SPE letter symbol	Computer letter symbol	Quantity	Dimensions
y		MFRV	mole fraction of a component in vapor phase	
R	N	MRF	molecular refraction	L^3
M		MWT	molecular weight (mass, relative)	m
M_L		MWTAVL	molecular weight of produced liquids, mole-weighted average	m
n	N	NMBM	moles, number of	
n_j	N_j	MOLJ	moles of component j	
n_{pj}	N_{pj}	MOLPJ	moles of component j produced, cumulative	
L	n_L	MOLL	moles of liquid phase	
V	n_v	MOLV	moles of vapor phase	
n_t	N_t	NMBMT	moles, number of, total	
M_L		MWTAVL	mole-weighted average molecular weight m of produced liquids	
R_{mc}	ρ_{mc}, r_{mc}	RESMC	mud-cake resistivity	mL^3tq^2
h_{mc}	d_{mc}, e_{mc}	THKMC	mud-cake thickness	L
R_{mf}	ρ_{mf}, r_{mf}	RESMF	mud-filtrate resistivity	mL^3tq^2
G_m	f_{Gm}	GMFM	mud geometrical factor (multiplier) (electrical logging)	
R_m	$\rho m, r_m$	RESM	mud resistivity	mL^3tq^2
G	f_G	GMF	multiplier (factor), geometrical (electrical logging)	
G_{an}	f_{Gan}	GMFAN	multiplier (factor), geometrical, annulus (electrical logging)	
G_{xo}	f_{Gxo}	GMFXO	multiplier (factor), geometrical, flushed zone (electrical logging)	
G_i	f_{Gi}	GMFI	multiplier (factor), geometrical, invaded zone (electrical logging)	
G_m	f_{Gm}	GMFM	multiplier (factor), geometrical, mud (electrical logging)	
G_p	f_{Gp}	GMFP	multiplier (factor), geometrical, pseudo (electrical logging)	
G_t	f_{Gt}	GMFT	multiplier (factor, geometrical, true (electrical logging)	
K	M	COE	multiplier or coefficient	various
C_L	c_L, n_L	CNTL	natural gas liquids or condensate content	various
In			natural logarithm, base e	
h_n	d_n, e_n	THKN	net pay thickness	L
N_N	N_n, C_N	NEUN	neutron count rate	I/t
n_N		NMBN	neutrons, density (number) of	
t_N	τ_N, t_n	NFL	neutron lifetime	1/t
N	$m_{\phi ND}$	SND	neutron porosity-density slope (absolute value)	L^3/m
N		NEU	neutron [usually with identifying subscript(s)]	various
g_c		GRVC	Newton's Second Law of Motion, conversion factor in	
a		DEC	nominal decline factor	
σ	s	XNL	nucleus cross section, microscopic	L^2
Z		ANM	number, atomic	
N		NUMQ	number, dimensionless, in general (always with identifying subscripts)	

PETROLEUM ENGINEERING: PRINCIPLES AND PRACTICE

Letter symbol	Reserve SPE Letter symbol	Computer letter symbol	Quantity	Dimensions
n	N	NMB	number (of variables, or components, or steps, or increments, etc.)	
n	N	NMB	number (quantity)	
M	m	NMBCP	number of compounding periods (usually per year)	
C_n	C	NMBC	number of components	
n_t	N_t	NMBM	number of moles, total	
N_{Re}		REYQ	number, Reynolds (dimensionless number)	
N	n	OIL	oil (always with identifying subscripts)	various
S_{wo}	S_{wb}	SATWO	oil band interstitial-water saturation	
c_o	$k_o, \kappa o$	CMPO	oil compressibility	Lt^2/m
ρ_o	D_o	DENO	oil density	m/L^3
δ_{ob}	F_{dob}	DPROB	oil displaced from burned volume, volume per unit volume of burned reservoir rock	
δ_{ou}	F_{dou}	DPROU	oil displaced from unburned volume, volume per unit volume of unburned reservoir rock	
k_o	K_o	PRMO	oil, effective permeability to	L^2
B_o	F_o	FVFO	oil formation volume factor	
B_{ob}	F_{ob}	FVFOB	oil formation volume factor at bubble point conditions	
R_s	F_{gs}, F_{gos}	GORS	oil, gas solubility in (solution gas-oil ratio)	
N	n	OILTI	oil in place in reservoir, initial	L^3
N_e		OILE	oil influx (encroachment) cumulative	L^3
ΔN_e	Δn_e	DELOILE	oil influx (encroachment) during an interval	L^3
e_o	i_o	ENCO	oil influx (encroachment) rate	L^3/t
λ_o		MOBO	oil mobility	$L^3 t/m$
N_p	n_p	OILP	oil produced, cumulative	L^3
ΔN_p	Δn_p	DELOILP	oil produced during an interval	L^3
q_o	Q_o	RTEO	oil production rate	L^3/t
q_{oD}	Q_{oD}	RTEOQ	oil production rate, dimensionless	
b_o	f_o, F_o	RVFO	oil reciprocal formation volume factor (shrinkage factor)	
N_{pa}	n_{pa}	OILPUL	oil recovery, ultimate	L^3
k_{ro}	K_{ro}	PRMRO	oil, relative permeability to	
S_o	ρ_o, s_o	SATO	oil saturation	
S_{og}	ρ_{og}, S_{og}	SATOG	oil saturation in gas cap, interstitial	
S_{or}	ρ_{or}, S_{or}	SATOR	oil saturation, residual	
γ_o	s_o, F_{os}	SPGO	oil specific gravity	
μ_o	v_o	VISO	oil viscosity	m/Lt
I		INC	operating cash income	M
I_a		INCA	operating cash income, after taxes	M
I		INCB	operating cash income, before taxes	M
O		XPO	operating expense	various
O_u		XPOU	operating expense per unit produced	M/L^3
∇^2			operator, Laplacian	
U	U_T, U_θ	HTCU	over-all heat transfer coefficient	$m/t^3 T$
E_R	η_R, e_R	EFFR	over-all reservoir recovery efficiency: volume of hydrocarbons recovered divided by volume of hydrocarbons in place at start of project ($E_R = E_p E_s E_o = E_v E_D$)	

SPE NOMENCLATURE AND UNITS

Letter symbol	Reserve SPE letter symbol	Computer letter symbol	Quantity	Dimensions
C_{O_2}	c_{O_2}	CNCO2	oxygen concentration (concentration of other various elements or compounds would be indicated as, C_{CO2}, C_N, etc.)	
e_{O2}	E_{O_2}	UTLO2	oxygen utilization	
$\bar{d}_p$	$\bar{D}_p$	DIAAVP	particle diameter, mean	L
L	$s, \ell\,script\,l$	LTH	path length, length, or distance	L
E_p	η, e_p	EFFP	pattern sweep efficiency (developed from areal efficiency by proper weighting for variations in net pay thickness, porosity and hydrocarbon saturation: hydrocarbon pore space enclosed behind the injected-fluid or heat front divided by total hydrocarbon pore space of the reservoir or project	
h_t	d_t, e_t	THKT	pay thickness, gross (total)	L
h_n	d_n, e_n	THKN	pay thickness, net	L
T	Θ	PER	period	t
k	K	PRM	permeability, absolute (fluid flow)	L^2
k_g	K_g	PRMG	permeability, effective, to gas	L^2
k_o	K_o	PRMO	permeability, effective, to oil	L^2
k_w	K_w	PRMW	permeability, effective, to water	L^2
μ	m	PRMM	permeability, magnetic	mL/q^2
k_g/k_o	K_g/K_o	PRMGO	permeability ratio, gas-oil	
k_w/k_o	K_w/K_o	PRMWO	permeability ratio, water-oil	
k_{rg}	K_{rg}	PRMRG	permeability, relative, to gas	
k_{ro}	K_{ro}	PRMRO	permeability, relative, to oil	
k_{rw}	K_{rw}	PRMRW	permeability, relative, to water	
P		NMBP	phases, number of	
μ	ν, σ	PSN	Poisson's ratio	
V_p	v_p	VOLP	pore volume $V_b - V_s$	L^3
V_{pD}	v_{pD}	VOLPQ	pore volume, dimensionless	
Q_i	q_i	FLUIQ	pore volumes of injected fluid, cumulative, dimensionless	
ϕ	f, ε	POR	porosity $(V_b - V_s)/V_b$	
ϕ_a	f_a, ε_a	PORA	porosity, apparent	
ϕ_e	f_e, ε_e	PORE	porosity, effective (V_{pe}/V_b)	
m		MXP	porosity exponent (cementation) (in an empirical relation between F_R and ϕ)	
ϕ_h	f_h, ε_h	PORH	porosity, hydrocarbon-filled, fraction or percent of rock bulk volume occupied by hydrocarbons	
I_ϕ	i_ϕ	PRX	porosity index	
$I_{\phi 1}$	$i_{\phi 1}$	PRXPR	porosity index, primary	
I_{ϕ_2}	i_{ϕ_2}	PRXSE	porosity index, secondary	
ϕ_{ne}	f_{ne}, ε_{ne}	PORNE	porosity, non-effective (V_{pne}/V_b)	
ϕ_{ig}	f_{ig}, ε_{ig}	PORIG	'porosity' (space), intergranular $(V_b - V_{gr})/V_b$	
ϕ_{im}	f_{im}, ε_{im}	PORIM	'porosity' (space), intermatrix $(V_b - V_{ma}/V_b)$	
ϕ_E	f_E, ε_E	POREX	porosity of experimental pack	
ϕ_R	f_r, ε_R	PORR	porosity of reservoir or formation	
ϕ_t	f_t, ε_t	PORT	porosity, total	

PETROLEUM ENGINEERING: PRINCIPLES AND PRACTICE

Letter symbol	Reserve SPE letter symbol	Computer letter symbol	Quantity	Dimensions
Φ	f	POT	potential or potential function various	
V	U	VLT	potential difference (electric)	mL^2/qt^2
E_p		ENGP	potential energy	mL^2/t^2
p_{bh}	P_{bh}	PRSBH	pressure, bottomhole	m/Lt^2
p	P	PRS	pressure	m/Lt^2
pa	P_a	PRSA	pressure, atmospheric	m/Lt^2
$\bar{p}$	$\bar{P}$	PRSAV	pressure, average or mean	m/Lt^2
$\bar{p}$	$\bar{P}_d$	PRSAVR	pressure, average, reservoir	m/Lt^2
p_{ws}	P_{ws}	PRSWS	pressure, bottomhole, at any time after shut-in	m/Lt^2
p_{wf}	P_{wf}	PRSWF	pressure, bottomhole flowing	m/Lt^2
p_{iwf}	P_{iwf}	PRSIWF	pressure, bottomhole flowing, injection well	m/Lt^2
p_w	P_w	PRSW	pressure, bottomhole general	m/Lt_2
p_{ws}	P_{ws}	PRSWS	pressure, bottomhole static	m/Lt^2
p_{ww}	P_{ww}	PRSWW	pressure, bottomhole (well), in water phase	m/Lt^2
p_{iws}	P_{iws}	PRSIWS	pressure, bottomhole static, injection well	m/Lt^2
p_b	P_s, P_s, P_b	PRSB	pressure, bubble-point (saturation)	m/Lt^2
P_c	P_c, p_c	PRSCP	pressure, capillary	m/Lt^2
p_{cf}	P_{cf}	PRSCF	pressure, casing flowing	m/Lt^2
p_{cs}	P_{cs}	PRSCS	pressure, casing static	m/Lt^2
p_c	P_c	PRSC	pressure, critical	m/Lt^2
p_d	P_d	PRSD	pressure, dew-point	m/Lt^2
p_D	P_D	PRSQ	pressure, dimensionless	
p_e	P_e	PRSE	pressure, external boundary	m/Lt^2
p_{ext}	P_{ext}	PRSXT	pressure, extrapolated	m/Lt^2
p_{wf}	P_{wf}	PRSWF	pressure, flowing bottomhole	m/Lt^2
p_{cf}	P_{cf}	PRSCF	pressure, flowing casing	m/Lt^2
p_{tf}	P_{tf}	PRSTF	pressure, flowing tubing	m/Lt^2
p_f	P_f	PRSF	pressure, front or interface	m/Lt^2
p_{tD}	P_{tD}	PRSTQQ	pressure function, dimensionless, at dimensionless time t_D	
p_i	P_i	PRSI	pressure, initial	m/Lt^2
p_{pc}	P_{pc}	PRSPC	pressure, pseudo-critical	m/Lt^2
p_{pr}	P_{pr}	PRSPRD	pressure, pseudo-reduced	m/Lt^2
p_r	P_r	PRSRD	pressure, reduced	
p_R	$\bar{P}$	PRSAVR	pressure, reservoir average	m/Lt^2
p_{sp}	P_{sp}	PRSSP	pressure, separator	m/Lt^2
p_{sc}	P_{sc}	PRSSC	pressure, standard conditions	m/Lt^2
p_{ws}	P_{ws}	PRSWS	pressure, static bottom-hole	m/Lt^2
p_{cs}	P_{cs}	PRSCS	pressure, static casing	m/Lt^2
p_{ts}	P_{ts}	PRSTS	pressure, static tubing	m/Lt^2
p_{tf}	P_{tf}	PRSTF	pressure, tubing flowing	m/Lt^2
p_{ts}	P_{ts}	PRSTS	pressure, tubing static	m/Lt^2
$I_{\phi 1}$	$i_{\phi 1}$	PRXPR	primary porosity index	
G_{Lp}	g_{Lp}	NGLP	produced condensate liquids, cumulative	L^3
Q_p		FLUP	produced fluids, cumulative (where Np and W_p are not applicable)	L^3
G_{Fp}	g_{Fp}	GASFP	produced free gas, cumulative	L^3
G_p	g_p	GASP	produced gas, cumulative	L^3
ΔG_p	Δg_p	DELGASP	produced gas during an interval	L^3

SPE NOMENCLATURE AND UNITS

Letter symbol	Reserve SPE letter symbol	Computer letter symbol	Quantity	Dimensions
G_{pE}	g_{pE}	GASPEX	produced gas from experimental tube run	L^3
G_{wgp}	g_{wgp}	GASWGP	produced gas, wet, cumulative	L^3
$\bar{\rho}_L$	$\bar{D}$	DENAVL	produced-liquid density, weight-weighted average	m/L^3
n_{pj}	N_{pj}	MOLPJ	produced moles of component j, cumulative	
N_p	n_p	OILP	produced oil, cumulative	L^3
ΔN_p	Δn_p	DELOILP	produced oil during an interval	L^3
W_p	w_p	WTRP	produced water, cumulative	L^3
Δw_p	Δw_p	DELWTRP	produced water during an interval	L^3
G_{wgp}	g_{wgp}	GASWGP	produced wet gas, cumulative	L^3
R	F_g, F_{go}	GOR	producing gas–oil ratio	
R_F	F_{gF}, F_{goF}	GORF	producing gas–oil ratio, free (free-gas volume/oil volume)	
F_{wo}		FACWO	producing water–oil ratio, instantaneous	
q_i	Q_i	RTEI	production rate at beginning of period	L^3/t
q_a	Q_a	RTEA	production rate at economic abandonment	L^3/t
q_D	Q_D	RTEQ	production rate, dimensionless	
q_g	Q_g	RTEG	production rate, gas	L^3/t
q_{gD}	Q_{gD}	RTEGQ	production rate, gas, dimensionless	
q_o	Q_o	RTEO	production rate, oil	L^3/t
q_{oD}	Q_{oD}	RTEOQ	production rate, oil, dimensionless	
q	Q	RTE	production rate or flow rate	L^3/t
$q_{\bar{p}}$	$Q_{\bar{p}}$	RTEPAV	production rate or flow rate at mean pressure	L^3/t
$\bar{q}$	$\bar{Q}$	RTEAV	production rate or flow rate, average	L^3/t
q_w	Q_w	RTEW	production rate, water	L^3/t
q_{wD}	Q_{wD}	RTEWQ	production rate, water, dimensionless	
Δt_{wf}	$\Delta\tau$	DELTIMWF	production time after well is opened to production (presure drawdown)	t
t_p	τ_p	TIMP	production time of well, equivalent, prior to shut-in (pseudo-time)	t
J	j	PDX	productivity index	$L^4 t/m$
P_k		PRAK	profit, annual net, over year k	M
f_{Pk}		PRAPK	profit, annual, over year k, fraction of unamortized investment	
P	P_t	PRFT	profit, total	M
$\propto$			proportional to	
J_s	j_s	PDXS	productivity index, specific	$L^3 t/m$
T_{pc}	θ_{pc}	TEMPC	pseudo-critical temperature	T
p_{pc}	P_{pc}	PRSPC	pseudo-critical pressure	m/Lt^2
G_p	f_{Gp}	GMFP	pseudo-geometrical factor (multiplier) (electrical logging)	
c_{pr}	K_{pr}, κ_{pr}	CMPPRD	pseudo-reduced compressibility	
p_{pr}	P_{pr}	PRSPRD	pseudo-reduced pressure	
E_{pSP}	Φ_{SP}	EMFP	pseudo-SP	mL^2/qt^2
T_{pr}	θ_{pr}	TEMPRD	pseudo-reduced temperature	T
t_p	τ_p	TIMP	pseudo-time (equivalent time well was on production prior to shut-in)	t
f_s	Q, x	QLTS	quality (usually of steam)	

Letter symbol	Reserve SPE letter symbol	Computer letter symbol	Quantity	Dimensions
Δr	ΔR	DELRAD	radial distance (increment along radius)	L
I	I_T, I_θ	HCTI	radiation heat transfer coefficient	$m/t^3 T$
r	R	RAD	radius	L
r_{wa}	R_{wa}	RADWA	radius, apparent or effective, of wellbore (includes effects of well damage or stimulation)	L
r_D	R_D	RADQ	radius, dimensionless	
r_e	R_e	RADE	radius, external boundary	L
r_H	R_H	RADHL	radius, hydraulic	L
r_d	R_d	RADD	radius of drainage	L
r_{wa}	R_{wa}	RADWA	radius of wellbore, apparent or effective (includes effects of well damage or stimulation)	L
r_s	R_s	RADS	radius of well damage or stimulation (skin)	L
r_w	R_w	RADW	radius, well	L
i_a		INJA	rate, air injection	L^3/t
i	k_i	RTE	rate: discount, effective profit, of return, reinvestment, etc; use symbol i with suitable subscripts	
q	Q	RTE	rate, flow or production	L^3/t
N_{GR}	N_γ, C_G	NGR	rate, gamma ray count	$1/t$
e_g	i_g	ENCG	rate, gas influx (encroachment)	L^3/t
i_g		INJG	rate, gas injection	L^3/t
q_g	Q_g	RTEG	rate, gas production	L^3/t
q_{gD}	Q_{gD}	RTEGQ	rate, gas production, dimensionless	
e	i	ENC	rate, influx (encroachment)	L^3/t
$\bar{x}$		MENES	random variable, mean value of x, estimated	
i		INJ	rate, injection	L^3/t
i		IRCE	rate, interest, effective compound (usually annual)	
i_M		IRPE	rate, interest, effective, per period	
j	r	IRA	rate, interest, nominal annual	
w	m	MRT	rate, mass flow	m/t
u	ψ	VELV	rate of flow or flux, per unit area (volumetric velocity)	L/t
Q	q, Φ	HRT	rate of heat flow	mL^2/t^3
i_r		RORI	rate of return (internal, true, or discounted cash flow) or earning power	
e_o	i_o	ENCO	rate, oil influx (encroachment)	L^3/t
q_o	Q_o	RTEO	rate, oil production	L^3/t
u	ψ	VELV	rate per unit area, flow (volumetric velocity)	L/t
q_{oD}	Q_{oD}	RTEOQ	rate, oil production, dimensionless	
q	Q	RTE	rate, production or flow	L^3/t
$q_{\bar{p}}$	$Q_{\bar{p}}$	RTEPAV	rate, production, at mean pressure	L^3/t
$\bar{q}$	$\bar{Q}$	RTEAV	rate, production, average	L^3/t
q_D	Q_D	RTEQ	rate, production, dimensionless	
q_s	Q_s	RTES	rate, segregation (in gravity drainage)	L^3/t
γ	e	SRT	rate, shear	$1/t$
v_b	V_b, U_b	VELB	rate (velocity) of burning-zone advance	L/t
e_w	i_w	ENCW	rate, water influx (encroachment)	L^3/t
i_w		INJW	rate, water injection	L^3/t

SPE NOMENCLATURE AND UNITS

Letter symbol	Reserve SPE letter symbol	Computer letter symbol	Quantity	Dimensions
q_w	Q_w	RTEW	rate, water production	L^3/t
q_{wD}	Q_{wD}	RTEWQ	rate, water production, dimensionless	
F_{aF}		FACAFU	ratio, air–fuel	various
F_s	F_d	DMRS	ratio, damage ('skin' conditions relative to formation conditions unaffected by well operations)	
δ	F_d	DPR	ratio, displacement	
δ_{ob}	F_{dob}	DPROB	ratio, displacement, oil from burned volume, volume per unit volume of burned reservoir rock	
δ_{ou}	F_{dou}	DPROU	ratio, displacement, oil from unburned volume, volume per unit volume of unburned reservoir rock	
δ_{wb}	F_{dwb}	DPRWB	ratio, displacement, water from burned volume, volume per unit volume of burned reservoir rock	
K	k, F_{eq}	EQR	ratio, equilibrium (y/x)	
R_F	F_{gF}, F_{goF}	GORF	ratio, free producing gas–oil (free-gas volume/oil volume)	
R_p	F_{gp}, F_{gop}	GORP	ratio, gas–oil, cumulative	
R_{si}	F_{gsi}	GORSI	ratio, gas–oil, initial solution	
k_g/K_o	K_g/k_o	PRMGO	ratio, gas–oil permeability	
R	F_g, F_{go}	GOR	ratio, gas–oil producing	
R_{sb}	F_{gsb}	GORSB	ratio, gas–oil, solution, at bubble-point conditions	
R_s	F_{gs}, F_{os}	GORS	ratio, gas–oil, solution (gas solubility in oil)	
M	F_λ	MBR	ratio, mobility, general ($\lambda_{\text{displacing}}/\lambda_{\text{displaced}}$)	
$M_{\bar{S}}$	M_{Dd}, M_{su}	MBRSAV	ratio, mobility, diffuse-front approximation $[(\lambda_D + \lambda_d)_{\text{swept}}/(\lambda_d)_{\text{unswept}}]$; D signifies displacing; d signifies displaced; mobilities are evaluated at average saturation conditions behind and ahead of front	
M	F_λ	MBR	ratio, mobility, sharp-front approximation (λ_D/λ_d)	
M_t	$F_{\lambda t}$	MBRT	ratio, mobility, total $[(\lambda_t)_{\text{swept}}/(\lambda_t)_{\text{unswept}}]$; 'swept' and 'unswept' refer to invaded and uninvaded regions behind and ahead of leading edge of a displacement front	
m	$F_F o, F_{go}$	MGO	ratio of initial reservoir free-gas volume to initial reservoir oil volume	
F		FAC	ratio or factor in general (always with identifying subscripts)	various
k_g/k_o	K_g/K_o	PRMGO	ratio, permeability, gas–oil	
R	F_g, F_{go}	GOR	ratio, producing gas–oil	
k_w/k_o	K_w/K_o	PRMWO	ratio, permeability, water–oil	
R_{sb}	F_{gsb}	GORSB	ratio, solution gas–oil, at bubble-point conditions	
R_s	F_{gs}, F_{gos}	GORS	ratio, solution gas–oil (gas solubility in oil)	
R_{si}	F_{gsi}	GORSI	ratio, solution gas–oil, initial	
F_{wF}		FACWFU	ratio, water–fuel	
F_{wop}		FACWOP	ratio, water–oil, cumulative	

PETROLEUM ENGINEERING: PRINCIPLES AND PRACTICE

Letter symbol	Reserve SPE letter symbol	Computer letter symbol	Quantity	Dimensions
k_w/k_o	K_w/K_o	PRMWO	ratio, water–oil permeability	
F_{wo}		FACWO	ratio, water–oil, producing, instantaneous	
X		XEL	reactance	ML^2/tq^2
k	r, j	RRC	reaction rate constant	L/t
$\mathscr{R}(z)$ script R			real part of complex number z	
b	f, F	RVF	reciprocal formation volume factor, volume at standard conditions divided by volume at reservoir conditions (shrinkage factor)	
b_g	f_g, F_g	RVFG	reciprocal gas formation volume factor	
b_{gb}	f_{gb}, F_{gb}	RVFGB	reciprocal gas formation volume factor at bubble-point conditions	
j	ω		reciprocal permeability	$1/L^2$
b_o	f_o, F_o	RVFO	reciprocal oil formation volume fator (shrinkage factor)	
E_R	η_R, e_R	EFFR	recovery efficiency, reservoir over-all; volume of hydrocarbons recovered divided by volume of hydrocarbons in place at start of project. ($E_R = E_p E_I E_D = E_v E_D$)	
G_{pa}	g_{pa}	GASPUL	recovery, ultimate gas	
p_r	P_r	PRSRD	reduced pressure	
T_r	θ_r	TEMRD	reduced temperature	
α		RED	reduction ratio or reduction term	
α_{SP}		REDSP	reduction, SP (general) due to shaliness	
R	N	MRF	refraction, molecular	
n	μ	RFX	refraction index	
α_{SPsh}		REDSH	reduction ratio, SP, due to shaliness	
A_r		AMPR	relative amplitude	
A		AWT	relative atomic mass (atomic weight)	
M		MWT	relative molecular weight (molecular weight)	
β	γ	BRGR	relative bearing	
γ	s, F_s	SPG	relative density (specific gravity)	
k_{rg}	K_{rg}	PRMRG	relative permeability to gas	
k_{ro}	K_{ro}	PRMRO	relative permeability to oil	
k_{rw}	K_{rw}	PRMRW	relative permeability to water	
t_2	ν_2	TIMAV	relaxation time, free-precession decay	t
t_1	τ_1	TIMRP	relaxation time, proton thermal	t
a	Fa	AIR	requirement, air	
a_E	F_{aE}	AIREX	requirement, unit air, in laboratory experimental run, volumes or air per unit mass of pack	L^3/m
a_R	F_{aR}	AIRR	requirement, unit air, in reservoir, volumes of air per unit bulk volume of reservoir rock	
G_{Fi}	g_{Fi}	GASFI	reservoir initial free-gas volume ($=mNB_{oi}$)	L^3
ϕ_R	f_R, ε_R	PORR	reservoir or formation porosity	
$\bar{p}$	$\bar{P}_R$	PRSAVR	reservoir pressure, average	m/Lt^2
E_R	η_R, e_R	EFFR	reservoir recovery efficiency, over-all; volume of hydrocarbons recovered divided by volume of hydrocarbons in place at start of project ($E_R = [sx] = E_v E_D$)	
V_{Rb}	v_{Rb}	VOLRB	reservoir rock burned, volume of	L^3

SPE NOMENCLATURE AND UNITS

Letter symbol	Reserve SPE letter symbol	Computer letter symbol	Quantity	Dimensions
V_{Ru}	v_{Ru}	VOLRU	reservoir rock unburned, volume of	L^3
T_R	θ_R	TEMR	reservoir temperature	T
S_{gr}	ρ_{gr}, s_{gr}	SATGR	residual gas saturation	
S_{hr}	ρ_{hr}, s_{hr}	SATHR	residual hydrocarbon saturation	
S_{or}	ρ_{or}, s_{or}	SATOR	residual oil saturation	
s_{wr}	ρ_{wr}, s_{wr}	SATWR	residual water saturation	
r	R	RST	resistance	ML^2/tq^2
R	ρ, r	RES	resistivity (electrical)	mL^3tq^2
R_{an}	ρ_{an}, r_{an}	RESAN	resistivity, annulus	mL^3tq^2
R_a	ρ_a, r_a	RESA	resistivity, apparent	mL^3tq^2
R_z	ρ_z, r_z	RESZ	resistivity, apparent, of the conductive fluids in an invaded zone (due to fingering)	mL^3tq^2
K_R	M_R, a, C	COER	resistivity factor coefficient, formation ($F_R\phi^m$)	mL^3tq^2
F_R		FACHR	resistivity factor, formation, equals R_0/R_w a numerical subscript to F indicates the R_w	
R_{xo}	ρ_{xo}, r_{xo}	RESXO	resistivity flushed zone (that part of the invaded zone closest to the wall of the borehole, where flushing has been the maximum)	mL^3tq^2
R_0	ρ_0, r_0	RESZR	resistivity, formation 100% saturated with water of resistivity R_w	mL^3tq^2
R_t	ρ_t, r_t	REST	resistivity, formation, true	mL^3tq^2
I_R	i_R	RSXH	resistivity index (hydrocarbon) equals R_t/R_0	
R_i	ρ_i, r_i	RESI	resistivity, invaded zone	mL^3tq^2
R_m	ρ_m, r_m	RESM	resistivity, mud	mL^3tq^2
R_{mc}	ρ_{mc}, r_{mc}	RESMC	resistivity, mud-cake	mL^3tq^2
R_{mf}	ρ_{mf}, r_{mf}	RESMF	resistivity, mud-filtrate	mL^3tq^2
R_{sh}	ρ_{sh}, r_{sh}	RESSH	resistivity, shale	mL^3tq^2
R_s	ρ_s, r_s	RESS	resistivity, surrounding formation	mL^3tq^2
R_w	ρ_w, r_w	RESW	resistivity, water	mL^3tq^2
V_u	R_u	GRRU	revenue, gross ('value'), per unit produced	M/L^3
V	R, V_t, R_t	GRRT	revenue, gross ('value'), total	M
N_{Re}		REYQ	Reynolds number (dimensionless number)	
c_f	k_f, κ_f	CMPF	rock or formation compressibility	Lt^2/m
C	c, n	CNC	salinity	various
S	ρ, s	SAT	saturation	
n		SXP	saturation exponent	
S_g	ρ_g, s_g	SATG	saturation, gas	
S_{gc}	ρ_{gc}, s_{gc}	SATGC	saturation, gas, critical	
S_{gr}	ρ_{gr}, s_{gr}	SATGR	saturation, gas, residual	
S_{og}	ρ_{og}, s_{og}	SATOG	saturation, interstitial-oil, in gas cap	
S_{wg}	ρ_{wg}, s_{wg}	SATWG	saturation, interstitial-water, in gas cap	
S_h	ρ_h, s_h	SATH	saturation, hydrocarbon	
S_{hr}	ρ_{hr}, s_{hr}	SATHR	saturation, residual hydrocarbon	
S_o	ρ_o, s_o	SATO	saturation, oil	
S_{or}	ρ_{or}, s_{or}	SATOR	saturation, oil, residual	

PETROLEUM ENGINEERING: PRINCIPLES AND PRACTICE

Letter symbol	Reserve SPE letter symbol	Computer letter symbol	Quantity	Dimensions
p_b	p_s, P_s, P_b	PRSB	saturation or bubble-point pressure	m/Lt^2
S_L	ρ_L, s_L	SATL	saturation, total (combined) liquid	
S_w	ρ_w, s_w	SATW	saturation, water	
S_{wc}	ρ_{wc}, s_{wc}	SATWC	saturation, water, critical	
S_{wi}	ρ_{wi}, s_{wi}	SATWI	saturation, water, initial	
S_{iw}	ρ_{iw}, s_{iw}	SATIW	saturation, water irreducible	
S_{wr}	ρ_{wr}, s_{wr}	SATWR	saturation, water, residual	
$I_{\phi 2}$	$i_{\phi 2}$	PRXSE	secondary porosity index	
q_s	Q_s	RTES	segregation rate (in gravity drainage)	L^3/t
p_{sp}	P_{sp}	PRSSP	separator pressure	m/Lt^2
t_{sh} script t	Δt_{sh}	TACSH	shale interval transit time	t/L
R_{sh}	ρ_{sh}, r_{sh}	RESSH	shale resistivity	mL^3tq^2
I_{shGR}	i_{shGR}	SHXGR	shaliness gamma-ray index $(\gamma_{\log} - \gamma_{cn})/(\gamma_{sh} - \gamma_{cn})$	
G	E_s	ELMS	shear modulus	m/Lt^2
$\dot{\gamma}$	$\dot{e}$	SRT	shear rate	$1/t$
A_s		AMPS	shear wave amplitude	various
b_o	f_o, F_o	RVFO	shrinkage factor (reciprocal oil formation volume factor)	
p_{ws}	P_{ws}	PRSWS	shut-in bottomhole pressure, at any time	m/Lt^2
Δt_{ws}	$\Delta \tau_{ws}$	DELTIMWS	shut-in time (time after well is shut in) (pressure buildup)	t
D_{SP}		DSCSP	single payment discount factor	
D_{SPC}		DSCSPC	single payment discount factor (constant annual rate)	
δ	r_s	SKD	skin depth (logging)	L
s	S, σ	SKN	skin effect	various
r_s	R_s	RADS	skin radius (radius of well damage or stimulation)	L
m	A	SLP	slope	various
M	$m_{\theta D}$	SAD	slope, interval transit time vs density (absolute value)	tL^2/m
N	$m_{\theta ND}$	SND	slope, neutron porosity vs density (absolute value)	L^3/m
$<$		LT	smaller than	
ρ_{sE}	D_{sE}	DENSEX	solid particles density of experimental rock	m/L^3
V_s	v_s	VOLS	solid(s) volume (volume of all formation solids)	L^3
ρ_{ma}	D_{ma}	DENMA	solids (matrix, grain) density	m/L^3
R_s	F_{gs}, F_{gos}	GORS	solubility, gas in oil (solution gas-oil ratio)	
R_{sw}		GWRS	solubility, gas in water	
R_{sb}	F_{gsb}	GORSB	solution gas–oil ratio at bubble-point conditions	
R_s	F_{gs}, F_{gos}	GORS	solution gas–oil ratio (gas solubility in oil)	
R_{si}	F_{gsi}	GORSI	solution gas–oil ratio, initial	
E_c	Φ_c	EMFC	SP, electrochemical component of	mL^2/t^2q
E_k	Φ_k	EMFK	SP, electrokinetic component of	mL^2/t^2q
E_{SP}	Φ_{SP}	EMFSP	SP (measured SP) (Self Potential)	mL^2/t^2q
E_{pSP}	Φ_{pSP}	EMFPSP	SP, pseudo	mL^2/t^2q
E_{SSP}	Φ_{SSP}	EMFSSP	SP, static (SSP)	mL^2/t^2q
L_s	s_s, ℓ_s script l	LENS	spacing (electrical logging)	L
s	σ	HERS	specific entropy	L^2/t^2T
γ	s, F_s	SPG	specific gravity (relative density)	
γ_g	s_g, F_{gs}	SPGG	specific gravity, gas	

SPE NOMENCLATURE AND UNITS

Letter symbol	Reserve SPE letter symbol	Computer letter symbol	Quantity	Dimensions
γ_o	s_o, F_{os}	SPGO	specific gravity, oil	
γ_w	s_w, F_{ws}	SPGW	specific gravity, water	
C	c	HSP	specific heat capacity (always with phase or system subscripts)	L^2/t^2T
γ	k	HSPR	specific heat capacity ratio	
I_s	i_s	IJXS	specific injectivity index	L^3t/m
J_s	j_s	PDXS	specific productivity index	L^3t/m
v	v_s	SPV	specific volume	L^3/m
F_{WV}	γ	WGTS	specific weight	mL^2/t^2
E_{SSP}	Φ_{SSP}	EMFSSP	SSP (static SP)	mL^2/t^2q
t_s	τ_s	TIMS	stabilization time of a well	t
σ		SDV	standard deviation of a random variable	
s		SDVES	standard deviation of a random variable, estimated	
p_{iws}	P_{iws}	PRSIWS	static bottom-hole pressure, injection well	m/Lt^2
p_{ws}	P_{ws}	PRSWS	static pressure, bottom-hole, at any time after shut-in	m/Lt^2
p_{cs}	P_{cs}	PRSCS	static pressure, casing	m/Lt^2
p_{ts}	P_{ts}	PRSTS	static pressure, tubing	m/Lt^2
r_s	R_s	RADS	stimulation or damage radius of well (skin)	L
ε	e, ε_n	STN	strain, normal and general	
γ	ε_s	STNS	strain, shear	
θ	θ_V	STNV	strain, volume	
Ψ		STR	stream function	various
σ	s	STS	stress, normal and general	m/Lt^2
τ	s_s	STSS	stress, shear	m/Lt^2
Σ		SUM	summation (operator)	
u	ψ	VELV	superficial phase velocity (flux rate of a particular fluid phase flowing in pipe; use appropriate phase subscripts)	L/t
q_{sc}	q_o, Q_{sc}	RTESC	surface production rate	L^3/t
σ	y, γ	SFT	surface tension, interfacial	m/t^2
R_s	ρ_s, r_s	RESS	surrounding formation resistivity	mL^3tq^2
k	κ	SUSM	susceptibility, magnetic	mL/q^2
T	θ	TEM	temperature	T
T_{bh}	θ_{BH}	TEMBH	temperature, bottomhole	T
T_c	θ_c	TEMC	temperature, critical	T
T_f	θ_f	TEMF	temperature, formation	T
g_T	g_h	GRDT	temperature gradient	T/L
T_{pc}	θ_{pc}	TEMPC	temperature, pseudo-critical	T
T_{pr}	θ_{pr}	TEMPRD	temperature, pseudo-reduced	T
T_r	θ_r	TEMRD	temperature, reduced	T
T_R	θ_R	TEMR	temperature, reservoir	T
T_{sc}	θ_{sc}	TEMSC	temperature, standard conditions	T
σ	y, γ	SFT	tension, surface (interfacial)	m/t^2
$\overline{\overline{X}}$			tensor of x	
k_h	λ	HCN	thermal conductivity (always with additional phase or system subscripts)	mL/t^3T
β	b	HEC	thermal cubic expansion coefficient	$1/T$
α	a, η_h	HTD	thermal or heat diffusivity	L^2/t
h	d, e	THK	thickness (general and individual bed)	L
h_t	d_t, e_t	THKT	thickness, gross pay (total)	L

Letter symbol	Reserve SPE letter symbol	Computer letter symbol	Quantity	Dimensions
h_{mc}	d_{mc}, e_{mc}	THKMC	thickness, mud-cake	L
h_t	d_t, e_t	THKT	thickness, pay, gross (total)	L
h_n	d_n, e_n	THKN	thickness, net pay	L
t	τ	TIM	time	t
Δt_{wf}	$\delta \tau_{wf}$	DELTIMWF	time after well is opened to production (pressure drawdown)	t
Δt_{ws}	$\Delta \tau_{ws}$	DELTIMWS	time after well is shut in (pressure build-up)	t
τ	τ_c	TIMC	time constant	t
τ_d	t_d	TIMD	time, decay (mean life) $(1/\lambda)$	t
t_d	τ_d	TIMD	time, delay	t
Δt	$\Delta \tau$	DELTIM	time difference (time period or interval, fixed length)	t
t_D	τ_D	TIMQ	time, dimensionless	
t_{Dm}	τ_{Dm}	TIMMQ	time, dimensionless at condition m	
t_s	τ_s	TIMS	time for stabilization of a well	t
t script t	Δt	TAC	time, interval transit	t/L
t_a script t	Δt_a	TACA	time, interval transit, apparent	t/L
t_f script t	Δt_f	TACF	time, interval transit, fluid	t/L
t_{ma} script t	Δt_{ma}	TACMA	time, interval transit, matrix	t/L
t_{sh} script t	Δt_{sh}	TACSH	time, interval transit, shale	t/L
t_{dN}		TIMDN	time, neutron decay (neutron mean life)	t
t_p	τ_p, t_{po}	TIMPO	time, pay-out (pay-off, pay-back)	t
Δt	$\Delta \tau$	DELTIM	time period or interval, fixed length	t
t_p	τ_p	TIMP	time well was on production prior to shut-in, equivalent (pseudo-time)	t
τ		TOR	tortuosity	
τ_e		TORE	tortuosity, electric	
τ_H		TORHL	tortuosity, hydraulic	
S_L	ρ_L, s_L	SATL	total (combined) liquid saturation	
s		HER	total entropy	$L^2/t^2 T$
λ_t	Λ	MOBT	total mobility of all fluids in a particular region of the reservoir, e.g., $(\lambda_o + \lambda_g + \lambda_w)$	$L^3 t/m$
M_t	$F_{\lambda t}$	MBRT	total mobility ratio $[(\lambda_t)_{swept}/(\lambda_t)_{unswept}]$; 'swept' and 'unswept' refer to invaded and uninvaded regions behind and ahead of leading edge of a displacement front	
h_t	d_t, e_t	THKT	total (gross) pay thickness	L
V	R, V_t, R_t	GRRT	total gross revenue ('value')	M
G	g	GASTI	total initial gas in place in reservoir	L^3
n	n_t, N_t	NMBM	total moles	
φ_t	f_t, ε_t	PORT	total porosity	
B_t	F_t	FVFT	total (two-phase) formation volume factor	
h	h_h, h_T	HTCC	transfer coefficient, convective heat	$m/t^3 T$
U	U_T, U_θ	HTCU	transfer coefficient, heat, over-all	$m/t^3 T$
I	I_T, I_θ	HTCI	transfer coefficient, heat, radiation	$m/t^3 T$
t script t	Δt	TAC	transit time, interval	t/L
t_a script t	Δt_a	TACA	transit time, apparent, interval	t/L
t_f script t	Δt_f	TACF	transit time, fluid interval	t/L
t_{ma} script t	Δt_{ma}	TACMA	transit time, matrix interval	t/L
t_{sh} script t	Δt_{sh}	TACSH	transit time, shale interval	t/L
$\mathscr{L}(y)$ script L			transform, Laplace of y $\int_0^\infty y(t)e^{-st}dt$	

SPE NOMENCLATURE AND UNITS

Letter symbol	Reserve SPE letter symbol	Computer letter symbol	Quantity	Dimensions
s			transform, Laplace, variable	
T	T	TRM	transmissivity, transmissibility	various
ρ_t	D_t	DENT	true density	m/L^3
R_t	ρ_t, r_t	REST	true formation resistivity	mL^3tq^2
G_t	f_{Gt}	GMFT	true geometrical factor (multiplier) (non-invaded zone) (electrical logging)	
p_{tf}	P_{tf}	PRSTF	tubing pressure, flowing	m/Lt^2
p_{ts}	P_{ts}	PRSTS	tubing pressure, static	m/Lt^2
F_B		FACB	turbulence factor	
B_t	F_t	FVFT	two-phase or total formation volume factor	
G_{pa}	g_{pa}	GASPUL	ultimate gas recovery	L^3
C_{uk}		INVUK	unamortized investment over year k	
P			undiscounted cash flow	M
V_{Ru}	v_{Ru}	VOLRU	unburned reservoir rock, volume of	L^3
a_E	F_{aE}	AIREX	unit air requirement in laboratory experimental run, volumes of air per unit mass of pack	L^3/m
a_R	F_{aR}	AIRR	unit air requirement in reservoir, volumes of air per bulk volume of reservoir rock	
C_m	c_m, n_m	CNCFU	unit fuel concentration (see symbol m)	various
R		RRR	universal gas constant (per mole)	mL^2/t^2T
e_{O_2}	E_{O_2}	UTLO2	utilization, oxygen	
z		VAL	valence	
y		MFRV	vapour phase, mole fraction of component	
V		MOLV	vapour phase, moles of	
L_v	λ_v	HLTV	vaporization, latent heat of	L^2/t^2
σ^2		VAR	variance of a random variable	
s^2		VARES	variance of a random variable, estimated	
$\vec{x}$			vector of x	
v	V, u	VEL	velocity	L/t
v	V, u	VAC	velocity, acoustic	L/t
v_a	V_a, u_a	VACA	velocity, acoustic apparent (measured)	L/t
v_f	V_f, u_f	VACF	velocity, acoustic fluid	L/t
v_{ma}	V_{ma}, u_{ma}	VACMA	velocity, matrix acoustic	L/t
v_{sh}	V_{sh}, u_{sh}	VACSH	velocity, shale acoustic	L/t
v_b	V_b, u_b	VELB	velocity (rate) of burning-zone advance	L/t
E_I	η_I, e_I	EFFI	vertical (invasion) efficiency: hydrocarbon pore space invaded (affected, contacted) by the injected-fluid or heat front divided by the hydrocarbon pore space enclosed in all layers behind the injected-fluid or heat front	
μ_a	η_a	VISA	viscosity, air	m/Lt
$\mu_{\bar{p}}$	$\eta_{\bar{p}}$	VISPAV	viscosity at mean pressure	m/Lt
μ	η	VIS	viscosity, dynamic	m/Lt
μ_g	η_g	VISG	viscosity, gas	m/Lt
μ_{ga}	η_{ga}	VISGA	viscosity, gas, at 1 atm	m/Lt
v	N	VSK	viscosity, kinematic	L^2/t
μ_o	η_o	VISO	viscosity, oil	m/Lt
μ_w	η_w	VISW	viscosity, water	m/Lt
V	v	VOL	volume	L^3
V_{bp}	v_{bp}	VOLBP	volume at bubble-point pressure	L^3
V_b	v_b	VOLB	volume, bulk	L^3

Letter symbol	Reserve SPE letter symbol	Computer letter symbol	Quantity	Dimensions
V_{bE}	v_{bE}	VOLBEX	volume, bulk, of pack burned in experimental run	L^3
V_e	V_{pe}, v_e	VOLG	volume, effective pore	L^3
V	f_v, F_v	VLF	volume fraction or ratio (as needed, use same subscripted symbols as for 'volumes'; note that bulk volume fraction is unity and pore volume fractions are ϕ])	various
G_{Fi}	g_{Fi}	GASFI	volume, free-gas, initial reservoir ($=mNb_{oi}$)	L^3
V_{gr}	v_{gr}	VOLGR	volume, grain (volume of all formation solids except shales)	L^3
V_{ig}	v_{ig}	VOLIG	volume, intergranular (volume between grains; consists of fluids and all shales) ($V_b - V_{gr}$)	L^3
V_{im}	v_{im}	VOLIM	volume, intermatrix (consists of fluids and dispersed shale) ($V_b - V_{ma}$)	L^3
V_{ma}	v_{ma}	VOLMA	volume, matrix (framework) (volume of all formation solids except dispersed shale)	
V_{ne}	V_{pne}, v_{ne}	VOLNE	volume, noneffective pore ($V_p - V_e$)	L^3
V_{Rb}		VOLRB	volume of reservoir rock burned	L^3
V_{Ru}		VOLRU	volume of reservoir rock unburned	L^3
V_M		VOLM	volume per mole (molal volume)	L^3
V_p	v_p	VOLP	volume, pore ($V_b - V_s$)	L^3
V_{pD}	v_{pD}	VOLPQ	volume, pore, dimensionless	
V_{shd}	v_{shd}	VOLSHD	volume, shale, dispersed	L^3
$V_{sh\ell}$ script l	$v_{sh\ell}$ script l	VSHLAM	volume, shale, laminated	L^3
V_{shs}	v_{shs}	VOLSHS	volume, shale, structural	L^3
V_{sh}	v_{sh}	VOLSH	volume, shale(s) (volume of all shales: structural and dispersed)	L^3
V_s	v_s	VOLS	volume, solid(s) (volume of all formation solids)	L^3
v	v_s	SPV	volume, specific	L^3/m
E_{Vb}	η_{Vb}, e_{Vb}	EFFVB	volumetric efficiency for burned portion only, *in situ* combustion pattern	
E_V	η_V, e_V	EFFV	volumetric efficiency: product of pattern sweep and invasion efficiencies	
q	Q	RTE	volumetric flow rate	L^3/t
q_{dh}	q_{wf}, q_{DH}, Q_{dh}	RTEDH	volumetric flow rate downhole	L^3/t
q_{sc}	q_σ, Q_{sc}	RTESC	volumetric flow rate, surface conditions	L^3/t
M		HSPV	volumetric heat capacity	m/Lt^2T
u	ψ	VELV	volumetric velocity (flow rate or flux, per unit area)	L/t
W	w	WTR	water (always with identifying subscripts)	various
c_w	k_w, κ_w	CMPW	water compressibility	Lt^2/m
ρ_w	D_w	DENW	water density	m/L^3
σ_{wb}	F_{wb}	DPRWB	water displaced from burned volume, volume per unit volume of burned reservoir rock	
C		WDC	water-drive constant	L^4t^2/m
C_L		WDCL	water-drive constant, linear aquifer	L^4t^2/m
k_w	K_w	PRMW	water, effective permeability to	L^2

SPE NOMENCLATURE AND UNITS

Letter symbol	Reserve SPE letter symbol	Computer letter symbol	Quantity	Dimensions
B_w	F_w	FVFW	water formation volume factor	
F_{wF}		FACWFU	water–fuel ratio	various
R_{sw}		GWRS	water, gas solubility in	
W	w	WTRTI	water in place in reservoir, initial	L^3
W_e	w_e	WTRE	water influx (encroachment), cumulative	L^3
ΔW_e	Δw_e	DELWTRE	water influx (encroachment) during an interval	L^3
e_w	i_w	ENCW	water influx (encroachment) rate	L^3/t
W_i	w_i	WTRI	water injected, cumulative	L^3
ΔW_i	Δw_i	DELWTRI	water injected during an interval	L^3
i_w		INJW	water injection rate	L^3/t
λ_w		MOBW	water mobility	$L^3 t/m$
k_w/k_o	K_w/K_o	PRMWO	water–oil permeability ratio	
F_{wop}		FACWOP	water–oil ratio, cumulative	
F_{wo}		FACWO	water–oil ratio, producing, instantaneous	
W_p	w_p	WTRP	water produced, cumulative	L^3
ΔW_p	Δw_p	DELWTRP	water produced during an interval	L^3
q_w	Q_w	RTEW	water production rate	L^3/t
q_{wD}	Q_{wD}	RTEWQ	water production rate, dimensionless	
k_{rw}	K_{rw}	PRMRW	water, relative permeability to	
R_w	ρ_w, r_w	RESW	water resistivity	$mL^3 t q^2$
S_w	ρ_w, s_w	SATW	water saturation	
S_{wc}	ρ_{wc}, s_{wc}	SATWC	water saturation, critical	
S_{wi}	ρ_{wi}, s_{wi}	SATWI	water saturation, initial	
S_{wo}	S_{wb}	SATWO	water saturation (interstitial) in oil band	
S_{wg}	ρ_{wg}, s_{wg}	SATWG	water saturation in gas cap, interstitial	
S_{iw}	ρ_{iw}, s_{iw}	SATIW	water saturation, irreducible	
S_{wr}	ρ_{wr}, s_{wr}	SATWR	water saturation, residual	
γ_w	S_w, F_{ws}	SPGW	water specific gravity	
μ_w	η_w	VISW	water viscosity	m/Lt
λ		WVL	wave length $(1/\sigma)$	L
σ	$\tilde{v}$	WVN	wave number $(1/\lambda)$	$1/L$
W	w, G	WGT	weight (gravitational)	m/Lt^2
$\bar{\rho}_L$	$\bar{D}_L$	DENAVL	weight-weighted average density of produced liquid	m/L^3
A		AWT	weight, atomic	m
M		MWT	weight, molecular	m
r_w	R_w	RADW	well radius	L
r_s	R_s	RADS	well radius of damage or stimulation (skin)	L
t_s	τ	TIMS	well stabilization time	t
r_{wa}	R_{wa}	RADWA	wellbore radius, effective or apparent (includes effects of well damage or stimulation	L
C_{wg}	c_{wg}, n_{wg}	CNTWG	wet-gas content	various
G_{wgp}	g_{wgp}	GASWGP	wet gas produced, cumulative	L^3
b	w	WTH	width, breadth, or (primarily in fracturing) thickness	L
W	w	WRK	work	mL^2/t^2
E	Y	ELMY	Young's modulus (modulus of elasticity)	m/Lt^2
d_i	d_I, D_i	DIAI	zone diameter, invaded, electrically equivalent	L
R_i	ρ_i, r_i	RESI	zone resistivity, invaded	$mL^3 t q^2$

B. Subscripts alphabetized by physical quantity

Subscript definition	Letter subscript	Reserve SPE subscript	Computer letter subscript
abandonment	a	A	A
acoustic	a	A, α alpha	A
activation log, neutron	NA	na	NA
active, activity, or acting	a		A
after taxes	a		A
air	a	A	A
air-fuel	aF		AFU
altered	a		A
amplitude log	A	a	A
angle, angular, or angular coordinate	θ theta		THE
anhydrite	anh		AH
anisotropic	ani		ANI
annulus apparent (from log readings; use tool description subscripts)	an	AN	AN
apparent (general)	a	ap	A
apparent wellbore (usually with wellbore radius)	wa		WA
areal	A		A
atmosphere, atmospheric	a	A	A
average or mean pressure	$\bar{p}$		PAV
average or mean saturation	$\bar{S}$	$\bar{s}, \bar{p}$ rho	SAV
band or oil band	b	B	B
bank or bank region	b		B
base	b	r, β beta	B
before taxes	b	B	B
bond log, cement	CB	cb	CB
borehole televiewer log	TV	tv	TV
bottom hole	bh	w, BH	BH
bottom-hole, flowing (usually with pressure or time)	wf		WF
bottom-hole, static (usually with pressure or time)	ws		WS
boundary conditions, external	e	o	E
breakthrough	BT	bt	BT
bubble	b		B
bubble-point conditions, oil at (usually with formation volume factor, B_{ob})	ob		OB
bubble-point conditions, solution at (usually with gas-oil ratio, R_{sb})	sb		SB
bubble point (saturation)	b	s, bp	B
bubble-point or saturation (usually with volume, V_{bp})	bp		B
bulk (usually with volume V_b)	b	B, t	B
burned in experimental tube run (usually with volume, V_{bE})	bE		BEX
burned or burning	b	B	B
burned portion of in situ combustion pattern, displacement from (usually with efficiency, E_{Db})	Db		DB
burned portion of in situ combustion pattern, volumetric of (usually with efficiency, E_{vb})	Vb		VB
burned reservoir rock	Rb		RB

SPE NOMENCLATURE AND UNITS

Subscript definition	Letter subscript	Reserve SPE subscript	Computer letter subscript
burned volume, oil from (usually with displacement ratio, δ_{ob})	ob		OB
burned volume water from (usually with displacement ratio, δ_{wb})	wb		WB
calculated	C	$calc$	CA
caliper log	C	c	C
capillary (usually with capillary pressure, P_c)	c	C	CP
capture	cap		C
carbon dioxide	CO_2		CO2
carbon monoxide	CO		CO
casing or casinghead	c	cg	CS
casing, flowing (usually with pressure)	cf		CF
casing, static (usually with pressure)	cs		CS
cement bond log	CB	cb	CB
chemical	c		C
chlorine log	CL	cl	CL
clay	cl	cla	CL
clean	cn	cln	CN
coil	C	c	C
compaction	cp		CP
compensated density log	CD	cd	CD
compensated neutron log	CN	cn	CN
component(s)	C		C
component j	j		J
component j produced (usually with moles, n_{pj})	pj		PJ
compressional wave	c	C	C
conditions for infinite dimensions	∞	INF	INF
conductive liquids in invaded zone	z		Z
constant	c	C	C
contact (usually with contact angle, θ_c)	c	C	C
contact log, microlog, minilog	ML	$m\ell$ script l	ML
convective			C
conversion (usually with conversion factor in Newton's law of motion, g_c)	c		C
core	c	C	C
corrected	cor		COR
critical	c	cr	CR
cumulative influx (encroachment)	e	i	E
cumulative injected	i		I
cumulative produced	p		P
cumulative produced free value (usually with gas, G_{Fp})	F_p		FP
cumulative produced liquid (usually with condensate, G_{Lp})	L_P		
damage or damaged (includes 'skin' conditions)	s	d	S
decay	d		D
deep induction log	ID	id	ID
deep laterolog	LLD	$\ell\ell d$ script ll	LLD
delay	d	δ delta	D

302 PETROLEUM ENGINEERING: PRINCIPLES AND PRACTICE

Subscript definition	Letter subscript	Reserve SPE subscript	Computer letter subscript
density	ρ rho		RHO
density log, compensated	CD	cd	CD
density log	D	d	D
depleted region, depletion	d	δ delta	D
dew-point	d		D
differential separation	d		D
differential temperature log	DT	dt	DT
diffusivity	η eta		ETA
dimensionless pore value (usually with volume V_{pD})	pD		PQ
dimensionless quantity	D		Q
dimensionless quantity at condition m	Dm		QM
dimensionless time	tD		TQ
dimensionless water	wD		WQ
dip (usually with angle, α_d)	d		D
diplog, dipmeter	DM	dm	DM
directional survey	DR	dr	DR
dirty (clayey, shaly)	dy	dty	DY
discounted value, present worth, or present value	PV	pv	PV
dispersed	d	D	D
dispersion	K	d	K
displaced	d	s,D	DD
displacement from burned portion of in situ combustion pattern (usually with efficiency, E_{Db})	Db		DB
displacement from unburned portion of in situ combustion pattern (usually with efficiency, E_{Du})	Du		DU
displacing or displacement (efficiency)	D	s, σ sigma	DN
dolomite	dol		DL
down-hole	dh	DH	DH
drainage (usually with drainage radius, r_d)	d		D
dual induction log	DI	di	DI
dual laterolog	DLL	dℓ script ll	DLL
earth	e	E	E
effective (or equivalent)	e		E
electric, electrical	e	E	E
electrochemical	c	ec	C
electrode	E	e	E
electrokinetic	k	ek	K
electrolog, electrical log, electrical survey	EL	el, ES	EL
electromagnetic pipe inspection log	EP	ep	EP
electron	el	ℓ script el	E
empirical	E	EM	EM
encroachment (influx), cumulative	e	i	E
entry	e	E	E
epithermal neutron log	NE	ne	NE
eqivalent	eq	EV	EV
estimated	E	est	ES
ethane	C_2		C2
experimental	E	EX	EX

SPE NOMENCLATURE AND UNITS

Subscript definition	Letter subscript	Reserve SPE subscript	Computer letter subscript
experimental value per mole of produced gas (usually with fuel consumption, m_{Eg})	Eg		EXG
external, outer boundary conditions	e	o	E
extrapolated	ext		XT
fast neutron log	NF	nf	NF
fill-up	F	f	F
finger or fingering	f	F	F
flash separation	f	F	F
flowing bottom-hole (usually with pressure or time)	wf		WF
flowing casing (usually with pressure)	cf		CF
flowing conditions, injection well (usually with pressure, p_{iwf})	iwf		IWF
flowing conditions, well (usually with time)	wf	f	WF
flowing tubing (usually with pressure)	tf		TF
fluid	f	fl	F
fluids in an invaded zone, conductive	z		Z
flushed zone	xo		XO
formation 100% saturated with water (used in R_0 only)	0 zero		ZR
formation (rock)	f	fm	F
formation, surrounding	s		S
fraction or fractional	f	r	F
fracture, fractured or fracturing	f	F	FR
free (usually with gas or gas–oil ratio quantities)	F	f	F
free fluid	F_f	f	FF
free value, cumulative produced, (usually with gas, G_{Fp})	F_p		FP
free value, initial (usually with gas, G_{Fi})	Fi		FI
front, front region, or interface	f	F	F
fuel, mass of (usually with fuel concentration, C_m)	m		FU
fuel (usually with fuel properties, such as ρ_F)	F		FU
gamma-gamma ray log	GG	gg	GG
gamma ray log	GR	gr	GR
gas	g	G	G
gas at atmospheric conditions	ga		GA
gas at bubble-point conditions	gb		GB
gas cap, oil in (usually with saturation, S_{og})	og		OG
gas cap, water in (usually with saturation, S_{wg})	wg		WG
gas, dimensionless	gD		GQ
gas–oil, solution (usually with gas–oil ratios)	s		S
gas–water, solution (usually with gas solubility in water, R_{sw})	sw		
geometrical	G		G
geothermal	G	T	GT
grain	gr		GR
grain (matrix, solids)	ma		MA
gravity meter log	GM	gm	GM
gross (total)	t	T	T
guard log	G	g	G
gypsum	gyp		GY
half	$1/2$		H

Subscript definition	Letter subscript	Reserve SPE subscript	Computer letter subscript
heat or thermal	h	T, θ theta	HT
heavy phase	HP	hp	HP
hole	h	H	H
horizontal	H	h	H
hydraulic	H		HL
hydrocarbon	h	H	H
hydrogen nuclei or atoms	H		HY
hydrocarbon, residual	hr		HR
hydrogen sulphide	H_2S		H2S
imbibition	I	ℓ script i	I
induction log, deep investigation	ID	id	ID
induction log	I	i	I
induction log, dual	DI	di	DI
induction log, medium investigation	IM	im	IM
infinite dimensions, conditions for	∞		INF
influx (encroachment), cumulative	e	i	E
initial conditions or value	i		I
initial free value (usually with gas, G_{Fi})	Fi		FI
initial solution (usually with gas–oil ratio, R_{si})	si		SI
initial value or conditions	i		I
injected, cumulative	i	I	I
injection, injected or injecting	i	inj	I
injection well, flowing conditions (usually with pressure, p_{iwf})	iwf		IWF
injection well, static conditions (usually with pressure, p_{iws})	iws		IWS
inner or interior	i	ι iota, ℓ script i	I
interface, front region, or front	f	F	F
interference	I	i, ℓ script i	I
intergranular	ig		IG
intermatrix	im		IM
internal	i	ι iota, ℓ script i	I
intrinsic	int		I
invaded	i	I	I
invaded zone	i	I	I
invaded zone, conductive liquids in an	z		Z
invasion (usually with invasion efficiency, E_I)	I	i	I
irreducible	i	ir, ι iota, ℓ script i	IR
jth component	j		J
jth component, produced	pj		PJ
junction	j		J
laminar	ℓ script l	L	LAM
laminated, lamination	ℓ script L	L	LAM
lateral (resistivity) log	L	ℓ script l	L
laterolog (add further tool configuration subscripts as needed)	LL	$\ell\ell$ script ll	LL
laterolog, dual	DLL	$d\ell\ell$ script ll	DLL
lifetime log, neutron, TDT	PNL	$n\ell$ script l	PNL
light phase	LP	ℓp script l	LP
limestone	ls	lst	LS
limiting value	lim		LM

SPE NOMENCLATURE AND UNITS

Subscript definition	Letter subscript	Reserve SPE subscript	Computer lltter subscript
linear, lineal	L	ℓ script l	L
liquid or liquid phase	L	ℓ script l	L
liquids, conductive, invaded zone	z		Z
liquid produced, cumulative (usually with condensate G_{Lp})	Lp		
location subscripts, usage is secondary to that for representing times or time periods	1,2,3, etc.		
log	LOG	log	L
lower	ℓ script l	L	L
magnetism log, nuclear	NM	nm	NM
mass of fuel (usually with fuel concentration, C_m)	m		FU
matrix (solids, grain)	ma		MA
matrix [solids, except (nonstructural) clay or shale]	ma		MA
maximum	max		MX
mean or average pressure	$\bar{p}$		PAV
mean or average saturation	$\bar{S}$	$\bar{s}, \bar{\rho}$ rho	SAV
medium investigation induction log	IM	im	IM
methane	C_1		C1
microlaterolog	MLL	mℓ script ll	MLL
microlog, minilog, contact log	ML	mℓ script l	ML
micro-seismogram log, signature log, variable density log	VD	vd	VD
minimum	min		MN
mixture	M	z,m	M
mobility	λ lambda	M	LAM
molal (usually with volume, V_M)	M		M
Mth period or interval	M	m	M
mud	m		M
mud cake	mc		MC
mud filtrate	mf		MF
net	n		N
neutron	N	n	N
neutron activation log	NA	na	NA
neutron lifetime log, TDT	PNL	nℓ script l	PNL
neutron log, compensated	CN	cn	CN
neutron log	N	n	N
neutron log, epithermal	NE	ne	NE
neutron log, fast	NF	nf	NF
neutron log, sidewall	SN	sn	SN
neutron log, thermal	NT	nt	NT
nitrogen	N_2		N2
noneffective	ne		NE
nonwetting	nw	NW	NW
normal	n		N
normal (resistivity) log (add numerical spacing to subscript to N; e.g., $N16$)	N	n	N
normalized (fractional or relative)	n	r,R	N
nth year, period, income, payment, or unit	n	N	N
nuclear magnetism log	NM	nm	NM

Subscript definition	Letter subscript	Reserve SPE subscript	Computer letter subscript
numerical subscripts (intended primarily to represent times or time periods; available secondarily as location subscripts or for other purposes)	1,2,3, etc.		
observed	OB		OB
oil at bubble-point conditions (usually with formation volume factor, B_{ob})	ob		OB
oil, dimensionless	oD		OQ
oil	o	N, n	O
oil from burned volume (usually with displacement ratio, δ_{ob})	ob		OB
oil from unburned volume (usually with displacement ratio, δ_{ou})	ou		OU
oil in gas cap (usually with saturation, S_{og})	og		OG
outer (external)	e	o	E
oxygen	O_2		O2
particle (usually with diameter, $\bar{d}_p$)	p		P
particular period, element, or interval	k	K	K
pattern (usually with pattern efficiency, E_p)	P		P
pay-out, pay-off, or pay-back	p	po	PO
permeability	k	K	K
phase or phases	P		P
pipe inspection log, electromagnetic	EP	ep	EP
pore (usually with volume, V_p)	p	P	P
pore value, dimensionless (usually with volume, V_{pD})	pD		PQ
porosity	ϕ phi	f, ε epsilon	PHI
porosity data	ϕ phi	f, ε epsilon	P
pressure, mean or average	$\bar{p}$		PAV
primary	1 one	p, pri	PR
produced	p	P	P
produced component j (usually with moles, n_{pj})	pj		PJ
produced, cumulative	p		P
produced free value, cumulative (usually with gas, G_{Fp})	F_p		FP
produced in experiment	pE		PEX
produced liquid, cumulative (usually with condensate, G_{Lp})	Lp		
produced water–oil (cumulative) (usually with cumulative water–oil ratio, F_{wop})	wop		WOP
production period (usually with time, t_p)	p	P	P
profit – unamortized investment	Pk		PK
proximity log	P	p	P
pseudo	p		P
pseudo-critical	pc		PC
pseudo-dimensionless	pD		PQ
pseudo-reduced	pr		PRD
pseudo-SP	pSP		PSP
radial	r	R	R
radius, radial, or radial distance	r	R	R
rate of return	r	R	R

SPE NOMENCLATURE AND UNITS

Subscript definition	Letter subscript	Reserve SPE subscript	Computer letter subscript
recovery (usually with recovery efficiency, E_R)	R		R
reduced	r		RD
reference	r	b, ρ rho	R
relative	r	R	R
reservoir	R	r	R
reservoir rock, burned	Rb		RB
reservoir rock, unburned	Ru		RU
residual	r	R	R
residual hydrocarbon	hr		HR
resistivity	R		R
resistivity log	R	r, ρ rho	R
Reynolds (used with Reynolds number only, N_{Re})	Re		
rock (formation)	f	fm	F
sand	sd	sa	SD
sandstone	ss	sst	SS
saturation, mean or average	$\bar{S}$	$\bar{s}$, $\bar{\rho}$ rho	SAV
saturation or bubble point	b	s	B
saturation or bubble point (usually with volume, V_{bp})	bp		BP
scattered, scattering	sc		SC
secondary	2 two	s,sec	SE
segregation (usually with segregation rate, q_s)	s	S, σ sigma	S
separator conditions	sp		SP
shale	sh	sha	SH
shallow laterolog	LLS	ℓs script ll	LLS
shear	s	τ tau	
shear wave	s	τ tau	S
sidewall	S	SW	SW
sidewall neutron log	SN	sn	SN
signature log, micro-seismogram log, variable density log	VD	vd	VD
silt	sl	slt	SL
single payment	sp		SP
skin (stimulation or damage)	s	S	S
slip or slippage	s	σ sigma	S
slurry ('mixture')	M	z,m	M
solid(s) (all formation solids)	s	σ sigma	S
solids in experiment	sE		SEX
solids (matrix, grain)	ma		MA
solution at bubble-point conditions (usually with gas–oil ratio, R_{sb})	sb		SB
solution in water (usually with gas solubility in water, R_{sw})	sw		
solution, initial (usually with gas–oil ratio, R_{si})	si		SI
solution (usually with gas–oil ratios)	s		S
sonde, tool	T	t	T
sonic velocity log	SV	sv	SV

Subscript definition	Letter subscript	Reserve SPE subscript	Computer letter subscript
SP	*SP*	*sp*	SP
spacing	*s*		L
specific (usually with *J* and *I*)	*s*		S
SSP	*SSP*		SSP
stabilization (usually with time)	*s*	*S*	S
standard conditions	*sc*	σ sigma	SC
static bottom-hole (usually with pressure or time)	*ws*		WS
static casing (usually with pressure)	*cs*		CS
static conditions, injection well (usually with pressure)	*iws*		IWS
static or shut-in conditions (usually with time)	*ws*	*s*	WS
static tubing (usually with pressure)	*ts*		TS
static well conditions (usually with time)	*ws*	*s*	WS
steam or steam zone	*s*	*S*	S
stimulation (includes 'skin' conditions)	*s*	*S*	S
stock-tank conditions	*st*		ST
storage or storage capacity	*S*	*s*, σ sigma	S
strain	ε epsilon	*e*	EPS
structural	*st*	*s*	ST
surface	*s*	σ sigma	S
surrounding formation	*s*		S
swept or swept region	*s*	*S*, σ sigma	S
system	*s*	σ sigma	S
TDT log, neutron lifetime log	*PNL*	*pn*/script l	PNL
televiewer log, borehole	*TV*	*tv*	TV
temperature	*T*	*h*, θ theta	T
temperature log	*T*	*t,h*	T
temperature log, differential	*DT*	*dt*	DT
thermal (heat)	*h*	*T*, θ theta	HT
thermal decay time (TDT) log	*PNL*	*pn*/script l	PNL
thermal neutron log	*NT*	*nt*	NT
time, dimensionless	*tD*		TQ
times or time periods	1,2,3, etc.		
tool-description subscripts: see individual entries such as 'amplitude log', 'neutron log,', etc.			
tool, sonde	*T*	*t*	T
total initial in place in reservoir	*ti*		TI
total (gross)	*t*	*T*	T
total, total system	*t*	*T*	T
transmissibility	*T*	*t*	T
treatment or treating	*t*	τ tau	T
true (opposed to apparent)	*t*	*tr*	T
tubing flowing (usually with pressure)	*tf*		TF
tubing or tubinghead	*t*	*tg*	T
tubing, static (usually with pressure)	*ts*		TS
turbulence (used with *F* only, F_B)	*B*		B
ultimate	*a*	*ul*	UL
unamortized	*u*	*U*	U
unburned	*u*		U
unburned portion of *in situ* combustion pattern displacement from (usually with efficiency, E_{Du})	*Du*		DU
unburned reservoir rock	*Ru*		RU

SPE NOMENCLATURE AND UNITS

Subscript definition	Letter subscript	Reserve SPE subscript	Computer letter subscript
unburned volume, oil from (usually with displacement ratio, δ_{ou})	ou		OU
unit	u	U	U
unswept or unswept region	u	U	U
upper	u	U	U
vaporization, vapour, or vapour phase	v	V	V
variable density log, micro-seismogram log, signature log	VD	vd	VD
velocity	v	V	V
velocity, sonic or acoustic log	SV	sv	SV
vertical	V	v	V
volumetric of burned portion of *in situ* combustion pattern (usually with efficiency, E_{vb})	Vb		VB
volume or volumetric	V	v	V
water	w	W	W
water, dimensionless	wD		WQ
water from burned volume (usually with displacement ratio, δ_{wb})	wb		WB
water-fuel	wF		WFU
water in gas cap (usually with saturation, S_{wg})	wg		WG
water–oil (usually with instantaneous producing water–oil ratio, F_{wo})	wo		WO
water–oil produced (cumulative) (usually with cumulative water–oil ratio, F_{wop})	wop		WOP
water, solution in (usually with gas solubility in water, R_{sw})	sw		SW
water-saturated formation, 100%	0 zero	zr	ZR
weight	W	w	W
well conditions	w		W
well, flowing conditions (usually with time)	wf	f	WF
well, static conditions (usually with time)	ws	s	WS
well, injection, flowing conditions (usually with pressure p_{iwf})	iwf		IWF
well, injection, static conditions (usually with pressure p_{iws})	iws		IWS
well, static conditions (usually with time)	ws		WS
wellbore, apparent (usually with wellbore radius, r_{wa})	wa		WA
wellhead	wh	th	WH
wet gas (usually with composition or content, C_{wg})	wg		WG
wet gas produced	wgp		WGP
wetting	w	W	W
Young's modulus, refers to	Y		Y
zero hydrocarbon saturation	0 zero	zr	ZR
zone, conductive fluids in an invaded	z		Z
zone, flushed	xo		XO
zone, invaded	i	I	I

Appendix 2
Solutions to Examples

Chapter 2

Solution 2.1

Although this problem should place probabilistic ranges on the given data and assumptions, it will be calculated deterministically.

We will assume that the combination of oil expelled from source rocks and trapped in potential structures represents some 8% of the converted source rocks, i.e.:

Oil converted for source rock $= 5 \times 4500 \times 12 \times 10^6 \text{ m}^3$
Trapped oil ($=$ OIP) $= 0.085 \times 4500 \times 12 \times 10^6 \text{ m}^3$
$= 2.16 \times 10^{10} \text{ m}^3$

Assuming an average formation volume factor of 1.4 rm^3/sm^3 this yields a stock tank oil in place of 1.54×10^{10} sm^3. For an assumed overall technical recovery factor of 0.35 this yields a recoverable reserve of

$$1.54 \times 10^{10} \times 0.35 = 5.4 \times 10^9 \text{ sm}^3$$

(This is equivalent to 34×10^9 STB.)

(N.B. The UK Government's 1983 'Brown Book' indicates a probable range of technically recoverable reserves between 11 and 23×10^9 STB, assuming an oil formation volume factor of 1.4 rm^3/sm^3.)

Chapter 3

Solution 3.1
Casing Design Example

(a) The buoyancy factor (BF) is given by

$$BF = \frac{SG_{steel} - SG_{fluid}}{SG_{steel}}$$

For the external system:

$$BF = \frac{7.84 - 1.92}{7.84} = 0.755$$

and for the internal fluid system:

$$BF = \frac{7.84 - 1.15}{7.84} = 0.853$$

SOLUTIONS TO EXAMPLES

The neutral point (NP) is thus the depth at which the string above is in tension and below in compression.

$$NP = 13000 \times BF$$
$$= 13000 \times 0.755$$
$$= 9820 \text{ ft}$$

This is rounded off to 9800 ft.

(b) For the design weight of casing (CWT) we have

$$CWT = \text{weight in air} \times BR$$

where the buoyancy ratio BR is given by

$$BR = \frac{\text{BF for outside mud system}}{\text{BF for internal fluid system}} = \frac{0.755}{0.853} = 0.885$$

(c) In the lower section we can check criteria:

(i) Collapse
The external mud gradient is $SG \times 0.433$ psi/ft

$$= 1.92 \times 0.433$$
$$= 0.831 \text{ psi/ft}$$

The collapse limit of the P-110 casing of the various weights is given from Table A3.1 as

$$\frac{9570}{0.831} = 11520 \text{ ft for 20 ppf casing, and}$$

$$\frac{11630}{0.831} = 14000 \text{ ft for 23 ppf casing}$$

∴ Use 23 ppf casing from bottom to 11 520 ft, that is $(13000 - 11520) = 1480$ ft
(NB no tension problem since neutral point is at 9800 ft.)

(ii) Burst check
Since a more dense mud is used outside the casing then the greatest internal:external pressure difference is at the top of each section.

At 11 520 ft, internal differential is:

(max surface pressure) + (internal fluid head) − (external fluid head)

Internal pressure gradient = $(SG \times 0.433) = 1.15 \times 0.433 = 0.498$ psi/ft

∴ $8000 + 11520 [0.498 - 0.831] = 4164$ psi

As burst pressure of 23 ppf casing is given as 11 780 psi no problem arises.

(iii) Joint strength calculation check
Since the entire section is below the neutral point, tension is not a problem so an API joint with long threads is sufficient.

(iv) Design weight for the section (CWT)

$$CWT = \text{Design length} \times \text{wt per foot} \times BR$$
$$= 1480 \times 23 \times 0.885$$
$$= 30125 \text{ lbs.}$$

(d) For the next section N-80, 23 ppf has the next highest collapse pressure to P-110, 20 ppf and can be set below the neutral point (see Table A3.1).

(i) Collapse limit = $\dfrac{8370}{0.831} = 10072$ ft

Rounding off, we can propose a section length of
$11520 - 10070 = 1450$ ft

(ii) Burst check

$8000 + 10070 [0.498 - 0.831] = 4647$ psi
no problem arises.

(iii) As we are below the neutral point no joint strength problem.

(iv) Design weight for this section

$1450 \times 20 \times 0.855 = 24\,795$ lb

Total weight calculated so far $= (30\,125 + 24\,795) = 54\,920$ lb.

(e) In the next section we might consider the use of P-11017 ppf but only a relatively short section could be used. It is considered more economical to design for N-80, 20 ppf.

(i) Collapse limit $= \dfrac{6930}{0.831} = 8339$ ft, round to 8340 ft

This is above the neutral point and therefore subject to the weight of casing above.

We calculate the ratio (R) for unit tensile stress to minimum yield strength using the ellipse of biaxial yield stress curve (Fig. A3.1) to obtain the percent of full collapse pressure that is appropriate. From Table A3.1 the plain end area (A) of 20 ppf N-80 is 5.828 in². For the minimum yield strength (Y_m) of 80 000 psi we have:

$$R = \dfrac{\text{weight in air of casing above neutral point}}{Y_m \cdot A}$$

Assume casing above neutral point is 20 ppf

$$R = \dfrac{20\,(9800 - D)}{80\,000\,(5.828)}$$

We have to choose D such that the reduction factor (F_R) correlated with R to obtain the effective collapse depth is consistent:

i.e. $\dfrac{6930}{0.831} \times F_R = f(R) = f\left\{\dfrac{20\,(9800 - D)}{80\,000\,(5.828)}\right\}$

This is solved by trial and we might choose D to be 7900 ft

$$R = \dfrac{20\,(9800 - 7900)}{80\,000\,(5.828)} = 0.0815$$

From Fig. A3.1 the value of F_R corresponding to 0.0815 is 0.956%

Collapse limit is $0.956 \times \dfrac{6930}{0.831} = 7972$ ft

We could converge a little better but might accept 7900 ft as a suitable depth, giving 2170 feet of casing required between 7900 and 10 070 ft.

(ii) Burst check for internal differential at 7900 ft

$= 8000 + 7900\,[0.498 - 0.831]$
$= 5369$ psi

This is within the tolerance of both 20 and 23 ppf N-80

(iii) Joint strength check
Section design weight $= (2170 \times 20 \times 885) = 38\,409$ lb
Total design weight $= 38\,409 + 54\,920$
$\qquad\qquad\qquad\ = 93\,329$ lb

We can see that the joint strengths of 20 and 23 ppf N-80 casing are both greater than the design weights (Table A3.1):

23 ppf : 251 000 lb
20 ppf : 214 000 lb

(f) In abnormal pressure wells, a depth can be reached where either collapse or burst may control. A design trial for the next section is made using 17 ppf N-80.

(i) Collapse check $\dfrac{5240}{0.831} = 6305$ ft

SOLUTIONS TO EXAMPLES

[Figure: Ratio of unit tensile stress to minimum yield strength (R) vs. fraction of full collapse pressure]

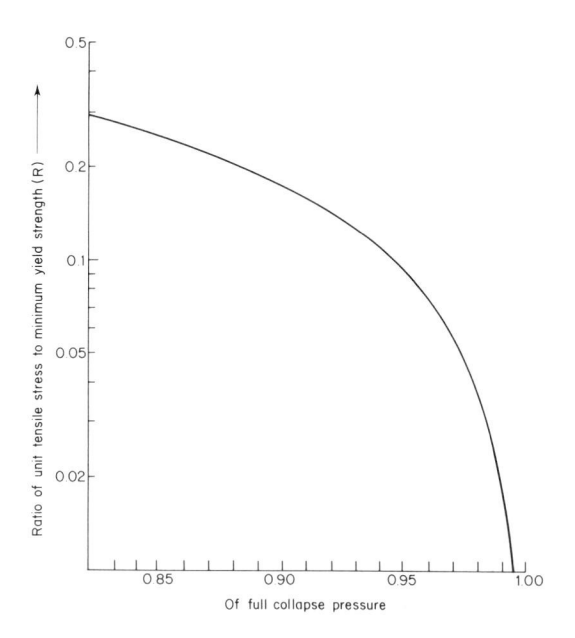

We can converge on a reduced setting depth of 5430 feet.
$$R = \frac{20(9800 - 7900) + 17(7900 - 5430)}{4.962\,(80\,000)} = 0.202, \text{ giving } F_R = 0.884$$

and a collapse limit of 5573 ft which is in tolerance.

The possible length of this section is thus $(7900 - 5430) = 2470$ ft

(ii) Burst check

Internal differential at 5430 ft

$$= 8000 + (5430\,[0.498 - 0.831])$$
$$= 6192 \text{ psi}$$

The burst strength of 17 ppf N-80 is quoted in Table A3.1 as 6180 psi.

We must check the depth at which burst governs, i.e. the depth equivalent to a burst strength of 6180 psi.

$$\text{Depth} = \frac{8000 - 6180}{0.831 - 0.498} = 5466 \text{ ft}$$

The depth that 17 ppf N-80 will withstand the internal pressure differential is below its allowable collapse depth and this grade cannot be used in this part of the design. We must therefore consider using 20 ppf N-80 as we know that this is collapse designed down to 7900 ft. The burst strength for this is 7400 psi.

$$\text{Depth} = \frac{8000 - 7400}{0.831 - 0.498} = 1802 \text{ ft, round up to 1820 ft}$$

This means that we could design a section of length $(7900 - 1800) = 6080$ ft

(iii) Joint strength check
Design weight for section is $(6080 \times 20 \times 0.885) = 107\,616$ lb
   Total weight is $107\,616 + 93\,329$
                     $= 200\,945$ lb

The joint strength for 20 ppf N-80 is given in Table A3.1 as 214 000 lb. We have so far designed 11 180 ft of the total well depth of 13 000 ft. The remaining 1820 ft are considered using P-110, 17 ppf grade casing.

(g) (i) Collapse check

$$R = \frac{20(9800 - 1820) + 17(1820)}{11\,000\,(4.962)} = 0.35$$

$$F_R = 0.78$$

Setting depth $= \dfrac{7000\,(0.78)}{0.831} = 6570$ ft

a proposed setting at 1820 ft is acceptable.

(ii) Burst check

Internal difference at top of string is 8000 psi (max) and Table A3.1 gives burst rating as 8500 psi, therefore design is acceptable.

(iii) Joint strength check
Design weight of section added $= 1820 \times 17 \times 0.885 = 27\,382$ lb

Total string weight $= 27\,382 + 200\,945$
Joint strength of P-110 L $= 247\,000$ lb   design is acceptable.

(h) We can summarize the design as follows:

Section	Length (ft)	Casing Grade
Surface – 1820	1820	17 ppf P-110 L
1820 – 10 070	8250	20 ppf N-80 L
10 070 – 11 520	1450	23 ppf N-80 L
11 520 – 13 000	1480	23 ppf P-110 L

It should be emphasized that this design is one of many combinations which may be acceptable and optimization in terms of economics is possible.

Solution 3.2

The average gradients give a pore pressure at 13 000 ft of

$13\,000 \times 0.455 = 5915$ psi

and a fracture pressure at 13 000 ft of

$13\,000 \times 0.80 = 10\,400$ psi

The minimum setting depth is given by equating, above 13 000 feet, the gas and fracture gradients to a common pressure. If the distance above 13 000 ft is D' then

$$P_g = 5915 - (0.1 \times D')$$
$$P_{fr} = 10\,400 - (0.8 \times D')$$

Setting $P_g = P_{fr}$ we have

$$D' = \frac{10\,400 - 5915}{0.8 - 0.1} = 6407 \text{ ft}$$

Minimum setting depth is $13\,000 - 6407 = 6593$ ft.

TABLE A3.1 Casing data for example (Grade N80-L/P110-L 5.5 in. OD.)

Weight (lb/ft)	Wall thickness (in)	Collapse incl. safety factor (psi)	Burst strength int. wk. press (incl. S.F.) psi	Joint strength (incl. S.F.) 1000 lb	Section area (in²)
17.0 P110	0.304	7 000	8 500	247	4.962
17.0 N80	0.304	5 240	6 180	174	4.962
20.0 P110	0.361	9 570	10 180	274	5.828
20.0 N80	0.361	6 930	7 400	214	5.828
23.0 P110	0.415	11 630	11 780	322	6.630
23.0 N80	0.415	8 370	8 570	251	6.630

Minimum yield strength $(Y_m) = 80\,000$ psi for N-80
$\phantom{\text{Minimum yield strength }(Y_m)} = 110\,000$ psi for P-110

SOLUTIONS TO EXAMPLES

Chapter 4

Solution 4.1

$$\text{API} = \frac{141.5}{\text{SG}} - 131.5$$

SG	API	SG	API
0.70	70.6	0.80	45.4
0.72	65.0	0.82	41.0
0.74	59.7	0.84	36.9
0.76	54.7	0.86	33.0
0.78	49.9	0.88	29.2
		0.90	25.72

NB API gravity is non-linear, inverse scale.
Water SG = 1.0; API = 10.

Solution 4.2

	Y_i	MW	Y_i MW	P_c	$Y_i P_{ci}$	T_c	$Y_i T_{ci}$
C_1	0.90	16	14.4	673	605.7	343	308.7
C_2	0.05	30	1.5	708	35.4	550	27.5
C_3	0.03	44	1.32	617	18.5	666	19.9
C_4	0.02	58	1.16	551	11.0	765	15.3
			(a) $\sum = 18.38$		(c) $\sum = 670.6$		(c) $\sum = 371.5$

(b) Specific gravity $= \dfrac{\text{MW}}{28.97} = \dfrac{18.38}{28.97} = 0.634$

Gas density $= \dfrac{m}{V} = \dfrac{MP}{RT} = \dfrac{18.38 \times 14.7}{10.732 \times 520} = 4.8 \times 10^{-2} \text{ lb ft}^3$

(d) At 2000 psia and 595°R

$$T_{pr} = \frac{595}{371.5} = 1.60$$

$$P_{pr} = \frac{2000}{670.6} = 2.98$$

(e) From graphs $z = 0.825$ (fig 4.7)

(f) Density $= \dfrac{MP}{zRT} = \dfrac{18.38 \times 2000}{0.825 \times 10.732 \times 595} = 6.977 \text{ lb ft}^3$

Gas gradient $= \dfrac{6.977}{144} \text{ psi/ft} = 0.0485 \text{ psi ft}^{-1}$

(g) $sB_g = \left(\dfrac{V_R}{V_s}\right) = \left(\dfrac{zT}{P}\right)_R = \left(\dfrac{P}{zT}\right)_s = \dfrac{0.825 \times 595 \times 14.696}{2000 \times 519.7} = 6.9 \times 10^{-3}$ vols/vol.

$$= 6.9 \times 10^{-3} \frac{10^3}{5.615} = 1.235 \text{ BBL/MSCF}$$

(h) From graphs, $\mu_1 = 0.0116$ (Fig. 4.8) and

Ratio $\dfrac{\mu}{\mu_o} = 1.3$ (Fig. 4.9)

Therefore $\mu = 0.015$ cp

316   PETROLEUM ENGINEERING: PRINCIPLES AND PRACTICE

(i) Compressibility
$$c_g = \frac{1}{p} - \frac{1}{z}\frac{dz}{dP}$$
$$= \frac{1}{P_{pc}}\left(\frac{1}{P_{pr}} - \frac{1}{z}\frac{dz}{dP_{pr}}\right)$$
$$= \frac{1}{P_{pc}}\left(\frac{P_{pc}}{P} - \frac{1}{z}\frac{dz}{dP_{pr}}\right)$$

from graph (Fig. 4.7) of z vs. reduced properties, by graphical differentiation

$$c_g = \frac{1}{670.6}\left(\frac{670.6}{2000} - \frac{1}{0.825} \times (-0.01)\right)$$

$$= 5.2 \times 10^{-4} \text{ psi}^{-1}$$

(j) At 4100 ft SS aquifer pressure would be $0.44 \times 4100 = 1804$ psi

Since gas has a smaller density than water, it will lie above water. At gas–water contact, pressures are equal. From the given data clearly this gas–water contact will be below 4100 ft. Let this extra distance be x ft. Assume too that density of gas is a constant over the distances concerned, and that the reservoir temperature is 135°F, thus the density takes the value calculated in (f), 6.977 lb ft^3 (or gradient 0.0485 psi ft^{-1}).

Pressure balance at gas–water contact:

$(4100 + x)\,0.44 = 2000 + 0.0485\,x$

$\therefore x = 196/0.3915 = 500$ ft

Therefore *gas–water* contact depth = 4600 ft SS

(k) From (j), gas–water contact is at 4600 ft SS, and pressure is $4600 \times 0.44 = 2024$ psi

Assuming the gas density remains constant for 1000 ft, pressure due to gas = $0.0485 \times 1000 = 48.5$ psi

Therefore pressure at crest of structure = $2024 - 48.5 = 1975.5$ psi

Therefore pressure of mud at this point will be = $1975.5 + 500 = 2475.5$ psi

Assuming the mud to be incompressible, let density of mud = ρ lbs/cu ft

Pressure exerted by mud at 3600 ft = $\dfrac{\rho}{144} \times 3600 = 2475.5$ psi

Therefore $\rho = 99.0$ lbs/cu ft

i.e. specific gravity of mud = 1.58

Solution 4.3

$c_g = 1/p = 1/1923 = 520 \times 10^{-6}$ psi^{-1}

(a) Total compressibility

$$c_T = c_f + c_o S_o + c_w S_w + c_g S_g$$
$$= 10^{-6}\left[5 + 0.45(10) + 0.24(3) + 0.31(520)\right]$$
$$= 171.5 \times 10^{-6} \text{ psi}^{-1}$$

(b) Effective hydrocarbon compressibility

$$c_{oe} = \frac{c_T}{1 - S_{wi}} = \frac{171.5 \times 10^{-6}}{0.76} = 225 \times 10^{-6} \text{ psi}^{-1}$$

SOLUTIONS TO EXAMPLES

Solution 4.4

(a) From graphs (Fig. 4.21, 4.22) or correlation equations for
API = 38°; GOR = 750; T = 175°F; and $\gamma_g = 0.7$:

bubble point pressure = 2800 psia

formation volume factor = 1.4 RB/STB

specific gravity of tank oil = $\dfrac{141.5}{131.5 + 38} = 0.834$

(b) Density of reservoir oil = $\left\{ \dfrac{\text{weight of oil and gas in solution}}{\text{volume of oil}} \right\}$ reservoir conditions

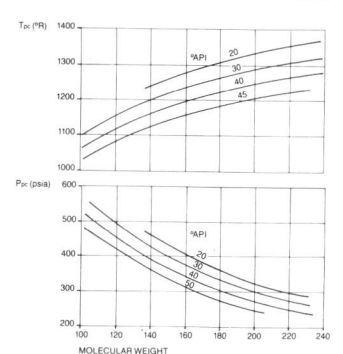

Fig. A4.1 Pseudo critical properties of hydrocarbon liquids

Weight of one barrel of water = $5.615 \times 62.4 = 350.4$ pounds (density of fresh water is 62.4 lb/ft³ and 5.615 cu ft = 1 barrel).

From specific gravity of tank oil, weight of one barrel of oil is $350.4 \times 0.834 = 292.2$ lb.

Avogadro's law states that 1 lb-mole of any ideal gas occupies 379.4 cu ft at 60°F and 14.7 psia.

∴ weight of gas which will dissolve in 1 STB of tank oil is given by the number of moles of gas times its molecular weight. The molecular weight of gas is the gas gravity × molecular weight of air
∴ weight of gas/STB = $(R_s/379.4) \times 0.7 \times 28.97$ lbs = $0.05345\, R_s$ lbs.

Volume of 1 STB oil at reservoir conditions = B_o BBL

∴ Density of reservoir condition oil = $\dfrac{[292.2] + [750 \times 0.053445]}{1.400}$ lbs/BBL

∴ SG = $\dfrac{\text{density at reservoir conditions}}{350.4} = 0.677$

The reservoir oil gradient is therefore 0.677×0.433 psi/ft where 0.433 is the fresh water gradient
∴ oil gradient = 0.293 psi/ft.

For an oil–water contact of 7000 ft SS the hydrostatic pressure is $7000 \times 0.465 = 3255$ psi.

The bubble point pressure is the pressure of oil saturated with gas in equilibrium at the gas–oil contact ∴ pressure at top of oil column = 2800 psi.

For constant oil gradient, height of oil zone = $\dfrac{8255 - 2800}{0.293} = 1550$ ft

∴ GOC = $7000 - 1550 = 5450$ ft SS

For a molecular weight of 180 and 38° API oil the liquid critical temperature is 1220°R and the liquid critical pressure is 310 psia

$T_{pr} = \dfrac{(460 + 175)}{1220} = 0.52$ and $P_{pr} = \dfrac{4000}{310} = 12.9$

The reduced compressibility from charts (Fig. A4.1) is given at this T_{pr}, P_{pr} condition as $c_R = 0.002$.

Since $c_R = c_o \cdot P_c$ then $c_o = \dfrac{0.002}{310} = 6 \times 10^{-6}$ psia^{-1}

From a constant oil compressibility between 2800 and 4000 psia

$$B_o = B_{ob}(1 - C_o \Delta P)$$
$$= 1.40 \left(1.0 - 6 \times 10^{-6}(4000 - 2800) \right)$$
$$= 1.389 \text{ RB/STB}$$

From graphs, viscosity of dead oil at reservoir conditions = 1.4 cP

∴ viscosity of reservoir crude = 0.6 cP.

Solution 4.5

From graph of system pressure vs. system volume the bubble point is estimated by inflexion at 2500 psi.

Liquid volume at standard conditions = 295 ml.

At 3000 psia liquid compressibility $c_o = -\frac{1}{V}\left(\frac{\partial V}{\partial P}\right)_T$

$$c_o = \frac{(404 - 410)}{(4000 - 2500)} \cdot \frac{1}{408} = 9.8 \times 10^{-6} \text{ psi}^{-1}$$

$$B_{o3000} \text{ psia} = \frac{408}{295} = 1.383 \text{ RB/STB}$$

$$B_{o2500} \text{ psia} = \frac{410}{295} = 1.390 \text{ RB/STB}$$

$$R_s = \frac{26.275}{295(10^{-3})} = 89.06 \text{ v/v} = 89.06(5.615) = 500 \text{ SCF/STB}$$

At 2000 psia

$$B_o = \frac{388}{295} = 1.315 \text{ RB/STB}; \quad B_t = \frac{430}{295} = 1.457 \text{ RB/STB}$$

$$R_s = \frac{21}{295 \times 10^{-3}} \times 5.615 = 400 \text{ SCF/STB}$$

∴ $B_t = B_o + (R_{si} - R_s) B_g$

∴ $B_g = \frac{B_t - B_o}{(R_{si} - R_s)} = \frac{(1.457 - 1.315)(295)}{(26.275 - 21.0)10^3} = 7.94 \times 10^{-3} \text{ v/v}$

$$z = \left(\frac{P1}{T1}\right) \cdot \left(\frac{T2}{P2}\right) \cdot \left(\frac{V1}{V2}\right) \cdot = \left(\frac{2000}{660}\right) \cdot \left(\frac{520}{14.7}\right) \cdot 7.94 \times 10^{-3} = 0.85$$

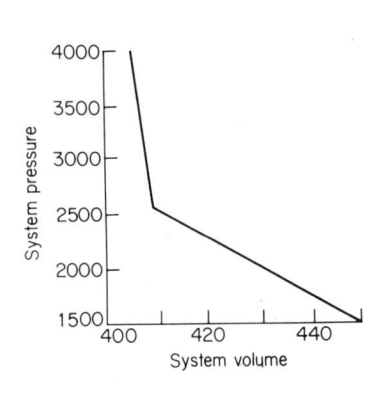

Chapter 5

Solution 5.1

F:	30	19.3	12.5	8.4	6.0
φ:	0.092	0.120	0.165	0.205	0.268

Plot either on log : log scales, or log F : log φ on coordinate scales.

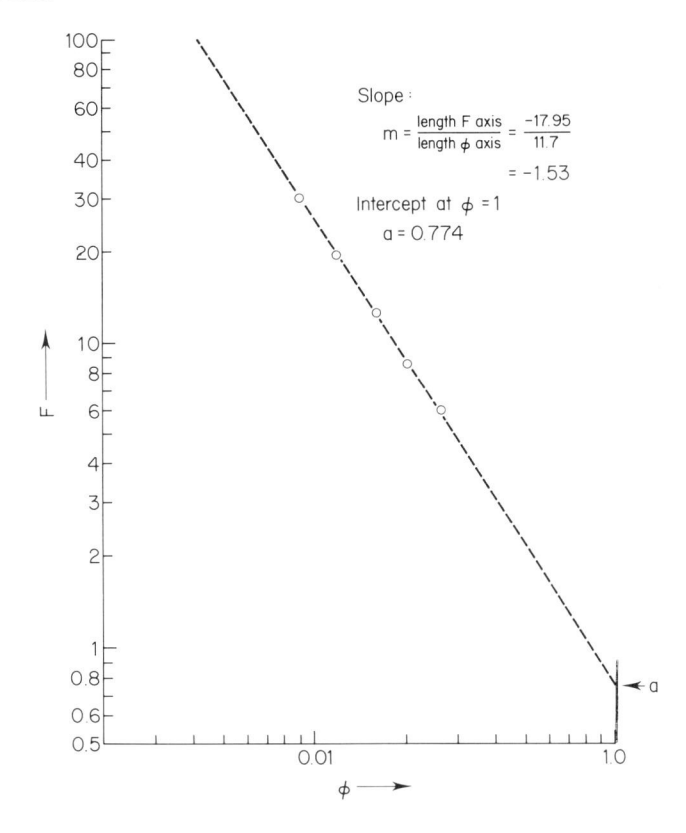

SOLUTIONS TO EXAMPLES

Fig. A5.1 F vs. ϕ

From plot $m = -1.53$
$\quad\quad\quad\; a = 0.774$

Substitute back into laboratory data to calculate check values of F.

Check	ϕ	0.092	0.120	0.165	0.205	0.268
Calculate	F	29.8	19.8	12.2	8.7	5.80

If the true resistivity is 1.29 Ωm and water resistivity is 0.056 Ωm then

$$F = \frac{R_o}{R_w} = \frac{1.29}{0.056} = 23.04$$

$\phi = 0.109$

If $I = \dfrac{1}{S_w^{\,n}}$ where $\exp n = 2$

$\quad R_t = 11.84\ \Omega\text{m} \quad\quad R_o = 1.29\ \Omega\text{m}$

then $I = \dfrac{11.84}{1.29} = 9.18$

$S_w = \left[\dfrac{1}{I}\right]^{0.5} = 0.330$

If $\exp = 1.8 \quad\quad S_w = 0.292$

If $\exp = 2.2 \quad\quad S_w = 0.365$

Solution 5.2

(a)

	Data values					Calculated values		
Log values	GR	FDC	SNP	C_{ILD}	R_{ILd}	$V_{SH_{GR}}$	$V_{SH_{D/N}}$	$\varphi_{D/N}$
Shale	102	2.52	29.0	1100	0.91	1.00	1.00	—
Zone A	52	2.22	22.5	150	6.67	0.39	0.00	0.26
B	72	2.37	20.5	350	2.86	0.63	0.31	0.14
C	20	2.20	21.0	4650	0.215	0.00	0.00	0.25

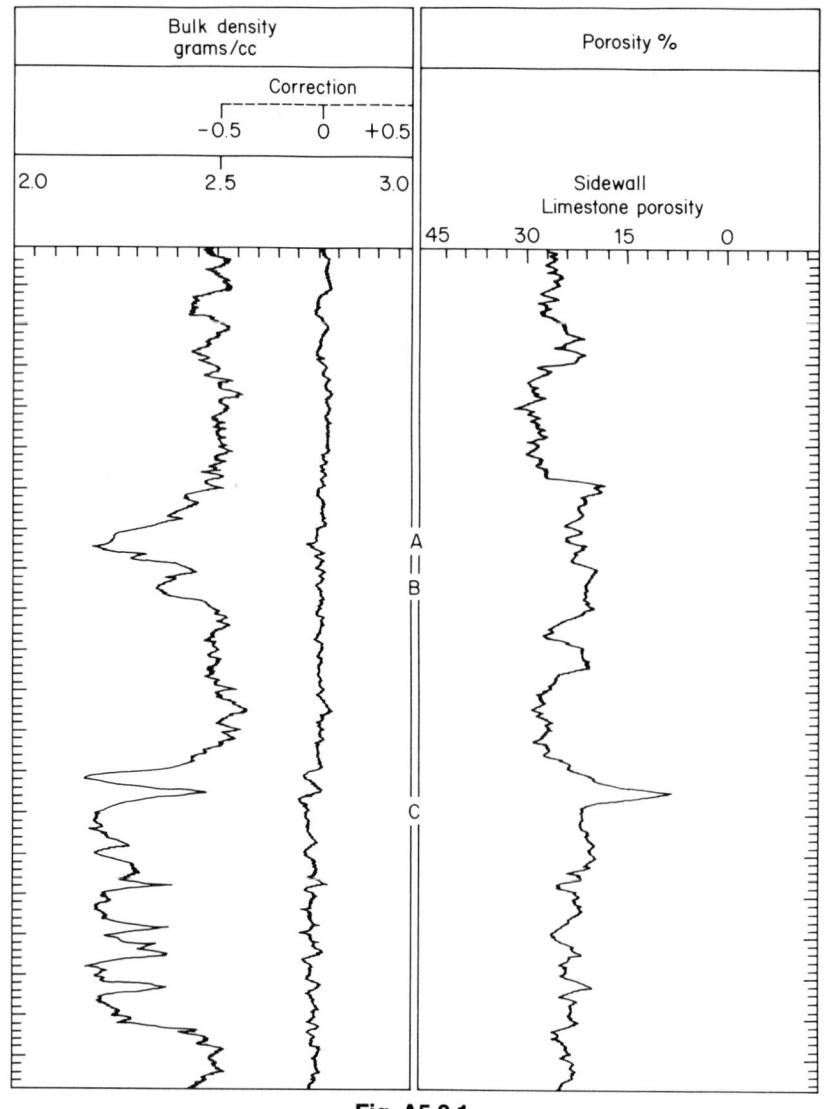

Fig. A5.2.1

SOLUTIONS TO EXAMPLES

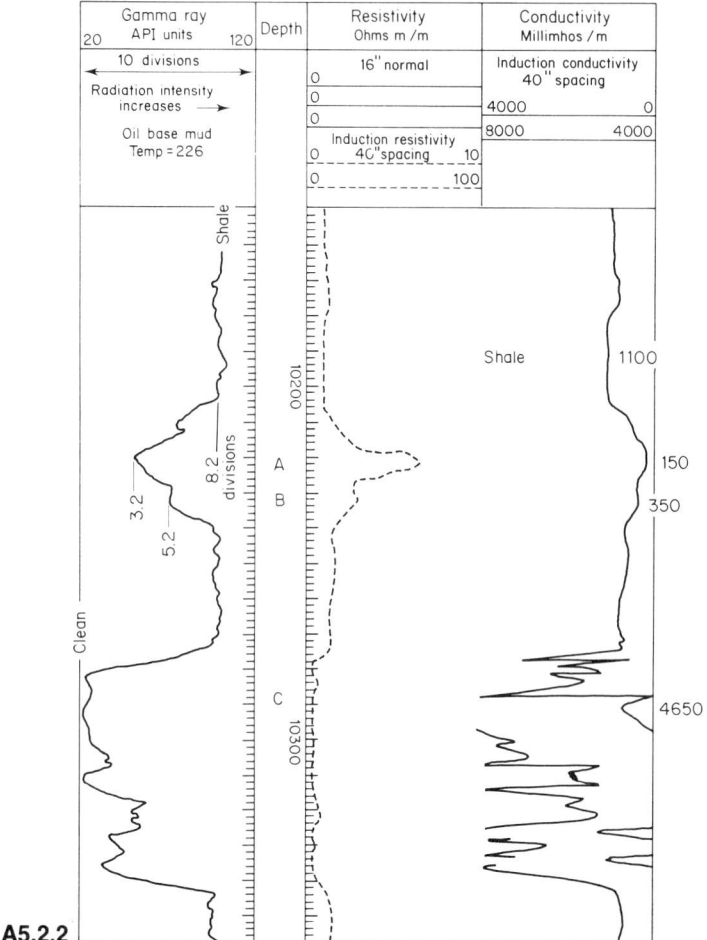

Fig. A5.2.2

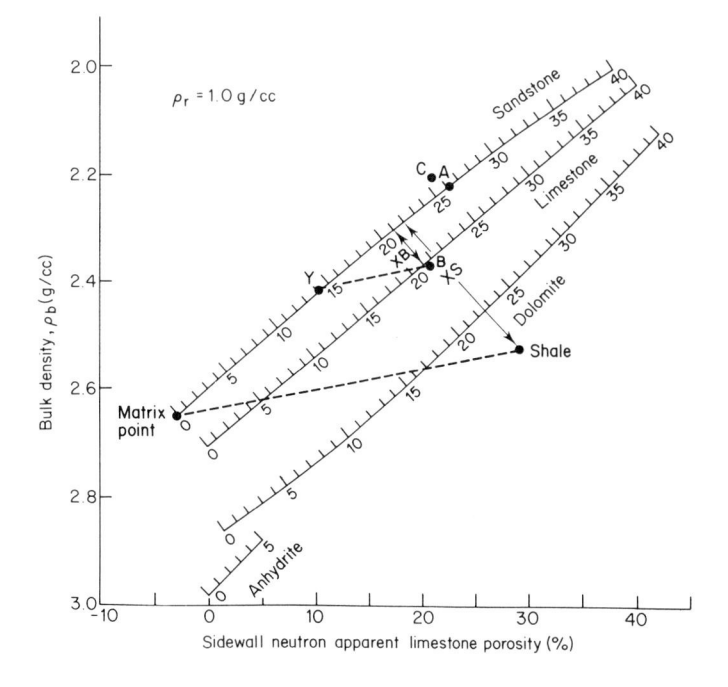

Fig. A5.2.3
Density/SNP crossplot.

(b) For zone C, point plots close to clean sandstone line with $\phi = 0.25$. Assuming C to be water bearing
$R_o = FRw = 1/\phi^2 \cdot \{R_w\}$
$R_w = \phi^2 R_o = 0.26^2 \times 0.215 = 0.0145$ (taking R_{ILd} as R_o)

(c) Shale values are listed above.

(d) $GR_{clean} = 20$, $GR_{shale} = 102$

$$V_{shGR} = \frac{GR - GR_{clean}}{GR_{shale} - GR_{clean}} = \frac{GR - 20}{82}$$

V_{shGR} values calculated are tabulated above.

(e) See Fig. A5.2.3 for shale point. Only level B shows a significant displacement from clean line. Graphically V_{sh} for zone B = $XB/XS = 1.25/4 = 0.31$.

(f) Taking the minimum shale indication (from D/N) gives only B as shaly. Presumably there are radioactive minerals in the sands (such as feldspar) so the GR overestimates shale content.

As above graphically for level B, $V_{sh} = 0.31$. The porosity is given by point Y on the clean sandstone line where BY is parallel to the matrix shale line, i.e. $\phi = 0.14$. The graphical construction is complicated by the curve on the sandstone line. More rigorously convert density and neutron values to sandstone matrix $\phi_D = 16.5$, $\phi_N = 24.2$. $\phi_{NSH} = 32$, $\phi_{DSH} = 7.5$, $\phi = \phi_N - V_{SH}\phi_{NSH}$, $\phi = \phi_D - V_{SH}\phi_{DSH}$

Solving the equations for unknown V_{SH}

$$V_{SH} = \frac{\phi_N - \phi_D}{\phi_{NSH} - \phi_{DSH}} = \frac{24.2 - 16.5}{32 - 7.5} = 0.31$$

$$\phi = \phi_N - V_{SH}\phi_{NSH} = 24.2 - 0.31 \times 7.5 = 0.14$$

(g) Saturation calculations

Level A all equations reduce to Archie ($V_{SH} = 0$)

$$\therefore \frac{1}{R_t} = \left(\frac{1}{FR_w}\right) \cdot S_w^2 \quad \therefore S_w = \sqrt{\left(\frac{FR_w}{R_t}\right)} = \frac{1}{\phi}\sqrt{\frac{R_w}{R_t}} = \frac{1}{0.26}\sqrt{\frac{0.0145}{6.67}} = 0.18$$

Level B with $n = 2$, $R_w = 0.0145$, $R_t = 2.86$, $R_{SH} = 0.91$, $V_{SH} = 0.31$, $\phi = 0.14$.

Archie
$$\frac{1}{R_t} = \left(\frac{1}{FR_w}\right) \cdot S_w^2 \quad \therefore 0.35 = 1.352 S_w^2$$
$$\therefore \underline{S_w = 0.51}$$

Simandoux
$$\frac{1}{R_t} = \left(\frac{1}{FR_w}\right) \cdot S_w^2 + \left(\frac{V_{SH}}{R_{SH}}\right) \quad \therefore 0.35 = 1.352 S_w^2 + 0.341$$
$$\therefore \underline{S_w = 0.082}$$

Modified Simandoux
$$\frac{1}{R_t} = \left(\frac{1}{FR_w}\right) \cdot S_w^2 + \left(\frac{V_{SH}}{R_{SH}}\right) \cdot S_w \quad \therefore 0.35 = 1.352 S_w^2 + 0.341 S_w$$
Solving quadratic + ve root only
$$\therefore \underline{S_w = 0.376}$$

SOLUTIONS TO EXAMPLES

Poupon and Leveaux (Indonesia)
$$\frac{1}{\sqrt{R_t}} = \frac{1}{\sqrt{FR_w}} S_w + \frac{V_{SH}^{(1-V_{SH}/2)}}{\sqrt{R_{SH}}} \cdot S_w$$
$\therefore 0.592 = 1.163 S_w + 390 S_w$
$\therefore \underline{S_w = 0.38}$

where
$\frac{1}{R_t} = \frac{1}{2.86} = 0.350 ; \frac{1}{\sqrt{R_t}} = 0.592$
$\frac{V_{SH}}{R_{SH}} = 0.341 ; V_{SH}^{(1-V_{SH}/2)} = 0.372 ; \frac{V_{SH}^{(1-V_{SH}/2)}}{\sqrt{R_{SH}}} = 0.390$
$\frac{1}{FR_w} = \frac{\phi^2}{R_w} = 1.352 ; \frac{1}{\sqrt{FR_w}} = 1.163$

Thus the modified Simandoux and Indonesia equations give similar S_w's which are less than the Archie S_w. The shale conductance in the basic Simandoux is already near to the measured conductance so the solution gives an unlikely optimistic value for a shaly sand.

Solution 5.3

Waxman and Thomas equation with $a = 1, m = 2, n = 2$
$$\frac{1}{R_t} = \frac{1}{FR_w} S_w^2 + \frac{BQ_v}{F} S_w$$
$$= \frac{1}{F} \left\{ \frac{1}{R_w} S_w^2 + BQ_v S_w \right\}$$

$BQ_v = 0.046 \times 0.3$ mho.cm^2.meq^{-1}.meq/cc
$\quad = 0.0138$ mho cm^{-1} or ohm^{-1} cm^{-1}
$\quad = 100 \times 0.0138 = 1.38$ ohm^{-1} m^{-1}

$\therefore \frac{1}{R_t} = \frac{1}{F} (10 S_w^2 + 1.38 S_w)$
$F = 1/\phi^2 = 1/0.26^2 = 14.79 \; R_t = 5$
$\therefore 0.2 = 0.0676 (10 S_w^2 + 1.38 S_w)$
$\quad = 0.676 S_w^2 + 0.0933 S_w$
$\therefore 0.676 S_w^2 + 0.0933 S_w - 0.2 = 0$

Solving the quadratic
$$S_w = \left(\frac{-0.0933 \pm \sqrt{[0.0933^2 - 4.676 \cdot (-0.2)]}}{2(0.676)} \right) = 0.4$$
(see Archie solution, $S_w = \sqrt{\frac{FR_w}{R_t}} = 0.544$)

Modified Simandoux model
$$\frac{1}{R_t} = \frac{1}{FR_w} S_w^2 + \frac{V_{SH}}{R_{SH}} \cdot S_w$$
Comparing with the Waxman Thomas equation

$$\frac{BQ_v}{F} = \frac{V_{SH}}{R_{SH}} \therefore V_{SH} = R_{SH}\frac{BQ_v}{F}$$

$$V_{SH} = \frac{1.5 \times 1.38}{14.79} = 0.140$$

i.e. it would take 14% shale with resistivity 1.5 ohm-m to get the same result as the Waxman Thomas equation

Basic Simandoux $\frac{1}{R_t} = \frac{1}{FR_w} S_w^2 + \frac{V_{SH}}{R_{SH}}$; $\therefore 0.2 = \frac{0.0676}{0.1} S_w^2 + \frac{0.14}{1.5}$

$0.2 = 0.676 S_w^2 + 0.0933$

$$S_w = \sqrt{\left[\frac{0.2 - 0.0933}{0.676}\right]} = 0.397$$

Solution 5.4

(a) Prove

From Darcy's law:
$$q = \frac{-kA}{\mu}\frac{\partial P}{\partial x}$$

Assuming Boyle's law:
$Q_{sc}P_o = qP$ and $P_o = 1$ atm.

Hence:
$$Q_{sc} = \frac{-kA}{\mu} P \frac{\partial P}{\partial x}$$

or $Q_{sc} = \frac{kA}{\mu}\frac{P_2^2 - P_1^2}{2L}$

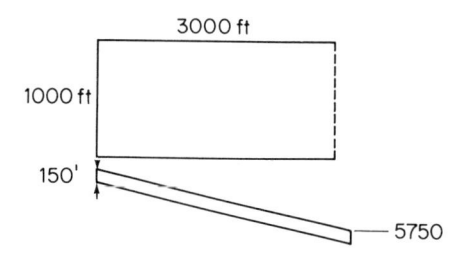

(b) $k = \frac{Q_{sc} \mu 2L}{A(P_1^2 - P_2^2)}$

$$= \frac{6.2 \times 2 \times 0.018 \times 2.54}{\pi \times 1.27^2 \left(\left(\frac{950}{760}\right)^2 - 1\right)}$$

$= 0.2$ D

Solution 5.5

The problem requires correction of pressure so that the linear Darcy law can be used. In field units:

$$q = 1.127 \times 10^{-3}\frac{kA}{\mu}\frac{\Delta P}{L} \text{ BBL/d}$$

Assuming average water gradient of 0.45 psi/ft (0.433 × 1.038) and referring to a HWC datum of 5250 ft SS, static pressure at the outcrop is:

$P_{5250} = 0.45 \times 5250 = 2362.5$ psi

But pressure = 1450 psi at 5250

Hence, $q = 1.127 \times 10^{-3} \times \frac{750 \times 3000 \times 65}{1} \times \frac{(2362.5 - 1450)}{52\,800}$

$q = 2848.5$ BBL/d

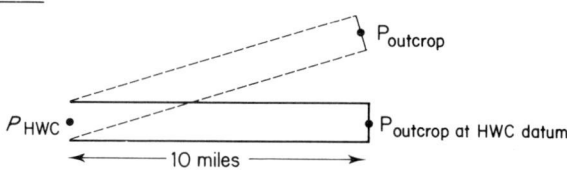

SOLUTIONS TO EXAMPLES

Solution 5.6

Using the equation:
$$k = \frac{Q_{sc} 2 \mu L}{A(P_1^2 - P_2^2)} = \frac{(6.4/60) \times 2 \times 0.018 \times 2.54}{\pi \, 1.27^2 \left(\left(\frac{861}{760}\right)^2 - 1^2\right)}$$

S_c for rate 1 = 0.0068 D = 6.8 mD
S_c for rate 2 = 6.02 mD
S_c for rate 3 = 5.0 mD

This is because of the Klinkenberg effect.

Plotting k against $1/P_{mean}$ gives k_L as $1/P_{mean} \to 0$ as 3 mD.

Solution 5.7

Assume cross-sectional area A.

$$q = -A\frac{dh}{dt}$$ where q is flow rate and h is current height measured from *bottom* of core plug.

Flow across core is:
$$q = \frac{-kA}{\mu}\frac{\Delta P}{L}$$

But ΔP = datum correction pressure difference, so:
$$q = \frac{-kA}{\mu}\frac{\rho g h}{L} = -A\frac{dh}{dt}$$

so $$-\int_{h_0}^{L}\frac{dh}{h} = \frac{k\rho g'}{\mu L}\int_0^t dt$$

or $$\log_e \frac{h_0}{h} = \frac{k\rho g'}{\mu L}t$$

so $$k = \frac{\mu L}{\rho g'}\frac{\triangle[\log(h_0/h)]}{\triangle t} = \frac{\mu L}{\rho g'}\left(\frac{\log_e \frac{h_0}{h_2} - \log_e \frac{h_0}{h_1}}{\Delta_t}\right)$$

Note: $\rho g'$ has to be in units such that $\rho g' h$ = atm.

Hence $$k = \frac{1 \times 2 \times 10^6}{1.02 \times 981} \times \frac{\log_e 84 - \log_e 15.5}{4500}$$

$$= 0.8 \text{D}$$

Note: a plot of $\log_e h$ against t would be best.

Solution 5.8

$$\rho_{oil} = 50 \text{ lb/ft}^3 = \frac{50}{144}\text{psi/ft} = 0.3472 \text{ psi/ft}$$

(a) Correct well pressures to 5750 ft = 1750 + 0.3472 × 750
$$= 2010.4 \text{ psi}$$

(b) Flowing gradient
$$q = \frac{kA}{\mu}\frac{\Delta P}{L} 1.127 \times 10^{-3}$$

$$\Delta P = \frac{q\mu L}{1.127 \times 10^{-3} \times k \times A} = \frac{1000 \times 1.135 \times 7 \times 3000}{1.127 \times 10^{-3} \times 150 \times 150 \times 1000}$$

$$= 94 \text{ psi}$$

So, $P_{owc} = 2010.4 + 94 = 2104.4$ psi

$PV_{res} = 3000 \times 1000 \times 150 \times \varphi$

Equating production $V_p = V2\bar{c} \cdot \Delta P$, then

$$PV_{aquifer} = \frac{3000 \times 1000 \times 150 \times \phi}{(2104.4 - 500) 3 \times 10^{-6}} = 93.5 \times 10^9 \times \phi \, \text{ft}^3$$

Solution 5.9

Darcy's equation $\frac{Q}{A} = -\frac{k}{\mu}\frac{dP}{dx}$ for non-compressible flow

(a) *Linear beds – parallel flow*

$Q = q_1 + q_2 + q_3$

Assume infinitely thin barriers between layers

$$Q = q_1 + q_2 + \ldots = k_1 A_1 \frac{\Delta P}{\mu L} + k_2 A_2 \frac{\Delta P}{\mu L} + \ldots$$

$$= k' A \frac{\Delta P}{\mu L}$$

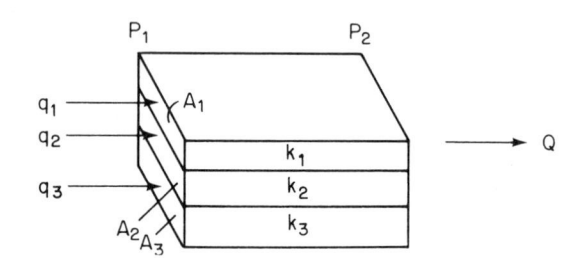

where k' is the apparent permeability and A the total area.

Hence $k'A = k_1 A_1 + k_2 A_2 + \ldots$

Therefore $k' = \dfrac{\sum\limits_{i=1}^{n} k_i A_i}{\sum\limits_{i} A_i}$

or if beds all same width $= \dfrac{\sum k_i h_i}{h_t}$

(b) *Series flow*

Assume equal areas

$A_1 = A_2 = \ldots$

$q_t = q_1 = q_2 = q_3 \ldots$

Now $P_1 - P_4 = (P_1 - P_2) + (P_2 - P_3) + (P_3 - P_4) \ldots$

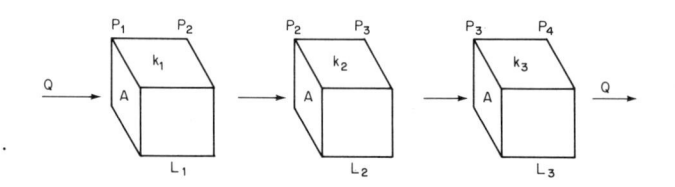

Using Darcy's law

$$q_t \frac{L}{A} \frac{\mu}{k'} = q_1 \frac{L_1}{A} \frac{\mu}{k_1} + q_2 \frac{L_2}{A} \frac{\mu}{k_1} + \ldots$$

Since flow rates, cross-sections and viscosities are equal in all beds

$$\frac{L}{k'} = \frac{L_1}{k_1} + \frac{L_2}{k_2} + \ldots$$

$$k' = \frac{\sum h_i}{\sum (L_i/k_i)}$$

(c) *Radial flow parallel*

From the figure, it is noted that the same terms appear in the radial flow network as in the linear system.

$$Q = \frac{2\pi k h (P_e - P_w)}{\mu \ln (r_e/r_w)}$$

e – external boundary
w – internal boundary

SOLUTIONS TO EXAMPLES

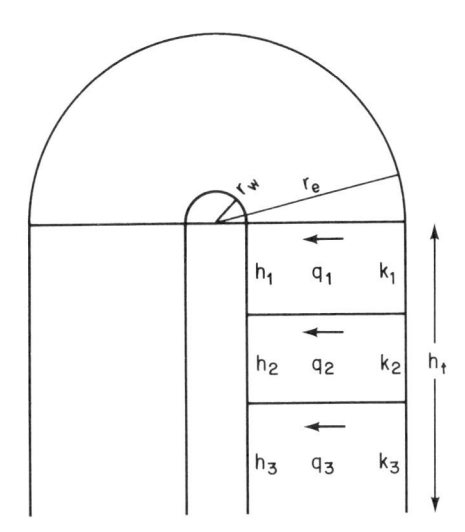

The only difference in the two systems is the manner of expressing the length over which the pressure drop occurs. All these terms are the same in each case.

Therefore $k' = \dfrac{\sum k_i h_i}{h_t}$

(d) Radial flow series

By same reasoning as in the linear case

$$k' = \dfrac{\ln (r_e/r_w)}{\sum\limits_{j=1}^{n} \dfrac{\ln (r_j/r_{j-1})}{k_j}}$$

Bed	Depth/ Length of bed	Horizontal permeability; mD
1	250	25
2	250	50
3	500	100
4	1000	200

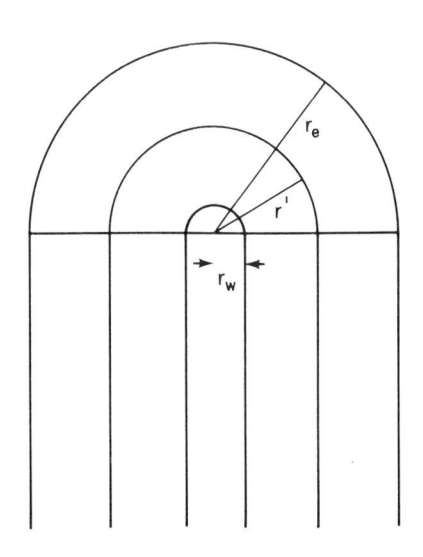

For radial systems, wellbore = 6″, and radius of effective drainage 2000′ and bed 1 is adjacent to wellbore.

Linear flow – parallel, and radial flow – parallel, take data lengths as bed depths and bed lengths and radii to be equal.

Linear flow in parallel

$$k' = \dfrac{250 \times 25 + 250 \times 50 + 500 \times 100 + 1000 \times 200}{2000} = 134.4 \text{ mD}$$

Radial flow in parallel

$$k' = \dfrac{\sum kh}{h_t}$$

$$k' = \dfrac{250 \times 25 + 250 \times 50 + 500 \times 100 + 1000 \times 200}{2000}$$

$$k' = \dfrac{268\ 750}{2000} = 134.4 \text{ mD}$$

Linear flow in series

$$k' = \frac{2000}{\frac{250}{25} + \frac{250}{50} + \frac{500}{100} + \frac{1000}{200}} = \frac{2000}{25} = 80 \text{ mD}$$

Radial flow in series

$$k' = \frac{\ln(2000/0.5)}{\frac{\ln 250/0.5}{25} + \frac{\ln 500/250}{50} + \frac{\ln 1000/500}{100} + \frac{\ln 2000/1000}{200}}$$

$$= 30.4 \text{ mD}$$

i.e. permeability near wellbore most important.

Chapter 6

Solution 6.1

P_c	0	4.4	5.3	5.6	7.6	10.5	15.7	35.0
S_w	100	100	90.1	82.4	60.0	43.7	32.2	29.8
h	0	33.3	40.2	42.4	57.5	79.6	119.0	265.3

$$\left(h = \frac{P_c}{(\rho_w - \rho_o)/144}\right)$$

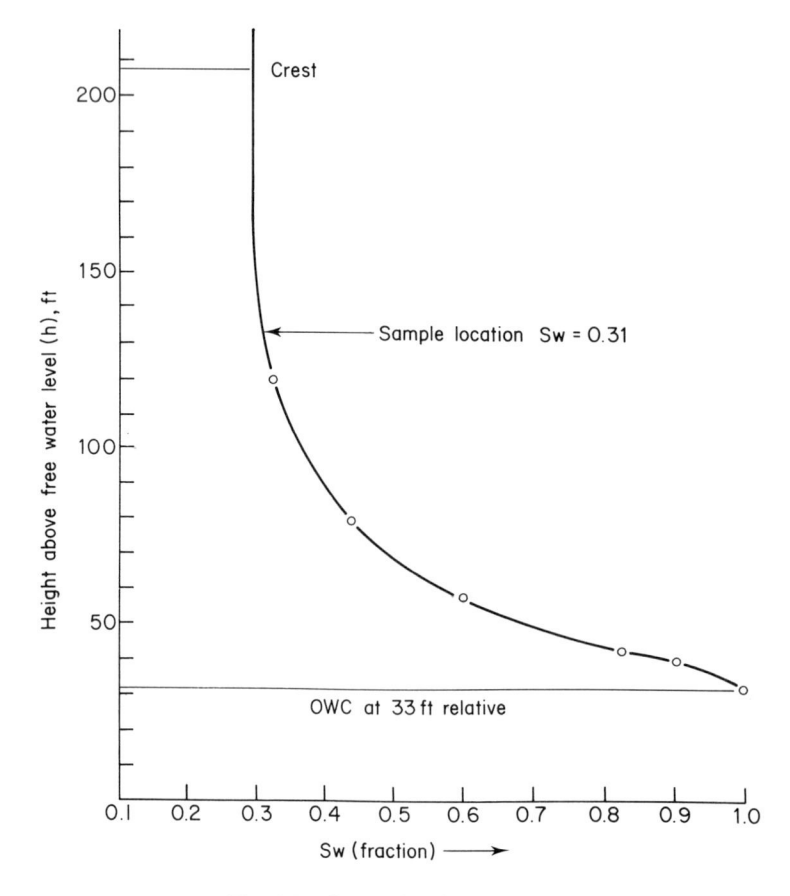

Fig. A6.1 Saturation distribution.

SOLUTIONS TO EXAMPLES

Note that the oil–water contact is at $S_w = 1.0$, not at $P_c = 0$.
At 100 ft above OWC, $S_w = 0.31$ (135 ft relative)

$$\bar{S}_w = \frac{\int S_w \, dh}{h}$$

From area under S_w against h curve:
$\bar{S}_w = 0.37$

Solution 6.2

S_w	100	100	90.1	82.4	60.0	43.7	32.2	29.8
$(P_c)_{\text{O-W}}$	0	4.4	5.3	5.6	7.6	10.5	15.7	35.0
$(P_c)_{\text{Hg}}$	0	65.1	78.4	82.9	112.5	155.4	232.4	518.0
$f(J) = P_c \sqrt{\frac{K}{\varnothing}}$	0	1534.4	1847.9	1954.0	2651.7	3662.8	5477.7	12209.0
$(P_c)_{\text{Hg}}$	0	110.7	133.3	140.9	191.2	264.1	395.0	880.4

for 25 mD
and $\varnothing = 0.13$

Solution 6.3

For the laboratory data $\sqrt{k/\phi}c = (150/0.22)^{0.5} = 26.11$ and using $J(S_w) = \frac{P_{c(sw)}}{\sigma \cos \theta} \sqrt{\frac{k}{\phi}}$ with $\sigma \cos \theta = 72$ dyne/cm the $J_{(sw)}$ vs S_w relationship is calculated.

$J(S_w) = 0.363 \, P_c (S_w)_{\text{lab}}$

S_w	1.0	1.0	0.9	0.8	0.7	0.6	0.5	0.4	0.3	0.2	0.2
$J_{(sw)}$	0	0.363	1.451	2.176	2.901	3.445	4.862	4.968	5.984	8.341	36.27

At reservoir conditions $P_{c(S_w)_{res}} = \dfrac{J_{(sw)} \, \sigma \cos \theta}{\sqrt{k/\phi}}$

for $\sigma \cos \theta = 26$ and $\sqrt{k/\phi} = 44.72$
$P_{c \, (S_w)_{res}} = 0.581 \, J(S_w)$ and the reservoir condition P_c curve is therefore calculated as

S_w	1.0	1.0	0.9	0.8	0.7	0.6	0.5	0.4	0.3	0.2	0.2
$P_{c(\text{res})}$	0	0.211	0.843	1.264	1.685	2.00	2.823	2.886	3.451	4.846	21.07

For the reservoir specific gravity of oil and water given

$\Delta \rho = (1.026 - 0.785)$
$= 0.241$

The relationship between capillary pressure and height H above FWL is, in the units required, $P_{c(Sw)} = 0.433 \, H \Delta \rho$

$\therefore H = \dfrac{P_{c(Sw)}}{0.104}$

Using the threshold value of $P_{c(Sw)} \, (= P_{ct})$ as the observed oil water contact, then

$H_{\text{owc}} = \dfrac{0.211}{0.104} = 2$ ft above the FWL

$H_{\text{TTZ}} = \dfrac{4.85}{0.104} = 46.5$ ft above the FWL

Chapter 7

Solution 7.1

From Darcy's law modified for effective permeability in horizontal linear flow

$$K_o(s) = \frac{q_o \mu_o L}{A \Delta P_o} \text{ and } K_w(s) = \frac{q_w \mu_w L}{A \Delta P_w}$$

Assuming zero capillary pressure ($P_c = 0 = P_o - P_w$) so $\Delta P_o = \Delta P_w = \Delta P$, and using Darcy units of cc/s for rate and atmospheres for ΔP, then:

$$K_e \text{ (md)} = \frac{q\mu}{\Delta P} \left[\frac{(4)(9)(1000)}{\pi (3.2)^2 \, 3600} \right]$$

For oil $K_o = \dfrac{q_o}{\Delta P}$ (9.14)

For water $K_w = \dfrac{q_w}{\Delta P}$ (5.0)

For $K_{ro} = \dfrac{K_o}{K_{o(cw)}}$ and $K_{rw} = \dfrac{K_w}{K_{o(cw)}}$

$K_{o(cw)} = \dfrac{90}{49.25} (9.14) = \underline{16.7 \text{ md}}$

S_w	K_{ro}	K_{rw}
15.0	1.0	0
19.8	0.452	0.017
25.1	0.30	0.025
32.1	0.20	0.049
41.0	0.12	0.075
54.9	0.05	0.156
68.1	0	0.249

These data are plotted in Fig. A 7.1

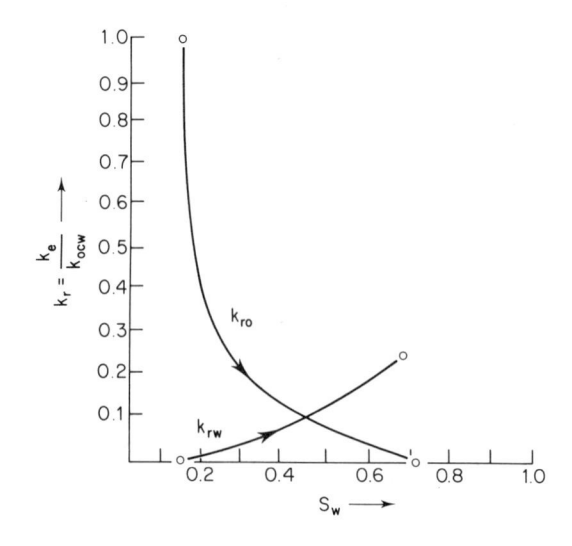

Fig. A7.1 Steady-state relative permeability.

SOLUTIONS TO EXAMPLES

Solution 7.2

For pressure maintenance, the oil rate in RB/D is

$10\,000 \times 1.2765 = 12\,765$ RB/D

The end points of the relative permeability curve are

$K_{ro}' = 0.9$ at $S_{wi} = 0.28$
$K_{rw}' = 0.7$ at $S_{or} = 0.35$

The ratio $\dfrac{\mu_w}{\mu_o}$ is then calculated from the given end point mobility ratio of 2.778.

Since $M' = \dfrac{k_{rw}'}{\mu_w} \cdot \dfrac{\mu_o}{k_{ro}'}$, then $\dfrac{\mu_w}{\mu_o} = \dfrac{k_{rw}'}{M' \, k_{ro}'} = \dfrac{0.7}{2.778\,(0.9)} = 0.28$

The fractional flow curve can now be calculated for the horizontal reservoir:

$$f_w = \dfrac{1}{1 + 0.28 \left\{\dfrac{k_{ro}}{k_{rw}}\right\}}$$

S_w	0.28	0.30	0.35	0.45	0.55	0.60	0.65
f_w	0	0.082	0.295	0.708	0.931	0.984	1.00

A line tangential to the fractional flow curve from $S_w = 0.28$ gives the tangent at $S_{wf} = 0.4$ ($f_w = 0.535$) and the intercept with $f_w = 1$ at $S_w = 0.505$. The gradient of this tangent $[df_w/dS_w]_{S_{wf}}$ is 4.44.

From Buckley-Leverett theory the constant rate frontal advance of the 40% saturation front is:

$$X_f/\text{day} = \dfrac{q(t)\,(5.615)}{(A)\,(\phi)} \left[\dfrac{df}{dS_w}\right]_{S_{wf}}$$

$$X_f/\text{day} = \dfrac{(12\,765)\,(5.615)\,(4.44)}{(5280)\,(50)\,(0.25)} = 4.82 \text{ ft/day}$$

For a system length of 5280 ft, breakthrough therefore occurs in 1095 days (= 3 years)
At year 4 the pore volume injected is

$$= \dfrac{4\,(365)\,(12\,765)\,(5.615)}{(5280)\,(50)\,(5280)\,(0.25)} = 0.3 \text{ PV}$$

and $df/ds \Big|_{S_{we}} = \dfrac{1}{0.3} = 3.33$

The tangent of gradient 3.33 to the fractional flow curve at saturations greater than frontal occurs at $S_{we} = 0.45$ (from a plot of $\dfrac{df_w}{dS_w}$ vs S_w).

At this saturation (S_{we}), $f_{we} = 0.71$, the reservoir condition water cut.

The average saturation remaining in the reservoir is given by the Welge equation as:

$$\bar{S}_w = S_{we} + \dfrac{f_{oe}}{[df_w/dS_w]_{S_{we}}}$$

$\bar{S}_w = 0.45 + \dfrac{(1 - 0.71)}{3.33}$ $\therefore \bar{S}_w = 0.537$

The recovery factor is thus:

$$RF = \dfrac{\bar{S}_w - S_{wi}}{1 - S_{wi}} = 0.36$$

Solution 7.3

The critical injection rate for gas is given in field units of SCF/D as:

$$q \text{ SCF/D} = \frac{4.9 \times 10^{-4} \, k \, k_{rg}' \, A \, (\gamma_g - \gamma_o) \sin \alpha}{\mu_g \, B_g \, (M-1)}$$

where B_g is in units of RB/SCF and α is negative for updip injection.

The density difference in terms of specific gravity is:

$$\Delta \gamma = \frac{17 - 48}{62.4} = -0.4968$$

$$M' = \frac{0.5}{0.028} \frac{(1.8)}{(0.9)} = 35.71$$

$$\sin(-10°) = -0.1736$$

$$\therefore q_{\text{crit}} = \frac{4.9 \times 10^{-4} \, (800) \, (0.5) \, (8000) \, (100) \, (-0.4968) \, (-0.1736)}{(0.028) \, (35.71 - 1) \, (7.5 \times 10^{-4})}$$

$$= 18.589 \text{ MMSCF/D}$$

The rate of injection proposed (15 MMSCF/D) is less than the critical rate and might almost lead to a stable displacement.

The oil rate expected prior to breakthrough is therefore:

$$Q_o = \frac{15 \times 10^6 \times 7.5 \times 10^{-4}}{1.125} = 10 \text{ MSTB/D}$$

Solution 7.4

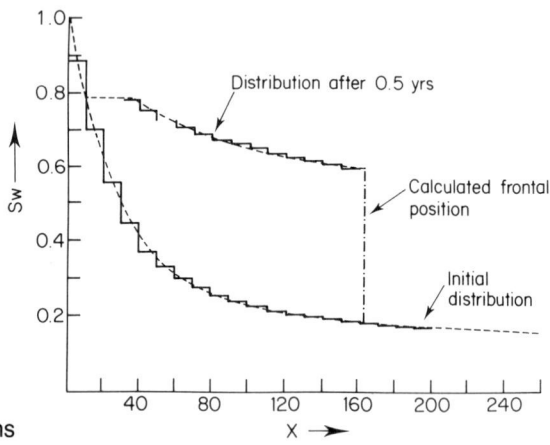

Fig. A7.2 Saturation distributions

From the given data the saturation is plotted as shown in Fig. A 7.2

$q_t = 9434$ rb/d   Dip $= 6°$ $\mu_o = 1.51$ cp
$h = 100'$ $k = 276$ mD $\mu_w = 0.83$ cp
$w = 8000'$ $\phi = 0.215$
$\Delta \gamma = 0.04$ $A = 800\,000$ ft^2

The fractional flow curve is calculated as follows:

$$f_w = \frac{1}{1 + \frac{\mu_w}{k_{rw}} \cdot \frac{k_{ro}}{\mu_o}} \left\{ 1 + 1.127 \times 10^{-3} \frac{k \, k_{ro} \, A}{q_t \, \mu_o} \left[-0.4335 \, \Delta \gamma \sin \alpha \right] \right\}$$

SOLUTIONS TO EXAMPLES

The results are shown in Fig. A 7.3

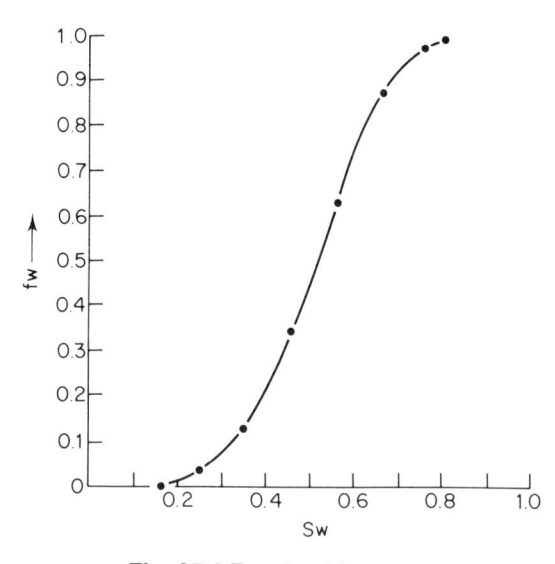

Fig. A7.3 Fractional flow curve.

Therefore:

S_w	0.16	0.25	0.35	0.45	0.55	0.65	0.75	0.79
f_w	0	0.036	0.127	0.344	0.64	0.88	0.98	1.0

Since there is no uniform saturation distribution initially a material balance solution is used:

$$(\Delta X)_{S_{wj}} = \frac{5.615\, q_t\, \Delta t}{\varphi A} \left[\frac{\Delta f_w}{\Delta S_w}\right]_{S_{wj}} = 0.308\, \Delta t \left[\frac{\Delta f_w}{\Delta S_w}\right]_{S_{wj}} \quad \text{for } \Delta t \text{ in days}$$

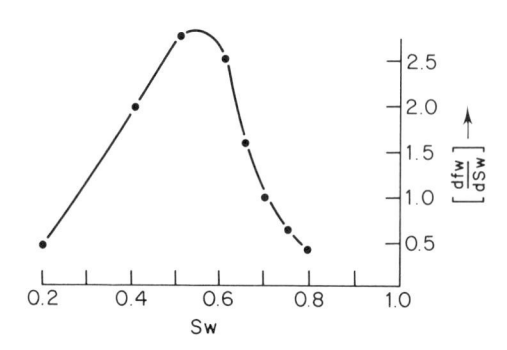

Fig. A7.4 Slopes of fractional flow curve.

The slope of the fractional flow curve as a function of saturation is plotted in Fig. A 7.4. Selecting saturations

For $S_w = 0.79$		For $S_w = 0.75$		For $S_w = 0.7$	
t (yrs)	X	t (yrs)	X	t (yrs)	X
0	10 ft	0	12 ft	0	15 ft
0.5	10 + 23.5 = 33.5	0.5	12 + 36.5 = 48.5	0.5	15 + 56.2 = 71.2
1.0	10 + 47 = 57	1.0	12 + 73.0 = 85	1.0	15 + 112.4 = 127.4
2.0	10 + 94 = 104	2.0	12 + 146 = 158	2.0	15 + 225 = 240

At 0.5 years the saturation distribution is shown on Fig. A7.2 and is represented in 10' increments.

$$\left(\text{Note:} \quad \frac{5.615 \, q_t \, \Delta t}{\phi A} = 56.22 \right)$$

Δx	$\sum \Delta x$	$\bar{S}_{wi}$	$\bar{S}_w \, (0.5 \, yr)$	$\Delta x \, (\bar{S}_w - \bar{S}_{wi})$	$\sum \Delta x \, (\bar{S}_w - \bar{S}_{wi})$
10	10	0.79	0.79	0	0
10	20	0.70	0.79	0.9	0.9
10	30	0.56	0.79	2.3	3.2
10	40	0.45	0.78	3.3	6.5
10	50	0.375	0.755	3.8	10.3
10	60	0.33	0.730	4.0	14.3
10	70	0.30	0.710	4.10	18.4
10	80	0.278	0.690	4.12	22.52
10	90	0.254	0.675	4.21	26.72
10	100	0.24	0.650	4.10	30.83
10	110	0.23	0.640	4.10	34.93
10	120	0.215	0.630	4.15	39.08
10	130	0.205	0.620	4.15	43.23
10	140	0.20	0.613	4.13	47.36
10	150	0.195	0.605	4.10	51.46
10	160	0.190	0.600	4.10	55.56
10	170	0.183	0.595	4.12	59.68

Interpolation $\therefore X_f = 160 + 10 \left\{ \dfrac{56.22 - 55.56}{59.68 - 55.56} \right\} = 161.6$ ft from OWC

From Fig. A 7.2, at $X_f = 161.6$ ft, $S_{wf} = 0.60$

Solution 7.5

For the particular example the problem reduces to the following tabulation, numbering layers n, from $n = 0$ to $n = N = 5$, bottom to top.

$$\bar{S}_{w_n} = \frac{0.7n + 0.15(5-n)}{5} \, ; \quad \bar{K}_{rw_n} = \frac{0.5 \sum_1^n k_j}{\sum_1^5 k_j} \, ; \quad \bar{K}_{ro_n} = 0.9 \frac{\sum_{n+1}^5 k_j}{\sum_1^5 k_j}$$

where: $\sum_1^5 k_j = 50 + 500 + 1500 + 2000 + 500 = 4550$ mD.

n	$\sum_1^n k_j$	$\sum_{n+1}^N k_j$	$\dfrac{\sum_1^n k_j}{\sum_1^5 k_j}$	$\dfrac{\sum_{n+1}^N k_j}{\sum_1^5 k_j}$	$\bar{K}_{rw_n}$	$\bar{K}_{ro_n}$	$\bar{S}_{w_n}$
0	0	4550	0	1.000	0	0.900	0.15
1	50	4550	0.0110	0.989	0.0055	0.8901	0.26
2	550	4000	0.1209	0.879	0.0605	0.7911	0.37
3	2050	2500	0.4505	0.5494	0.2253	0.4940	0.48
4	4050	500	0.8901	0.1099	0.4451	0.0989	0.59
5	4550	0	1.00	0	0.50	0	0.70

The resultant pseudo-relative permeability is plotted as $\bar{S}_{w_n}$ vs $\bar{K}_{rw_n}$ and $\bar{K}_{ro_n}$

Chapter 8

Solution 8.1

Using the relationship $h + 139 = 164/\sinh x$ the saturation vs height relation is calculated as follows:

X (frac)	0.33	0.40	0.50	0.60	0.70	0.80	0.90	1.0
$\sinh x$	0.3360	0.4108	0.5211	0.6367	0.7586	0.8881	1.0265	1.1752
h (ft)	349	260.2	175	118	77	45	20.8	0.55

Fig. A 8.1 shows the plot of water saturation and porosity as a function of depth. Fig. A. 8.2 shows the plot of isopach value vs area contained within the contour. In the absence of a planimeter to measure area use metric graph paper in a simplified approach. Take 50 ft intervals from base to crest. Count squares to determine volume for each interval. Assign appropriate value of φ and S_w for each interval (1cm square = 2500 acre ft).

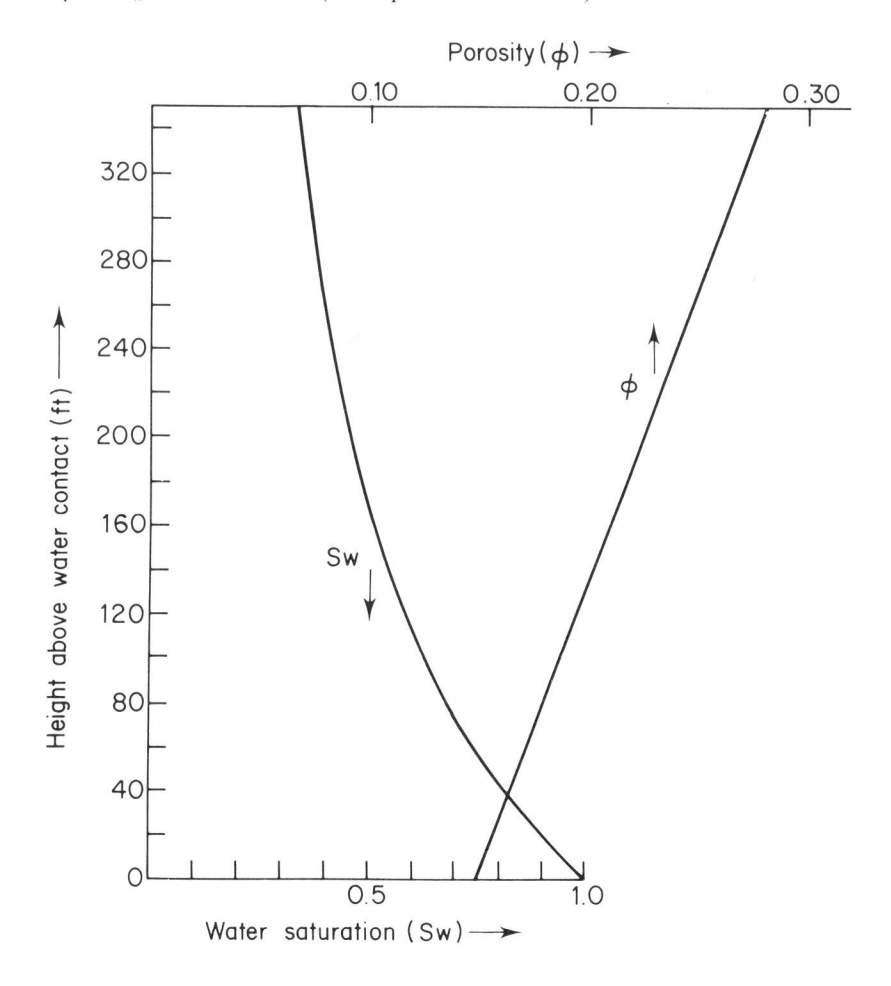

Fig. A8.1 S_w and ϕ vs. depth.

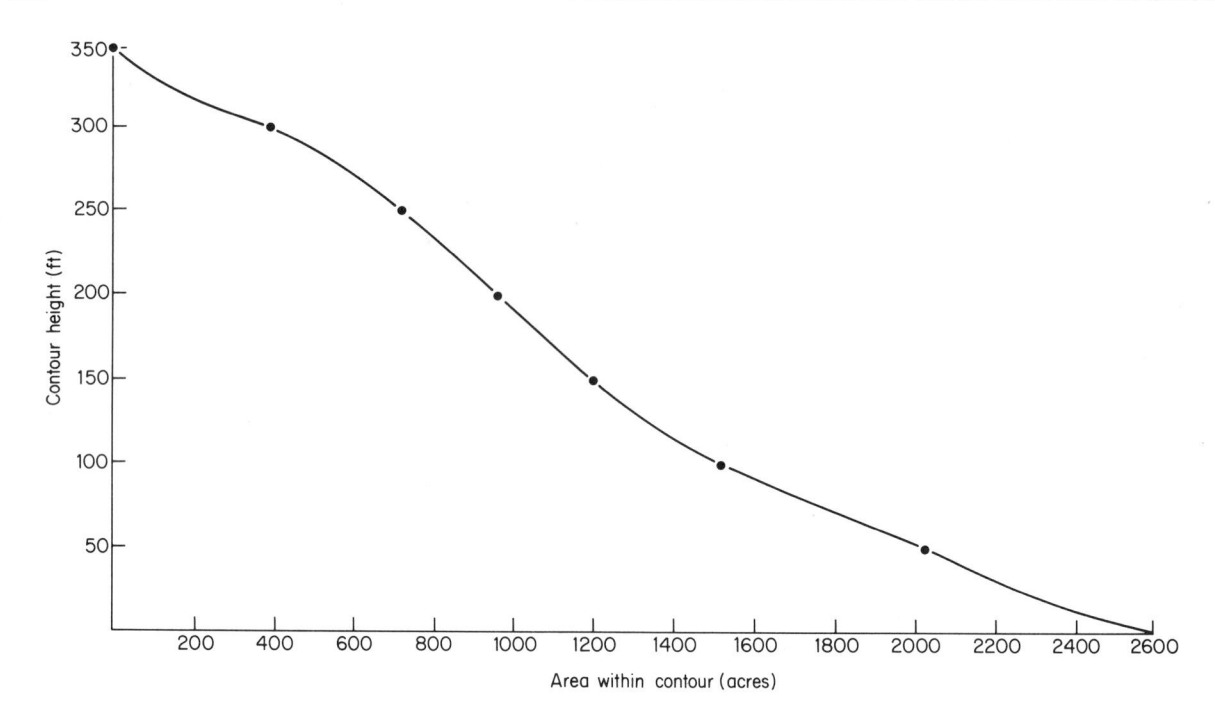

336   PETROLEUM ENGINEERING: PRINCIPLES AND PRACTICE

Interval	No. of squares	Gross rock volume (10^3 acre ft)	Saturation (S_w)	Porosity (φ)	Hydrocarbon (volume × 10^6 BBLs)
0–50	46	115.00	0.875	0.160	16.73
50–100	35	87.50	0.70	0.178	36.25
100–150	26.7	66.75	0.57	0.197	43.87
150–200	21.5	53.75	0.49	0.215	45.72
200–250	17.0	42.50	0.43	0.234	43.98
250–300	11.3	28.25	0.39	0.252	33.69
300–350	3.0	7.50	0.36	0.271	10.09
		$\sum = 401.250$			$\sum = 230.326$

Hydrocarbon in place = 230 326 250 BBLs reservoir oil
                       = 170 × 10^6 BBLs stock tank/oil

Solution 8.2

The oil in place at stock tank conditions is evaluated using the relationship

$$N = \frac{7758 \, A_h \varphi \, S_o}{B_{oi}}$$

where N is in STB
       A is in acres
       h is in feet
       φS_o is a fraction
       B_{oi} is in RB/STB

The recoverable reserve is $N.(RF)$ where RF is the recovery factor (fraction). Deterministically, the minimum, 'most likely', and maximum values are calculated as:

minimum	43 × 10^6 STB
'most likely'	116 × 10^6 STB
maximum	274 × 10^6 STB

SOLUTIONS TO EXAMPLES

The distribution functions of the reservoir parameters are shown in Fig. A 8.3. These data are interrogated randomly using a Monte Carlo approach in the recoverable reserve calculation. The resulting cumulative frequency greater than a given value plot is shown in Fig. A 8.4. The values associated with the 90%, 50% and 10% levels are as follows:

at 90% the recoverable reserve is at least 72×10^6 STB
at 50% the recoverable reserve is at least 120×10^6 STB
at 10% the recoverable reserve is at least 185×10^6 STB

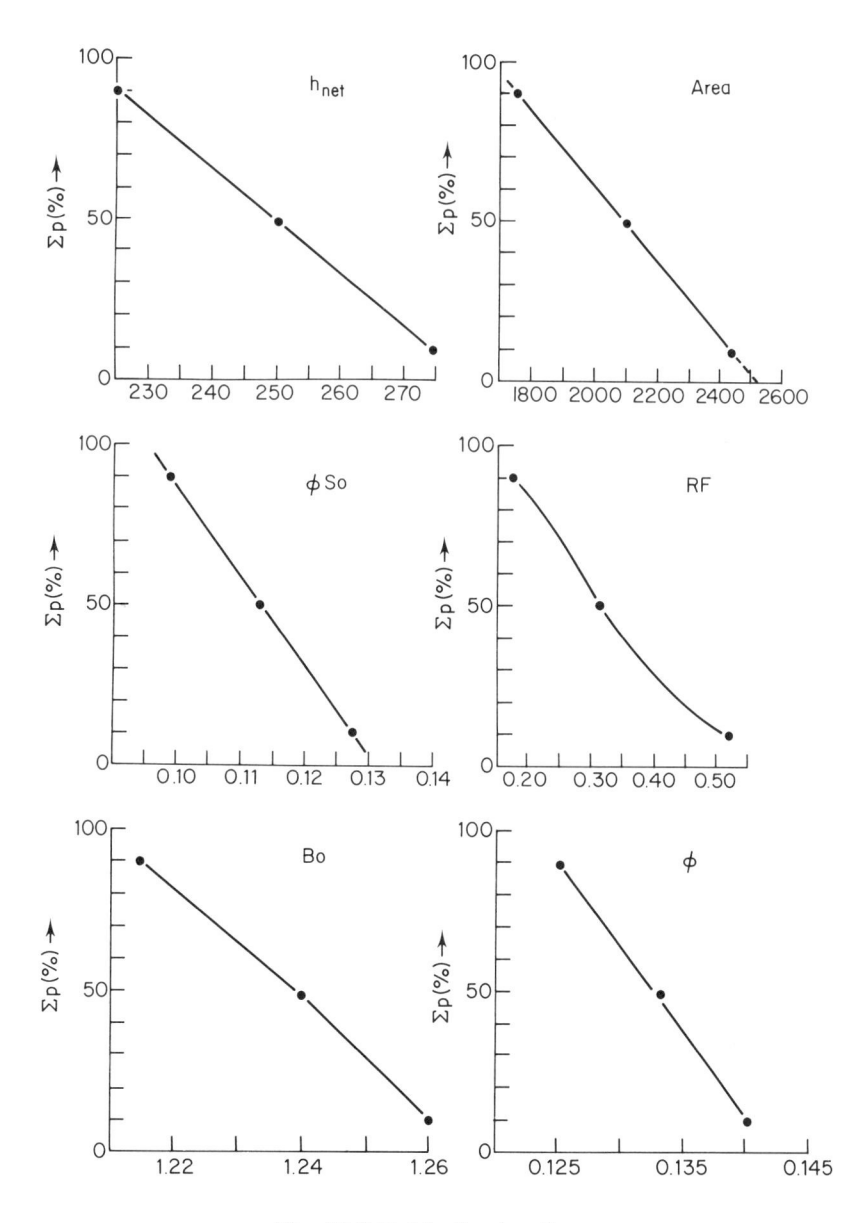

Fig. A8.3 Distribution functions.

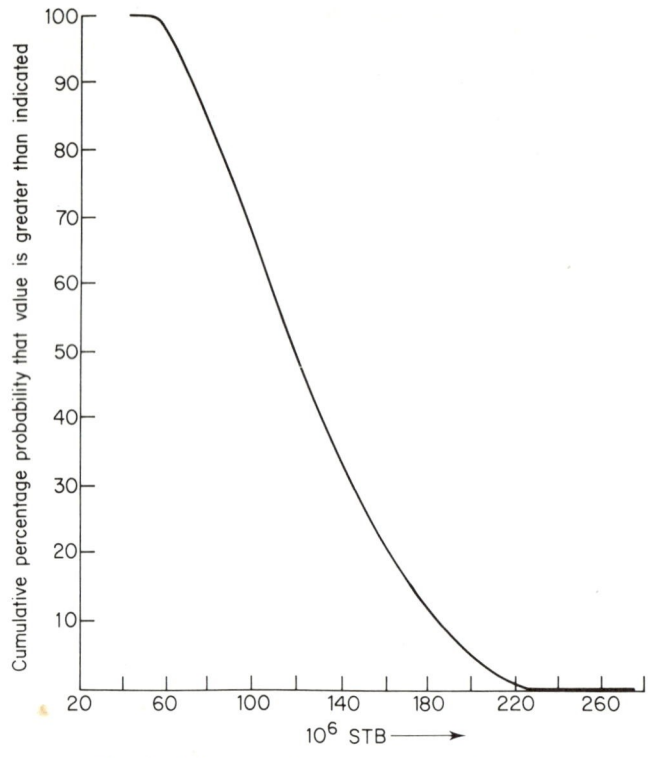

Fig. A8.4 Recoverable reserves distribution.

Chapter 9

Solution 9.1

$$t_D = \frac{K_t}{\phi\mu c r^2}$$

with (a) $t_D = 1481$
(b) $t_D = 14815$
(c) $t_D = 7.4 \times 10^{-3}$

Solution 9.2

$$p_i - p = \frac{q\mu}{4\pi Kh}\left[-E_i\left(-\frac{\phi\mu c r^2}{4Kt}\right)\right]$$

For (a) $x = \dfrac{\phi\mu c r^2}{4K_t} = 4.2 \times 10^{-3}$

as x is small

$-E_i(-x) = -0.5772 - \log_e x$
$= 4.895$

Hence $\Delta P = 22.72$ atmospheres

For (b) $x = 0.4375$ From graph $-E_i(-x) = 0.62$

Hence $\Delta P = 2.875$ atmospheres

For (c) $x = 0.49$ From graph $-E_i(-x) = 0.55$

Hence $\Delta P = 64$ atmospheres

SOLUTIONS TO EXAMPLES

Solution 9.3

From the plot shown in Fig. A 9.1, of P_{wf} vs $\log_{10} t$

$m = 18$ psi/cycle

Then $Kh = \dfrac{162.6\,(500)\,(0.5)\,(1.7535)}{18}$

$= 3960$ mD ft

$K_o = \dfrac{3960}{60} = 66$ mD

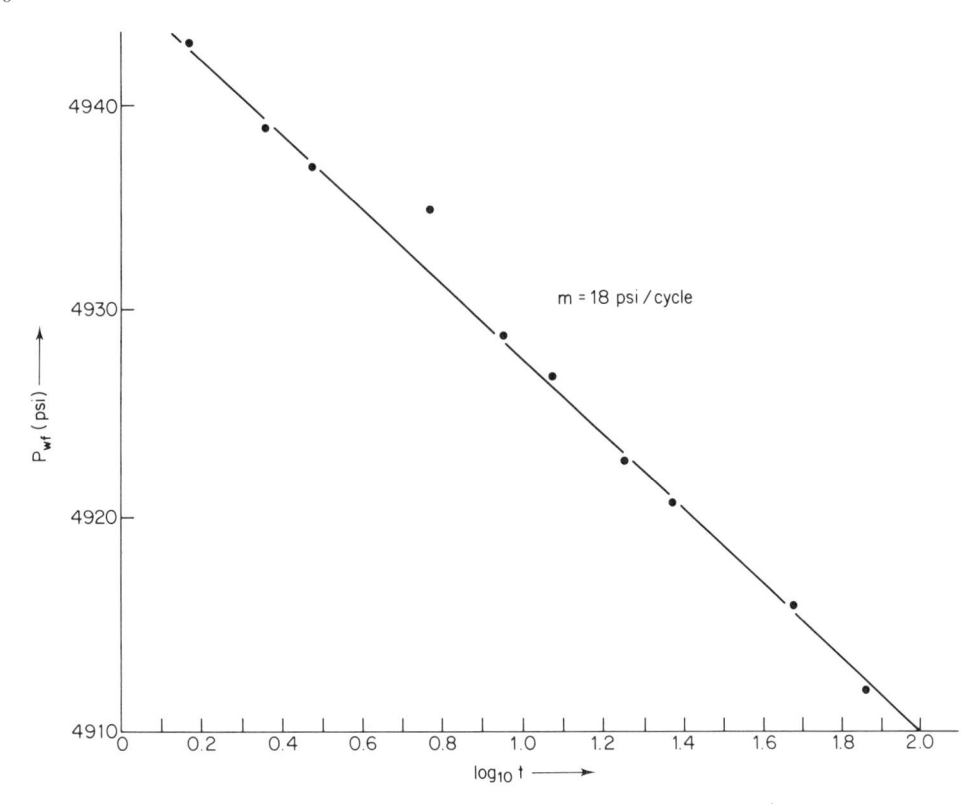

Fig. A9.1 P_{wf} vs $\log_{10} t$.

Solution 9.4

From a graph of P vs $\left\{\log \dfrac{t+\Delta t}{\Delta t}\right\}$ with the points in the table calculated, the slope is determined as 21.7 psi/cycle ($= m$).

For a reservoir rate q of 500 (1.454) rb/d ($=$ 727 rb/d)

Then, $kh = \dfrac{162.6 q \mu}{m} = 3800$ mD.ft

For $h = 120$ ft then $K_o = 32$ mD.

The value of P_{1h}, corresponding to a Horner time function of 3.16 is 4981 psi

$S = 1.151 \left\{ \dfrac{4981 - 4728}{21.7} - \log_{10} \dfrac{32}{(0.135)(0.7)(17 \times 10^{-6})(0.5)^2} + 3.23 \right\}$

$= +7$

$\Delta P_s = 0.87\, m\, S = 132$ psi

Efficiency $= \dfrac{4981 - 4728 - 132}{4981 - 4728} = 0.5$ (approx.)

Data points for Horner plot on semi-log paper, P vs $\dfrac{t+\Delta t}{\Delta t}$ or a plot of p vs $\log\left[\dfrac{t+\Delta t}{\Delta t}\right]$ on linear scales are as follows:

$t = 60 \times 24 = 1440$ hours

Time (h)	$\dfrac{t+\Delta t}{\Delta t}$	$\log_{10}\{(t+\Delta t)/\Delta t\}$	P (psi)
0.25	5761	3.76	4967
0.5	2881	3.46	4974
1.0	1441	3.16	4981
1.5	961	2.98	4984
2.0	721	2.88	4987
3.0	481	2.68	4991
6.0	241	2.38	4998
9.0	161	2.21	5002
18.0	81	1.91	5008
36.0	41	1.61	5014
48.0	31	1.49	5017

Solution 9.5

Examination of the data shows that: $\Delta P/\text{day} \approx 3$ psi

Assuming $1 - S_w \approx 0.7$

We have $NB_{oi} = N_p B_o/(c_o)_e\, \Delta P$

and $(c_o)_e = 15 \times 10^{-6}/0.7 = 21.4 \times 10^{-6}$

and $N_p = 500$ b/d. For $B_o = B_{oi}$ then

$N = \dfrac{500}{21.4 \times 10^{-6} \times 3} = 7.8 \times 10^6$ BBL

Solution 9.6

Rate	Q (MSCF/D)	(ΔP^2) total
1	7 290	42 181
2	16 737	126 120
3	25 724	237 162
4	35 522	391 616

Assume t_{flow} prior to build up is 4.5 hours:

Time since shut in	$\log_{10}\dfrac{t+\Delta t}{\Delta t}$	P
1	0.7404	2509.7
1.5	0.6201	2510.7
2	0.5119	2511.3
2.5	0.4472	2511.7
3	0.3979	2512.1
4	0.3274	2512.5
5	0.2788	2513.0
6	0.2430	2513.2

Slope $= 7$ psi/cycle from Horner plot

Hence $Kh = \dfrac{162.6\,(q\,B_g)\,\mu}{m}$

$B_g = \dfrac{0.00504zT}{P}$ BBL/scf $= 0.00103$

$Kh \approx 14\,500 \quad K = 72$ mD

Now $P_e^2 - P_w^2 = \dfrac{1424\,\mu\,zTQ}{Kh}\left\{\ln 0.606\, r_e/r_w + S_1\right\}$

$= 0.855\, Q\, \{8.93 + S_1\}$

Or $S_1 = \dfrac{P_e^2 - P_w^2}{0.855\, Q} - 8.93$

Rate	S_1	Q
1	-2.16	7290
2	-0.12	
3	$+1.85$	
4	$+3.96$	35 522

SOLUTIONS TO EXAMPLES

$D = \Delta S/\Delta Q = 2.16 \times 10^{-4}$

$S = -3.7$

$\beta = \dfrac{Dh\mu r_w}{2.225 \; 10^{-15} \, K\gamma_g} = 2.865 \times 10^9$

$\beta_{theoretical} = \dfrac{48\,211}{\phi^{5.5}\sqrt{K}} = 1.80 \times 10^9$

This is order of magnitude agreement.

The inertial pressure term $\Delta P^2_{inertial}$ is calculated from B as follows:

$B = \dfrac{3.16 \times 10^{-12} \gamma_g Tz\beta}{h^2 \, r_w}$

$= 0.000185$

Hence $(\Delta P^2)_{inertial}$ is as follows:

Rate	Q(MSCF/D)	$(\Delta P^2)_{inertial}$	$(\Delta P^2)_{total}$
1	7 290	9 851	42 181
2	16 737	51 928	126 120
3	25 724	122 665	237 162
4	35 522	233 906	391 616

Comparison between the numbers shows that at high rates the inertial drop is over half the total drop, and that in this case only the inertial drop is close to the total drop of the previous rate.

The AOF plot is shown in Fig. A 9.2 and when ΔP^2 is equal to P_e^2 ($6.32 \times 10^6 \text{psi}^2$) then $Q_{AOF} = 220 \times 10^6$ SCF/d

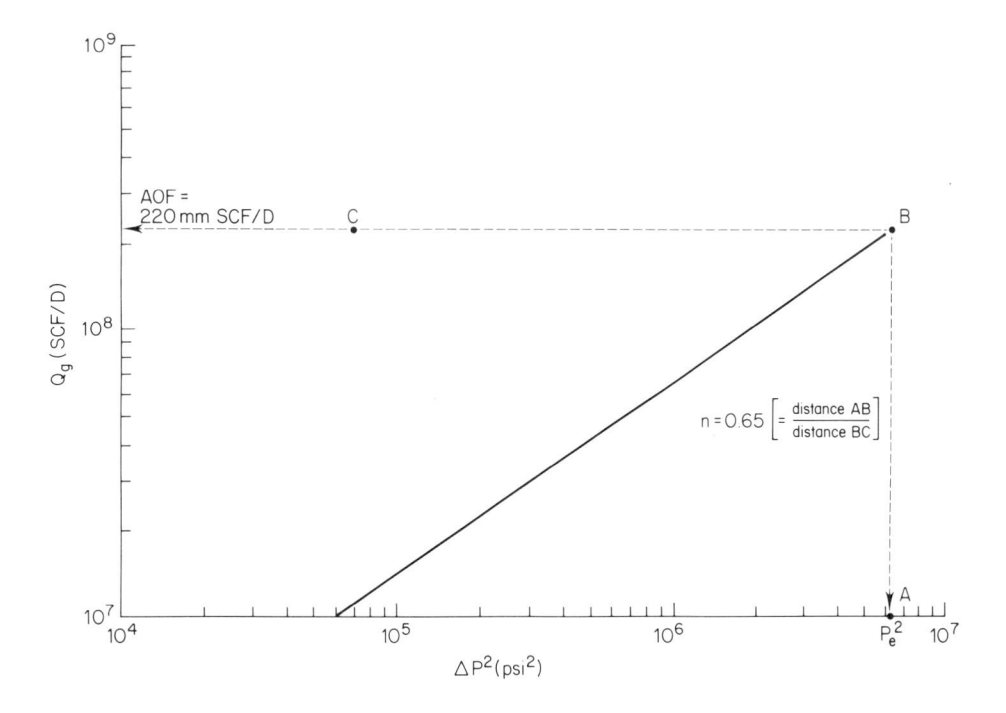

Fig. A9.2 AOF determination.

Chapter 10

Solution 10.1

Volume of reservoir = $V = 100 \times (5280)^2 \times 500$ cu.ft

Volume of reservoir available for fluid = $(1 - S_w) \phi V = V_r$
$$= 0.65 \times 0.12 \times V$$

Volume of gas at standard conditions of 14.7 psi and 60°F:

$$\left(\frac{PV}{zT}\right)_1 = \left(\frac{PV}{zT}\right)_2$$

$$V_{sc} = \frac{2000}{14.7} \frac{V_r}{0.825} \frac{520}{595} = 15.6 \times 10^{12} \text{ SCF}$$

(1) Assume no water influx,

Initial moles in place $n_i = \dfrac{P_i V_r}{z_i RT_r} = \left(\dfrac{PV_i}{RT}\right)_{std}$

V_i – gas in place measured at standard conditions.

Abandonment moles left in place $n_a = \dfrac{P_a V_r}{z_a R T_r} = \left(\dfrac{PV_a}{RT}\right)_{Std}$

Gas recovered $\Delta n = \dfrac{V_r}{RT_r}\left(\dfrac{P_i}{z_i} - \dfrac{P_a}{z_a}\right) = \dfrac{P_s}{RT_s} \Delta V$

Recoverable gas measured at standard conditions

$$= \frac{V_r T_s}{T_r P_s}\left(\frac{P_i}{z_i} - \frac{P_a}{z_a}\right)$$

At 500 psi, reduced pressure $P_{pr} = \dfrac{500}{670.6} = 0.75 \qquad T_{pr} = 1.6 \qquad z = 0.94$

Therefore recoverable gas $= \dfrac{V_r T_s P_i}{T_r P_s z_i}\left(1 - \dfrac{z_i}{z_a}\dfrac{P_a}{P_i}\right)$

$$= 15.6 \times 10^{12}\left(1 - \frac{0.825}{0.94} \cdot \frac{500}{2000}\right)$$

$$= 15.6 \times 10^{12} (1 - 0.219)$$

$$= 12.2 \times 10^{12} \text{ SCF}$$

Recovery factor $= \dfrac{12.2}{15.6} = 78\%$

Solution 10.2

Radial flow of oil $q_o = \dfrac{2\pi k_o h}{\mu_o B_o} \dfrac{\Delta P_o}{\log_e \dfrac{r_e}{r_w}}$

Radial flow of gas $q_g = \dfrac{2\pi k_g h}{\mu_g B_g} \dfrac{\Delta P_g}{\log_e \dfrac{r_e}{r_w}}$

and if the capillary pressure gradient is negligible, and the pressure drop over the same radii are considered,

$$\frac{q_g}{q_o} = \frac{k_g \mu_o B_o}{k_o \mu_g B_g}$$

SOLUTIONS TO EXAMPLES

To this must be added the gas evolved from solution in the oil.
The total measured gas–oil ratio will then be:

$$= \frac{k_g \, \mu_o \, B_o}{k_o \, \mu_g \, B_g} + R_s$$

For the figures given:

$$= \frac{(96)(0.8)(1.363)}{(1000)(0.018)(0.001162)} + 500$$

$$= 5005 + 500$$

$$= 5505 \text{ SCF/STB}$$

Solution 10.3

1. $N = \dfrac{N_p \, B_t + B_g(R_p - R_{si}) - (W_e - B_w \, W_p)}{B_t - B_{oi}}$

$W_e = N_p \, B_t + B_g(R_p - R_{si}) - N(B_t - B_{oi})$

(i) At cumulative 1.715×10^6 BBL ($P = 1600$)

$W_{e1} = (1.715 \times 10^6)[1.437 + 0.0015(878 - 690)] - 14.5 \times 10^6[1.437 - 1.363]$

$= 1.875 \times 10^6$

(ii) At cumulative 3.43×10^6 BBL ($P = 1300$)

$W_{e2} = (3.43 \times 10^6)[1.594 + 0.0019(996 - 690)] - 14.5[1.594 - 1.363]$

$= 4.112 \times 10^6$

At $P = 1000$ estimated water influx $= 6.375 \times 10^6$ (from trend)

$N_p = \dfrac{N(B_t - B_{oi}) + W_e}{B_t + B_g(R_p - R_{si})}$

$= \dfrac{14.5(1.748 - 1.363) \times 10^6 + 6\,375\,000}{1.748 + 0.0025(1100 - 690)}$

$= 4.312 \times 10^6$ BBL

Solution 10.4

Total hydrocarbon in place $= \tfrac{1}{3}\pi \, r^2 h \phi \, (1 - S_w)$

$= \pi \dfrac{9}{3}(5280)^2 \dfrac{750 \times 0.17 \times 0.76}{5.615} = 4.54 \times 10^9$ BBL

Since bubble-point is 1850 psi, this must be pressure at any gas–oil contact.
Elevation of gas–oil contact above oil–water contact is:

$$\dfrac{(1919 - 1850)\,144}{43.4} = 229 \text{ ft}$$

This is less than hydrocarbon column so gas–oil contact exists at 4031 ft SS

Height of gas zone $= 750 - 229 = 521$ ft

Ratio gas/total $= \dfrac{r_1^2 h_1^2}{r_2^2 h_2} = \dfrac{h_1^3}{h_2^3} = \left(\dfrac{520}{750}\right)^3 = 0.34$

Therefore, oil in place $= \dfrac{0.66 \times 4.54 \times 10^9}{1.363} = 2.198 \times 10^9$ STB

$m \simeq 0.5 \ (\simeq 1/3 \div 2/3)$

Material balance

$$N = \frac{N_p[B_t + B_g(R_p - R_{si})] - W_e + W_p B_w}{(B_t - B_{oi}) + \frac{m B_{oi}}{B_{gi}}(B_g - B_{gi})}$$

At 1600 psi:

$$B_t + B_g(R_p - R_{si}) = 1.437 + 0.00150(1100 - 690)$$
$$= 2.0520$$

$$B_t - B_{oi} + \frac{m B_{oi}}{B_{gi}}(B_g - B_{gi}) = 1.437 - 1.363 + \frac{0.5\,(1.363)}{0.00124}(0.0015 - 0.00124)$$
$$= 0.0740 + 0.1429$$
$$= 0.2169$$

$$W_e = (2.052 \times 3.1 \times 10^8 + 31 \times 10^6) - 2.198 \times 10^9 \times 0.2169$$
$$= 1.904 \times 10^8 \text{ BBL}$$

At 1300 psi:

$$B_t + B_g(R_p - R_{si}) = 1.594 + 0.0019(1350 - 690)$$
$$= 2.8480$$

$$B_t - B_{oi} + \frac{m B_{oi}}{B_{gi}}(B_g - B_{gi}) = 1.594 - 1.363 + \frac{0.5\,(1.363)}{0.00124}(0.0019 - 0.00124)$$
$$= 0.5937$$

$$W_e = 5.5 \times 10^8 \times 2.8480 + 55 \times 10^6 - 2.198 \times 10^9 \times 0.5937$$
$$= 3.164 \times 10^9$$

This is not simply linear with pressure but extrapolation is reasonably straightforward and water influx at 1000 psi is estimated at 3.75×10^9 BBL

$$B_t + B_g(R_p - R_{si}) = 1.748 + 0.0025(1800 - 690)$$
$$= 4.523$$

$$B_t - B_{oi} + \frac{m B_{oi}}{B_{gi}}(B_g - B_{gi}) = 1.748 - 1.363 + \frac{(0.5)1.363}{0.00124}(0.00250 - 0.00124)$$
$$= 1.0775 \text{ (denominator term)}$$

$$N_p = \frac{N \times \text{denom.}) + W_e - W_p B_w}{B_t + B_g(R_p - R_{si})}$$
$$= \frac{2.198 \times 10^9 \times 1.0775 + 3.75 \times 10^8 - 63 \times 10^6}{4.5230}$$
$$= 5.926 \times 10^8$$
$$= 590 \times 10^6 \text{ STB}$$

Solution 10.5

Gas cap : oil zone ratio $m = \dfrac{GB_{gi}}{NB_{oi}} = \dfrac{120.7 \times 10^9 \times 6.486 \times 10^{-4}}{300 \times 10^6 \times 1.3050} = 0.2$

From *PVT* data the values of B_o, R_s and B_g at 4300 psi can be estimated by linear interpolation as:

$B_o = 1.228$ RB/STB; $R_s = 338$ SCF/STB; $B_g = 7.545 \times 10^{-4}$ RB/SCF

SPE NOMENCLATURE AND UNITS   345

From production data the value of R_p is calculated as G_p/N_p to give the following table.

Time	P(psi)	$\sum W_{inj}(10^6 BBL)$	$N_p(10^6 STB)$	$R_p(SCF/STB)$	$R_s(SCF/STB)$
1.1.80	5000	0	0	0	500
1.1.81	4300	0	21.9	550	338
1.1.82	4250	25.55	43.8	600	325

Using the relationship $F = N(E_T) + W_e + W_{inj}B_{winj}$ the following is calculated where:
$E_T = mE_g + E_o + E_{fw}$

	Units	1.1.81	1.1.82
(a) $F = N_p B_o + (R_p - R_s)B_g$	10^6 RB	30.28	62.41
(b) $E_o = (B_o - B_{oi}) + (R_{si} - R_s) B_g$	RB/STB	0.0420	0.0435
(c) $E_g = B_{oi}\left[\dfrac{B_g}{B_{gi}} - 1\right]$	RB/STB	0.2131	0.2302
(d) $E_{fw} = (1 + m) B_{oi} \Delta P \left[\dfrac{c_w S_w + c_f}{1 - S_w}\right]$	RB/STB	0.0061	0.0065
(e) $E_T = mE_g + E_o + E_{fw}$	RB/STB	0.0887	0.0960
(f) $W_e = F - N(E_T) - W_{inj} B_{winj}$	10^6 BBL	3.67	8.06

Solution 10.6

The dimensionless radius ratio is:
$$re_D = \frac{r \text{ aquifer}}{r \text{ oil zone}} = \frac{81\,000}{9000} = 9$$

The dimensionless time t_D is related to real time by:
$$t_D = \frac{2.309\, k\, t\,(\text{years})}{\phi \bar{c}\, \mu\, r_o^2} = \frac{2.309\,(707t)}{(0.18)(7\times 10^{-6})(0.4)(9000)^2} = 40t$$

The instantaneous pressure drops which at the start of each year are equivalent to the continuous pressure declines are:

$$\Delta P_0 = \frac{P_i - P_1}{2} = \frac{5870 - 5020}{2} = 425 \text{ psi}$$

$$\Delta P_1 = \frac{P_i - P_2}{2} = \frac{5870 - 4310}{2} = 780 \text{ psi}$$

$$\Delta P_2 = \frac{P_1 - P_3}{2} = \frac{5020 - 3850}{2} = 585 \text{ psi}$$

The aquifer constant is:

$U = 1.119\, f\phi h\, \bar{c}\, r_o^2$

$U = 1.119 \times 1 \times 0.18 \times 200 \times (7 \times 10^{-6}) \times (9000)^2$

$U = 22\,841$ BBL/psi

From tables or charts for dimensionless influx at $re_D = 9$ we have:

t_D	$W_D(t_D)$
40	21
80	29
120	34

From $W_e = U \sum_{j=0}^{j=n-1} \Delta PW_D (T_D - t_{Dj})$

$W_{e1} = 22\,841\,[425\,(21)] = 203.9 \times 10^6$ BBL

$W_{e2} = 22\,841\,[425\,(29) + 780\,(21)] = 655.7 \times 10^6$ BBL

$W_{e3} = 22\,841\,[425\,(34) + 780\,(29) + 585\,(21)] = 1127.3 \times 10^6$ BBL

Chapter 11

Solution 11.1

$$PI = \frac{1}{\mu_o} \left[\frac{0.00708\,k\,k_{ro}\,h}{\ln \frac{r_e}{r_w} - 0.75 + S} \right]$$

For $r_e = 1500$ ft
$r_w = 0.5$ ft
$S = +4$
$K_{ro} = 0.6$
$h = 100$ ft
$k = 1325$ mD

$$PI = \frac{50}{\mu_o}$$

∴

μ_o	0.5	5	50	500	5000
PI	100	10	1	0.1	0.01

Solution 11.2

The injectivity index is given in field units by:

$$II = \frac{0.00708\,k\,k_{rw}\,h}{\mu_w \left[\ln \frac{r_e}{r_w} - 0.75 + S \right]}$$

Assuming all other factors equal then

$$\frac{II_{cold}}{II_{hot}} = \frac{(\mu_w)_{hot}}{(\mu_w)_{cold}} = \frac{0.35}{0.55} = 0.64$$

Solution 11.3

Use is made of the plot in Fig. 11.4 which correlates areal sweep efficiency E_A as a function of end point mobility ration (M') for different fractional injection volumes, V_D.

$$M' = \frac{K_w'}{\mu_w} \cdot \frac{\mu_o}{K_o'} = \frac{0.4}{0.4} \cdot \frac{3.4}{0.85} = 4$$

The volume of injected fluid, in reservoir barrels, after 10 years is:

$10 \times 365.25 \times 53\,000 \times 1.005$

$= 1.945 \times 10^8$ RB

SOLUTIONS TO EXAMPLES

The displaceable pore volume ($= PV (1-S_{or} - S_{wi})$) is given in reservoir barrels as follows:

$$\frac{(4 \times 5280)(1 \times 5280)(98)(0.25)}{5.615}[1 - 0.3 - 0.3]$$

$$= 1.946 \times 10^8 \text{ RB}$$

$$V_D = \frac{1.945 \times 10^8}{1.946 \times 10^8} \simeq 1$$

From Fig. 11.4 the value of E_A corresponding to $M' = 4$ and $V_D = 1$ is 0.7

Solution 11.4

For stable cone formation

$$\Delta \Phi' = g' X (\rho_w - \rho_o)$$

For $\Delta \Phi'$ (in psi), and cone height X (in feet) and density difference as specific gravities then

$$\Delta \Phi' = \frac{62.4}{144}(1.01 - 0.81) \, 50$$

$$= 4.33 \text{ psi}$$

Chapter 12

Solution 12.1

(a) In field units

$$U = \frac{1.25 \,(4000)}{70 \,(1500)} = 0.0476 \text{ B/D} - \text{ft}^3$$

The viscous-gravity force ratio is calculated from

$$R_{v\text{-}g} = \frac{2050 \, U \, \mu_o \, L}{(\rho_o - \rho_s) \, \bar{k} h}$$

$$= \frac{2050 \,(0.0476)(0.5)(1500)}{(0.8 - 0.4)(130)(70)} = 20$$

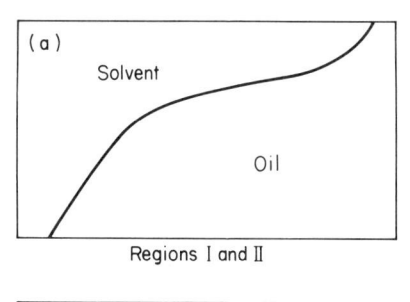

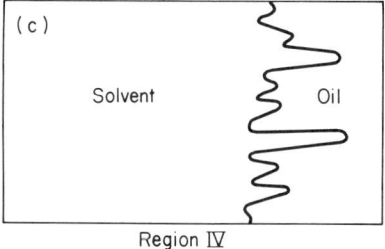

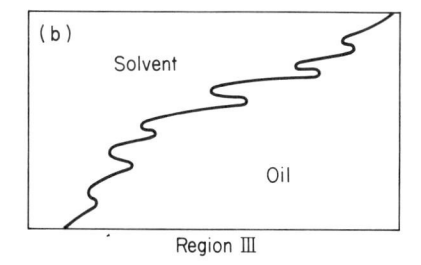

Region I : Single gravity override tongue

Region II : Single tongue but sweepout independent of $R_{V\text{-}G}$ for given M

Region III : Transition region with secondary fingers below main tongue

Region IV : Multiple fingers with sweepout independent of $R_{V\text{-}G}$ for given M

Fig. A12.1 Displacement regimes.

For a mobility ratio, M represented by μ_o/μ_s (= 25), Figure A 12.2 shows a breakthrough sweep efficiency of about 15% and a flow dominated by gravity tonguing. (Fig A 12.1)

(b) In field units

$$U = \frac{1.25\,(1000)}{30\,(2000)} = 0.0208 \text{ B/D} - \text{ft}^2$$

The viscous gravity force ratio requires an approximation of permeability as:

$$\bar{k} = \sqrt{K_v \cdot K_h}$$
$$\therefore \bar{k} = ((1)(3))^{0.5} = 1.73 \text{ md}$$

Then

$$R_{v\text{-}g} = \frac{2050\,(0.0208)\,(0.36)\,(2000)}{(0.75 - 0.64)\,(1.73)\,(30)}$$
$$= 5378$$

For a mobility ratio of M ($\mu_o/\mu_s = 0.36/0.055 = 6.55$), Figures A 12.1 and A 12.2 show a breakthrough sweepout efficiency of around 50% and a flow dominated by viscous fingering.

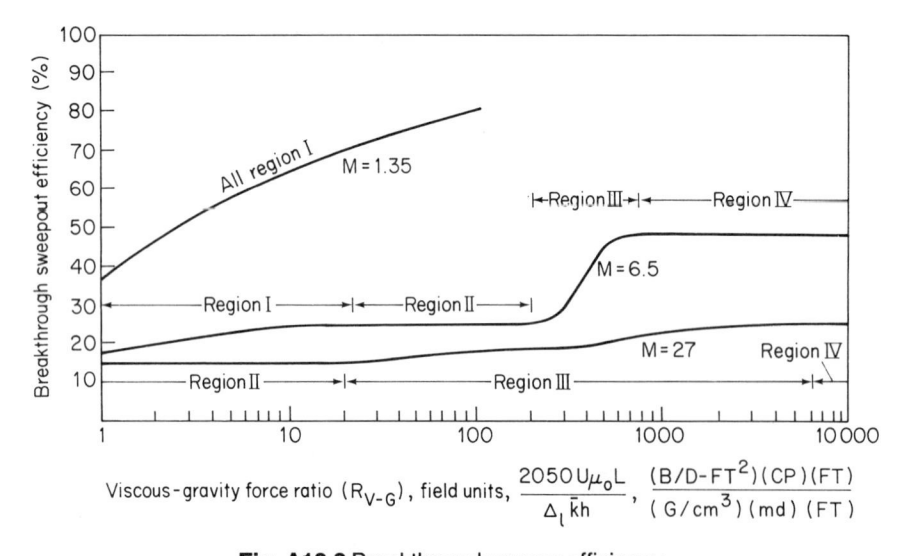

Fig. A12.2 Breakthrough sweep efficiency.

Solution 12.2

The tie lines for the system join the equilibrium compositions of systems A and B in the two phase region. The compositions are plotted in Figure A12.3

(a) The critical point (*CP*) is estimated where the limiting tie line becomes tangential to the phase envelope and has the composition, wt%, 21% surfactant, 67% oil; 12% brine.

(b) The point with the composition 4% surfactant and 77% oil is given on Figure A12.3 as point A. From the slope of tie lines in this region the equilibrium phase compositions are A_1 and A_2 with weight percents estimated as:

A_1 10% oil; 10% surfactant; 80% brine
A_2 97% oil; 2% surfactant; 1% brine

For an original 200 g mixture containing

8g surfactant, 154 g oil, 38 g brine

SOLUTIONS TO EXAMPLES 349

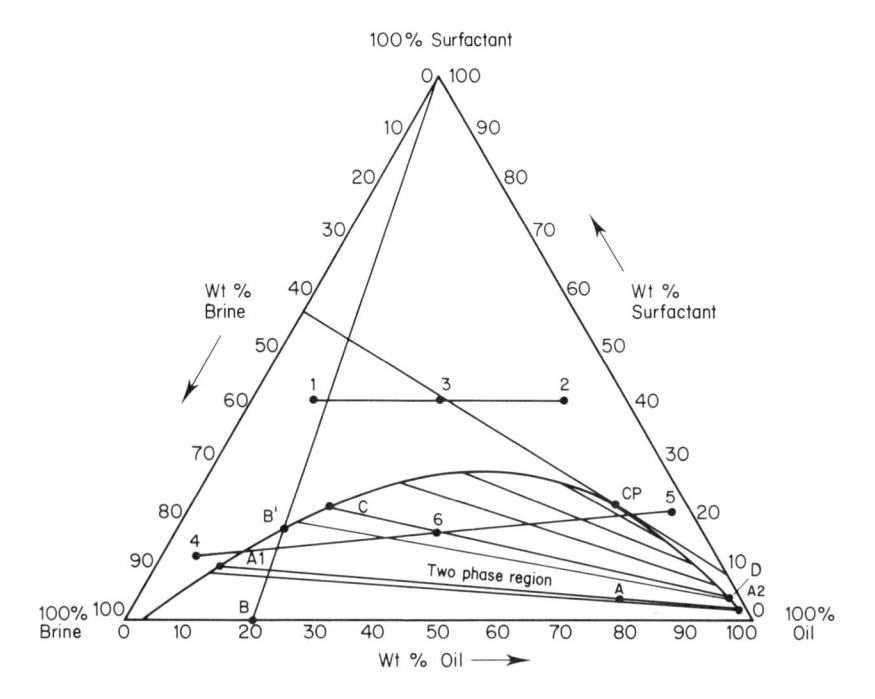

Fig. A12.3 Ternary diagram.

The tie line ratios give:

wt of A_1 phase $3/13 \times 200 = 46$ g
wt of A_2 phase $10/13 \times 200 = 154$ g

∴ Composition of A_1 = 4.6 g oil
　　　　　　　　　　4.6 g surfactant
　　　　　　　　　　36.8 g brine

∴ Composition of A_2 = 149.5 g oil
　　　　　　　　　　3.0 g surfactant
　　　　　　　　　　1.5 g brine

(c) On Figure A 12.3 the composition 20% oil and 80% brine is shown at location B. A line from B to the 100% surfactant point leaves the two phase region at location B′, having a composition oil 16.5%, surfactant 17.5%, brine 66%. The oil + brine weight is 100 g and would constitute 82.5% of the mixture, so surfactant needed is 0.175 (100/0.825) = 21.2 g.

(d) On Figure A 12.3, location 1 is 10% oil, 40% surfactant and location 2 is 50% oil, 40% surfactant. They are in a single phase region and the resulting mixture contains 30% oil, 40% surfactant and 30% brine, as denoted by position 3.

(e) On Figure A 12.3, location 4 is 12% surfactant, 5% oil and location 5 is 20% surfactant, 77% oil.

The mixture weight is 200 g and contains 41% oil, 16% surfactant, and 43% brine. It is shown as location 6. The mixture is in the two phase region and equilibrates to compositions C and D on the equilibrium tie line through location 6. The compositions are:

C: 58% brine; 21.5% oil, 20.5% surfactant (146 g total)
D: 94% oil; 5% surfactant; 1% brine (54 g total)

Solution 12.3

For conventional production

$$\frac{q_{inj}}{\Delta P} = \frac{0.003541\, k_o h}{\mu_o \left[\ln\left(\frac{208.71\, A^{0.5}}{r_w}\right) - 0.964\right]}$$

$$\frac{q_{inj}}{150} = \frac{700}{\mu_o \left[\ln\left(\frac{208.71\,(3)}{0.5}\right) - 0.964\right]} = \frac{(0.003541\,(1000)(60))}{} = 161 \text{ rb/d For 9 acre spacing and a 200 psi differential.}$$

For thermal stimulation and steam injection a 5 fold improvement in flow resistance between producers and injectors would lead to rates around 800 b/d. To determine the steady state production/injection time at which such rates will lead to 50% of the pattern volume being occupied by steam we can conduct the following analysis:

The cumulative heat injected into the reservoir, Q_i, can be calculated from heat injection rate, using the mass rate of injection W_i

$$\frac{Q_i}{t} = w_i \left[C_w\,\Delta T + f_{sdh}\,L_{Vdh} \right]$$

$$= q_{inj}\,(5.615)(62.4) \left[C_w\,(380 - 100) + 0.75\,(845) \right]$$

The average specific heat, C_w, over the temperature range $380 - 100°F$ is given by:

$$C_w = \frac{h_{w(Ts)} - h_{w(Tres)}}{T_s - T_{res}} = \frac{355 - 69}{380 - 100} = 1.02 \text{ Btu/lb m - degF}$$

$$\therefore Q_i = t \cdot q_{inj} \cdot 322118 \text{ Btu}$$

The ratio of latent heat to total energy injected, f_{hv} is calculated from:

$$f_{hv} = \left\{ 1 + \frac{C_w\,\Delta T}{f_{sdh}\,L_{vdh}} \right\}^{-1} = \left\{ 1 + \frac{(1.02\,(380 - 100))}{0.75\,(845)} \right\}^{-1}$$

$$= 0.689$$

Figure A 12.4 can now be used to estimate the thermal efficiency of the steam zone, E_{hs}, at different values of dimensionless time, t_D. The values of t_D are given from:

$$t_D = 4t \left[\frac{M_s}{M_R}\right]^2 \frac{\alpha_s}{h^2}$$

$$= 4t \left[\frac{45}{35}\right]^2 \left[\frac{0.75}{(60)^2}\right]$$

$$= 0.00138t \text{ days or } 0.504\,t \text{ years}$$

The following table may now be constructed using $f_{hv} = 0.689$ on Fig A 12.4.

t (yr)	t (days)	t_D	E_{hs}	$E_{hs} \cdot t$
1.0	365.25	0.5	0.64	233.8
1.5	547.9	0.75	0.59	323.3
2.0	730.5	1.0	0.56	409.1
2.5	913.1	1.25	0.52	474.8

The volume of a steam zone, V_s, is in general given by:

$$V_s = \frac{Q_i\,E_{hs}}{43560\,M_R\,\Delta T}$$

SOLUTIONS TO EXAMPLES

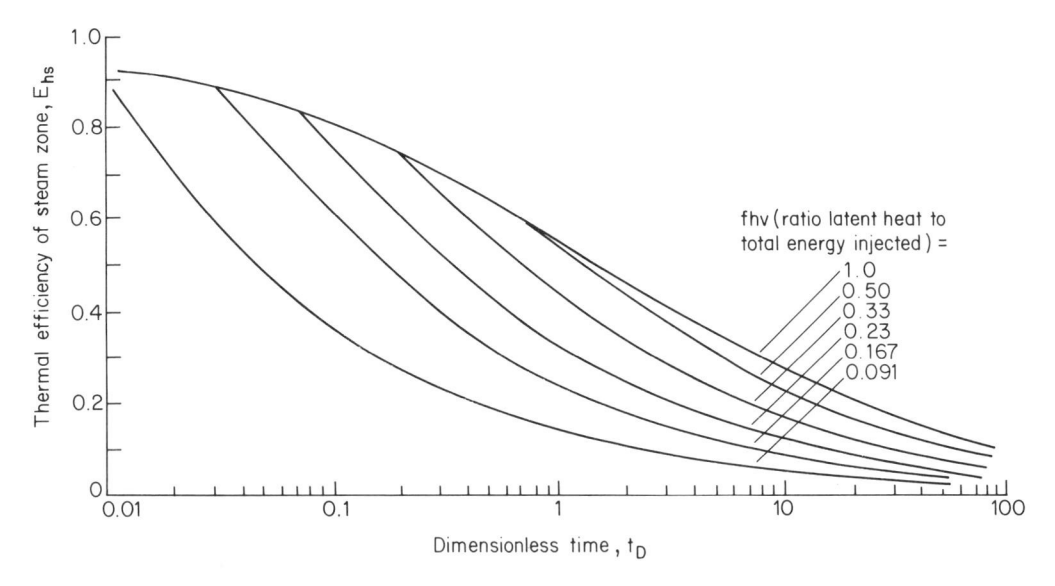

Fig. A12.4 Thermal efficiency

For the case of 50% steam volume in the pattern of area A acres then
$$Q_i = 0.5 \, Ah \left\{ \frac{43560 \, M_R \, \Delta T}{E_{hs}} \right\}$$

Equating values of Q_i we obtain the relationship
$$322118 \, q_{inj} \cdot t = \frac{0.5 \, (9) \, (60) \, (43560) \, (35) \, (280)}{E_{hs}}$$

where t is in days

That is $\quad q_{inj} = \dfrac{357817.5}{E_{hs} \cdot t}$

The injection rates needed to provide 50% pattern volume of steam at the following times are therefore as shown in the following table.

t (yr)	q_{inj} (rb/D)
1.0	1531
1.5	1107
2.0	875
2.5	753

These data may be further evaluated in terms of steam injection equipment capacity and project economics.

Solution 12.4

The wet condensate gas volume is obtained from the volumetric calculation:
$$V_{sc} = \frac{A \, h_n \, \phi \, (\bar{S}_g)}{B_{gi}}$$

In terms of standard cubic feet this is:
$$V_{sc} = \frac{1}{B_{gi}} [\pi (3 \times 5280)^2 \, 300 \, (0.18) \, (0.75)]$$

$$V_{sc} = \frac{3.1937 \times 10^{10}}{B_{gi}} \text{ SCF}$$

In order to find B_{gi} we need the super compressibility factor z which can be obtained from Fig 4.7 using the reservoir condition molecular weight or gas gravity.

The oil molecular weight is given by

$$M_o = \frac{44.3 \rho_L}{(1.03 - \rho_L)}$$

Now $\rho_L = \frac{141.5}{API + 131.5} 0.75$

$\therefore M_o = 119$

The weight associated with a stock tank barrel of liquid is given by:

$$W = (5.615 \times 62.4 = 0.75) + \frac{5000 (0.58) (28.97)}{379.4}$$

$= 262.78 + 221.44$

$= 484.22$

The number of moles associated with this weight is

$$n = \frac{5000}{379.4} + \frac{(62.4)(0.75)(5.615)}{119}$$

$n = 13.18 + 2.21$

$n = 15.39$

$\therefore MW_{(res)} = \frac{W}{n} = \frac{484.22}{15.39} = 31.46$

and $\gamma_{g(res)} = \frac{MW_{(res)}}{28.97} = \frac{31.46}{28.97} = 1.086$

i.e. $\gamma_{g(res)} = 1.09$

From Fig 4.7, $P_{pc} = 620$ and $T_{pc} = 465$

From reservoir datum conditions

$$P_{pr} = \frac{4500}{620} = 726 \text{ and } T_{pr} = \frac{670}{465} = 1.44$$

So, from Fig 4.7 $z = 0.925$

Then:

$$B_{gi} = \frac{(0.02829)(0.925)(670)}{4500}$$

$= 3.8962 \times 10^{-3}$ RCF/SCF

$$V_{sc} = \frac{3.1937 \times 10^{10}}{3.8962 \times 10^{-3}}$$

$= 8.197 \times 10^{12}$ SCF

The dry gas volume

$$G = \left[8.197 \times 10^{12}\right]\left[\frac{5000/379.4}{15.39}\right]$$

$G = (8.197 \times 10^{12})(0.8563)$

$G = 7.019 \times 10^{12}$ SCF

Similarly the oil volume

$$N \times \frac{V_{sc}}{R} = \frac{8.197 \times 10^{12}}{5000}$$

$N = 1.639 \times 10^9$ STB

Chapter 13

Solution 13.1

Using the relationship that the depth equivalent of the total head is equal to the sum of the depth equivalents of the well head pressure and the well depth, then:

$$D_T = D_{whp} + D_{well}$$

(a) From Fig. A 13.1 at a well head pressure of 400 psi then D_{whp} = 3700 ft. Since D_{well} = 6000 ft then D_T = 9700 ft. At the GOR of 200 scf/stb the pressure at a depth equivalent of 9700 ft is read as 2400 psi.

(b) From Fig. A 13.2 at the bottom hole pressure of 1200 psi and GOR of 500 scf/stb the depth equivalent D_T, is read as 8900 ft. Since D_{well} is 5000 ft then D_{whp} is 3900 ft. The well head pressure is read from the graph at 3900 ft as 360 psi.

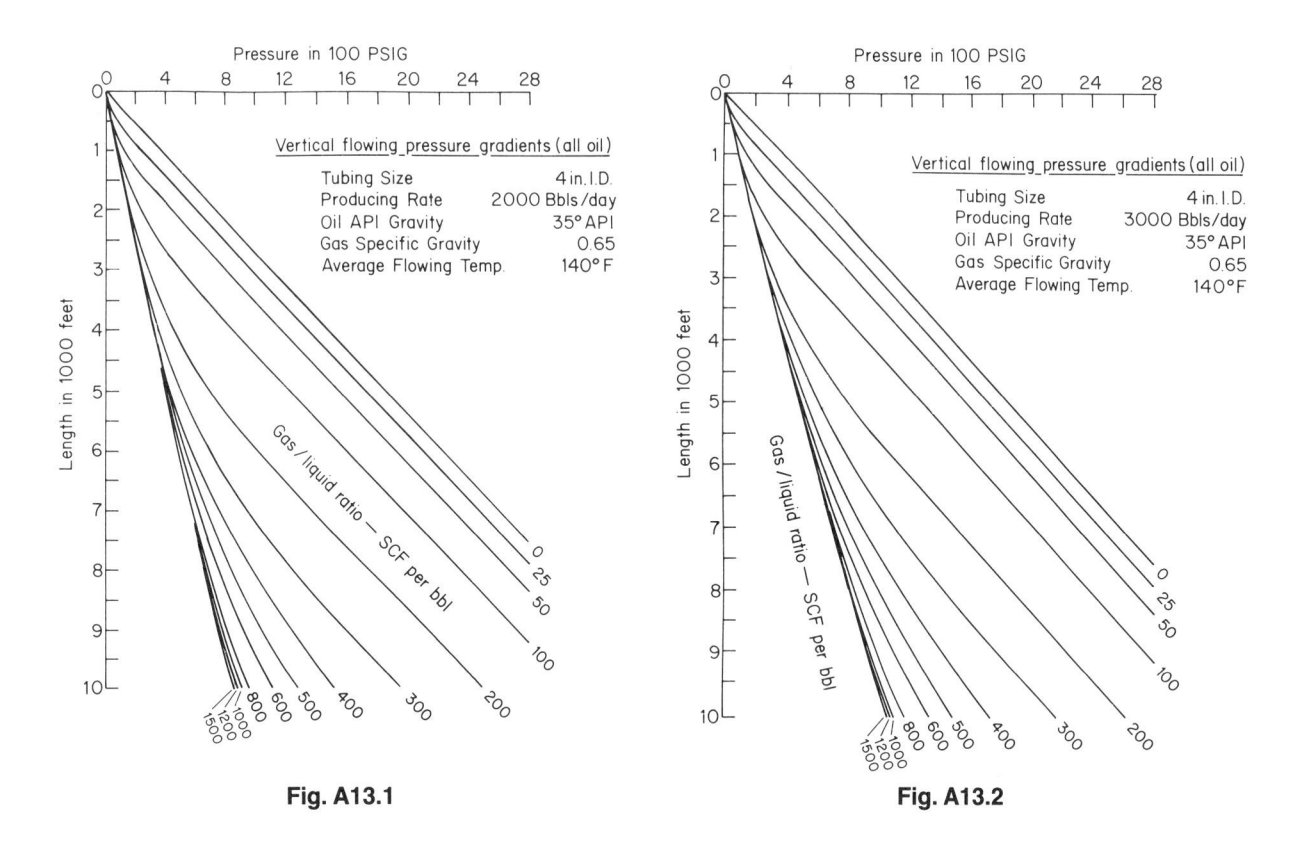

Fig. A13.1

Fig. A13.2

Solution 13.2

The maximum production rate q_{max} can be evaluated using the Vogel relationship, with $\bar{p}$, the static pressure, i.e.

$$q/q_{max} = 1 - 0.2 \left[\frac{P}{\bar{P}}\right] - 0.8 \left[\frac{P}{\bar{P}}\right]^2$$

$$= 1 - 0.2 \left[\frac{1500}{2600}\right] - 0.8 \left[\frac{1500}{2600}\right]^2$$

$$= 0.619$$

therefore, $q_{max} = \dfrac{3315}{0.619} = 5355$ b/d

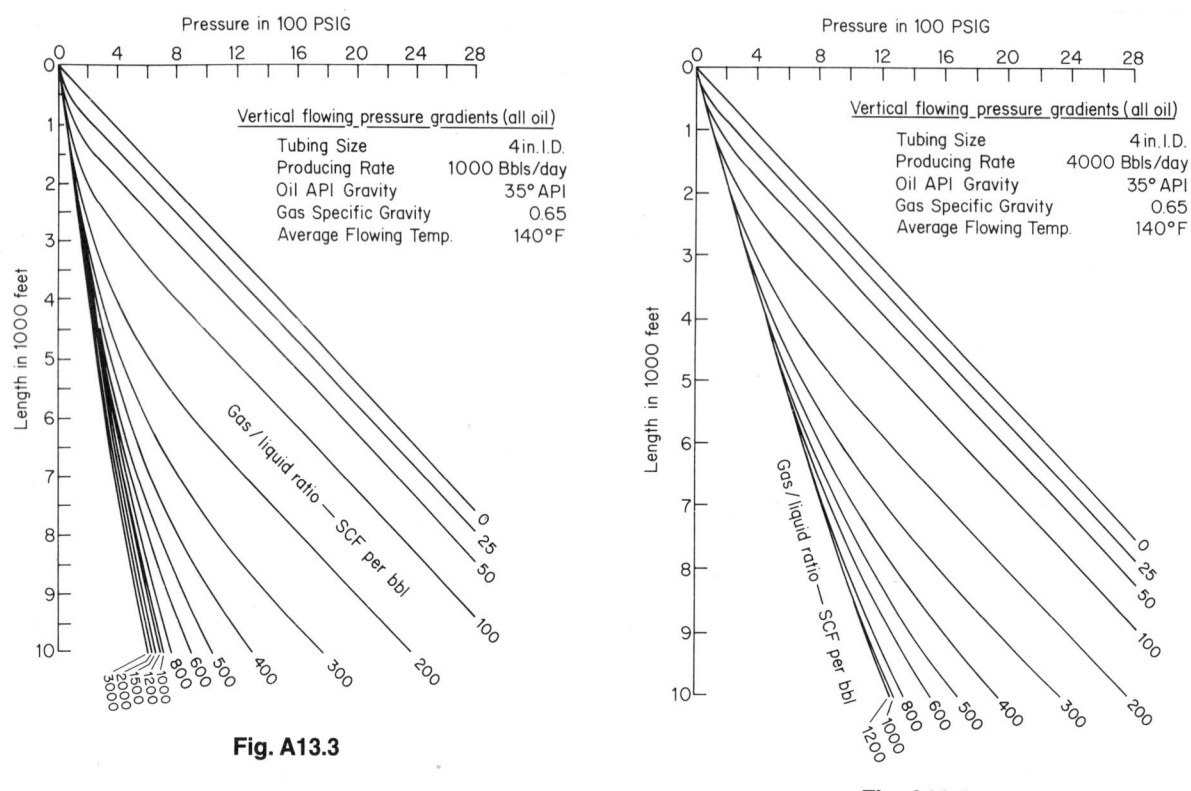

354   **PETROLEUM ENGINEERING: PRINCIPLES AND PRACTICE**

Fig. A13.3

Fig. A13.4

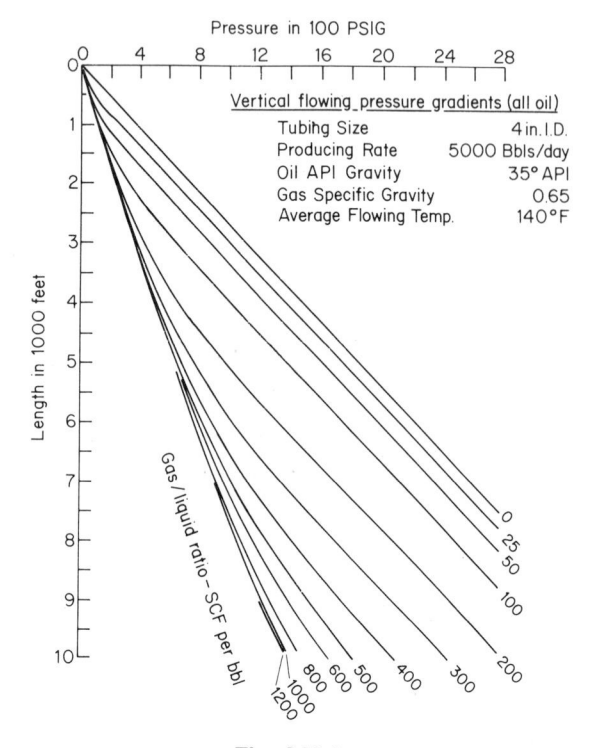

Fig. A13.5

SOLUTIONS TO EXAMPLES

From Fig. A 13.1 to A 13.5 the different vertical flowing pressure gradient curves at different rates are found for 4 in. tubing and a GOR of 200 SCF/STB. The total head depth is obtained as the sum of the well depth and the depth equivalent to a tubing head pressure of 400 psig. The flowing bottom hole pressure equivalent to the total head depth is recorded as a function of flow rate. It can be seen that the bottom hole pressure is essentially independent of rate at this condition and is 2200 psi.

Hence
$$q = q_{max}\left[1 - 0.2\left(\frac{2200}{2600}\right) - 0.8\left(\frac{2200}{2600}\right)^2\right]$$
$$= 1400 \text{ b/d}$$

Solution 13.3

For a residence time of 3 min. the volume of oil in the separator will be:
$$V_o = \frac{(1000)(3)}{(24)(60)} = 2.083 \text{ m}^3$$

At 40°C and 20 bar the volumetric rate of associated gas will be
$$\frac{V_g}{t} = \frac{(1000)(95)}{(24)(60)(60)} \frac{(313.15)}{(273.15)} \frac{(1)}{(20)} = 0.06303 \text{ m}^3/\text{s}$$

At separator conditions the gas density ρ_g is given by
$$\rho_g = 1.272 \, (0.75)(20) \frac{(273.15)}{(313.15)}$$
$$= 16.682 \text{ kg/m}^3$$

The maximum velocity equation is then used:
$$U_{max} = 0.125 \left[\frac{796 - 16.682}{16.682}\right]^{0.5} \text{ m/s}$$
$$= 0.8544 \text{ m/s}$$

Since cross-sectional area = volume rate/velocity then for an interface half way up the separator we have:
$$\frac{\pi D^2}{(2)(4)} = \frac{0.06303}{0.8544}$$
$$\therefore D = 0.4334 \text{ m}$$

Total volume of the separator is thus twice the oil volume for an interface half way up the separator $\therefore$
$$V_{sep} = 2V_o$$
$$= 4.166 \text{ m}^3$$

Design length for $L/D = 3$ gives
$$3D = L = \frac{(4.166)(4)}{\pi D^2}$$
$$\therefore D^3 = \frac{(4.166)(4)}{3\pi}$$
$$\therefore D = 1.209 \text{ m}$$
and $L = 3.627$ m

Design length for $L/D = 4$ gives
$$D^3 = \frac{(4.166)(4)}{4\pi}$$
$$\therefore D = 1.099$$
and $L = 4.396$ m

In practice the separator design would be based on a standard size selected to be nearest the size calculated.

Index

Abandonment pressure 159
absolute permeability 102
AFE (authorisation for expenditure) document 23, **24–5**
Amerada gauge 147, 148
API (American Petroleum Institute) gravity and oil density 14
aquifer characteristics
   correlation with model 167
   determination of 165–6
aquifers and pressure change 165
areal sweep efficiency *176*, 182–3

Back pressure equation 143–4, 221
barrel 14
bedforms, grain size and stream power 242
biocides and injection water 229
biopolymers 197
black oil reservoir modelling, uncertainties in 246–7
black oil systems 42
blow-out 35
blow-out preventers *34–5*
blowdown 210
Boltzmann transformation 134
bond number 191
BOPs *see* blow-out preventers
bottlenecks 219
bottom-hole sampling 52
Boyle's law method and grain volume 73
Brent Sand reservoirs 10, 11
brine disposal 186
bubble-point 41, 51, 53, 54, 55, 159, 220, 221
bubble-point pressure 52, 54–5, 56–7, 160, 163, 221
   in volatile oil reservoirs 211
Buckley–Leverett theory 105
Buckley–Leverett/Welge technique 107, 109

Capillary number 191, 193
capillary pressure 93
   and residual fluids 111–12
   defined 92
capillary pressure data (given rock type, correlation 99
capillary pressure hysteresis 97–8
capillary suction pressure *see* imbibition wetting phase threshold pressure
carbon dioxide in miscible displacement 195, 196
casing a well, reasons for **25**
casing eccentricity 35–6
casing selection *27*
   main design criteria 28
casings 23, 25, 26, 28
caustic solutions 196
cementation problems 35–6
chemical flood processes 196–200
choke assembly 146
Christmas tree 36
coalescer 227
Coates and Dumanoir equation 86
combination drive material balance equation 166
compaction drive 161
complete voidage replacement 173
completion 28, *29*
completion for production (permanent, normal) 36
composite cores 111
compressibility 42–3, 55
Compton scattering 76
conceptual models 233, 245
condensate analysis **208**
condensate reservoirs and liquid drop-out 208
condensate systems 42
condensing gas drive 194–5
cone height, critical 182
coning 181–2
core analysis
   and permeability distribution 83–4
   routine 69–71, 81
   presentation of results **70**, 71

Footnote: Numbers in italic indicate figures; Numbers in bold indicate tables

358   PETROLEUM ENGINEERING: PRINCIPLES AND PRACTICE

core data and palaeogeographical reconstruction 237–8
    and recognition of sand body type 238
core-derived data 68
core floods and surfactant testing 200
core for special core analysis 67, 68
core length and imbibition processes 110–11
core log *64*, 68
core plug experiments, concern over
    laboratory-derived data 113–14
core plugs 68
    analysis on 65
    and effective permeability 109
    and fluid saturation 93–4
    and oil saturation 193
    and permeability 81
    and porosity 72
    and residual saturation 174
core porosity, compaction corrected 131
core preservation 67–8
core recovery, fluids for 31
Coregamma surface logger 68
cores 62
    composite 111
    correlation with wireline logs *63*, 65, 75
    data obtainable from *63*
    diversity of information available 64
    and geological studies 68–9
    and heavy oil reservoirs 202
    residual fluid saturation
        determination 69
coring
    the case for 65
    conventional and oriented 66
    of development wells 65–6
    of exploration wells 65
coring decisions 64–6
coring mud systems 66–7
corresponding states, law of 44–5, 47
Cricondenbar 41, 42
Cricondentherm 41
critical displacement rate 177
critical displacement ratio 112
critical gas (equilibrium) saturation 159
critical production rate (coning) 182
crude oil
    flow of in wellbore 221, 223
    metering of 229
    processing 226–8
cushion 147
cuttings logs 31
cyclic steam stimulation *205*

Darcy (def.) 79
Darcy's equation 79
data acquisition during drilling 30–1
datum correction 79–80
deltaic environments, division of 238, *240*, **241, 242**
deltaic models, use of 238–43
deltaic system model 242, *244*
demulsifiers and heavy oil processing 228
depositional processes and reservoir rocks 7
dew-point 41
dew-point locus 42
diamond coring 33
differential liberation at reservoir temperature 53
displacement calculations, validation of relative permeability data for 113–14

displacement principles 173–5
drawdown testing 138
drill bits 22, 32–3
drill collars 23
drill stem testing 145
    testing tools and assemblies 145–7
drilling, turbine versus rotary 33
drilling costs 23, **24, 25**
drilling fluid *see* drilling mud
drilling logs 30
drilling mud pressure, excessive 29
drilling muds 22–3
    control of 28–9
    main constituents 67
drilling muds and cements, rheology of 29–30
drilling optimization 32–3
drilling, special problems in cementation problems 35–6
    pressure control and well kicks 34–5
    stuck pipe and fishing 33–4
drillstring 23
drive mechanisms 159
dry gas reservoirs 41–2
dual porosity systems 71, 73
    and gravity drainage 164–5

Early (transient) time solution 138
economic factors and oil production rates 180
effective permeability 102
    and wettability 108
enhanced oil recovery schemes and uncertainty 247
equity, distribution of, petroleum reservoirs 130–1
exploration well drilling 7, *8*

Faults, identification of 238
faults (in-reservoir), effect on injection/production well locations 180
field processing 224
filtration, injection water treatment 229
flash liberation at reservoir temperature 52–3
flash separation tests 53–4
flooding efficiency ratio 110
flow equations, linear and radial 80–1
flow string 145
fluid contacts 12–13
    multiple *12*
fluid flow in porous media 78–9
fluid pairs 93
fluid pressure and overburden load 11–12
fluid pressures, hydrocarbon zone 12–13
fluid saturation, laboratory measurements and relationship with reservoir systems 93–6
fluids, recovery of by depletion **211**
Forcheimer equation 143
formation breakdown pressure 30
formation density logs and interpretation of porosity 202–3
formation density tool response 75–6
formation factor *see* formation resistivity factor
    formation interval tester (FIT) 148
formation resistivity factor 74
formation tester (FT) 148
formation volume factor 14, 55
    two-phase 55–6
formation volume factors B 49–51
formation waters 14
fractional flow 104–6
    analysis methods 105–6
    effect of dip angle and wettability 175, *177*
    free water level (FWL) 12, 95

INDEX

Gas cap expansion drive 163–4
gas compressibilities 48–9
gas condensate, critical properties of 210
gas condensate and volatile oil reservoirs,
    uncertainties in 247
gas condensate reservoirs 207–11
    production methods for 209–11
gas deviation factor Z 46, 47
gas expansion during production 157
gas flow and gradient 159
gas flow and permeability 81
gas flow rate, measurement of 150, 229
gas formation volume factor 157
gas formation volume factor B_g 49–50
gas properties 45
gas recycling, gas condensate reservoirs 210
gas reinjection 186
gas reservoirs, recovery from 157–9
gas viscosities 47–8
gas-kicks 12
gas–oil ratio 14, 51–2, 54, 159
gas–oil systems and relative permeability 103–4
gas well testing 143–5
gases, behaviour of 43–4
gases, flow of in wellbore 221
geological model, development of 237–8
geothermal gradient and hydrocarbon generation 7, 9
geothermal gradient and reservoir temperature 13
GOR *see* gas–oil ratio
grain density 71
grain volume and Boyle's law method 73
gravity drainage and dual porosity systems 164–5
gravity segregation and recovery efficiencies 164–5
gravity stabilization and reservoir dip 175

Head loss in wellbores 221
heavy crude oil
    characteristics of UKCS heavy crude oils **201**
    general classification 200
    Yen classification 200, *201*
heavy oil processing 228
heavy oil recovery 200–2
heavy oil reservoirs
    examples of **201**
    permeability increase and production improvement 204
    production characteristics of 203–4
    properties of 202–3
    and thermal energy 204–7
    and uncertainty 247
heavy oil systems and thermal energy addition 204
HKW (highest known water) *12*, 13
homogeneous reservoirs and coning 181–2
Horner analysis 13
hydrates 224
hydrocarbon accumulation and sedimentary basins 7
hydrocarbon accumulations and formation waters 14
hydrocarbon exploitation, types of interactions *16*
hydrocarbon field 7
hydrocarbon generation and geothermal gradient 7, 9
hydrocarbon pore thickness (HPT) 126–7
hydrocarbon pore volume maps 126–7
hydrocarbon properties **47**
hydrocarbon recovery, improved 191–211
hydrocarbon reservoir fluids **15**
hydrocarbon systems
    volumetric and phase behaviour 40–1
    applications to field systems 41–2

hydrocarbon volume in place calculations 127–8
hydrocarbons, migration of (modelled) 93–4
hydrocarbons (commercial reservoirs),
    geological characteristics 62
hydrostatic gradient, regional 10–11

Ideal gas law (and modification) 43
imbibition processes
    and core length 110–11
    liquid 104
imbibition wetting phase threshold pressure 97
in-place volume 122
inflow performance relationship, 220
    dimensionless, for oil wells 220–1
    for gas wells 221
injection fluids, compatibility with reservoir fluids 183–4
injection fluids, quality of 183–6
injection water, viscosity of 184
injection water treatment 229
injectivity index 174
insert bits 33
isobaric thermal expansion coefficient 43
isocapacity maps 126
isochores 124
isochronal testing 144
isoliths 124
isopachs 124
isoporosity maps 125
isosaturation lines 99
isosaturation maps 126
isothermal compressibility 43
isothermal retrograde condensation 42

Kay's rule 45
kelly 23
kick 34–5
Kimmeridge Clay 7, 9
Klinkenberg correction 81, *82*
Kolmogorov–Smirnoff test 84

Lasater correlation (bubble-point pressure) *55*
leak off tests 30
Leverett J–function correlation 99
light oil processing 226
    foaming problems 227–8
    separator design considerations 227
    wax problems 228
line source solution (fluid flowing in a porous medium) 134–5
    development of 135–6
liquid drop out 208
liquids systems, generalized correlations 54–8
lithofacies representation 125
LKO (lowest known oil) *12*, 13
low interfacial tension (IFT) systems 193

Material balance, reservoirs with water encroachment or water
    injection 165–8
material balance calculations
    generation of data 52
    sources of error 168–9
material balance equation 158
    combination drive 166
    gas cap expansion drive 163–4
    solution gas drive 161–3
material balance residual oil saturation 174
mathematical models 233–4

mercury injection and porosimetry 73, 96, *97*
meters 229
microemulsion 198
middle (late transient) time solution 139
miscible displacement mechanisms 194–5
miscible displacement processes 193
miscible floods 194
    applications 195–6
    examples 196
miscible fluids, properties of **195**
mobility ratio 104–5, 107, 175, *176*
    and polymers 197
modelling of reservoirs 130–1
models 233–4
mole (def.) 44
Monte Carlo
    approach, probabilistic estimation 127
    technique and recoverable reserves estimate 130
movable hydrocarbon formula (MHV) 130
mud cake 36
mud circulation system *22, 23*
mud composition, general limitations on 67
mud logging 30–1
mud systems, bland (unreactive) and core recovery 31–2, 67
multicomponent systems, phase behaviour 41
multimodal porosity 78
multirate data, analysis of 144–5
multiphase flow, equations of 234–5

Natural gas
    calorific value 226
    dehydration 224–5
    onshore processing 225–6
    sales specification 224
    sweetening 225
natural gas processing 224–6
nitrogen in miscible displacement 195, 196
non–wetting phase fluid 94
non–wetting phase saturation 102
North Sea, heavy oil reservoirs 202
North Sea, hydrocarbon fields
    Beryl field 196
    Brent field 196
    Buchan field *37*
    Dunlin field *131, 178*
    Forties field *249*
    Fulmar field 249–51
    Magnus field *184*
    Maureen field *187*
    Montrose reservoir (RFT data) *151*
    Murchison field *125*
    Rough gas field *123, 124, 126, 127*
    Statfjord field 196, 245, 246
    Thistle oil reservoir 122, *123, 125*
North Sea, oil correlations, recent 56–8
North Sea, reservoirs, fluid choice for miscible displacement 196
North Sea, reservoirs and surfactants 198, 199

ODT (oil down to) 13
offshore production/injection system, principle components of 184, *185*, 186
offshore system 21
oil bank formation 195
oil density 14
oil flow rate, measurement of 150
oil formation factor B_o 51

oil saturation, local, influences on 191
oil viscosity 56
oil–water contact (OWC) 96, 98–9
oil–water systems and relative permeability 102–3
open–hole tests 145
optimal salinity 198
orifice meters 229
overpressure 11, 12

Packer 146
Peng and Robinson equation 44
permeabilities, averaging of 83
permeability 7, 78–86
    and critical displacement ratio 112
    anistropy 82–3
    distributions 83–4
    improvement 193–4
    laboratory determination of 81–2
    ratios 104–5
    variation, effects of 106–8
permeameter 81
petroleum
    migration of 9–10
    origin and formation of 7
    recovery 5
petroleum engineering
    function of 1
    problem solving in **3**
phase (def.) 14
phase inversion temperature (PIT) 198
physical models 233
piston displacement, stratified reservoirs 107–8
planimeter 124, 127
polyacrylamides 197
polymer fluids 193
polymer systems and adsorption 197
pool *see* reservoir
pore fluid pressures **11**
pore pressure, significance in drilling and well completion 26, 28
pore size distribution 96–7
pore space characteristics and equilibrium saturation distribution 92–3
pore volume compressibility 160
    of reservoir rocks **203**
poro–perm data, validity of 242
porosity 7, 71–8
    and permeability, relationship between 84–6
    cut-off 124
    distributions 77–8
    logs 75–7
    main logging tools for 75
    measurement of 72–3
potential gradient 174
pressure (abnormal) and d-exponent 25–6
pressure build–up analysis 139–40
pressure build–up (testing) 149
pressure control and well kicks 34–5
pressure decline, rates of 137
pressure depletion 210
pressure drawdown and reservoir limit testing 142–3
pressure equilibrium, static system 12
pressure gauges 137, 147
    (downhole), characteristics of **136**
pressure gradients and heterogeneity of reservoir pore space 129
pressure maintenance 173
pressure regimes, abnormal 11–12
primary recovery, oil reservoirs 159–64

INDEX

probabilistic estimation 127–8, *129, 130*
produced fluids and offshore processing 184–6
produced water treatment 228
producing rates (well inflow equations/pressure loss calculations) 174–5
production costs, significance of 1, 3
production engineering, and well performance 220–1
production engineering described 218
production operations, influencing factors 218–29
production rate effects 180–2
production rates, technical and economic factors 219
production system 218–19
production testing 150–1
productivity index (PI) 245
 and inflow performance 220
pseudo–critical temperatures and pressures 45–7
pseudo–relative permeability in dynamic systems 115
pseudo–relative permeability functions 177, *178*, 243, 245
 static 115–16
pseudo–relative permeability
 relationships and thicker sands 107
PVT analysis 52–4
PVT relationships, single and multicomponent systems 40–1

Radial equations in practical units 136
radial flow in a simple system 134–5, *137*
recombination sampling 52
recovery efficiency, water reservoirs 168
recovery factors and reserves 128–30
recovery string 34
recovery targets 191
Redlich–Kwong equation 44
relative permeability 102–4, 106–7
 effect of temperature *204*
relative permeability
 data, laboratory determination of 109–11
 from correlations 112–13
 improvement, heavy oil reservoirs 204
relative spreading concept 93
repeat formation tester (RFT) 148–50
reservoir behaviour in production engineering 220–1
reservoir condition
 material balance techniques 160
 volumetric balance techniques 160–1
reservoir data, sources 14–15, **17**
reservoir (def.) 7
reservoir description in modelling 237–45
 uncertainty in 245–7
reservoir development, costs of 3, *4*
reservoir dip angle 175, 177
reservoir flow rate, effect of 181
reservoir fluid properties,
 measurement and prediction of 43–9
reservoir fluids
 and compressibility 42–3
 nature of 14
 properties of 40–58
reservoir geometry and continuity 180, 238–45
reservoir heterogeneity 177–80
reservoir mapping and cross–section interpretation 245–6, *247*
reservoir modelling
 analysis and data requirements 237
 application in field development 248–51
 concepts in 233–48
reservoir performance analysis 157–68
reservoir pore volume and change in fluid pressure 42–3
reservoir pressures 10–12

reservoir rocks, characteristics of 62–86
 pore volume compressibility **203**
reservoir simulation modelling 233–7
reservoir simulation and vertical communication 243, 245
reservoir temperatures 13
reservoirs 7–18
 areal extent of 122–4
residual oil 53, 191
 influence of recovery mechanism 191, 193
residual oil saturation **192**
 average 174
 and material balance 174
 measurement of 191, *192*
residual saturations 111–112
resistivity factor *see* formation resistivity factor
resistivity index 74
retrograde condensation 208
reverse circulating sub 146
rotary table 23

Safety joints and jars 147
salinity and water viscosity 56
samplers 147
sand body continuity 180
 importance of 238, *239–40*
sand body type
 effect on injected water and oil displacement 178–80
 recognition of 238
saturation distributions in reservoir intervals 98–9
saturation gradients 164
saturation pressure *see* bubble–point pressure
scribe shoe 66
sea water as injection water 184
seawater floods (continuous) and low surfactant concentration 199–200
secondary recovery and pressure maintenance 173–86
secondary recovery techniques 173
sedimentary basins
 and hydrocarbon accumulation 7
 origin of 7
 worldwide *2*
segregated displacement *177*
sensitivity studies 246–7
shaliness, effect of 13
Shinoda diagrams 198
simulators
 applications 235
 classification of 235, **236**
single component systems, phase behaviour 40–1
skin effect 140–2
 negative factors 142
skin zone 194
slabbing 68
solution gas drive, analysis by material balance 159–63
solution gas–oil ratio 53, 54, 55
Standing–Katz correlations 46, 47
Standing's data (bubble–point correlation) *55*
STB (stock tank barrel) 14
steady state permeability tests 110
steam flooding *205*
steam properties 206, *207*
steamdrive analysis, example data requirements **207**
Stiles technique 107–8
stock tank oil 54
 and retrograde condensation 208
stock tank oil in place and equity

determination 130
stock tank units 14
stock tank volume 53
Stratapax bits 33
stratified reservoir analysis 106
stripping 191
structure contour maps 122
stuck pipe and fishing 33–4
summation of fluids and porosity 72–3, *74*
superposition technique 140
surfactant concentration (low) and continuous seawater floods 199–200
surfactant flooding 198–200
surfactant phase systems 197–8
surfactant processes 197–200
surfactants 193
   synthetic 199
sweetening, natural gas 225

Tester valve 146
thermal energy 204–7
thermal injection processes 204–6
thickness maps 124
threshold capillary pressure (reservoir rocks) 95
threshold pressure 94
traps (structural and stratigraphic) 10
tricone bits 32, 33
trip gas 34
turbine meters 229

Ultimate recovery formula *see* movable hydrocarbon formula
uncertainty in reservoir model description 245–8
unitization 130–1
universal gas constant, values of **43**
unsteady state relative permeability tests 109–10
USA, heavy oil resource distribution **202**

Van der Laan method (volume in place) 128
vaporizing gas drive 194, *195*
vapour phase 42
vertical bed resolution 76
vertical permeability variation and fractional flow curve 177
vertical pressure logging 148–50
Viking Graben area (N North Sea) 10
Vogel dimensionless IPR 220–1
volatile oil reservoirs 211
volatile oil systems 42
volumetric balance techniques 160
vugular carbonates and whole core analysis 69

Walther's law of facies 238
water drive and gas condensate reservoirs 209, 210
water drive reservoirs 167
   recovery efficiency of 168
water formation factor B_w 50–1
water influx 165, 166
water influx, gas reservoir 158–9
water injection 166, 178
water saturation distribution, homogeneous reservoir *96*
water viscosity 56
waterflooding 178, *179*, 180
Welge analysis 106
Welge's equations 174
well arrangements, dipping reservoirs 181
well classification **20**
well description log 31, *32*
well drilling operations 20–3
well locations and patterns 182–3
well performance, radial flow analysis of 134–51
well productivity improvement 193–4
well test methods, applications of analytical solutions 136–9
well test procedures 145–50
   data analysis 147–8
well testing and pressure analysis 150–1
well/reservoir responses, different reservoir systems *139*
wellbore, altered zone 141
wellbore flow 221–3
wellbore inflow equations 174
wellsite controls and core recovery 68
wettability 175
   change in 67, 196
   degree of 93
wettability control, *in situ* 112
wettability effects 108
wettability preference 93
wetting phase fluid 93
wetting phase saturation 94
wetting preference 175
wireline logs, correlation with cores *63*, 65, 75
wireline testing 148–50
WUT (water up to) 13

Xanthan gums 197

Zonation 99, 131, 242, 243, *245*
   Forties reservoir *249*
   and geological core study 68–9
   and histogram analysis 84
   and permeability distributions 84